AF767409

Smart Food Packaging Systems

Smart Food Packaging Systems

Innovations and Technology Applications

Edited by

Avik Mukherjee
Central Institute of Technology Kokrajhar
Kokrajhar, Assam, India

Santosh Kumar
Central Institute of Technology Kokrajhar
Kokrajhar, Assam, India

Manjusri Misra
University of Guelph
Guelph, Ontario, Canada

Amar K. Mohanty
University of Guelph
Guelph, Ontario, Canada

WILEY

This edition first published 2025
© 2025 John Wiley & Sons Ltd

All rights reserved, including rights for text and data mining and training of artificial technologies or similar technologies. No part of this publication may be reproduced, stored in a retrieval system, or transmitted, in any form or by any means, electronic, mechanical, photocopying, recording or otherwise, except as permitted by law. Advice on how to obtain permission to reuse material from this title is available at http://www.wiley.com/go/permissions.

The right of Avik Mukherjee, Santosh Kumar, Manjusri Misra, and Amar K. Mohanty to be identified as the authors of the editorial material in this work has been asserted in accordance with law.

Registered Office(s)
John Wiley & Sons, Inc., 111 River Street, Hoboken, NJ 07030, USA
John Wiley & Sons Ltd, The Atrium, Southern Gate, Chichester, West Sussex, PO19 8SQ, UK

For details of our global editorial offices, customer services, and more information about Wiley products visit us at www.wiley.com.

Wiley also publishes its books in a variety of electronic formats and by print-on-demand. Some content that appears in standard print versions of this book may not be available in other formats.

Trademarks: Wiley and the Wiley logo are trademarks or registered trademarks of John Wiley & Sons, Inc. and/or its affiliates in the United States and other countries and may not be used without written permission. All other trademarks are the property of their respective owners. John Wiley & Sons, Inc. is not associated with any product or vendor mentioned in this book.

Limit of Liability/Disclaimer of Warranty
While the publisher and authors have used their best efforts in preparing this work, they make no representations or warranties with respect to the accuracy or completeness of the contents of this work and specifically disclaim all warranties, including without limitation any implied warranties of merchantability or fitness for a particular purpose. No warranty may be created or extended by sales representatives, written sales materials or promotional statements for this work. This work is sold with the understanding that the publisher is not engaged in rendering professional services. The advice and strategies contained herein may not be suitable for your situation. You should consult with a specialist where appropriate. The fact that an organization, website, or product is referred to in this work as a citation and/or potential source of further information does not mean that the publisher and authors endorse the information or services the organization, website, or product may provide or recommendations it may make. Further, readers should be aware that websites listed in this work may have changed or disappeared between when this work was written and when it is read. Neither the publisher nor authors shall be liable for any loss of profit or any other commercial damages, including but not limited to special, incidental, consequential, or other damages.

Library of Congress Cataloging-in-Publication Data Applied for:
Hardback ISBN: 9781394189564

Cover Design: Wiley
Cover Image: © Mark Agnor/Shutterstock

Set in 9.5/12.5pt STIXTwoText by Straive, Pondicherry, India

Printed and bound by CPI Group (UK) Ltd, Croydon, CR0 4YY

C9781394189564_260924

This book is dedicated to future of sustainable food packaging.

Contents

About the Editors *xvii*
List of Contributors *xxi*
Preface *xxv*
Acknowledgments *xxvii*

1 Introduction to Active Food Packaging System *1*
Sweety Kalita, Amar K. Mohanty, Manjusri Misra, Avik Mukherjee, and Santosh Kumar
1.1 Introduction *1*
1.2 Types of Active Food Packaging *3*
1.2.1 Active Scavenging/Absorbing System *3*
1.2.1.1 Ethylene Scavenger *3*
1.2.1.2 Oxygen Scavenger *4*
1.2.1.3 Moisture Absorber *6*
1.2.1.4 Ultraviolet (UV) Light Absorbers *7*
1.2.1.5 Flavor Absorbers *10*
1.2.2 Active Releasing or Emitting Systems *10*
1.2.2.1 Carbon Dioxide Emitter *10*
1.2.2.2 Antimicrobial Agent Emitter/Antimicrobial Packaging *11*
1.2.2.3 Antioxidant Emitters *13*
1.2.2.4 Ethanol Emitters *16*
1.2.2.5 Sulfur Dioxide Emitters *17*
1.2.2.6 Flavor Releasers *17*
1.3 Forms of Active Packaging Systems *18*
1.3.1 Sachets *18*
1.3.2 Polymeric Films *19*
1.3.3 Other Forms *20*
1.4 Active Packaging and Food Distribution *21*
1.5 Regulatory and Market Status *22*
1.6 Conclusion and Future Perspective *23*
References *23*

2 Additives in Active Food Packaging Systems *39*
Debarshi Nath, Amar K. Mohanty, Avik Mukherjee, Santosh Kumar, and Manjusri Misra
2.1 Introduction *39*
2.2 Techniques to Incorporate Active Agents Into Packaging Systems *42*
2.2.1 Solution Casting *42*
2.2.2 Coating and Spraying *42*
2.2.3 Melt Blending *42*
2.3 Metal Oxide Nanoparticles (NPs) Used in Food Packaging *43*
2.3.1 Zinc Oxide *43*
2.3.2 Titanium Dioxide *44*
2.3.3 Silicon Dioxide *45*
2.4 Active Agents Derived from Organic Sources *45*
2.4.1 Polyphenols *45*
2.4.2 Phenolic Acids *46*
2.4.3 Flavonoids *46*
2.4.4 Tannins *47*
2.4.5 Polyphenolic Amides *47*
2.4.6 Other Polyphenols *48*
2.4.6.1 Curcumin *48*
2.4.6.2 Resveratrol *48*
2.5 Antimicrobial Peptides *49*
2.5.1 Nisin *50*
2.5.2 Pediocin *50*
2.6 Essential Oils *50*
2.6.1 Essential Oil Components *52*
2.6.1.1 Cinnamaldehyde *52*
2.6.1.2 Eugenol *52*
2.6.1.3 Carvacrol *53*
2.6.1.4 Thymol *53*
2.7 Polymer Nanoparticles *54*
2.7.1 Chitosan *55*
2.7.2 Alginate *56*
2.7.3 Gelatin *56*
2.8 Ecotoxicology *57*
2.9 Conclusion *61*
CRediT Author Contributions *62*
Conflicts of Interest *62*
Acknowledgments *62*
References *63*

3 Antimicrobial Food Packaging Systems *73*
Jigyasa, Jaspreet Kaur Rajput, and Smriti Dogra
3.1 Introduction *73*
3.2 Foodborne Pathogens and Their Control *74*

3.3 Antimicrobial Active Agents *74*
3.3.1 Antimicrobials from Plant Origin *74*
3.3.2 Antimicrobials from Animal Origin *81*
3.3.2.1 Chitosan *81*
3.3.2.2 Defensin *82*
3.3.2.3 Lactoferrin *83*
3.3.2.4 Lactoperoxidase *83*
3.3.2.5 Lysozyme *83*
3.3.2.6 Pleurocidin *84*
3.3.2.7 Protamine *84*
3.3.3 Antimicrobials from Microbial Origin *84*
3.3.3.1 Bacteriophages *84*
3.3.3.2 Natamycin *85*
3.3.3.3 Nisin *85*
3.3.3.4 Reuterin *85*
3.3.4 Antimicrobials Nanomaterials *86*
3.4 Technologies of Incorporation of Antimicrobials in Food Packaging System *86*
3.4.1 Ionic or Covalent Bonding Between Antimicrobial and Food Packaging System *86*
3.4.2 Antimicrobial Sachets in Packaging System *87*
3.4.3 Antimicrobial-Coated Films *87*
3.5 Release of Antimicrobials from Food Packaging *87*
3.6 Applications of Antimicrobial Active Packaging *88*
3.6.1 Fruits and Vegetables *88*
3.6.2 Bakery and Other Food Products *90*
3.7 Conclusion and Future Perspective *91*
References *92*

4 Oxygen Scavengers in Active Food Packaging Systems *103*
Arihant Ahuja, Dimple Singh, Varsha Singh, Anamika Singh, Satish Kumar Bachala, and Vibhore Kumar Rastogi
4.1 Introduction *103*
4.2 Oxygen Scavengers *104*
4.2.1 Metal-Based Scavenging System *106*
4.2.2 Bio-Based Scavenging System *108*
4.2.3 Enzyme-Based Scavenging System *110*
4.2.4 Polymer-Based Scavenging System *112*
4.2.5 Microorganism-Based Oxygen Scavenger System *115*
4.3 Loading of Oxygen Scavengers in Active Packaging *116*
4.3.1 Sachets or Labels *116*
4.3.2 Incorporation Techniques *117*
4.3.2.1 Coatings *117*
4.3.2.2 Blending *118*
4.3.2.3 Multilayer Structure *119*
4.3.2.4 Immobilization *119*

4.4 Applications of Oxygen Scavengers in Active Packaging *120*
4.4.1 Fruits and Vegetables *120*
4.4.2 Dairy Products *120*
4.4.3 Muscle Foods *121*
4.4.4 Bakery and Other Food Products *122*
4.5 Conclusion and Future Perspective *122*
Acknowledgments *123*
References *123*

5 Carbon Dioxide (CO_2) Scavengers and Emitters in Food Packaging *131*
Narender Raju Panjagari and Deepak Kumar
5.1 Introduction *131*
5.2 CO_2 Scavengers and Emitters *133*
5.2.1 CO_2 Scavengers *133*
5.2.2 CO_2 Emitters *134*
5.3 Mechanistic Principles of CO_2 Scavenger *134*
5.3.1 Chemical Absorbers *135*
5.3.2 Physical Absorbers *136*
5.4 Mechanism of CO_2 Emitters *137*
5.4.1 CO_2 Emitters *137*
5.4.2 Active Valves for CO_2 Emission or Removal to Environment *138*
5.5 Incorporation of CO_2 Scavengers and Emitters in Active Food Packaging *139*
5.6 CO_2 Emitters in Active Food Packaging *139*
5.6.1 Antimicrobial Applications *139*
5.6.2 Antioxidant Applications *141*
5.7 Application of CO_2 Scavenging and Emitting Systems in Active Packaging *141*
5.7.1 Fruits and Vegetables *141*
5.7.2 Dairy Products *142*
5.7.3 Muscle Foods *143*
5.7.4 Bakery and Other Food Products *144*
5.8 Conclusion *144*
References *145*

6 Ethylene Scavengers for Active Packaging of Fresh Horticultural Produce *149*
Gurbuz Gunes
6.1 Introduction *149*
6.2 Biosynthesis Pathway and Effects of Ethylene on Ripening of Fruits and Vegetables *150*
6.2.1 Ethylene Biosynthesis *150*
6.2.2 Effects of Ethylene on Ripening *152*
6.3 Types of Ethylene Scavengers *153*
6.3.1 Potassium and Sodium Permanganate *154*
6.3.2 Titanium Oxide *154*
6.3.3 Novel Oxidizers *155*
6.3.4 Activated Carbon *155*
6.3.5 Zeolite *155*

6.3.6 Clays *156*
6.3.7 Metal–Organic Frameworks *156*
6.3.8 Methylcyclopropene *156*
6.4 Principles and Mechanisms of Ethylene Scavenging *156*
6.5 Advances and Applications of Ethylene-Scavenging Systems *157*
6.5.1 Banana *159*
6.5.2 Mango *159*
6.5.3 Tomato *160*
6.5.4 Apple *160*
6.5.5 Others *161*
6.6 Conclusion and Future Perspective *162*
References *162*

7 Moisture-Absorbent Food Packaging Systems and the Role of Chitosan *169*
Srasti Yadav and Pradip Kumar Dutta
7.1 Introduction *169*
7.2 Classification of Moisture Absorbents *171*
7.2.1 Inorganic Absorbents *171*
7.2.2 Organic Absorbents *173*
7.2.3 Polymeric Absorbents *174*
7.3 Mechanisms of Moisture Absorption *176*
7.4 Incorporation of Moisture Absorbents in Food Packaging *176*
7.5 Moisture-Absorbent-Based Active Food Packaging *178*
7.6 Chitosan: A Better Substrate in Food Packaging *179*
7.7 Applications of Moisture Absorbents in Food Packaging *183*
7.7.1 Fruits and Vegetables *183*
7.7.2 Muscle Foods *183*
7.7.3 Dry Foods *184*
7.7.4 Other Foods *184*
7.8 Conclusion and Future Perspective *185*
References *186*

8 Off-Flavor Absorbents in Active Food Packaging *195*
Rishabh Goyal, Het Paraskumar Bhadania, Sushil Kumar Singh, and Poonam Singha
8.1 Introduction *195*
8.2 Flavor Absorbents *196*
8.2.1 Direct Flavor Absorbents *198*
8.2.1.1 Activated Carbon *198*
8.2.1.2 Zeolites *198*
8.2.1.3 Resins and Polymers *201*
8.2.1.4 Enzymes *201*
8.2.1.5 Acid Compounds *201*
8.2.1.6 Metal *202*
8.2.2 Indirect Off-Flavor Absorbent *202*
8.2.2.1 Oxygen Absorbers *203*

8.2.2.2 Ethylene Scavengers *203*
8.2.2.3 Moisture Absorbers *203*
8.3 Mechanisms of Off-Flavor Absorbents *203*
8.3.1 Adsorption *204*
8.3.2 Chemical and Enzymatic Reaction *206*
8.3.2.1 Oxidation and Reduction *206*
8.3.2.2 Hydrolysis and Enzymatic Reactions *206*
8.3.2.3 Hydrophobic and Hydrophilic Interactions *206*
8.3.2.4 Electrostatic Interaction *207*
8.3.2.5 Chelation *207*
8.3.3 Diffusion *207*
8.4 Off-Flavor Absorbent-Based Active Food Packaging *208*
8.4.1 Sulfide Scavengers *209*
8.4.2 Amine Scavengers *209*
8.4.3 Aldehyde Scavengers *210*
8.5 Applications of Off-Flavor Absorbing System in Food Packaging *210*
8.5.1 Fruits and Vegetable Products *210*
8.5.2 Meat and Meat Products *211*
8.5.3 Milk and Milk Products *212*
8.5.4 Fatty Foods *213*
8.6 Conclusion and Perspective *213*
References *214*

9 Biopolymer-Based Nanocomposites for Active Food Packaging *219*
Soubani Baidya and R. Anandalakshmi
9.1 Introduction *219*
9.2 Biopolymers *220*
9.2.1 Natural Biopolymers *221*
9.2.2 Synthetic Biopolymers *221*
9.2.3 Biopolymers from Microbial Fermentation *221*
9.3 Nanomaterials and Their Functional Properties *222*
9.3.1 Antimicrobial Property of Nanomaterials *224*
9.3.2 Ethylene and Oxygen-Scavenging Property of Nanomaterials *224*
9.3.3 Antioxidant Property of Nanomaterials *225*
9.3.4 Flavor-Releasing Property of Nanomaterials *225*
9.3.5 UV-Blocking Property of Nanomaterials *226*
9.4 Biopolymer-Nanocomposites Active Packaging Systems *226*
9.4.1 Starch-Based Active Nanocomposite System *227*
9.4.2 Cellulose-Based Active Nanocomposite System *228*
9.4.3 Chitosan-Based Active Nanocomposite System *229*
9.4.4 Gelatin-Based Active Nanocomposite System *230*
9.4.5 Casein and Whey Protein-Based Active Nanocomposite System *230*
9.4.6 Polylactic Acid-Based Active Nanocomposite System *232*
9.4.7 Other Biopolymer-Based Active Nanocomposite System *232*
9.5 Application of Nanocomposites in Food Packaging Systems *234*

9.5.1 Fruits and Vegetables *234*
9.5.2 Dairy Products *239*
9.5.3 Muscle Foods *240*
9.5.4 Bakery and Other Products *241*
9.6 Regulation and Safety Aspects *242*
9.7 Conclusion and Perspective *243*
References *244*

10 Introduction to Intelligent Food Packaging *257*
Krishna Kamal Hazarika, Pallabi Hazarika, and Swapnali Hazarika
10.1 Introduction *257*
10.2 Types of Intelligent Packaging *258*
10.3 Indicators *259*
10.3.1 Time–Temperature Indicators *259*
10.3.1.1 Chemical Time and Temperature Indicators *260*
10.3.1.2 Physical Time–Temperature Indicator *261*
10.3.1.3 Time–Temperature Indication for Biological Systems *262*
10.3.2 Freshness Indicators *262*
10.3.3 Food Quality Indicators *263*
10.3.3.1 Oxygen and Carbon Dioxide *263*
10.3.3.2 pH Change *263*
10.3.3.3 Humidity *265*
10.3.4 Gas and Volatile Indicators *265*
10.3.5 pH Indicators *266*
10.3.5.1 Meat Products *267*
10.3.5.2 Seafood and Fish Products *267*
10.3.5.3 Milk and Dairy Products *268*
10.3.6 Leak Indicators *268*
10.3.6.1 O_2 Indicators *269*
10.3.6.2 CO_2 Indicators *269*
10.3.6.3 Indicators of Microbial Growth *270*
10.3.7 Thermochromic Ink *270*
10.4 Sensors *271*
10.4.1 Gas Sensors *271*
10.4.2 Humidity Sensors *271*
10.5 Radio Frequency Identification *272*
10.6 3D Printing and Intelligent Packaging *272*
10.7 Legal Aspects of Intelligent Packaging *273*
10.8 Conclusion and Perspective *274*
References *274*

11 Time–Temperature Indicators (TTIs) for Monitoring Food Quality *281*
Kandeepan Gurunathan
11.1 Introduction *281*
11.2 Types of TTIs *283*

11.2.1 Chemical TTIs *284*
11.2.1.1 Polymerization-Based TTIs *284*
11.2.1.2 Photochromic-Based TTIs *284*
11.2.1.3 Redox Reaction-Based TTIs *284*
11.2.1.4 Diffusion-Based TTIs *285*
11.2.1.5 Nanoparticle-Based TTIs *285*
11.2.2 Enzymatic TTIs *285*
11.2.3 Microbial TTIs *287*
11.2.4 Other Innovative TTIs *289*
11.2.4.1 TTIs Based on Thermochromic Polymer/Dye Blends *289*
11.2.4.2 TTIs Based on Photonic Lattice Change *289*
11.2.4.3 Time–Temperature Sensors Based on Electronics *289*
11.3 Applications of TTIs for Monitoring Food Quality *290*
11.3.1 Fruits and Vegetables *290*
11.3.2 Dairy Products *290*
11.3.3 Muscle Foods *291*
11.3.4 Other Foods *292*
11.4 Assessment of Applicability and Commercialization *292*
11.4.1 Assessment of Applicability of TTI *292*
11.4.2 Commercialization of TTI *293*
11.5 Conclusion and Perspective *298*
References *299*

12 Carbon Dioxide Sensors/Indicators-Based Intelligent Food Packaging *305*
Nitin Gupta, Neelisetty Sesha Sai Baba, Dimple Singh, Ayyan P. Borkakoty, Vinayak Srivastava, and Vibhore Kumar Rastogi
12.1 Introduction *305*
12.2 Types of CO_2 Sensors *306*
12.2.1 Conventional CO_2 Sensors *306*
12.2.2 Innovative Optical CO_2 Sensors *307*
12.3 Applications of CO_2 Sensors/Indicators in Monitoring Food Quality *311*
12.3.1 Fruits and Vegetables *311*
12.3.2 Dairy Products *314*
12.3.3 Muscle Foods *317*
12.3.4 Other Foods *322*
12.3.4.1 Grains *322*
12.3.4.2 Beverages *324*
12.3.4.3 Kimchi *324*
12.4 Safety Aspects of CO_2 Sensors *325*
12.5 Conclusion and Perspective *325*
Acknowledgment *326*
References *326*

13 Food Quality Indicator-Based Intelligent Food Packaging *337*
K. P. Chaithra, B. Gayathri, P. S. Aiswarya, and T. P. Vinod
13.1 Introduction *337*
13.2 Food Quality Indicators *338*
13.2.1 Humidity Indicators *338*
13.2.2 Gas Indicators/Detectors *340*
13.2.3 pH Change Indicators/Detectors *340*
13.2.4 Specific Chemical Indicators/Detectors *342*
13.3 Intelligent Packaging (IP) System for Monitoring Food Quality *343*
13.3.1 Humidity-Based IP *345*
13.3.2 O_2-Based IP *345*
13.3.3 CO_2-Based IP *346*
13.3.4 Microbial-Based IP *347*
13.3.5 pH-Based IP *349*
13.4 Conclusion and Perspective *351*
References *351*

14 Freshness Indicator in Intelligent Food Packaging System *361*
Satti Venu Gopala Kumari, Kannan Pakshirajan, and G. Pugazhenthi
14.1 Introduction *361*
14.2 Substance Releases Indicating Food Freshness *362*
14.2.1 Volatile Base Nitrogen *363*
14.2.2 Carbon Dioxide *363*
14.2.3 Hydrogen Sulfide *363*
14.2.4 Other Characteristic Substance Releases *364*
14.3 Types of Freshness Indicators *364*
14.3.1 Nature of Support *364*
14.3.1.1 Synthetic Polymer-Based Indicators *364*
14.3.1.2 Natural Polymer-Based Indicators *365*
14.3.2 Nature of Chromogenic Agents *365*
14.3.2.1 Chemical Chromogen-Based Indicators *365*
14.3.2.2 Natural Chromogen-Based Indicators *366*
14.4 Fabrication of FIs *368*
14.5 Application of FI in Monitoring Food Quality *368*
14.5.1 Fruits and Vegetables *368*
14.5.2 Dairy Products *372*
14.5.3 Muscle Foods *374*
14.5.4 Other Foods *376*
14.5.5 Commercial Availability of FIs *377*
14.6 Conclusion and Future Perspective *377*
References *378*

15 Biopolymer-Based Nanocomposites for Intelligent Food Packaging *387*
Himakshi Baishya, Sweety Kalita, Avik Mukherjee, and Santosh Kumar
15.1 Introduction *387*
15.2 Bio-Nanocomposite-Based Intelligent Packaging Films/Labels *390*
15.3 Natural Active Compounds-Based Intelligent Packaging Systems *392*
15.4 Bio-Nanocomposite-Based Color Sensor *393*
15.5 Food Packaging Applications of Bio-Nanocomposite Intelligent Films/Labels *397*
15.5.1 Fruits and Vegetables *397*
15.5.2 Dairy Products *398*
15.5.3 Muscle Foods *399*
15.6 Conclusion and Perspective *404*
References *404*

16 Life Cycle Assessment of Smart Food Packaging Systems *415*
Sweety Kalita, Debarshi Nath, Ehsan Pesaranhajiabbas, Manjusri Misra, Amar K. Mohanty, Avik Mukherjee, and Santosh Kumar
16.1 Introduction *415*
16.2 Biodegradability in the Circular Economy Framework *417*
16.3 Sustainability of Food Packaging: Rationales, Importance, and Approaches *418*
16.3.1 Approaches to Achieve Sustainability in Packaging *420*
16.3.2 Use of Biodegradable Materials *420*
16.3.2.1 Circular Economy and Recycling *421*
16.3.2.2 Reducing and Lightweighting *422*
16.3.2.3 Regulatory Initiatives and Role of Customers *422*
16.4 Case Studies on Life Cycle Assessment of Smart Packaging Systems *423*
16.5 Conclusion *424*
References *425*

Index *431*

About the Editors

Dr. Avik Mukherjee

Avik Mukherjee is an Associate Professor at the Department of Food Engineering and Technology, Central Institute of Technology Kokrajhar (Deemed-to-be-University, MHRD, Govt. of India), Assam, India. He earned his MS and PhD degrees, majoring in Food Science, at the University of Minnesota, USA, and then worked as a Postdoctoral Fellow at the Center for Meat Safety and Quality, Department of Animal Sciences, Colorado State University, USA. Dr. Mukherjee has 18 years of teaching and research experience in the areas of microbial food safety, sustainable and novel food preservation and packaging. He has published almost 40 international peer-reviewed journals, coauthored and/or co-edited 8 books and 13 book chapters, and his research works have produced more than 30 conference papers. Dr. Mukherjee's research area includes the application of natural and sustainable food preservation and packaging technologies, including the application of nanotechnology to ensure safe and healthy foods, and the utilization of food processing waste(s) for biofuel production. Dr. Mukherjee has also shown academic and administrative leadership skills serving his Institutes as Head of the Department for more than nine years, and as Dean for more than five years.

Dr. Santosh Kumar

Dr. Santosh Kumar is currently working as an Assistant Professor in the Department of Food Engineering and Technology, Central Institute of Technology Kokrajhar, India, since 2011. He earned his MTech and PhD degrees from the Department of Food Technology and Biochemical Engineering (FTBE), Jadavpur University, Kolkata, India, and then worked as Postdoctoral Fellow at KTH Royal Institute of Technology, Sweden, during 2018–2019. Dr. Kumar has more than 12 years of teaching and research experience in Sustainable Food Systems, particularly in Active and Intelligent Food Packaging, Edible Films and Coatings, and Agro-food Waste Valorization. Dr. Kumar has authored/edited more than 65 publications including

45 peer-reviewed journal papers, 3 books, and 15 book chapters. Dr. Kumar has received several awards including Overseas Associateship Award by the Department of Biotechnology, Government of India. He is a professional member of many scientific societies and institutions, including the Royal Society of Chemistry, London. He also serves as an Editorial Board member of many reputed journals including "Sustainable Chemistry and Pharmacy" published by Elsevier. Dr. Kumar has shown academic and administrative leadership skills by serving his institutions in many positions, including as the Head of the Department for about five years.

Dr. Manjusri Misra

Dr. Manjusri Misra is a professor and Tier 1 Canada Research Chair (CRC) in Sustainable Biocomposites in the School of Engineering and holds a joint appointment in the Department of Plant Agriculture at the University of Guelph. As well, she is the Research Program Director of the Bioeconomy Panel for the Ontario Agri-Food Innovation Alliance, a program between the Ontario Ministry of Agriculture and Rural Affairs (OMAFRA) and the University of Guelph. Dr. Misra's research directly contributes to the UN's Sustainable Development Goals. She is a Fellow of the Royal Society of Chemistry (UK), the American Institute of Chemical Engineers (AIChE), and the Society of Plastic Engineers (SPE). She was an editor of the CRC Press volume, "Natural Fibers, Biopolymers and Biocomposites," Taylor & Francis Group, Boca Raton, FL (2005); American Scientific Publishers volume "Packaging Nanotechnology", Valencia, California (2009); "Polymer Nanocomposites", Springer (2014); "Biocomposites: Design and Mechanical Performance", Woodhead Publishing (2015); "Fiber Technology for Fiber-Reinforced Composites", Woodhead Publishing (2017); and "Nanomaterials from Renewable Resources for Emerging Applications (Emerging Materials and Technologies)", CRC Press, Taylor & Francis Group (2023).

Dr. Amar K. Mohanty

Dr. Amar K. Mohanty is a Professor, Distinguished Research Excellence Chair in Sustainable Materials, and the Director of the Bioproducts Discovery and Development Centre. He is a Full Professor in two departments – Department of Plant Agriculture and the School of Engineering at the University of Guelph, Canada. Dr. Mohanty is the *Editor-in-Chief* of *Composites Part C* (ELSEVIER). He is a Fellow of the *Society of Plastic Engineers* (SPE), *Royal Society of Canada (RSC)*, the *American Institute of Chemical Engineers* (AIChE), the *Royal Society of Chemistry* (UK), and the *Indian Institute of Chemical Engineers* (IIChE). He is one of the most cited researchers worldwide with 498 peer-reviewed journal papers, 71 patents (awarded/applied), and over 160 plenary and keynote presentations. He has received many awards, including the prestigious *Miroslaw Romanowski Medal* for his significant scientific contributions to the resolution of environmental problems from the *Royal Society of Canada*. He also received the *Synergy Award for Innovation* from the *Natural Sciences and Engineering Research Council of Canada* (NSERC), the *Andrew Chase Forest Products Division Award* from the AIChE, *J. L White Innovation Award* from the International *Polymer Processing Society* (PPS), and the *Lifetime Achievement Award* from the *BioEnvironmental Polymer Society* (BEPS), USA. His research encompasses sustainable polymeric materials, circular economy, environmental sustainability, waste plastic valorization, 3D printing, AI for sustainable materials development, biodegradable plastics, biocarbon-based composites all targeted in reducing greenhouse gas emission.

List of Contributors

Arihant Ahuja
Department of Paper Technology
Indian Institute of Technology Roorkee
Roorkee, Uttarakhand
India

P. S. Aiswarya
Department of Chemistry
Christ University
Bangalore, Karnataka
India

R. Anandalakshmi
Department of Chemical Engineering
Indian Institute of Technology Guwahati
Guwahati, Assam
India

Satish Kumar Bachala
Department of Paper Technology
Indian Institute of Technology Roorkee
Roorkee, Uttarakhand
India

Soubani Baidya
Department of Chemical Engineering
Indian Institute of Technology Guwahati
Guwahati, Assam
India

Himakshi Baishya
Department of Food Engineering and Technology
Central Institute of Technology Kokrajhar
Kokrajhar, Assam
India

Het Paraskumar Bhadania
Department of Chemical Engineering
Institute of Chemical Technology-Mumbai
IOC Campus
Bhubaneswar, Odisha
India

Ayyan P. Borkakoty
Department of Paper Technology
Indian Institute of Technology Roorkee
Roorkee, Uttarakhand
India

K. P. Chaithra
Department of Chemistry
Christ University
Bangalore, Karnataka
India

Smriti Dogra
Department of Chemistry
Dr B R Ambedkar National Institute of Technology
Jalandhar, Punjab
India

Pradip Kumar Dutta
Department of Chemistry
Motilal Nehru National Institute of Technology
Allahabad, Prayagraj
India

B. Gayathri
Department of Chemistry
Christ University
Bangalore, Karnataka
India

Rishabh Goyal
Department of Food Process Engineering
National Institute of Technology Rourkela
Odisha
India

Gurbuz Gunes
Department of Food Engineering
Faculty of Chemical and Metallurgical Engineering
Istanbul Technical University
Istanbul
Turkey

Nitin Gupta
Wood K Plus - Competence Center for Wood Composites and Wood Chemistry
St. Veit an der Glan
Austria

Kandeepan Gurunathan
Packaging Lab
ICAR-National Meat Research Institute
Hyderabad, Telangana
India

Krishna Kamal Hazarika
Chemical Engineering Group and Centre for Petroleum Research Division
CSIR-North East Institute of Science and Technology
Jorhat, Assam
India

Academy of Scientific and Innovative Research (AcSIR)
Ghaziabad, Uttar Pradesh
India

Pallabi Hazarika
Chemical Engineering Group and Centre for Petroleum Research Division
CSIR-North East Institute of Science and Technology
Jorhat, Assam
India

Academy of Scientific and Innovative Research (AcSIR)
Ghaziabad, Uttar Pradesh
India

Swapnali Hazarika
Chemical Engineering Group and Centre for Petroleum Research Division
CSIR-North East Institute of Science and Technology
Jorhat, Assam
India

Academy of Scientific and Innovative Research (AcSIR)
Ghaziabad, Uttar Pradesh
India

Jigyasa
Department of Chemistry
Maharishi Markandeshwar University
Mullana, Ambala, Haryana
India

Sweety Kalita
Department of Food Engineering and Technology
Central Institute of Technology Kokrajhar
Kokrajhar, Assam
India

Deepak Kumar
Dairy Technology Division
Food Packaging Laboratory
ICAR-National Dairy Research Institute
Karnal, Haryana
India

Santosh Kumar
Department of Food Engineering and Technology
Central Institute of Technology Kokrajhar
Kokrajhar, Assam
India

Satti Venu Gopala Kumari
Department of Chemical Engineering
Indian Institute of Technology Guwahati
Guwahati, Assam
India

Manjusri Misra
School of Engineering, Thornbrough Building
University of Guelph
Guelph, Ontario
Canada

Department of Plant Agriculture
Bioproducts Discovery and Development Center
Crop Science Building, University of Guelph
Guelph, Ontario
Canada

Amar K. Mohanty
School of Engineering, Thornbrough Building
University of Guelph
Guelph, Ontario
Canada

Department of Plant Agriculture
Bioproducts Discovery and Development Center
Crop Science Building, University of Guelph
Guelph, Ontario
Canada

Avik Mukherjee
Department of Food Engineering and Technology
Central Institute of Technology Kokrajhar
Kokrajhar, Assam
India

Debarshi Nath
School of Engineering, Thornbrough Building
University of Guelph
Guelph, Ontario
Canada

Department of Plant Agriculture
Bioproducts Discovery and Development Center
Crop Science Building, University of Guelph
Guelph, Ontario
Canada

Kannan Pakshirajan
Department of Biosciences and Bioengineering
Indian Institute of Technology Guwahati
Guwahati, Assam
India

Narender Raju Panjagari
Dairy Technology Division
Food Packaging Laboratory
ICAR-National Dairy Research Institute
Karnal, Haryana
India

Ehsan Pesaranhajiabbas
Department of Plant Agriculture
Bioproducts Discovery and Development Center, Crop Science Building
University of Guelph
Guelph, Ontario
Canada

School of Engineering
Thornbrough Building
University of Guelph
Guelph, Ontario
Canada

G. Pugazhenthi
Department of Chemical Engineering
Indian Institute of Technology Guwahati
Guwahati, Assam
India

Centre for Sustainable Polymers
Indian Institute of Technology Guwahati
Guwahati, Assam
India

Jaspreet Kaur Rajput
Department of Chemistry
Dr B R Ambedkar National Institute of Technology
Jalandhar, Punjab
India

Vibhore Kumar Rastogi
Department of Paper Technology
Indian Institute of Technology Roorkee
Roorkee, Uttarakhand
India

Neelisetty Sesha Sai Baba
Department of Paper Technology
Indian Institute of Technology Roorkee
Roorkee, Uttarakhand
India

Anamika Singh
Department of Paper Technology
Indian Institute of Technology Roorkee
Roorkee, Uttarakhand
India

Dimple Singh
Department of Paper Technology
Indian Institute of Technology Roorkee
Roorkee, Uttarakhand
India

Poonam Singha
Department of Food Process Engineering
National Institute of Technology Rourkela
Odisha
India

Sushil Kumar Singh
Department of Food Process Engineering
National Institute of Technology Rourkela
Odisha
India

Varsha Singh
Department of Paper Technology
Indian Institute of Technology Roorkee
Roorkee, Uttarakhand
India

Vinayak Srivastava
Department of Paper Technology
Indian Institute of Technology Roorkee
Roorkee, Uttarakhand
India

T. P. Vinod
Department of Chemistry
Christ University
Bangalore, Karnataka
India

Srasti Yadav
Division of Chemistry, School of Basic Sciences
Galgotias University
Greater Noida, Uttar Pradesh
India

Preface

The project *Smart Food Packaging Systems: Innovations and Technology Applications* was born sometime in the year 2022 inspired by successful completion of our first book project with John Wiley & Sons. The idea of such a book stemmed from the emerging interest and emphasis on safe and healthy food ensured by sustainable food packaging preferred by consumers, food industries, and all other stakeholders. As more and more user-friendly active and intelligent food packaging systems are being developed, the replacement of hazardous synthetic inputs with natural, biodegradable, and/or edible materials is an important step towards the development of sustainable food packaging systems. This book is a complete and systematic account of the contemporary developments in the area of smart food packaging, including applications of bionanocomposites and natural active/intelligent agent(s) as sustainable alternatives to synthetic plastic polymers and synthetic chemicals, respectively. It is a unique reading and resource material for academicians, professionals of food packaging and other related industries, research scholars, graduate students, entrepreneurs, and people working in the area of food preservation, food packaging, food shelf life enhancement, environmental safety, human health, sustainability, and circular economy. Also, the book includes detailed discussions on intelligent agents of natural origins that add functionalities and communicate quality status of packaged food to consumers.

This book consists of sixteen (16) chapters divided into two parts: the first nine chapters covering different types of active food packaging systems incorporated with antimicrobial, scavengers of oxygen, ethylene, carbon dioxide emitters, absorbents of moisture, off-flavor, and various different nanomaterials, and the next six chapters elaborating on major types of intelligent packaging systems, e.g., time-temperature-indicators, freshness indicators, gas sensors, food quality indicators, such as humidity indicators, pH indicators, indicators of specific chemical, etc. The final chapter of the book gives a synoptic description of the life cycle assessment of smart food packaging systems including their contribution towards achieving circularity in our day-to-day livelihood.

It has been a great pleasure and a privilege to collaborate with the contributing authors representing almost a dozen academic Institutes/Universities of international repute in India and abroad. The success of this project depends on the usefulness of the book to potential readers across the entire spectra of food packaging and sustainability in academia, industries, businesses, and government bodies. The reader's enthusiastic acceptance, appreciation, comments, and critiques will inspire us to venture into more such collaborative projects in the future. We could not have achieved success in our efforts without the support, cooperation, and understanding of our peers, friends, and families, which stems from our unwavering dedication and commitment towards the successful completion of this book.

Avik Mukherjee
Santosh Kumar
Manjusri Misra
Amar K. Mohanty

Acknowledgments

We take this opportunity to express our sincere gratitude to the research grant from the All India Council for Technical Education (AICTE), Govt. of India, for partially funding this book project under the Research Promotion Scheme for North Eastern Region (RPS-NER) of India. We are also thankful for the partial funding support by (i) the Ontario Agri-Food Innovation Alliance – Bioeconomy for Industrial Uses Research Program, (ii) the Ontario Research Fund, Research Excellence Program; Round 11 (ORF-RE11) Ontario Ministry of Colleges and Universities, (iii) the Natural Sciences and Engineering Research Council of Canada (NSERC), and (iv) Canada Research Chair (CRC) program. Moreover, it was a great pleasure and privilege to work with John Wiley & Sons, the publisher of the book.

Avik Mukherjee
Santosh Kumar
Manjusri Misra
Amar K. Mohanty

1

Introduction to Active Food Packaging System

Sweety Kalita[1], Amar K. Mohanty[2,3], Manjusri Misra[2,3], Avik Mukherjee[1], and Santosh Kumar[1]

[1] *Department of Food Engineering and Technology, Central Institute of Technology Kokrajhar, Kokrajhar, Assam, India*
[2] *School of Engineering, Thornbrough Building, University of Guelph, Guelph, Ontario, Canada*
[3] *Department of Plant Agriculture, Bioproducts Discovery and Development Center, Crop Science Building, University of Guelph, Guelph, Ontario, Canada*

1.1 Introduction

The realm of food packaging is rapidly growing within scientific research and industry driven by rising global demands, and the continuous, stringent updates and amendments of food safety regulations. Food packaging is crucial in safeguarding food quality and shelf life. Shifts in consumer lifestyles have heightened interest in fresh, high-quality, and clean products with a prolonged shelf life, which has consequently prompted the necessity for advanced packaging technology [1–3]. Furthermore, with the passage of time, the scope of food packaging has progressed from basic containment to intricate systems that actively engage with the packaged food, giving rise to active food packaging systems. Active packaging has become essential across diverse sectors of the food industry [4]. Active packaging systems were first developed in the 1980s, and are now integrated with active compounds to retain or enhance quality aspects and safety of the packaged food [5]. Depending on the active components added into the active packaging, they are designed to reduce respiration, inhibit microbial growth, and/or mitigate moisture migration by releasing active agent(s), absorbing moisture or gases, blocking certain deteriorations, or buffering pH alterations [6, 7]. Broadly, active packaging falls into two main categories: nonmigratory, which involves scavengers that remove/absorb undesirable substances from headspace of the packaging without any release of an active agent into the packaged food (called an active scavenging/absorbing system), and migratory packaging, which includes emitters that facilitate sustained release of the active compounds inside the package (called an active emitting/releasing system) (Figure 1.1) [4, 8, 9]. Active agents are incorporated into packaging systems as separate entities (for example, a pad or sachet), combined with polymers (e.g., composite film or coating), or applied to the film or packaging surface (e.g., by coating a layer of active agent on the packaging material). Furthermore, the active agent

Smart Food Packaging Systems: Innovations and Technology Applications, First Edition.
Edited by Avik Mukherjee, Santosh Kumar, Manjusri Misra, and Amar K. Mohanty.
© 2025 John Wiley & Sons Ltd. Published 2025 by John Wiley & Sons Ltd.

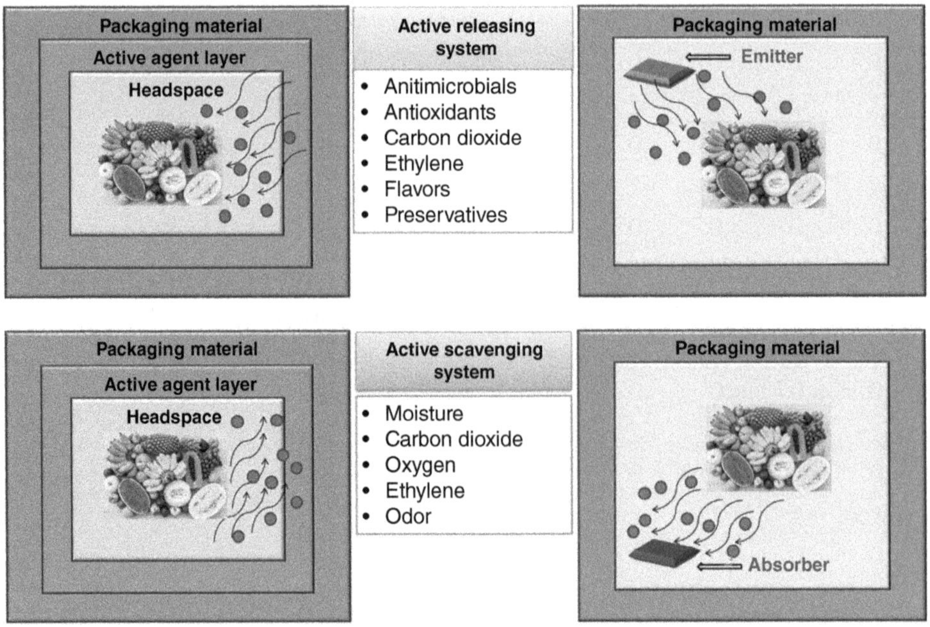

Figure 1.1 Active scavenging/absorbing and active emitting/releasing systems.

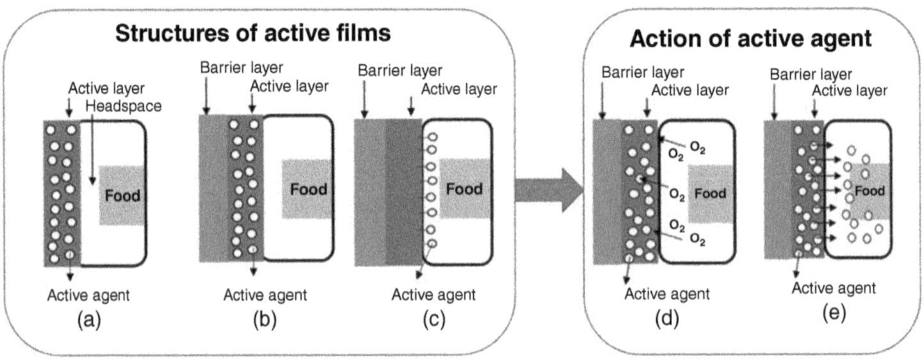

Figure 1.2 (a) Mono-layered film with an active agent. (b) Bi-layered film, in which the inner layer contains an active agent. (c) Bi-layered film having immobilized active substance on the film surface. (d) Active scavenging. (e) Active emitting systems. *Source:* Reference [10], 2022/MDPI/CC-BY-4.0.

either firmly fixes or immobilizes onto the film surface. The flexible active films can be mono-, bi-, or multilayered (Figure 1.2). Continuous technological progress is driving the evolution of novel active packaging systems, leading to elevated levels of food safety and quality, along with decreased waste and enhanced sustainability [11].

Food packaging made of biodegradable polymers is a sustainable alternative to the synthetic plastic-based packaging, and the former is gaining increased popularity due to consumer awareness about the detrimental effects of petroleum-based plastic packaging on human and environmental health [12]. Moreover, many natural compounds (e.g., certain secondary metabolites of some plants and microorganisms) act as antimicrobial and/or antioxidant active agents that can be blended into biopolymer-based active food packaging systems [12, 13]. Polysaccharides, such as chitosan, cellulose, and starch, proteins such as gelatin, whey, and casein, and lipids such as beeswax, carnauba wax, and shellac wax are being studied extensively for the development of active packaging [14–16]. In this chapter, an in-depth exploration of active packaging systems is presented, covering gas and moisture scavengers, ethylene absorbers, antioxidant-releasing systems, CO_2 emitters, and antimicrobial packaging. This chapter also reviews scientific research highlighting the benefits and challenges regarding the application of these novel systems for the packaging of perishable foods.

1.2 Types of Active Food Packaging

1.2.1 Active Scavenging/Absorbing System

Active scavenging/absorbing food packaging systems effectively scavenge/remove undesirable components/gases that are either present in the package or emanate from the food item into the packaging headspace. Ethylene scavengers for fresh fruits and vegetables, oxygen scavengers (OSs) for cut fruits and vegetables, fish, meat, edible oil, and moisture absorbers for milk powder, processed tea leaves, biscuits, crackers, and chips are effective for shelf-life extension of packaged products [17]. The various active scavenging systems are explained in the subsequent discussions.

1.2.1.1 Ethylene Scavenger

Ethylene (C_2H_4), a plant hormone, intricately regulates diverse physiological functions within plants, such as growth, ripening, and senescence with both positive and negative consequences. Its positive impact involves expediting the ripening process in fresh produce, while its negative effects lead to a diverse array of rapid undesirable changes [18, 19]. The process of fruit ripening generates ethylene, aldehydes, and other gases, amplifying the ripening progression. Ethylene influences ripening through two distinct forms: through endogenous ethylene, produced by the plant itself, and through external ethylene, originating from neighboring crops, automotive emissions, and polymers [20]. Ethylene also possesses the capability to trigger the expression of ripening-associated genes via signal transduction pathways [21]. Ethylene production can be controlled by preventing its synthesis, oxidation and by its absorption during the handling and transportation of fresh produce [20]. Notably, ethylene exerts pronounced effects on the quality of fruits and vegetables with even minute concentrations, as low as 0.1 μl/l, significantly impacting their growth and development [22]. Consequently, the implementation of ethylene control measures, such as the use of ethylene scavengers, becomes indispensable for curbing product losses and safeguarding food quality by delaying postharvest ripening processes [23]. Ethylene scavengers can effectively reduce ethylene levels in packaging, chemically or physically, as the gas is highly reactive due to its double bond [9]. Chemical methods typically employ potassium permanganate ($KMnO_4$) embedded in materials such as porous sachets, which oxidize ethylene into acetate, and ethanol without affecting the produce. However, $KMnO_4$ cannot be used in direct contact with food due to its toxicity and lack of efficacy in high-moisture environments [24, 25]. Other alternatives for ethylene removal include metal oxides, layer silicates, zeolites, nanoparticles, and activated carbon, which can be incorporated into packaging materials or provided inside the packaging in sachets [9, 19, 26]. These active materials/scavengers fall into two main categories: absorbers, which physically trap ethylene, and scavengers, which chemically react with ethylene. They can be incorporated in the film or in the form of sachets, and are commonly used in packaging for fruits and vegetables. The reduction of ethylene gas in the vicinity of fresh produce postpones fruit maturation by slowing down metabolic activities, resulting in a substantial extension of the shelf life of the produce [17, 22, 27]. Numerous scientific studies have explored the application of ethylene scavengers in active food packaging. Notably, Joung et al. [28] used encapsulated potassium permanganate (P)-based ethylene scavenger in halloysite nanotubes (HNTs) and revealed a reduction in ethylene production and respiration rates, as well as a delay in loss of firmness and color changes for 21 days in cherry tomatoes wrapped with the developed nanocomposite film. Furthermore, Kaewklin et al. [29] reported that the use of chitosan and TiO_2 nanocomposite films as an ethylene scavenger for postharvest management of climacteric fruits resulted in ethylene photodegradation, potentially contributing to the delay of the ripening process, and their shelf-life extension. Upadhyay et al. [30] developed an optimal ethylene scavenging film using a nanocomposite of corn starch (CS) and gum acacia (GA) with 20% sepiolite that demonstrated effectiveness in packaging broccoli florets for six days at 23 °C storage. Moreover, the cellulose nanofiber/TiO_2 nanotubes-Cu_2O-based film exhibited the capability to scavenge ethylene gas generated in the headspace during storage, resulting in delayed tomato discoloration, softening, and weight loss [31]. Another noteworthy development in the realm of ethylene scavengers involves innovative films containing nano-clay and nano-silica that have been incorporated with $KMnO_4$, which demonstrated an ability to extend the shelf life of bananas by up to 15 days when stored under ambient conditions [32]. An innovative $KMnO_4$-based ethylene scrubber, in combination with protonated montmorillonite (PMMT) and modified atmosphere packaging, effectively extended the shelf life of blueberries by 25 days at 2 °C and by 14 days at 10 °C [33].

1.2.1.2 Oxygen Scavenger

Intrusion of oxygen (O_2) within food packaging catalyzes the rapid oxidation of fats, oils, pigments, and vital nutrients resulting in food deterioration. Moreover, the presence of O_2 offers a conducive environment for the proliferation of microorganisms including aerobic bacteria, yeasts, and molds. Oxidative deteriorations result in undesirable traits, such as unpalatable odors, displeasing flavors, alterations in color, and degradation of essential nutrients. In response to this challenge, innovations such as packaging in vacuum and modified atmospheric conditions have achieved a reduction of in-package oxygen levels to 0.5–2% v/v. It is noteworthy that even within this range, deterioration may persist in the packaged food. In contrast, OSs offer a reliable solution by lowering oxygen concentrations to less than 0.1% (v/v), resulting in a far more impactful extension of the product's shelf life [1, 34]. These scavengers are rapid and high-capacity oxygen absorbers (OAs) that function through differential partial pressure, which facilitates the removal of oxygen from both the in-package environment and the packaged food product itself through diffusion [9, 35]. Iron powder sachets enclosed within small, oxygen-permeable bags are the most widely utilized OSs for commercial purposes. They function by oxidation of iron and ferrous salts, which react with water from the food to create a process that hydrates the iron within the packaging, irreversibly converting it to a stable oxide [36, 37]. The combination of modified atmospheric packaging (MAP) (e.g., 70% CO_2 and/or 30% N_2) and OSs containing Fe/FeO_2 effectively enhanced the shelf life of chicken thigh meat by hindering microbial growth and preventing lipid and protein oxidation for up to nine days [38]. Despite their effectiveness in diverse food applications, iron-containing sachets have drawbacks, such as accidental ruptures, a necessity for additional packaging, and an unsuitability for liquid food. Alternatively, iron nanoparticles, immobilized yeast, enzymes (such as glucose oxidase), ascorbic acid, photosensitive dyes, and unsaturated fatty acids (such as linoleic and linolenic acids) can be effective OAs [9, 34, 37]. Recent publications on alternative oxygen scavenging systems have highlighted nanoscale OSs featuring zerovalent iron particles within a silicon matrix. These nanoscale iron particles demonstrated a 10-fold increase in oxygen scavenging rates compared to that of commercially available iron-based scavengers within polyethylene or polypropylene (PP) polymer matrices [39]. In a study, a fast oxygen scavenge system was developed by Faas et al. [40] where they engineered an oxygen scavenging film utilizing palladium to mitigate lipid oxidation in linseed oil, the findings indicate that the application of this scavenging film in conjunction with MAP yielded a substantial reduction in lipid oxidation in linseed oil when compared to conventional MAP with 2 v/v O_2. Another metallic oxygen scavenger using titanium oxide nanotubes (TONT) demonstrated significantly higher oxygen uptake rates than that of the commercial iron-based scavengers, suggesting that TONT scavengers can be suitable as oxygen indicators for the secure packaging of highly oxygen-sensitive products, as they can switch from dark blue to yellowish brown color in the presence of oxygen [41]. Titanium dioxide (TiO_2) blended in hydroxyl-terminated polybutadiene (HTPB), low-density polyethylene (LDPE), and polylactic acid (PLA) revealed increased oxygen absorption by the composite [42]. Another study on coating mesoporous silica nanospheres with oxygen scavenging materials (iron or titania-based) has received a rapid response and enhanced adsorption efficiency, especially in dry conditions, in which activation by moisture is undesirable [43].

Researchers have now moved toward developing organic OSs in oxygen-absorbing films, as traditional metallic agents have limitations for packaging production. These organic agents may involve low-molecular-weight oligomers in polymer films with designed structures that react with oxygen, and thus provide better dispersion and compatibility [44]. Lee et al. [45] introduced a nonmetallic scavenger, i.e. activated carbon and sodium L-ascorbate, which showed higher oxygen scavenging efficiency (2.2 times) than Fe powder-based scavengers. These ascorbic acid-based systems, when applied to the packaging of raw meatloaf, effectively prevented the oxidation of lipid and lactic acid bacteria, yeasts, and mold growth [45]. Another study developed oxygen scavenging films by incorporating 10% or 20% pyrogallic acid, a phenolic compound known for its potent oxygen scavenging ability, in an LDPE-based film altered with adding sodium carbonate. The developed composite films were used for storing peeled garlic for 30 days at 5 and 25 °C, and the results showed reduced fungal decay, aerobic count, and weight loss in the product at both storage conditions [46]. The diffusion and interaction between oxygen and pyrogallol in the polymers are primarily controlled by factors such as molecular mobility, reaction rates, and film microstructures, as demonstrated by Promsorn and Harnkarnsujarit [47], who developed a thermoplastic starch-based composite film loaded with pyrogallol and linear LDPE. Gallic acid acts as a moisture-activated oxygen scavenger in degradable polyester films (polybutylene adipate co-terephthalate (PBAT)) and polybutylene succinate (PBS), among which PBAT shows smooth surface topography and structure, while the introduction of gallic acid in PBS reduced homogeneity leading to void spaces and an increase in surface roughness [48]. Numerous other reports have documented the successful incorporation of natural oxygen scavenging agents, such as gallic acid and polyisoprene into polymeric films through various fabrication methods [49–52].

1.2.1.3 Moisture Absorber

Condensation frequently occurs in packaged foods, particularly for those with high water vapor pressure due to the metabolic processes of fats and carbohydrates common in fruits and vegetables. Temperature variations between the outside and inside of the package lead to the formation of water droplets inside the package or on the surface of the food. The presence of water droplets negatively affects the package's appearance, reduces consumer appeal, and contributes to increased surface mold growth, thereby shortening food shelf life [1]. Moreover, storage conditions, especially relative humidity, profoundly affect the quality of dry food items. Insufficient humidity within packages can significantly degrade the quality, as increased moisture content heightens susceptibility to microbial growth and triggers undesirable alterations in appearance and texture, leading to a reduced shelf life. Conversely, for products with elevated water activity, such as fish, meat, fruits, and vegetables, maintaining controlled humidity proves advantageous in preventing dehydration, however, these packaged products may experience excess liquid due to drip loss and temperature fluctuations during storage [9, 53, 54]. A practical method for managing excess moisture within a package is incorporating a moisture absorber to regulate the water activity of the product, thus inhibiting microbial growth [55]. Moisture absorbers are typically available as sachets, films, trays, or pads, in which the materials designed for moisture absorption/scavenging are typically enclosed to control exudates from food products or to control the desired humidity inside the package [56]. Moisture absorbers are hygroscopic substances or desiccants such as silica gel, clays, synthetic crystalline sieves (e.g., potassium, zeolite, calcium alumina silicate, and sodium), and humectant salts (e.g., NaCl, $MgCl_2$, or compound such as sorbitol). The absorption capacity of desiccants is determined by the water vapor sorption isotherm. Desiccants are typically incorporated inside the packages as sachets, microporous bags, or combined with pads [9, 57]. A recent study showcased the enhancement of desiccants within the PP medium by adding different molecular weight polyethylene glycol and molecular sieves (MS) [58]. Notably, polyethylene glycol with molecular weight of 8000 facilitated uniform microphase separation in the composite because of its enhanced chemical compatibility with MS during the channeling structure formation, leading to substantial moisture absorption. Although most commercial manufacturers use desiccants for high-moisture foods, in recent years, a growing interest in superabsorbent composites, including polymer-based materials, is evident among packaging film industries [9, 57].

Researchers are particularly interested in food-grade organic absorbers such as fructose and sorbitol [44]. Bovi et al. [56] introduced a Fruitpad, which included fructose to enhance the overall moisture absorption capacity for the packaging of fresh strawberries. A Fruitpad containing 30% fructose demonstrated 0.94 g of moisture absorption (highest) per gram of pad at 100% RH and 20 °C [56]. Cellulose is a hygroscopic substance that exists naturally and is widely utilized in the manufacturing of moisture-absorbing films for food packaging. This polymer, and its derivatives such as carboxymethyl cellulose (CMC), can absorb moisture on the surface or within the bulk of the amorphous regions [59–61]. In the industrial production of moisture absorbers, inorganic materials, such as calcium oxide (CaO), silica gel, and calcium chloride in sachets for adsorption, play a crucial role. These materials employ both physical and chemical processes to effectively reduce the relative humidity inside the package or the collected liquid at the bottom of the food container. Various other minerals such as potassium chloride (KCl), potassium carbonate (K_2CO_3), and bentonite have also served as moisture absorbers in food packaging applications [1, 9]. In a recent study, a moisture absorber was developed using rice husk and $CaCl_2$, and the resulting sachet was applied to the packaging of coconut palm sugar granules (CSG). The results showed that the rice husk-$CaCl_2$ composite with a concentration of 15% $CaCl_2$ exhibited the highest moisture absorption capability maintaining the moisture content of CSG at approximately 2% under conditions with a relative humidity of 70% [62]. Furthermore, the moisture sorption isotherm behavior of the rice husk-$CaCl_2$ composite desiccants displayed a J-shaped curve of equilibrium moisture content with boundary zones A_{aw} 0.2 and B_{aw} 0.67, which indicated that the rice husk desiccant could effectively absorb water vapor at both low and moderate humidity levels [63]. In general, polymer-based packaging materials, such as sorbitol containing glycerin, incorporated with uniformly dispersed magnesium chloride, can effectively control the humidity in packaged foods by reducing condensed water and stabilizing relative humidity within the packaging headspace. However, a challenge in manufacturing such materials is the increased volume of a salt solution in comparison to its dry crystal. To address this, the polymeric film needs an appropriate porous structure to accommodate the salt solution, which functions inside the film through water vapor absorption [64, 65]. Sängerlaub et al. [66] observed that PP film containing dispersed NaCl increased porosity that facilitates enhanced diffusion of the salt boosting its water vapor absorption rates. The authors further proposed that a similar positive effect could be achieved through the thermoforming process. Consequently, thermoforming was found to elevate the packaging material's porosity from 0.28 to 0.63 [64]. In a recent study, various equilibrium-modified atmosphere packaging (EMAP) systems such as cellulose, PLA, and polyethylene terephthalate (PET) with PP film as moisture absorbers were tested for cape gooseberry fruits. Stored at 6 °C and 75% RH, fruits in PLA trays without sachets had a maximum shelf life of 42 days due to the higher water vapor permeability preventing condensation and reducing dehydration compared to cellulose packages [67]. Prior studies have proven the efficacy of polymeric moisture absorbers, including PVA/*Uncaria Gambir* (UG) extract film [68], HDPE bags with a moisture absorber containing triple-refined super-activated white DMF and PCP-free silica gel [69], and TEMPO-oxidized cellulose nanocrystal (CNC) film [70].

1.2.1.4 Ultraviolet (UV) Light Absorbers

Active packaging incorporated with UV light barriers is a notable alternative to traditional packaging that involves integrating light absorbers or coatings onto packaging materials to diminish photo-oxidation, a major cause of degradation of light-sensitive food such as muscle products and beverages. Aluminum/metalized foil is recognized as the top material for blocking UV and visible light. Despite its effectiveness, it has drawbacks related to recyclability, cost, and transparency. However, recently studied UV absorbers address these concerns when selectively choosing and applying them to suitable polymers and considering their specific functions [37, 71]. Inorganic UV-absorbing materials, mainly composed of metal oxide particles, include titanium dioxide (TiO_2), iron oxide (Fe_2O_3), and zinc oxide (ZnO) which utilize "pigment theory" through band gap absorption and light scattering pathways to attenuate light. The band gap property enables semi-conductive metal oxide particles to absorb photons when their energy matches or exceeds the band gap. Additionally, the high refractive index of metal oxide particles contributes to light blocking through reflection, refraction, and scattering, influenced by their internal crystal structure [71]. In a recent research work, PLA biopolymer-based film incorporated with 5 wt% TiO_2 nanoparticles was found to block UV radiation effectively [72]. Another contemporary study developed a nanocomposite using dapsone-capped TiO_2 nanoparticles (DAP-TiO_2-NPs) incorporated into biodegradable polyvinyl alcohol (PVA) films, the results revealed that the films containing DAP-TiO_2-NPs had excellent UV-shielding properties, blocking almost all UV bands and up to 98% of visible light even at lower nanofiller ratios [73]. Composite films, formed with 2% graphene oxide (GO) nanosheets and regenerated cellulose exhibited notable UV-shielding capabilities, blocking 66.7% of UVA, 54.2% of UVB, and showed outstanding visible light transmittance [74]. Organic UV absorbers, especially phenolic compounds, are either colorless or light yellowish in color with a high absorption coefficient in the UV range. Several research publications focused on organic UV light absorbers have effectively showcased their research findings that phenolic absorbers possess an excellent capacity to absorb and convert light energy into less harmful forms, along with excellent photo-stability [75–79]. These compounds typically incorporate intramolecular hydrogen bonds, such as O–H–O bridges or O–H–N bridges, which effectively dissipate energy, transforming incident light energy into low thermal energy through photo-physical reactions. This mechanism helps postpone the formation of free radicals in the early stages of degradation [71]. A film of poly (methyl methacrylate) combined with lignin (L) from coconut shell demonstrated UV-blocking properties, showed lower transparency (approximately 26%) compared to pristine films, transmitted about 79% visible light at 550 nm, and exhibited excellent ability toward UV light absorption [80]. Shikinaka et al. [81] reported on a film composed of lignosulfonate and clay minerals that demonstrated outstanding UV protection properties, providing 99% UVA protection. Additionally, hybrid UV absorbers can be developed by either incorporating an inorganic UV absorber into the organic polymer matrix or by polymerizing an organic monomer to envelop the inorganic phase of the UV absorber [71]. For instance, Lei et al. [82] developed catechol-functionalized chitosan (C-CS)/PVA composite films for active food packaging through a solution blending method in a neutral aqueous solution. The UV transmittance (at 280 nm) of C-CS/PVA films decreased by 67.6% with a 10 wt% C-CS content, while maintaining transparency in the visible range compared to pure PVA film [82]. Another study produced durable and transparent bioplastics through the blending of cellulose and naringin at various ratios, and the results exhibited outstanding transparency and highly effective UV-blocking capabilities of the developed films, particularly against UV-A and UV-B radiation [83]. Table 1.1 compiles diverse studies on hybrid UV absorber-based composites intended for application in food packaging.

Table 1.1 Hybrid UV light absorber-based food packaging systems.

Active substances	Film-forming polymer/biopolymer	Benefits	Reference
Extract of cocoa bean shell and zinc oxide nanoparticles	Pectin-based active film	The incorporation of the extract and nanoparticles significantly improved UV barrier properties by 98%, a crucial enhancement for UV-sensitive packaged foods.	[84]
Methylene blue (dye) and cloisite (clay)	Polylactic acid film (PLA)	PLA films, with a thickness of 100 μm and containing dye and clay demonstrated 100% UV protection performance, slightly surpassing that of PLA-Cloisite.	[85]
Grape syrup (GS)	Polylactic acid film (PLA)	PLA-GS films demonstrated UV absorption from 250 to 400 nm, as indicated by UV–Vis results due to O–H, C=C, and C=O functional groups.	[86]
Date palm syrup (DPS)	Starch films	UV–Vis spectroscopy revealed effective UV absorption for these films, attributed to C=O, C=C, and O–H functional groups.	[87]
Uncaria gambir (UG) extract and boric acid (BA)	Poly(vinyl alcohol) film	The PVA/UG/BA film absorbs around 98% of UV light at 400 nm without compromising transparency.	[68]
Waste tea residue carbon dots	Poly(vinyl alcohol) film	Higher amount of WTR-CDs in PVA films enhanced UV light protection, with 100% blockage of UV-C and UV-B at 230–280 nm, and 280–315 nm, respectively.	[88]
Cellulose acetate and phenyl salicylate	Polylactic acid (PLA) film	The composite films completely absorbed UV-C and UV-B regions, while achieving over 95% UV absorption in the UV-A range.	[89]
Lignin (L)	Cellulose nanofibril film (CNF)	The lignin-cellulose nanofibril films completely blocked UV rays in range of 290–400 nm.	[90]
Citric acid-modified lignin nanoparticles (LNP)	Poly(vinyl alcohol) nanocomposite film	The UV-blocking test showed excellent performance, with over 97.5% UV absorption.	[91]
Cellulose nanocrystals (CNCs) and diethyl ferulate (DEF)	Poly(vinyl alcohol) film	PVA films containing 20 wt% CNC-DEF provided total UV blockage (100%).	[92]
Lignocellulose particles (extracted from waste hemp hurd)	Poly(vinyl alcohol) film	The hemp/PVA films show enhanced UV protection mainly in the UVA range.	[93]
ZnO nanoparticles	Poly(vinyl alcohol) film	ZnO-blended films had a notable UV shielding below 380 nm.	[94]

1.2.1.5 Flavor Absorbers

Unpleasant aromas and flavors can deteriorate sensory attributes of foods, significantly influencing their consumer acceptability. Undesirable aromas can emanate from food due to various physiological and biochemical changes, e.g. amines produced by oxidation of protein in foods such as fish, rancid flavored aldehydes, and ketones resulting from oxidation of fatty acids in foods such as fried foods, bakery, and fruit juices. Additionally, packaging materials, particularly plastic components such as polyolefins, may contribute to undesirable odors. Flavor absorbers such as clays, zeolites, and activated carbon are designed to eliminate undesirable aromas within the package headspace [95, 96]. A typical flavor-absorbing package is constructed from an impermeable plastic such as PE or PET with sufficient thickness, featuring a respiratory gas-permitting port. The package includes a sachet containing a blend of nickel and charcoal for odor absorption and to counteract unpleasant odors, e.g. for alkaline volatile amines such as trimethylamine emanating from fish protein breakdown, citric acid is an effective odor absorber. ANICO Co. Ltd., Japan, manufactures the ANICO bags, which are food packaging films incorporated with ferrous salt and organic acids, such as ascorbic or citric acid, to oxidize free amines in the food they contain [97]. A recent US patent discloses a flavor absorbent pad with activated carbon, which acts as an antimicrobial agent and is also aimed at minimizing confinement odor in vacuum-packaged food products. The activated carbon not only traps odorous compounds emanating from spoiled foods but also effectively inhibits microbial spoilage of the packaged food [98]. Various other commercial flavor absorption food packaging systems include sachets such as MINIPAX and STRIPPAX (Multisorb Technologies, USA), designed to absorb off-flavors due to the formation of mercaptans and H_2S in certain spoiled foods during storage and distribution. The Profresh brand is promoted as a masterbatch for freshness preservation and malodor control in polymeric packages such as polyethylene and polystyrene, in which ADI50 (composition undisclosed) is incorporated as the active component to absorb ethylene, ethyl alcohol, ethyl acetate, and H_2S from within the headspace [99].

1.2.2 Active Releasing or Emitting Systems

1.2.2.1 Carbon Dioxide Emitter

CO_2 positively impacts the initial freshness of food and prevents oxidative deteriorations such as color and odor-related spoilage. Its antimicrobial activities work against aerobic bacteria, yeasts, and molds, making it useful for shelf life enhancement and quality retention, particularly in muscle foods and non-climacteric fruits. Higher concentrations of CO_2 in a MAP system effectively reduce microbial growth by influencing their lag and logarithmic growth phases through alterations in cytoplasmic pH, inhibiting bacterial enzyme synthesis, and affecting bacterial cell membranes [7, 37, 100]. Generally, maintaining oxygen levels below, and CO_2 levels above, atmospheric levels effectively preserves and extends the freshness of foods [95, 101]. MAP targets an optimal headspace gas volume to food product volume ratio (*g*/*f* ratio) between 2/1 and 3/1 for effective CO_2 dissolution and reduced package deformation. Higher *g*/*f* ratios may lead to larger packages, inefficient distribution, increased resource usage, and environmental impact. Decreasing the *g*/*f* ratio necessitates higher CO_2 amounts for equivalent antimicrobial effects. Continuous CO_2 refill by a CO_2

emitter is vital to prevent package deformation due to alteration(s) in pressure [37]. Carbon dioxide CO_2 emitters in packaging permit the gas to permeate through the film, reducing respiration rates and inhibiting microbial growth in the packaged fruits and vegetables. An effective commercial MAP strategy for highly perishable foods includes a dual-function system with both an oxygen scavenger and a CO_2 emitter. AgelessR, Verifrais, and FreshPaxR are some commercially successful examples of such MAP systems using materials such as ferrous carbonate or a mixture of an organic acid (e.g., citric or ascorbic acid) and sodium bicarbonate [102, 103]. The reaction begins when water within the food dissolves these components reducing the pH and triggering the production of CO_2, as sodium bicarbonate acts as a buffering salt that shifts towards un-dissociated carbonic acid and CO_2 formation [9]. A common method is to encase specified amounts and ratios of citric acid dry powder and sodium bicarbonate in an absorber pad. Positioned beneath the food product, this pad functions as both a CO_2 emitter and liquid absorber. The advantage of such a system lies in its ability to adjust the ratio of active ingredients for compatibility with the pH of the specific food product that is packaged [37]. Hansen et al. [100] studied vacuum or MAP packaging of Atlantic cod loins with 60% CO_2 to 40% N_2 ratio using CO_2 emitter pad. The study showed that adding a CO_2 emitter to vacuum-packaged and MAP cod loins enhanced microbial and sensory quality. Vacuum packaging with a CO_2 emitter achieved quality preservation comparable to MAP (gas-to-product ratio of 1.5/1). The CO_2 emitter in MA packages absorbed CO_2 in fish tissue, reducing its partial pressure in the MA package headspace compared to packages without the emitter [100]. Holck et al. [104] reported that the emitter effectively compensated for CO_2 absorbed by chicken fillets in a MAP system with 100% CO_2. This led to a significant reduction in drip loss (from 7.5% to 2.5% weight loss) in the packaged fillets due to a decrease in package collapse and physical pressure on the fillet [104]. Pettersen et al. [105] found that packaging fresh reindeer meat in a MAP system (60% CO_2, 40% N_2), with or without a CO_2 emitter, extended the sensory shelf life of the meat product to 21 days, compared to 17 days with vacuum packaging. Additionally, the packaging strategy with a CO_2 emitter led to a significantly reduced drip loss from 3% weight loss with an emitter, compared to 1% weight loss without an emitter [105]. Extensive research has examined the CO_2-releasing system in food packaging, exploring the impact of additional active agents such as ascorbic acid and ferrous carbonate as emitters on quality, shelf life, and efficacy of modified atmosphere packaging [37].

1.2.2.2 Antimicrobial Agent Emitter/Antimicrobial Packaging

Antimicrobial agents are used to inhibit and/or slow down microbial growth, which is the major cause of food spoilage. In general, a high degree of free water (i.e., high water activity) in food products is a leading factor in spoilage, which may be further exacerbated by insufficient processing and compromised packaging integrity. Traditional methods of preservation, such as thermal processing, including aseptic canning, drying, freezing, and an addition of synthetic preservatives, help mitigate these quality deteriorations effectively; however, thermal processing and an application of synthetic chemical preservatives are increasingly avoided by consumers due to their health concerns [97, 106, 107]. Active packaging with antimicrobial agents, including natural antimicrobial compounds applied in the form of emitting sachets or absorbent pads, enables the reduction or elimination of preservatives in food products [108, 109]. Most commercial antimicrobial packaging systems such as Bactiblock®, Biomaster®, Food-touch®, Surfacine®, AgIon®, IonPure®, and Sanic Films utilize active compounds such as glucose oxidase, chlorine dioxide, silver zeolite, ethanol vapor, triclosan, and nisin. They are applied in the form of sachets, bioactive coating agents, and as active compounds in packaging films and typically on products such as fruits and vegetables, bread and bakery products, cheese and other dairy products, and dried muscle food products. Active and antimicrobial absorbent pads are particularly effective for the packaging of muscle/meat products as these help to remove exuded liquids, thereby hindering microbial growth [110]. The literature on antimicrobial packaging encompasses a range of agents, including metal oxides such as ZnO, MgO, and metal ions such as silver, copper, essential oils (EOs) such as oregano, pimento, thyme, clove, plant extracts such as grape seed, pomegranate peel/rind, cinnamon bark, garlic, oregano, bioactive components, polysaccharides (chitosan), peptides (nisin, lactoferrin), enzymes (peroxidase, lysozyme), and synthetic antimicrobial agents (ammonium salts, EDTA, propionic acid, and benzoic acid) [111–113]. Metals, known for their antimicrobial properties, remain widely applicable in active food packaging. Among them, silver and its compounds, including metallic silver, silver ions, and silver nanoparticles exhibit highly effective antimicrobial action at low concentrations and demonstrate minimal systemic toxicity toward consumers [114]. Roy and Rhim [115] found that starch and agar-based composite packaging films incorporated with silver nanoparticles, exhibited enhanced physical properties, including mechanical strength and water vapor barrier, along with notable antibacterial activities against *Escherichia coli* and *Listeria monocytogenes*. Metals and metal oxides such as Cu, Zn, and TiO_2, including their nanoforms, have also demonstrated promising antimicrobial activities. In 2019, Amjadi et al. developed a bio-nanocomposite packaging using gelatin, chitosan nanofiber (10%), and ZnO NPs (30 nm size at the rate of 5%) that exhibited potent antimicrobial activity against foodborne pathogens including *Staphylococcus aureus*, *E. coli*, and *Pseudomonas aeruginosa* in chicken fillets and cheese samples during storage [116]. Another study revealed that an antimicrobial active packaging film blending semolina flour, ZnO, and kaolin for mozzarella cheese storage displayed robust antimicrobial activity against bacteria (*E. coli* and *S. aureus*), yeast (*Candida albicans*), and mold (*Aspergillus niger*) [117]. EOs derived from plants, such as clove, oregano, coriander, lemongrass, basil, and grape seed extract, along with the major active components isolated from them, e.g. carvacrol, eugenol, and thymol in active food packaging, have demonstrated significant efficacy in inhibiting microbial growth in various perishables. However, a challenge with EOs is the requirement of high concentrations to achieve effective antimicrobial activities in packaged or preserved food, which may potentially impact the organoleptic qualities of the food products [118]. A composite film poly (lactide)-poly(butylene adipate-co-terephthalate) (PLA-PBAT) incorporated with clove and thyme oils inhibited *E. coli* by 93.43% and 82.30%, respectively. Moreover, the clove oil composite film eliminated *S. aureus* (reducing their counts from 6.5 log CFU/ml to 0 log CFU/ml) and demonstrated UV-blocking properties [119]. Another study revealed that carvacrol-loaded PLA films enhanced the shelf life of ground beef, demonstrating a strong antimicrobial effect [120]. Apart from the aforementioned active agents, biomacromolecules, such as peptides (nisin), enzymes (lysozyme), and polysaccharides (chitosan, cellulose), are also studied for their antimicrobial activities in food. Nisin, which is the only permitted bacteriocin peptide, serves as a natural food preservative in various food products [121]. Recently,

Chang et al. [122] demonstrated that the incorporation of EDTA-nisin in PLA-chitosan films enhanced the antimicrobial activities of the packaging film, extending the shelf life of packaged fish fillet at 25 °C or 4 °C. EDTA disrupts the stability of the cell structure of Gram-negative bacteria by binding metal cations to nisin creating pores on the cell surface [122]. Lysozyme, a single-peptide enzyme acts as a food preservative by hydrolyzing the β-1,4-glucosidic linkages in the peptidoglycan layers of Gram-positive bacteria cell walls. However, it is less effective against Gram-negative bacteria due to the presence of a lipopolysaccharide layer preventing the exposure of their cell wall to lysozyme [123]. Recently, the incorporation of lysozyme nanofibers into pullulan films conferred antimicrobial activity against *S. aureus* [124]. Chitosan, derived from chitin, is widely researched in food packaging for its broad-spectrum antimicrobial activity due to the amino group, however, such effects differ between Gram-positive and Gram-negative bacteria, as highlighted in a recent review [125]. Chitosan-based composite films and coatings incorporated with various active agents such as silicon dioxide nanoparticles [126], titanium dioxide [127], and EOs [128, 129] have been studied extensively. The publications listed in Table 1.2 represent antimicrobial systems involving EOs, organic acids, biomacromolecules, metals, and the combination of multiple agents.

1.2.2.3 Antioxidant Emitters

Reactive free radicals and nonradicals, such as superoxide, peroxyl, hydroxyl, and alkoxyl cause oxidative damages such as a deterioration in texture, color, development of off-flavors, aldehyde formation, and a loss of nutritional value in food. These oxidative reactions are major contributors to the reduced shelf life of food, especially meat, poultry, seafood, cut fruits, vegetables, juices, edible oils, and fats. To address these food deteriorations, antioxidant packaging systems can be incorporated with active sachets or pads that contain certain compounds that retain or release gases or vapors, thus minimizing or inhibiting oxidative damage to food products [145]. Numerous synthetic and natural antioxidants, such as butylated hydroxytoluene (BHT) and butylated hydroxyanisole (BHA), have been extensively employed in these active packaging systems to inhibit oxidative deterioration in packaged foods. However, the addition of synthetic antioxidants such as BHA and BHT in food products raises health concerns, limiting their applications. Thus, there is a rising preference for natural antioxidants such as plant extracts, polyphenols, EOs, and tocopherols in active food packaging [37, 146]. An alternative approach involves incorporating active antioxidant agent(s) in packaging films, instead of a direct addition to the food. The antioxidant(s) are gradually released from the film creating a protective atmosphere in and around the packaged food item, effectively delaying oxidative reactions, resulting in a lengthened shelf life of the packaged/coated food [147]. In a recent investigation conducted by Minh et al. [148], it was demonstrated that LDPE bags incorporated with BHA effectively mitigated lipid oxidation in shrimp during frozen storage. Also, the integration of BHA and BHT into a PP matrix enhanced the mechanical strength, barrier characteristics, and antioxidant properties of the developed packaging film [149]. As a promising substitute for conventional plastic food packaging, a recent strategy involves biopolymer-based active packaging made of semi-refined carrageenan with cellulose nanofiber, in which BHA was incorporated, leading to improved solubility, opacity, thickness, moisture content, and antioxidant activity of the film [147]. Natural antioxidants such as tocopherol,

Table 1.2 Antimicrobial-based packaging systems.

Antimicrobial active agent	Film-forming polymer/ biopolymer	Food sample	Benefits	Reference
Clove essential oil (CEO) and montmorillonite (MMT)	Soy protein isolate (SPI)	Bluefin tuna	Clay blending in soy protein enhanced sustained release of active components, exhibiting prolonged antimicrobial effects.	[130]
Lactoferrin and lysozyme	Polyethylene terephthalate (PET)	Salmon	Lysozyme/lactoferrin-coated PET demonstrated effectiveness in reducing H_2S-releasing bacteria during storage at elevated temperatures.	[131]
ZnO nanoparticles	Polylactic acid (PLA)	Apple	The nano-blend film exhibits superior performance in inhibiting microbial growth, including bacteria, yeast, and fungi.	[132]
Oregano essential oil (OEO)	Pectin	Tomatoes	Fungal decay and oxidative deterioration were prevented.	[133]
Cinnamaldehyde	PLA/ polybutylene adipate (PBAT)	Meat analogues	Hybrid film effectively inhibited *E. coli* and *S. aureus* growth during storage at 4 °C.	[134]
ε-poly-(L-lysine) (ε-PL)	Carboxymethyl cellulose/ montmorillonite (MMT)	Strawberries	ε-PL incorporated CMC/ MMT films had antimicrobial activities against *S. aureus*, *E. coli*, *Botrytis cinerea*, and *Rhizopus oligosporus*.	[135]
Chitosan, nisin, potassium sorbate, and silver-substituted zeolite	Low-density polyethylene (LDPE)	Chicken drumsticks	Reduced thiobarbituric acid reactive substance (TBARS) and microbial spoilage.	[136]
Thyme essential oil	Zein nanofiber	Strawberries	TEO-loaded zein nanofiber film markedly reduced total bacterial counts, fungi, and yeast over a 15-day storage period.	[137]
Nisin-silica liposomes	Chitosan	Cheddar cheese	Composite was effective against *L. monocytogenes* without damaging sensory quality of the cheese.	[138]

Table 1.2 (Continued)

Antimicrobial active agent	Film-forming polymer/ biopolymer	Food sample	Benefits	Reference
Oxidized regenerated cellulose	Poly(ε-caprolactone) (PCL)	Sliced salami	50% reduction in *Listeria monocytogenes* growth observed over 14 days at 4 °C.	[139]
Nisin	Chitosan/ cellulose film	Cheese	Films containing 1 mg/ml of nisin completely eliminated *L. monocytogenes* on white cheese surfaces during 14 days of storage at 4 °C.	[140]
ZnO nanoparticles	carboxymethyl cellulose/okra mucilage	Chicken breast	The bio-nanocomposite demonstrated significant antimicrobial efficacy, exhibiting a large inhibition zone against *S. aureus,* and reducing lactic acid bacteria (LAB) counts in stored chicken breast.	[141]
ZnO nanoparticles, *Zataria multiflora* Boiss essential oil, and *Menthe piperita* L. essential oil	Polylactic acid (PLA)	Otolithesruber fish	Nanocomposite film exhibited effective antibacterial action against *E. coli, Pseudomonas aeruginosa, Salmonella enterica, Bacillus cereus*, and *S. aureus.*	[142]
Ag/TiO_2	High-density polyethylene (HDPE)	Bread	The microbial study of packaged bread demonstrated stability against yeast, mold, *Bacillus subtilis*, and *Bacillus cereus* counts.	[143]
Nisin	Chitosan/ nano-silicon dioxide	Blueberry	Chitosan/nano-silicon dioxide, supplemented with nisin, effectively inhibited microbial populations of molds/ yeast and mesophiles.	[144]

quercetin, catechin, thymol, ferulic, and ascorbic acid, along with plant and fruit extracts such as oregano, rosemary, mint, pomegranate peel, green tea, and EOs from lemongrass, clove, thyme, and ginger have gained popularity in active food packaging applications [150–153]. Antioxidant compounds are categorized based on their mechanisms of action i.e. primary (chain-breaking) antioxidants acting as free radical scavengers, and secondary (preventive) antioxidants including UV absorbers, OS, singlet oxygen quenchers, and metal chelators. Secondary antioxidants are advantageous as they can reduce or prevent oxidation reactions, while primary antioxidants react with free radicals, converting them into relatively stable products, and hindering further initiation or propagation of the reactions. Free radical scavengers, the most extensively studied antioxidants, donate hydrogen to react with free radicals, creating stable radicals that impede initiation or propagation of the deteriorating reactions. Examples include synthetic agents such as BHA, BHT, TBHQ, propyl gallate, and natural antioxidants such as plant extracts and EOs [154, 155]. Yang et al. [156] demonstrated that antioxidant films incorporated with green tea extract improved the oxidative stability of packaged pork meat products. Additionally, Lorenzo et al. [157] found that multilayer films with oregano essential oil (OEO) were more effective in stopping lipid oxidation in foal meat in MAP than those incorporated with green tea extract. Carrizo et al. [158] introduced a unique method using sealable multilayer films, where green tea extract in the laminating adhesive indirectly shielded packaged food (peanuts and cereals covered with chocolate) against oxidation. A recent study developed an active antioxidant sachet using PVA and CS with coconut shell extract and sapiolite clay. This formulation demonstrated enhanced antioxidant activity, contributing to improved oxidative stability of packaged soybean oil [159]. Besides primary antioxidants, secondary additives such as chelating agents, UV absorbers, and singlet oxygen quenchers are employed in active packaging to decrease the rate of oxidation in the packaged food products [154, 160–165].

1.2.2.4 Ethanol Emitters

Ethanol is a recognized traditional preservative valued for its robust antimicrobial activity and it is commonly applied either by direct spraying on the product or packaging or by incorporating it into the package through ethanol emitters (EEs). From the consumer perspective, ethanol stands out as a preferable preservative compared to other chemical alternatives largely due to its ability to reduce water activity and its notable anti-staling properties, which are particularly advantageous for bakery products. EEs are employed to release ethanol vapor into the headspace of the package, effectively inhibiting microbial growth in the packaged food [1, 7, 166, 167]. Several commercial ethanol-releasing systems are available (e.g., Ethicap and Oitech), which utilize ethanol adsorption in silica as the emitter with a high and uncontrolled volatilization rate. However, off-flavor development is a major challenge in the application of EEs. To mitigate this challenge, certain sachet emitters incorporated with added flavor(s), such as vanilla, can effectively conceal the alcohol odor in the package headspace [7, 168, 169]. Furthermore, Mu et al. [170] devised a controlled-release ethanol emitter to address drawbacks of the conventional ones. They mixed ethanol with varying concentrations of sodium stearate, creating a translucent ethanol gel, which was then combined with diatomite to enhance the strength. The resulting white powder ethanol emitter showed controlled release rates. The developed emitter was applied to the active packaging of Chinese bayberry fruits and demonstrated effectiveness in reducing decay, maintaining firmness, and inhibiting malondialdehyde formation [170]. Additionally, several studies have reported on synergistic effects of combining EEs and OAs. Latou et al. [171] studied efficacy of an ethanol emitter combined with an oxygen absorber (EE + OA) as a substitute for synthetic preservatives in prolonging the shelf life of sliced wheat bread. Results indicated that the developed synergistic effects extended the shelf life of the product from 4 to 30 days compared to 24 days for the EE system alone [171]. Janjarasskul et al. [172] showed combined effects of ethanol emitter and oxygen absorber on shelf life extension of sponge cake when wrapped with polyvinylidene chloride (PVDC)-coated nylon or pristine nylon.

1.2.2.5 Sulfur Dioxide Emitters

The incorporation of sulfur dioxide (SO_2) in certain foods such as tomatoes, cabbage, coconuts, peas, potatoes, carrots, and beverages, such as wine, has proven to be beneficial in mitigating oxidative discoloration during storage. SO_2-based emitters typically function through metabisulfite hydrolysis mechanism and the reaction of calcium sulfite with moisture [1, 7]. Besides SO_2, sulfite salts, such as sodium sulfite, sodium bisulfite, sodium metabisulfite, potassium sulfite, and copper sulfite, are employed to impede both enzymatic and nonenzymatic browning throughout the processing and storage phases. Codex Alimentarius categorizes sulfites as a significant allergen and strictly regulates the incorporation doses of SO_2 in foodstuffs. Concentrations exceeding $10\,mg\,l^{-1}$ are considered allergenic. The presence of SO_2 must be indicated on the product label [173]. For example, gelatin-based multifunctional composite films were fabricated and incorporated with varying amounts of copper sulfide nanoparticles (CuSNPs), which markedly improved the film's UV-blocking ability, mechanical strength, water vapor barrier properties, and antibacterial activities against *E. coli* and *L. monocytogenes* [174]. In another study, sulfur nanoparticles were synthesized and incorporated into calcium cross-linked alginate films resulting in improved tensile strength, water vapor barrier properties, and bacteriostatic activities against *E. coli* and *L. monocytogenes* [175]. The incorporation of sulfur quantum dots in gelatin/agar composite films improved their UV protection, while maintaining their appearance. The resulting composite film also exhibited antioxidant activity, a moderate bactericidal effect against *L. monocytogenes*, and inhibitory action against *E. coli* [176]. The researcher extended their work by further incorporating sulfur quantum dots into the alginate film that demonstrated notable antioxidant and robust antibacterial activities against bacteria (*E. coli* and *L. monocytogenes*) and fungi (*A. niger* and *P. chrysogenum*). The developed film efficiently controlled the growth of mold in wrapped bread for up to 14 days [177].

1.2.2.6 Flavor Releasers

To elaborate further on the context of previous sections, oxygen availability inside food packaging may lead to various spoilage mechanisms, particularly lipid oxidation, causing flavor deterioration, especially in oil-rich products. This process results in oxidative rancidity marked by unpleasant odors and flavors due to free radicals generated during oxidation. As the post-processing and postharvest storage of food extends, there is a further accumulation of microbial metabolites. Protein and lipid breakdown products in the package headspace, resulting in putrid and unpleasant odors, thus reducing consumer acceptability. To address such quality loss, flavor releasers can be employed as these can maintain and/or enhance sensory characteristics and overall consumer acceptability by emitting pleasant flavors into the food product [7, 178]. Flavor-releasing compounds are used through direct contact with food and packaging material or an indirect system, the latter being commonly applied to solid food, and includes the diffusion of aroma compounds within the package, desorption at the package/headspace interface, and adsorption into the food. An example is a commercial product named PolyIFF® by International Flavor and Fragrances Inc. The direct method is utilized for delivering aroma and nonvolatile flavor compounds, especially in beverage products involving five steps. The first being the diffusion of aroma/nonvolatile compounds in the packaging material. Second, the dissolution at the package/liquid interface. Third, desorption at the package/headspace interface. Fourth, dispersion/diffusion at the liquid/headspace interface, and fifth, desorption at the liquid/headspace interface, and dispersion of aroma/nonvolatile compounds in the headspace. An example is the chocolate-flavored milk bottle by a commercial product named Add Master [145]. Aromatic EO are notable plant extracts with diverse biological activities and are gaining attention as potential replacements for chemical preservatives in food preservation and packaging applications [179]. Generally Recognized as Safe (GRAS), these oils are used as flavoring agents in various foods, including bakery items, ice cream, beverages, and sweets [178]. Their volatile nature allows them to be vapor-released into the headspace from packaging material with or without direct food contact, diffusing toward the food surface where microbial growth occurs. However, controlled release is essential to optimize concentration and effectiveness, and hence carrier-biopolymer matrices effectively trap EOs or aroma compounds for regulated release [180, 181]. In a recent study, researchers formulated a slow-release PLA/PHA active packaging system incorporated with OEO and examined its impact on the fillet of a pufferfish's flavor and quality. The carvacrol released from the hybrid film to the pufferfish was measured at 9.70 mg/kg. This controlled release contributed to extending the shelf life of the fish fillets by an additional two to three days, when stored at 4 °C [182]. Another study demonstrated that the efficacy of an LDPE and PP-based active bilayer film when incorporated with attapulgite (AT) and loaded with *Allium sativum* essence oil (AEO) for the preservation of large yellow croaker fillets stored at 4 ± 1 °C. The inclusion of AEO controlled by AT resulted in a slow release extending the shelf life of large yellow croaker fillets by five days. This finding suggests that the controlled release of AEO facilitated by AT contributes to the maintenance of freshness and prolongs the shelf life of large yellow croaker [183].

1.3 Forms of Active Packaging Systems

1.3.1 Sachets

Sachets, originating in Japan in the late 1970s, have emerged as a popular choice for food packaging applications, which aim to extend the freshness of perishables without introducing additives or chemical preservatives into the packaged food item [184]. The industrial use of active packaging, particularly in sachet form, encompasses various systems, such as OS and absorbers, carbon dioxide emitters and absorbers, ethylene absorbers and emitters, moisture absorbers, odor removers, and antimicrobial agents [37]. Early-generation sachets were designed for moisture regulation, oxygen absorption, carbon dioxide control, and moisture absorption. In contrast, the latter generation offers a broader spectrum of functionalities, including emission or absorption of flavors and odors, controlled release of antioxidants, ethylene gas modulation, and antimicrobial agents. These advancements come in diverse formats, capacities, and triggering mechanisms [23]. The most used OSs are iron-based materials, in which powder is kept in a sachet within the packaging system. In various food applications, such as meat, poultry, coffee, baked goods, and dried foods, iron-based sachets have demonstrated satisfactory performance. Oxycatch and Freshcare are some of the commercially available sachets-based OSs on the market [1]. However, sachet-based applications present drawbacks, including the risk of accidental ruptures leading to the unintentional consumption of its content(s). To address this constraint, alternative solutions can be developed by incorporating OSs into polymeric packaging films or containers, providing a more controlled and versatile approach [36, 37]. Furthermore, antimicrobial packaging employs antimicrobial agents embedded in small sachets or pads to ensure continuous release of the active compound(s) [179, 184]. These sachets, classified as nonvolatile or volatile, play a crucial role in preventing microbial growth in bakery products. Bakery applications benefit from the use of volatile compounds, such as OEO, as they can efficiently penetrate the food matrix without direct contact with packaging materials, effectively inhibiting bacterial and fungal growth in the product [185, 186]. Additionally, Ju et al. [187] developed antibacterial sachets with eugenol and citral and reported that the sachets extended the shelf life of bread without compromising flavor. In a study by Hernandez et al. [188] active sachets incorporated with antifungal thymol (T) and an ethylene scavenger ($KMnO_4$-loaded sapiolite) resulted in the highest (≥91%) fungal inhibition on tomatoes. In another study, the utilization of starch-based microcellular foam sachets infused with peppermint essential oil demonstrated increased enzyme activity and reduced malondialdehyde levels, contributing positively to the extended shelf life of strawberries and exhibited efficiency in reducing microbial load [189].

1.3.2 Polymeric Films

Active packaging engages in functions beyond containment and physical protection of the food product with a focus on the interaction between food and packaging materials, especially films and containers. Surface contamination during handling and transportation is a key contributor to food spoilage. To counteract such deterioration, packaging films incorporated with active agents have been a preferred mode/form of active packaging. Controlled release using active components in packaging films is deemed more effective than bulk application with sachets [101]. Various active agents, particularly antimicrobials, antioxidants, and enzymes are commonly integrated into films or coatings to facilitate active packaging of food materials. For example, poly (vinyl alcohol)/ethyl cellulose/tea polyphenol (PVA/EC-TP) composite nanofibrous films showed strong antioxidant and antimicrobial activities, and extended the shelf life of pork by three days [190]. Furthermore, a food packaging film composed of ethylene-vinyl acetate copolymer incorporated with anthocyanin, ZnO nanoparticles, rosemary extract, and modified montmorillonite (MMT) provided the composite film with antimicrobial, antioxidant, and air barrier properties [191]. Blended films of poly (butylene adipate-co-terephthalate) and thermoplastic cassava starch incorporated with titanium dioxide (TiO_2) nanoparticles exhibited effective oxygen scavenging activity and resulted in slower darkening/color change in banana fruit resulting in an extended shelf life [192]. Numerous other works have employed antioxidant and antimicrobial-based active agents in polymeric form for food packaging systems [143, 193–195]. Moreover, the biopolymer-based films fabricated from polysaccharides, proteins, and lipids, either pristine or in combination, can also be incorporated with nanoparticles, antioxidants, antimicrobial agents, and nutraceuticals, thereby inducing their beneficial effects into the packaged product [196]. For example, Mohammadi et al. [197] created a biodegradable film using whey protein isolate incorporated with chitosan nanofiber and cinnamon essential oil, which resulted in reduced water content and water solubility, a slight increase in water vapor permeability, and enhanced antibacterial activity of the films. In a recent investigation, gelatin/agar films infused with clove bud essential oil demonstrated elevated UV light barrier properties, enhanced water resistance, and robust antioxidant activity, which show potential applications in active food packaging systems [198]. Additionally, the incorporation of copper-based metal-organic frameworks (Cu-MOFs) and red cabbage anthocyanin into gelatin/poly(vinyl alcohol) composite films has enhanced tensile strength, UV protection, and water resistance. 3 wt% Cu-MOFs and anthocyanins imparted excellent antioxidant properties and antibacterial activity, and these synergistic effects preserved the freshness of shrimp for a longer period [199]. Numerous other studies have demonstrated that the incorporation of active agents on/in biopolymer-based food packaging films offers active and smart packaging solutions for various food commodities [200–202].

1.3.3 Other Forms

Perishable food products, including meat, poultry, and fresh fruits and vegetables often produce exudates during storage. These liquids are not only detrimental to both sensory properties and the microbial quality of the packaged food but also signify nutrition and weight loss in the product. To mitigate this deterioration, absorbent pads, typically positioned at the base of the food packaging trays or containers, have been effective in absorbing the liquid exudates during storage (Figure 1.3) [184]. In a recent study, lemon essential oil (LEO) was absorbed by thermally stable and porous vermiculite (VML) to create the LEO/VML complex, which was then combined with konjac glucomannan-grafted-poly(acrylic acid)/polyvinyl alcohol (KGM-g-PAA/PVA) composite to form a bacteriostatic water-absorbing pad. The designed pad demonstrated a sustained and controlled release of LEO effectively inhibiting the growth of *E. coli* during storage and extended the shelf life of chilled pork by three days [203]. If the migration rate of antimicrobial agents surpasses the growth rate of target microorganisms, depletion may occur before the intended storage period ends. Conversely, if the release rate is too slow, microorganisms may proliferate before achieving sufficient antimicrobial activity. Therefore, the controlled release of EOs can be attained by incorporating EOs into biopolymeric matrix and covering the food surface by coating, wrapping, spraying, or impregnating, to reduce or prevent the movement of gases and moisture [106, 179]. In a recent investigation, camellia oil was identified as a potential active agent for dispersion in a konjac glucomannan/Kappa-carrageenan matrix through an emulsification method. The results showed that the developed emulsion coating significantly prolonged the shelf life of chicken meat for up to 10 days [204]. The challenge with biopolymeric-based active packaging is the evolving properties of materials, especially the active compounds within the biopolymer matrix. To mitigate this, nanocarriers can shield active compounds, improving their shelf life and efficacy with controlled release. Encapsulated active compounds are released from the

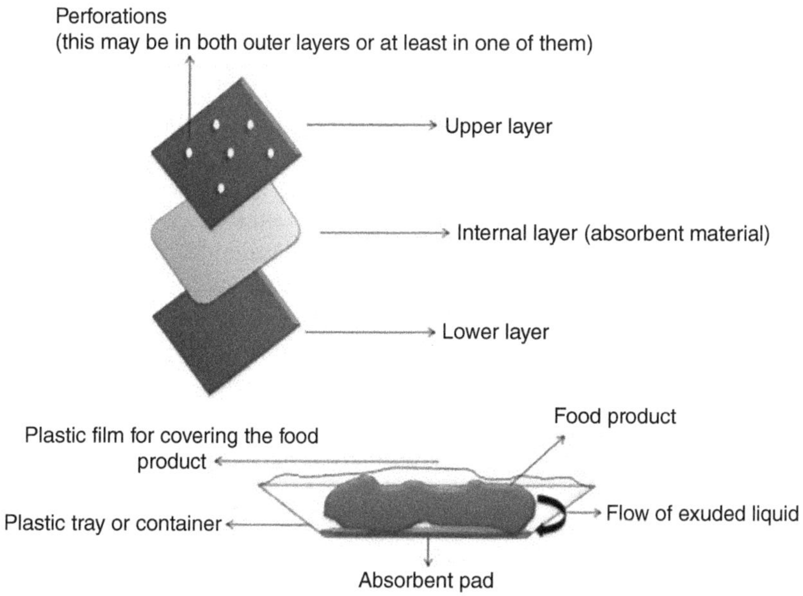

Figure 1.3 Architecture of the absorbent pad. *Source:* Reproduced from Otoni et al. [184]/ with permission of Elsevier.

biopolymer matrix to the food through two processes, migration and diffusion. Release rates are notably influenced by nanocarrier particle size, with smaller nanocarriers showing higher release rates [205–208]. For example, a cinnamon essential oil nanoemulsion was incorporated into pullulan-based active films to assess its impact on structure and properties. The film with the smallest droplet size (60 nm) and uniform distribution exhibited the highest cinnamon retention and bacteriostatic ability against both *E. coli* and *Staphylococcus aureus* [209]. Another novel active food packaging film was developed using pullulan/ethyl cellulose nanofiber loaded with cinnamaldehyde that exhibited enhanced flexibility, hydrophobicity, and antimicrobial activity against both *E. coli* and *S. aureus* [210]. On the other hand, cardboard is a commonly employed packaging material for fresh fruits and vegetables, often coated with waterproof lacquers (water-based emulsions) to enhance its mechanical properties. This is essential as fruits and vegetables are stored at high relative humidity to minimize water loss. In a study, a cardboard tray with water-based essential oil coating and β-cyclodextrins was tested for preserving cherry tomatoes stored at 8 °C for 24 days. The EO continued to be released until day 16 and the developed packaging maintained the tomato quality until day 24 [211].

1.4 Active Packaging and Food Distribution

Food loss and wastage pose significant challenges amid the continuous global population growth. According to FAO, a staggering 1.3 billion tons of food products are lost or wasted each year at various stages across the entire food supply chain, from farm to fork [212–214]. Internationally, the issue of food wastage has been acknowledged through the implementation of the United Nation's sustainable development goals (SDG). SDG 12.3 aims to reduce per capita global food wastage at the consumer level, during production and supply chains, including reducing postharvest losses by 50% by the year 2030 [215]. Generally, the loss of fresh agricultural produce occurs during handling, transportation, and storage due to the deterioration of quality, including impacts on nutritional and sensory quality [216]. Examining the food supply chain reveals multiple stages and causes of food wastage. Packaging of food, typically introduced at the postharvest stage, contributes significantly to the reduction of food wastage, however, post-packaging food losses can still occur due to damaged packaging, surpassing the expiration dates, and inadequate stock management. Following distribution, food stored in the retail sector often experiences losses due to spoilage, improper storage conditions, and/or inappropriate handling. Finally, after the consumption of a packaged food item, its packaging is subsequently disposed of, further adding to its environmental impact [217, 218]. In such cases, advanced technologies, such as temperature-controlled storage, energy-efficient systems, innovative packaging materials, designs, and smart monitoring systems, are crucial to minimizing postharvest food loss. Among these technologies, advanced packaging techniques, including active packaging and intelligent packaging, have been effective tools in reducing food wastage, ensuring food safety, and meeting consumer expectations [212]. Electronic components for intelligent packaging, such as electrochemical sensors, E-tongues, and E-noses, have gained significant momentum. However, these devices are often complex and expensive, posing challenges for integration into commercial food packaging. Active packaging, in which active/functional agent(s) are incorporated for the release and/or absorption of food deteriorating agents has emerged as a reliable solution to effectively tackle food loss and wastage [212, 219].

1.5 Regulatory and Market Status

Active materials can only be introduced to the market if they effectively serve their intended purpose and comply with the relevant legislative requirements. Unlike conventional materials, active packaging materials may alter food quality. However, these changes must comply with food regulations. Any modifications should not conceal signs of food spoilage or mislead consumers [8]. The criteria for active materials and articles designed for food contact, including food packaging, are outlined in the European Parliament and Council (EC) Regulation No. 1935/2004 dated October 27, 2004, concerning materials and articles intended for contact with food. This regulation stipulates that active materials for food contact are formulated to release active agents slowly into food or absorb substances present in it. Changes in the composition or sensory characteristics of food are permissible only if they align with the relevant community provisions on food [220]. Conversely, specific guidelines governing the utilization of active packaging are delineated in the Commission Regulation (EC) No. 450/2009 dated May 29, 2009, addressing active and intelligent materials and articles intended for food contact. This regulation addresses concerns regarding the safe application of materials within active packaging, establishes criteria for the commercialization of such materials, and outlines the European Commission's sanctioned list of permissible substances for use in active components. The regulation further delineates the criteria for substance inclusion, labeling rules, and mandates the presence of a declaration of conformity along with supporting documentation [221]. In summary, Regulation No. 1935/2004 enables the introduction of active packaging into the European Union market, and Regulation No. 450/2009 serves as the legal framework addressing substances or groups of substances acting as active or intelligent components [222].

The global active packaging market for the food and beverage industries is fast growing due to the increased consumer demand for fresh and high-quality packaged food, consumer convenience, and the emphasis on extending the shelf life of food products coming from manufacturers. The global active packaging market is expected to experience a moderate Compound Annual Growth Rate (CAGR) of 9.49% throughout the forecast period. As of 2023, the market is valued at US$ 26.13 billion and is predicted to exceed US$ 64.71 billion by 2033. Despite the market's growth, potential hindrances are anticipated. Government regulations in developing economies restricting the use of synthetic plastic in convenience stores might impede market expansion through 2033. Furthermore, a comprehensive analysis indicates that the high cost of active packaging is a significant constraint for the current market during the forecast period [223].

1.6 Conclusion and Future Perspective

Active packaging has emerged as a valuable solution for minimizing food loss and wastage by extending their shelf life using controlled-release active agents or by the absorption of undesirable components. Active packaging offers clear advantages, however, certain unresolved issues hinder its widespread adoption. Concerns include potential hazards from new materials and technologies that may impact product quality and human health. Addressing these challenges requires a deeper understanding of the governing principles and mechanisms, optimization of active agents, and the development of effective systems with minimal side effects. Despite the potential of active packaging, the market currently offers a limited number of products with active packaging compared to the extensive literature on the subject. Further research on packaging materials, particularly in limiting migration and assessing food protection, is essential for future progress toward sustainable food preservation and packaging.

References

1 Soltani Firouz, M., Mohi-Alden, K., and Omid, M. (2021). A critical review on intelligent and active packaging in the food industry: research and development. *Food Research International* 141: 110113.
2 Chauhan, P., Chauhan, E.S., and Chauhan, U. (2022). Role of active packaging in the food industry: a review. *Journal of Pharmaceutical Research International* 34: 20–32.
3 Deshmukh, R.K., Hakim, L., and Gaikwad, K.K. (2023). Active packaging materials. *Current Food Science and Technology Reports* 1: 123–132.
4 Jiang, Y., Zhang, Y., and Deng, Y. (2023). Latest advances in active materials for food packaging and their application. *Foods* 12: 4055.

5 Ahmed, M.W., Haque, M.A., Mohibbullah, M. et al. (2022). A review on active packaging for quality and safety of foods: current trends, applications, prospects and challenges. *Food Packaging and Shelf Life* 33: 100913.

6 Krasniewska, K., Galus, S., and Gniewosz, M. (2020). Biopolymers-based materials containing silver nanoparticles as active packaging for food applications – a review. *International Journal of Molecular Sciences* 21: 698.

7 Kuswandi, B. and Jumina (2020). Active and intelligent packaging, safety, and quality controls. *Fresh-Cut Fruits and Vegetables* 243–294.

8 Wyrwa, J. and Barska, A. (2017). Innovations in the food packaging market: active packaging. *European Food Research and Technology* 243: 1681–1692.

9 Yildirim, S., Rocker, B., Patterson, M.K. et al. (2017). Active packaging applications for food. *Comprehensive Reviews in Food Science and Food Safety* 17: 165–199.

10 Azevedo, A.G., Barros, C., Miranda, S. et al. (2022). Active flexible films for food packaging: a review. *Polymers* 14: https://doi.org/10.3390/polym14122442.

11 Stoleru, E., Irimia, A., and Butnaru, E. (2021). Bio-based bioplastics in active food packaging. In: *Bioplastics for sustainable development* (ed. M. Kuddus and Roohi), 347–379. Springer: Springer Nature Singapore Pte Ltd.

12 Kumar, S., Basumatary, I.B., Mukherjee, A. et al. (2022). An overview of natural biopolymers in food packaging. In: *Biopolymer-Based Food Packaging: Innovations and Technology Applications* (ed. S. Kumar, A. Mukherjee, and J. Dutta), 1–28. Wiley.

13 Moeini, A., Germann, N., Malinconico, M. et al. (2021). Formulation of secondary compounds as additives of biopolymer-based food packaging: a review. *Trends in Food Science & Technology* 114: 342–354.

14 Majumder, R., Das, A., Mukherjee, A. et al. (2022). Biopolymers. In: *Biopolymer-Based Food Packaging: Innovations and Technology Applications* (ed. S. Kumar, A. Mukherjee, and J. Dutta), 29–65. Wiley.

15 Basumatary, I.B., Kumar, S., Mitra, A. et al. (2023). Edible film and coating on food products: progress in deposition methods. In: *Comprehensive Materials Processing* (ed. S. Hashmi), 244–261. Elsevier.

16 Purkayastha, M.D. and Kumar, S. (2022). Protein-based films and coatings. In: *Biopolymer-Based Food Packaging: Innovations and Technology Applications* (ed. S. Kumar, A. Mukherjee, and J. Dutta), 178–224. Wiley.

17 Gaikwad, K.K., Singh, S., and Negi, Y.S. (2019). Ethylene scavengers for active packaging of fresh food produce. *Environmental Chemistry Letters* 18: 269–284.

18 Zhu, Z., Zhang, Y., Zhang, Y. et al. (2019). Preparation of PAN@TiO(2) nanofibers for fruit packaging materials with efficient photocatalytic degradation of ethylene. *Materials* 12 (6): 896.

19 Álvarez-Hernández, M.H., Artés-Hernández, F., Ávalos-Belmontes, F. et al. (2018). Current scenario of adsorbent materials used in ethylene scavenging systems to extend fruit and vegetable postharvest life. *Food and Bioprocess Technology* 11: 511–525.

20 Keller, N., Ducamp, M.N., Robert, D. et al. (2013). Ethylene removal and fresh product storage: a challenge at the frontiers of chemistry. Toward an approach by photocatalytic oxidation. *Chemical Reviews* 113: 5029–5070.

21 Jiang, T., Zheng, X., Li, J. et al. (2011). Integrated application of nitric oxide and modified atmosphere packaging to improve quality retention of button mushroom (*Agaricus bisporus*). *Food Chemistry* 126: 1693–1699.

22 Wei, H., Seidi, F., Zhang, T. et al. (2021). Ethylene scavengers for the preservation of fruits and vegetables: a review. *Food Chemistry* 337: 127750.

23 Janjarasskul, T. and Suppakul, P. (2017). Active and intelligent packaging: the indication of quality and safety. *Critical Reviews in Food Science and Nutrition* 58: 808–831.

24 Awalgaonkar, G., Beaudry, R., and Almenar, E. (2020). Ethylene-removing packaging: basis for development and latest advances. *Comprehensive Reviews in Food Science and Food Safety* 19: 3980–4007.

25 Álvarez-Hernández, M.H., Martínez-Hernández, G.B., Avalos-Belmontes, F. et al. (2019). Potassium permanganate-based ethylene scavengers for fresh horticultural produce as an active packaging. *Food Engineering Reviews* 11: 159–183.

26 Tas, C.E., Hendessi, S., Baysal, M. et al. (2017). Halloysite nanotubes/polyethylene nanocomposites for active food packaging materials with ethylene scavenging and gas barrier properties. *Food and Bioprocess Technology* 10: 789–798.

27 Sadeghi, K., Lee, Y., and Seo, J. (2019). Ethylene scavenging systems in packaging of fresh produce: a review. *Food Reviews International* 37: 155–176.

28 Joung, J., Boonsiriwit, A., Kim, M. et al. (2021). Application of ethylene scavenging nanocomposite film prepared by loading potassium permanganate-impregnated halloysite nanotubes into low-density polyethylene as active packaging material for fresh produce. *LWT* 145: 111309.

29 Kaewklin, P., Siripatrawan, U., Suwanagul, A. et al. (2018). Active packaging from chitosan-titanium dioxide nanocomposite film for prolonging storage life of tomato fruit. *International Journal of Biological Macromolecules* 112: 523–529.

30 Upadhyay, A., Kumar, P., Kardam, S.K. et al. (2022). Ethylene scavenging film based on corn starch-gum acacia impregnated with sepiolite clay and its effect on quality of fresh broccoli florets. *Food Bioscience* 46: 101556.

31 Riahi, Z., Ezati, P., Rhim, J.W. et al. (2022). Cellulose nanofiber-based ethylene scavenging antimicrobial films incorporated with various types of titanium dioxide nanoparticles to extend the shelf life of fruits. *ACS Applied Polymer Materials* 4: 4765–4773.

32 Ebrahimi, A., Khajavi, M.Z., Mortazavian, A.M. et al. (2021). Preparation of novel nano-based films impregnated by potassium permanganate as ethylene scavengers: an optimization study. *Polymer Testing* 93: 106934.

33 Álvarez-Hernández, M.H., Martínez-Hernández, G.B., Avalos-Belmontes, F. et al. (2019). An innovative ethylene scrubber made of potassium permanganate loaded on a protonated montmorillonite: a case study on blueberries. *Food and Bioprocess Technology* 12: 524–538.

34 Dey, A. and Neogi, S. (2019). Oxygen scavengers for food packaging applications: a review. *Trends in Food Science & Technology* 90: 26–34.

35 Awulachew, M.T. (2022). A review of food packaging materials and active packaging system. *International Journal of Health Policy Planning* 1: 28–35.

36 Dobrucka, R. and Przekop, R. (2019). New perspectives in active and intelligent food packaging. *Journal of Food Processing and Preservation* 43: 14194.

37 Vilela, C., Kurek, M., Hayouka, Z. et al. (2018). A concise guide to active agents for active food packaging. *Trends in Food Science & Technology* 80: 212–222.

38 Demirhan, B. and Candogan, K. (2017). Active packaging of chicken meats with modified atmosphere including oxygen scavengers. *Poultry Science* 96: 1394–1401.

39 Foltynowicz, Z., Bardenshtein, A., Sängerlaub, S. et al. (2017). Nanoscale, zero valent iron particles for application as oxygen scavenger in food packaging. *Food Packaging and Shelf Life* 11: 74–83.

40 Faas, N., Rocker, B., Smrke, S. et al. (2020). Prevention of lipid oxidation in linseed oil using a palladium-based oxygen scavenging film. *Food Packaging and Shelf Life* 24: 100488.

41 Tulsyan, G., Richter, C., and Diaz, C.A. (2017). Oxygen scavengers based on titanium oxide nanotubes for packaging applications. *Packaging Technology and Science* 30 (6): 251–256.

42 Kordjazi, Z. and Ajji, A. (2021). Development of TiO_2 catalyzed HTPB based oxygen scavenging films for food packaging applications. *Food Control* 121: 107639.

43 He, Y., Hu, X., Xu, M. et al. (2021). Mesoporous silica nanosphere-based oxygen scavengers. *Microporous and Mesoporous Materials* 327: 111426.

44 Gaikwad, K.K., Singh, S., and Lee, Y.S. (2018). Oxygen scavenging films in food packaging. *Environmental Chemistry Letters* 16: 523–538.

45 Lee, J.S., Chang, Y., Lee, E.-S. et al. (2018). Ascorbic acid-based oxygen scavenger in active food packaging system for raw meatloaf. *Journal of Food Science* 83: 682–688.

46 Singh, S., Gaikwad, K.K., and Lee, Y.S. (2019). Development and application of a pyrogallic acid-based oxygen scavenging packaging system for shelf life extension of peeled garlic. *Scientia Horticulturae* 256: 108548.

47 Promsorn, J. and Harnkarnsujarit, N. (2022). Pyrogallol loaded thermoplastic cassava starch based films as bio-based oxygen scavengers. *Industrial Crops and Products* 186: 115226.

48 Sonchaeng, U., Promsorn, J., Bumbudsanpharoke, N. et al. (2022). Polyesters incorporating gallic acid as oxygen scavenger in biodegradable packaging. *Polymers* 14: 5296.

49 Ahn, B.J., Gaikwad, K.K., and Lee, Y.S. (2016). Characterization and properties of LDPE film with gallic-acid-based oxygen scavenging system useful as a functional packaging material. *Journal of Applied Polymer Science* 133: 44138.

50 Gaikwad, K.K., Singh, S., Shin, J. et al. (2020). Novel polyisoprene based UV-activated oxygen scavenging films and their applications in packaging of beef jerky. *LWT* 117: 108643.

51 Singh, G., Singh, S., Kumar, B. et al. (2020). Active barrier chitosan films containing gallic acid based oxygen scavenger. *Journal of Food Measurement and Characterization* 15: 585–593.

52 Iribarren, E., Wongphan, P., Bumbudsanpharoke, N. et al. (2023). Thermoplastic agar blended PBAT films with enhanced oxygen scavenging activity. *Food Bioscience* 54: 102940.

53 Murmu, S.B. and Mishra, H.N. (2018). Selection of the best active modified atmosphere packaging with ethylene and moisture scavengers to maintain quality of guava during low-temperature storage. *Food Chemistry* 253: 55–62.

54 Rux, G., Mahajan, P.V., Linke, M. et al. (2016). Humidity-regulating trays: moisture absorption kinetics and applications for fresh produce packaging. *Food and Bioprocess Technology* 9: 709–716.

55 Choi, H.Y. and Lee, Y.S. (2013). Characteristics of moisture-absorbing film impregnated with synthesized attapulgite with acrylamide and its effect on the quality of seasoned laver during storage. *Journal of Food Engineering* 116: 829–839.

56 Bovi, G.G., Caleb, O.J., Klaus, E. et al. (2018). Moisture absorption kinetics of FruitPad for packaging of fresh strawberry. *Journal of Food Engineering* 223: 248–254.

57 Gaikwad, K.K., Singh, S., and Ajji, A. (2018). Moisture absorbers for food packaging applications. *Environmental Chemistry Letters* 17: 609–628.

58 Jeong, Y., Sadeghi, K., and Seo, J. (2023). Effects of molecular weight of polyethylene glycol on the moisture-absorption capacity and morphology of the channeling structure in desiccant composite. *Food Packaging and Shelf Life* 37: 101086.

59 Abral, H., Ariksa, J., Mahardika, M. et al. (2020). Transparent and antimicrobial cellulose film from ginger nanofiber. *Food Hydrocolloids* 98: 105266.

60 Wang, J., Gardner, D.J., Stark, N.M. et al. (2017). Moisture and oxygen barrier properties of cellulose nanomaterial-based films. *ACS Sustainable Chemistry & Engineering* 6: 49–70.

61 Pinming, C., Sukgorn, N., Suhatcho, T. et al. (2016). Humidity sensor using carboxymethyl cellulose hydrogel membrane. In: *13th International Conference on Electrical Engineering/Electronics, Computer, Telecommunications and Information Technology (ECTI-CON)*, 1–3. Chiang Mai, Thailand: IEEE.

62 Agriawati, D.P., Warsiki, E., Iskandar, A. et al. (2019). Application alkali treatment rice husk – $CaCl_2$ as moisture absorber of coconut palm sugar granule. *IOP Conference Series: Earth and Environmental Science* 347: 012062.

63 Warsiki, E., Agriawati, D., Noor, E. et al. (2021). Isotherm moisture sorption of composite desiccant made from rice husk biomass. *IOP Conference Series: Earth and Environmental Science* 749: 012011.

64 Sängerlaub, S., E. Lehmann, and K. Muller, et al., Porosity, thickness distribution and water vapour sorption of thermoformed foamed PP films with dispersed sodium chloride for humidity regulation *Food Packaging and Shelf Life*, 2020. 24. 100482.

65 Chen, C.-W., Xie, J., Yang, F.X. et al. (2018). Development of moisture-absorbing and antioxidant active packaging film based on poly(vinyl alcohol) incorporated with green tea extract and its effect on the quality of dried eel. *Journal of Food Processing and Preservation* 42: 13374.

66 Sängerlaub, S., Seibel, K., Miesbauer, O. et al. (2018). Functional properties of foamed and/or stretched polypropylene-films containing sodium chloride particles for humidity regulation. *Polymer Testing* 65: 339–351.

67 Garavito, J., Mendoza, S.M., and Castellanos, D.A. (2022). Configuration of biodegradable equilibrium modified atmosphere packages, including a moisture absorber for fresh cape gooseberry (*Physalis peruviana* L.) fruits. *Journal of Food Engineering* 314: 110761.

68 Rahmadiawan, D., Abral, H., Railis, R.M. et al. (2022). The enhanced moisture absorption and tensile strength of PVA/Uncaria gambir extract by boric acid as a highly moisture-resistant, anti-UV, and strong film for food packaging applications. *Journal of Composites Science* 6 (11): 110337.

69 Asfaq and Chand, K. (2020). Effect of moisture absorber and high-density polyethylene bags on shelf life of edible coated jaggery cubes during storage. *Sugar Tech* 22: 1130–1137.

70 Guo, X., Liu, L., Hu, Y. et al. (2018). Water vapor sorption properties of TEMPO oxidized and sulfuric acid treated cellulose nanocrystal films. *Carbohydrate Polymers* 197: 524–530.

71 Kwon, S., Orsuwan, A., Bumbudsanpharoke, N. et al. (2018). A short review of light barrier materials for food and beverage packaging. *Korean Journal of Packaging Science and Technology* 24: 141–148.

72 Mohr, L.C., Capelezzo, A.P., Baretta, C. et al. (2019). Titanium dioxide nanoparticles applied as ultraviolet radiation blocker in the polylactic acid bidegradable polymer. *Polymer Testing* 77: 105867.

73 Moustafa, H., Karmalawi, A.M., and Youssef, A.M. (2021). Development of dapsone-capped TiO2 hybrid nanocomposites and their effects on the UV radiation, mechanical, thermal properties and antibacterial activity of PVA bionanocomposites. *Environmental Nanotechnology, Monitoring & Management* 16: 100482.

74 Zhang, X.F., Song, L., Wang, Z. et al. (2020). Highly transparent graphene oxide/cellulose composite film bearing ultraviolet shielding property. *International Journal of Biological Macromolecules* 145: 663–667.

75 Wu, W., Liu, T., Deng, X. et al. (2019). Ecofriendly UV-protective films based on poly(propylene carbonate) biocomposites filled with TiO(2) decorated lignin. *International Journal of Biological Macromolecules* 126: 1030–1036.

76 Lee, E., Song, Y., and Lee, S. (2017). Crosslinking of lignin/poly(vinyl alcohol) nanocomposite fiber webs and their antimicrobial and ultraviolet-protective properties. *Textile Research Journal* 89: 3–12.

77 Nguyen, T.M., Lim, Y.W., Jana, S. et al. (2023). Low-migration macromolecular benzophenones as UV absorbers for food packaging polymers. *ACS Applied Polymer Materials* 5: 6932–6940.

78 Zhang, Y., Haque, A., and Naebe, M. (2021). Lignin-cellulose nanocrystals from hemp hurd as light-coloured ultraviolet (UV) functional filler for enhanced performance of polyvinyl alcohol nanocomposite films. *Nanomaterials* 11: 3425.

79 Bian, H., Chen, L., Dong, M. et al. (2021). Natural lignocellulosic nanofibril film with excellent ultraviolet blocking performance and robust environment resistance. *International Journal of Biological Macromolecules* 166: 1578–1585.

80 Avelino, F., de Oliveira, D.R., Mazzetto, S.E. et al. (2019). Poly(methyl methacrylate) films reinforced with coconut shell lignin fractions to enhance their UV-blocking, antioxidant and thermo-mechanical properties. *International Journal of Biological Macromolecules* 125: 171–180.

81 Shikinaka, K., Nakamura, M., Navarro, R.R. et al. (2019). Non-flammable and moisture-permeable UV protection films only from plant polymers and clay minerals. *Green Chemistry* 21: 498–502.

82 Lei, Y., Mao, L., Yao, J. et al. (2021). Improved mechanical, antibacterial and UV barrier properties of catechol-functionalized chitosan/polyvinyl alcohol biodegradable composites for active food packaging. *Carbohydrate Polymers* 264: 117997.

83 Guzman-Puyol, S., Hierrezuelo, J., Benítez, J.J. et al. (2022). Transparent, UV-blocking, and high barrier cellulose-based bioplastics with naringin as active food packaging materials. *International Journal of Biological Macromolecules* 209: 1985–1994.

84 Mellinas, A.C., Jimenez, A., and Garrigos, M.C. (2020). Pectin-based films with cocoa bean shell waste extract and ZnO/Zn-NPs with enhanced oxygen barrier, ultraviolet screen and photocatalytic properties. *Foods* 9: 1572.

85 Mahmoodi, A., Ghodrati, S., and Khorasani, M. (2019). High-strength, low-permeable, and light-protective nanocomposite films based on a hybrid nanopigment and biodegradable PLA for food packaging applications. *ACS Omega* 4: 14947–14954.

86 Nikvarz, N., Khayati, G.R., and Sharafi, S. (2020). Preparation of UV absorbent films using polylactic acid and grape syrup for food packaging application. *Materials Letters* 276: 128187.

87 Nikvarz, N., Khayati, G.R., and Sharafi, S. (2021). Bio-based ultraviolet protective packaging film preparation using starch with incorporated date palm syrup. *Materials Chemistry and Physics* 270: 124794.

88 Patil, A.S., Waghmare, R.D., Pawar, S.P. et al. (2020). Photophysical insights of highly transparent, flexible and re-emissive PVA @ WTR-CDs composite thin films: a next generation food packaging material for UV blocking applications. *Journal of Photochemistry and Photobiology A: Chemistry* 400: 112647.

89 Liang, L. and Chen, H. (2022). Development and characterization of biodegradable ultraviolet protective and antibacterial polylactic acid-cellulose acetate film modified by phenyl salicylate. *International Journal of Biological Macromolecules* 211: 85–93.

90 Dou, J., Vuorinen, T., Koivula, H. et al. (2021). Self-standing lignin-containing willow bark nanocellulose films for oxygen blocking and UV shielding. *ACS Applied Nano Materials* 4: 2921–2929.

91 He, X., Luzi, F., Hao, X. et al. (2019). Thermal, antioxidant and swelling behaviour of transparent polyvinyl (alcohol) films in presence of hydrophobic citric acid-modified lignin nanoparticles. *International Journal of Biological Macromolecules* 127: 665–676.

92 Mendoza, D.J., Browne, C., Raghuwanshi, V.S. et al. (2021). Phenolic ester-decorated cellulose nanocrystals as UV-absorbing nanoreinforcements in polyvinyl alcohol films. *ACS Sustainable Chemistry & Engineering* 9: 6427–6437.

93 Zhang, Y., Remadevi, R., Hinestroza, J.P. et al. (2020). Transparent ultraviolet (UV)-shielding films made from waste hemp hurd and polyvinyl alcohol (PVA). *Polymers* 12: 1190.

94 Channa, I.A., Ashfaq, J., Gilani, S.J. et al. (2022). UV blocking and oxygen barrier coatings based on polyvinyl alcohol and zinc oxide nanoparticles for packaging applications. *Coatings* 12: 897.

95 Bhardwaj, A., Alam, T., and Talwar, N. (2019). Recent advances in active packaging of agri-food products: a review. *Journal of Postharvest Technology* 7: 33–62.

96 Gutierrez-Pacheco, M.M., Gutierrez-Pacheco, S.L., Ortega-Ramirez, L.A. et al. (2016). *Active Packaging in Postharvest Management Approaches for Maintaining Quality of Fresh Produce* (ed. M.W. Siddiqui, J.F.A. Zavala, and C.-A.A. Hwang). Springer: Springer International Publishing Switzerland.

97 Ahmed, I., Lin, H., Zou, L. et al. (2017). A comprehensive review on the application of active packaging technologies to muscle foods. *Food Control* 82: 163–178.

98 Versteylen, S., Stoll, B.N., and Michaels, M.H. (2015). *Absorbent Pad to Reduce Confinement Odor in Food Packages*. United States: IFI CLAIMS Patent Services.

99 Dar, B.N. and Mir, S.A. (2020). *Emerging Technologies for Shelf-Life Enhancement of Fruits*. New York: Apple Academic Press.

100 Hansen, A.Å., Moen, B., Rødbotten, M. et al. (2016). Effect of vacuum or modified atmosphere packaging (MAP) in combination with a CO_2 emitter on quality parameters of cod loins (*Gadus morhua*). *Food Packaging and Shelf Life* 9: 29–37.

101 Singh, A.K., Ramakanth, D., Kumar, A. et al. (2021). Active packaging technologies for clean label food products: a review. *Journal of Food Measurement and Characterization* 15: 4314–4324.

102 Kerry, J.P. (2014). New packaging technologies, materials and formats for fast-moving consumer products. In: *Innovations in Food Packaging* (ed. J.H. Han), 549–584. Academic Press.

103 Coma, V. (2008). Bioactive packaging technologies for extended shelf life of meat-based products. *Meat Science* 78: 90–103.

104 Holck, A.L., Pettersen, M.K., Moen, M.H. et al. (2014). Prolonged shelf life and reduced drip loss of chicken filets by the use of carbon dioxide emitters and modified atmosphere packaging. *Journal of Food Protection* 77: 1133–1141.

105 Pettersen, M.K., Hansen, A.Å., and Mielnik, M. (2014). Effect of different packaging methods on quality and shelf life of fresh reindeer meat. *Packaging Technology and Science* 27: 987–997.

106 Mousavi Khaneghah, A., Hashemi, S.M.B., and Limbo, S. (2018). Antimicrobial agents and packaging systems in antimicrobial active food packaging: an overview of approaches and interactions. *Food and Bioproducts Processing* 111: 1–19.

107 Raghunath, A. and Perumal, E. (2017). Metal oxide nanoparticles as antimicrobial agents: a promise for the future. *International Journal of Antimicrobial Agents* 49: 137–152.

108 Yildirim, S. and Röcker, B. (2021). Sustainable antimicrobial packaging technologies. In: *Sustainable Food Packaging Technology* (ed. A. Athanassiou), 323–348. WILEY-VCH GmbH.

109 Biji, K.B., Ravishankar, C.N., Mohan, C.O. et al. (2015). Smart packaging systems for food applications: a review. *Journal of Food Science and Technology* 52: 6125–6135.

110 Paidari, S. and Ibrahim, S.A. (2021). Potential application of gold nanoparticles in food packaging: a mini review. *Gold Bulletin* 54: 31–36.

111 Basumatary, I.B., Mukherjee, A., and Kumar, S. (2023). Chitosan-based composite films containing eugenol nanoemulsion, ZnO nanoparticles and Aloe vera gel for active food packaging. *International Journal of Biological Macromolecules* 242: 124826.

112 Yu, D., Basumatary, I.B., Kumar, S. et al. (2023). Chitosan modified with bio-extract as an antibacterial coating with UV filtering feature. *International Journal of Biological Macromolecules* 230: 123145.

113 Devi, L.S., Mukherjee, A., Dutta, D. et al. (2023). Carnauba wax-based sustainable coatings for prolonging postharvest shelf-life of citrus fruits. *Sustainable Food Technology* 1 (3): 415–425.

114 Kim, I., Viswanathan, K., Kasi, G. et al. (2020). ZnO nanostructures in active antibacterial food packaging: preparation methods, antimicrobial mechanisms, safety issues, future prospects, and challenges. *Food Reviews International* 38: 537–565.

115 Roy, S. and Rhim, J.-W. (2022). Starch/agar-based functional films integrated with enoki mushroom-mediated silver nanoparticles for active packaging applications. Food. *Bioscience* 49: 101867.

116 Amjadi, S., Emaminia, S., Nazari, M. et al. (2019). Application of reinforced ZnO nanoparticle-incorporated gelatin bionanocomposite film with chitosan nanofiber for packaging of chicken fillet and cheese as food models. *Food and Bioprocess Technology* 12: 1205–1219.

117 Jafarzadeh, S., Rhim, J.-W., Alias, A.K. et al. (2019). Application of antimicrobial active packaging film made of semolina flour, nano zinc oxide and nano-kaolin to maintain the quality of low-moisture mozzarella cheese during low-temperature storage. *Journal of the Science of Food and Agriculture* 99: 2716–2725.

118 Tomic, A., Sovljanski, O., and Erceg, T. (2023). Insight on incorporation of essential oils as antimicrobial substances in biopolymer-based active packaging. *Antibiotics* 12: 1473.

119 Sharma, S., Barkauskaite, S., Duffy, B. et al. (2020). Characterization and antimicrobial activity of biodegradable active packaging enriched with clove and thyme essential oil for food packaging application. *Foods* 9: 1117.

120 Wang, L., Heising, J., Fogliano, V. et al. (2020). Fat content and storage conditions are key factors on the partitioning and activity of carvacrol in antimicrobial packaging. *Food Packaging and Shelf Life* 24: 100500.

121 Popa, E.E., Miteluţ, A.C., Râpă, M. et al. (2022). Antimicrobial active packaging containing nisin for preservation of products of animal origin: an overview. *Foods* 11: 3820.

122 Chang, S.H., Chen, Y.J., Tseng, H.J. et al. (2021). Applications of Nisin and EDTA in food packaging for improving fabricated chitosan-polylactate plastic film performance and fish fillet preservation. *Membranes* 11: 852.

123 Aziz, M. and Karboune, S. (2016). Natural antimicrobial/antioxidant agents in meat and poultry products as well as fruits and vegetables: a review. *Critical Reviews in Food Science and Nutrition* 58: 486–511.

124 Silva, N.H.C.S., Vilela, C., Almeida, A. et al. (2018). Pullulan-based nanocomposite films for functional food packaging: exploiting lysozyme nanofibers as antibacterial and antioxidant reinforcing additives. *Food Hydrocolloids* 77: 921–930.

125 Wang, H., Qian, J., and Ding, F. (2018). Emerging Chitosan-Based Films for Food Packaging Applications. *Journal of Agricultural and Food Chemistry* 66: 395–413.

126 Bi, F., Zhang, X., Liu, J. et al. (2020). Development of antioxidant and antimicrobial packaging films based on chitosan, D-α-tocopheryl polyethylene glycol 1000 succinate and silicon dioxide nanoparticles. *Food Packaging and Shelf Life* 24: 100503.

127 Mesgari, M., Aalami, A.H., and Sahebkar, A. (2021). Antimicrobial activities of chitosan/titanium dioxide composites as a biological nanolayer for food preservation: a review. *International Journal of Biological Macromolecules* 176: 530–539.

128 Wang, Y., Xia, Y., Zhang, P. et al. (2016). Physical characterization and pork packaging application of chitosan films incorporated with combined essential oils of cinnamon and ginger. *Food and Bioprocess Technology* 10: 503–511.

129 Hafsa, J., Smach, M.A., Khedher, M.R.B. et al. (2016). Physical, antioxidant and antimicrobial properties of chitosan films containing Eucalyptus globulus essential oil. *LWT- Food Science and Technology* 68: 356–364.

130 Echeverria, I., López-Caballero, M.E., Gómez-Guillén, M.C. et al. (2018). Active nanocomposite films based on soy proteins-montmorillonite- clove essential oil for the preservation of refrigerated bluefin tuna (*Thunnus thynnus*) fillets. *International Journal of Food Microbiology* 266: 142–149.

131 Rollini, M., Nielsen, T., Musatti, A. et al. (2016). Antimicrobial Performance of Two Different Packaging Materials on the Microbiological Quality of Fresh Salmon. *Coatings* 6: 6.

132 Li, W., L. Li, and Y. Cao, et al., Effects of PLA film incorporated with ZnO nanoparticle on the quality attributes of fresh-cut apple. *Nanomaterials*, 2017. 7. 207.
133 Rodriguez-Garcia, I., Cruz-Valenzuela, M.R., Silva-Espinoza, B.A. et al. (2016). Oregano (*Lippia graveolens*) essential oil added within pectin edible coatings prevents fungal decay and increases the antioxidant capacity of treated tomatoes. *Journal of the Science of Food and Agriculture* 96: 3772–3778.
134 Wang, L., et al., Preservation of soy protein-based meat analogues by using PLA/PBAT antimicrobial packaging film. *Food Chemistry*, 2022. 380: p. 132022.
135 He, Y., X. Fei, and H. Li, Carboxymethyl cellulose-based nanocomposites reinforced with montmorillonite and ε-poly-l-lysine for antimicrobial active food packaging. *Journal of Applied Polymer Science*, 2019. 137. 48782.
136 Soysal, Ç., Bozkurt, H., Dirican, E. et al. (2015). Effect of antimicrobial packaging on physicochemical and microbial quality of chicken drumsticks. *Food Control* 54: 294–299.
137 Ansarifar, E. and Moradinezhad, F. (2021). Preservation of strawberry fruit quality via the use of active packaging with encapsulated thyme essential oil in zein nanofiber film. *International Journal of Food Science & Technology* 56: 4239–4247.
138 Cui, H.Y., Wu, J., Li, C.Z. et al. (2016). Anti-listeria effects of chitosan-coated nisin-silica liposome on Cheddar cheese. *Journal of Dairy Science* 99: 8598–8606.
139 Aydemir Sezer, U., Sanko, V., Yuksekdag, Z.N. et al. (2016). Use of oxidized regenerated cellulose as bactericidal filler for food packaging applications. *Cellulose* 23: 3209–3219.
140 Divsalar, E., Tajik, H., Moradi, M. et al. (2018). Characterization of cellulosic paper coated with chitosan-zinc oxide nanocomposite containing nisin and its application in packaging of UF cheese. *International Journal of Biological Macromolecules* 109: 1311–1318.
141 Mohammadi, H., Kamkar, A., Misaghi, A. et al. (2019). Nanocomposite films with CMC, okra mucilage, and ZnO nanoparticles: Extending the shelf-life of chicken breast meat. *Food Packaging and Shelf Life* 21: 100330.
142 Heydari-Majd, M., Ghanbarzadeh, B., Shahidi-Noghabi, M. et al. (2019). A new active nanocomposite film based on PLA/ZnO nanoparticle/essential oils for the preservation of refrigerated *Otolithes ruber* fillets. *Food Packaging and Shelf Life* 19: 94–103.
143 Mihaly Cozmuta, A., Peter, A., Cozmuta, L.M. et al. (2014). Active packaging system based on Ag/TiO_2 nanocomposite used for extending the shelf life of bread. Chemical and microbiological investigations. *Packaging Technology and Science* 28: 271–284.
144 Eldib, R., Khojah, E., Elhakem, A. et al. (2020). Chitosan, nisin, silicon dioxide nanoparticles coating films effects on blueberry (*Vaccinium myrtillus*) quality. *Coatings* 10: 962.
145 Emanuel, N. and Sandhu, H.K. (2019). Food packaging development: recent perspective. *Journal of Thin Films, Coating Science Technology and Application* 6: 13–29.
146 Chacha, J.S., Ofoedu, C.E., and Xiao, K. (2022). Essential oil-based active polymer-based packaging system: a review of its effect on the antimicrobial, antioxidant, and sensory properties of beef and chicken meat. *Journal of Food Processing and Preservation* 46: 16933.
147 Yahaya, W.A.W. Chik, S.M.S.T., and Azman, N.A.M., et al. (2023). Mechanical properties and antioxidant activity of carrageenan-cellulose nanofiber incorporated butylated hydroxyanisole as active food packaging. *Materials Today: Proceedings*, August 2023, Malaysia

148 Minh, N.P., Nhi, T.T.Y., Lap, T.T. et al. (2019). Quality of frozen black tiger shrimp (Penaeus monodon) under a low-density polyethylene bag coextruded with butylated hydroxyanisole. *Journal of Pharmaceutical Sciences and Research* 11: 1483–1486.

149 Fasihnia, S.H., Peighambardoust, S.H., Peighambardoust, S.J. et al. (2020). Migration analysis, antioxidant, and mechanical characterization of polypropylene-based active food packaging films loaded with BHA, BHT, and TBHQ. *Journal of Food Science* 85: 2317–2328.

150 Valdés, A., Mellinas, A.C., Ramos, M. et al. (2015). Use of herbs, spices and their bioactive compounds in active food packaging. *RSC Advances* 5: 40324–40335.

151 Ganiari, S., Choulitoudi, E., and Oreopoulou, V. (2017). Edible and active films and coatings as carriers of natural antioxidants for lipid food. *Trends in Food Science & Technology* 68: 70–82.

152 Deshmukh, R.K. and Gaikwad, K.K. (2022). Natural antimicrobial and antioxidant compounds for active food packaging applications. *Biomass Conversion and Biorefinery* 14: 2190–6823.

153 Chang, B.P., Trinh, B.M., Tadele, D.T. et al. (2023). Natural antioxidant and antimicrobial agents and processing technologies for the design of active food packaging polymers. *Polymer Reviews* 63: 961–1013.

154 Tian, F., Decker, E.A., and Goddard, J.M. (2013). Controlling lipid oxidation of food by active packaging technologies. *Food & Function* 4: 669–680.

155 Islam, R.U., Khan, M.A., and Islam, S.U. (2017). Plant derivatives as promising materials for processing and packaging of meat-based products – focus on antioxidant and antimicrobial effects. *Journal of Food Processing and Preservation* 41: 12862.

156 Yang, H.-J., Lee, J.H., Won, M. et al. (2016). Antioxidant activities of distiller dried grains with solubles as protein films containing tea extracts and their application in the packaging of pork meat. *Food Chemistry* 196: 174–179.

157 Lorenzo, J.M., Batlle, R., and Gómez, M. (2014). Extension of the shelf-life of foal meat with two antioxidant active packaging systems. *LWT- Food Science and Technology* 59: 181–188.

158 Carrizo, D., Taborda, G., Nerín, C. et al. (2016). Extension of shelf life of two fatty foods using a new antioxidant multilayer packaging containing green tea extract. *Innovative Food Science & Emerging Technologies* 33: 534–541.

159 Tanwar, R., Gupta, V., Kumar, P. et al. (2021). Development and characterization of PVA-starch incorporated with coconut shell extract and sepiolite clay as an antioxidant film for active food packaging applications. *International Journal of Biological Macromolecules* 185: 451–461.

160 Ahmed, S. and Ikram, S. (2016). Chitosan and gelatin based biodegradable packaging films with UV-light protection. *Journal of Photochemistry and Photobiology, B: Biology* 163: 115–124.

161 Lin, Z. and Goddard, J. (2018). Photo-curable metal-chelating coatings offer a scalable approach to production of antioxidant active packaging. *Journal of Food Science* 83: 367–376.

162 Lin, Z., Roman, M.J., Decker, E.A. et al. (2016). Synthesis of iminodiacetate functionalized polypropylene films and their efficacy as antioxidant active-packaging materials. *Journal of Agricultural and Food Chemistry* 64: 4606–4617.

163 Tian, F., Decker, E.A., and Goddard, J.M. (2013). Controlling lipid oxidation via a biomimetic iron chelating active packaging material. *Journal of Agricultural and Food Chemistry* 61: 12397–12404.

164 Vilela, C., Pinto, R.J.B., Coelho, J. et al. (2017). Bioactive chitosan/ellagic acid films with UV-light protection for active food packaging. *Food Hydrocolloids* 73: 120–128.

165 Rizzo, V., Lombardo, S., Pandino, G. et al. (2021). Active packaging-releasing system with foeniculum vulgare essential oil for the quality preservation of ready-to-cook (RTC) globe artichoke slices. *Foods* 10: 517.

166 De Pilli, T. (2020). Development of a vegetable oil and egg proteins edible film to replace preservatives and primary packaging of sweet baked goods. *Food Control* 114: 107273.

167 Axel, C., Zannini, E., and Arendt, E.K. (2017). Mold spoilage of bread and its biopreservation: A review of current strategies for bread shelf life extension. *Critical Reviews in Food Science and Nutrition* 57: 3528–3542.

168 Melini, V. and Melini, F. (2018). Strategies to extend bread and gf bread shelf-life: from sourdough to antimicrobial active packaging and nanotechnology. *Fermentation* 4: 9.

169 Bodbodak, S. and Rafiee, Z. (2016). Recent trends in active packaging in fruits and vegetables. In: *Eco-Friendly Technology for Postharvest Produce Quality*, 77–125.

170 Mu, H., Gao, H., Chen, H. et al. (2017). A novel controlled release ethanol emitter: preparation and effect on some postharvest quality parameters of Chinese bayberry during storage. *Journal of the Science of Food and Agriculture* 97: 4929–4936.

171 Latou, E., Mexis, S.F., Badeka, A.V. et al. (2010). Shelf life extension of sliced wheat bread using either an ethanol emitter or an ethanol emitter combined with an oxygen absorber as alternatives to chemical preservatives. *Journal of Cereal Science* 52: 457–465.

172 Janjarasskul, T., Tananuwong, K., Kongpensook, V. et al. (2016). Shelf life extension of sponge cake by active packaging as an alternative to direct addition of chemical preservatives. *LWT- Food Science and Technology* 72: 166–174.

173 Pisoschi, A.M., Pop, A., Gajaila, I. et al. (2020). Analytical methods applied to the assay of sulfur-containing preserving agents. *Microchemical Journal* 155: 104681.

174 Roy, S. and Rhim, J.-W. (2021). Gelatin-based film integrated with copper sulfide nanoparticles for active packaging applications. *Applied Sciences* 11: 6307.

175 Priyadarshi, R., Kim, H.-J., and Rhim, J.-W. (2021). Effect of sulfur nanoparticles on properties of alginate-based films for active food packaging applications. *Food Hydrocolloids* 110: 106155.

176 Priyadarshi, R., Riahi, Z., Rhim, J.W. et al. (2021). Sulfur quantum dots as fillers in gelatin/agar-based functional food packaging films. *ACS Applied Nano Materials* 4: 14292–14302.

177 Riahi, Z., Priyadarshi, R., Rhim, J.W. et al. (2022). Alginate-based multifunctional films incorporated with sulfur quantum dots for active packaging applications. *Colloids and Surfaces B: Biointerfaces* 215: 112519.

178 Dutta, D. and Sit, N. (2022). Application of natural extracts as active ingredient in biopolymer based packaging systems. *Journal of Food Science and Technology* 60 (7): 1888–1902.

179 Ju, J., Chen, X., Xie, Y. et al. (2019). Application of essential oil as a sustained release preparation in food packaging. *Trends in Food Science & Technology* 92: 22–32.

180 Sharma, S., Barkauskaite, S., Jaiswal, A.K. et al. (2021). Essential oils as additives in active food packaging. *Food Chemistry* 343: 128403.

181 Wicochea-Rodriguez, J.D., Chalier, P., Ruiz, T. et al. (2019). Active food packaging based on biopolymers and aroma compounds: how to design and control the release. *Frontiers in Chemistry* 7: 398.

182 Zheng, H., Tang, H., Yang, C. et al. (2022). Evaluation of the slow-release polylactic acid/polyhydroxyalkanoates active film containing oregano essential oil on the quality and flavor of chilled pufferfish (*Takifugu obscurus*) fillets. *Food Chemistry* 385: 132693.

183 Dong, Z., Luo, C., Guo, Y. et al. (2019). Characterization of new active packaging based on PP/LDPE composite films containing attapulgite loaded with *Allium sativum* essence oil and its application for large yellow croaker (*Pseudosciaena crocea*) fillets. *Food Packaging and Shelf Life* 20: 100320.

184 Otoni, C.G., Espitia, P.J.P., Avena-Bustillos, R.J. et al. (2016). Trends in antimicrobial food packaging systems: emitting sachets and absorbent pads. *Food Research International* 83: 60–73.

185 Passarinho, A.T.P., Dias, N.F., Camilloto, G.P. et al. (2014). Sliced bread preservation through oregano essential oil-containing sachet. *Journal of Food Process Engineering* 37: 53–62.

186 Quiles, J.M., Manyes, L., Luciano, F. et al. (2015). Influence of the antimicrobial compound allyl isothiocyanate against the Aspergillus parasiticus growth and its aflatoxins production in pizza crust. *Food and Chemical Toxicology* 83: 222–228.

187 Ju, J., Xie, Y., Yu, H. et al. (2020). Synergistic inhibition effect of citral and eugenol against *Aspergillus niger* and their application in bread preservation. *Food Chemistry* 310: 125974.

188 Álvarez-Hernández, M.H., Martínez-Hernández, G.B., Castillejo, N. et al. (2021). Development of an antifungal active packaging containing thymol and an ethylene scavenger. Validation during storage of cherry tomatoes. *Food Packaging and Shelf Life* 29: 100734.

189 Amiri, A., Sourestani, M.M., Mortazavi, S.M.H. et al. (2022). Fabrication of the antimicrobial sachet by encapsulation of peppermint essential oil in active packaging of strawberry fruit. *Journal of Food Processing and Preservation* 46: 17181.

190 Yang, Y., Shi, Y., Cao, X. et al. (2021). Preparation and functional properties of poly(vinyl alcohol)/ethyl cellulose/tea polyphenol electrospun nanofibrous films for active packaging material. *Food Control* 130: 108331.

191 Eskandarabadi, S.M., Mahmoudian, M., Farah, K.R. et al. (2019). Active intelligent packaging film based on ethylene vinyl acetate nanocomposite containing extracted anthocyanin, rosemary extract and ZnO/Fe-MMT nanoparticles. *Food Packaging and Shelf Life* 22: 100389.

192 Phothisarattana, D., Wongphan, P., Promhuad, K. et al. (2021). Biodegradable poly(butylene adipate-co-terephthalate) and thermoplastic starch-blended TiO_2 nanocomposite blown films as functional active packaging of fresh fruit. *Polymers* 13: 4192.

193 Ahmed, J., Mulla, M., Arfat, Y.A. et al. (2018). Compression molded LLDPE films loaded with bimetallic (Ag-Cu) nanoparticles and cinnamon essential oil for chicken meat packaging applications. *LWT- Food Science and Technology* 93: 329–338.

194 Moura, L.E., de Souza, C.O., de Oliveira, E.A.S. et al. (2018). Bioactive efficacy of low-density polyethylene films with natural additives. *Journal of Applied Polymer Science* 135: 46461.

195 Srisa, A. and Harnkarnsujarit, N. (2020). Antifungal films from trans-cinnamaldehyde incorporated poly(lactic acid) and poly(butylene adipate-co-terephthalate) for bread packaging. *Food Chemistry* 333: 127537.

196 Akyuz, L., Kaya, M., Ilk, S. et al. (2018). Effect of different animal fat and plant oil additives on physicochemical, mechanical, antimicrobial and antioxidant properties of chitosan films. *International Journal of Biological Macromolecules* 111: 475–484.

197 Mohammadi, M., Mirabzadeh, S., Shahvalizadeh, R. et al. (2020). Development of novel active packaging films based on whey protein isolate incorporated with chitosan nanofiber and nano-formulated cinnamon oil. *International Journal of Biological Macromolecules* 149: 11–20.

198 Roy, S. and Rhim, J.-W. (2021). Gelatin/agar-based functional film integrated with pickering emulsion of clove essential oil stabilized with nanocellulose for active packaging applications. *Colloids and Surfaces A: Physicochemical and Engineering Aspects* 627: 127220.

199 Riahi, Z., Khan, A., Rhim, J.W. et al. (2023). Gelatin/poly(vinyl alcohol)-based dual functional composite films integrated with metal-organic frameworks and anthocyanin for active and intelligent food packaging. *International Journal of Biological Macromolecules* 249: 126040.

200 Deshmukh, R.K., Akhila, K., Ramakanth, D. et al. (2022). Guar gum/carboxymethyl cellulose based antioxidant film incorporated with halloysite nanotubes and litchi shell waste extract for active packaging. *International Journal of Biological Macromolecules* 201: 1–13.

201 Zheng, H., Chen, X., Li, L. et al. (2023). Development of gelatin-based active packaging and its application in bread preservation. *Journal of Renewable Materials* 11: 3693–3709.

202 Martiny, T.R., Raghavan, V., Moraes, C.C. et al. (2020). Bio-based active packaging: carrageenan film with olive leaf extract for lamb meat preservation. *Foods* 9: 1759.

203 Li, X., Xiao, N., Xiao, G. et al. (2021). Lemon essential oil/vermiculite encapsulated in electrospun konjac glucomannan-grafted-poly (acrylic acid)/polyvinyl alcohol bacteriostatic pad: Sustained control release and its application in food preservation. *Food Chemistry* 348: 129021.

204 Zhou, X., Zong, X., Zhang, M. et al. (2021). Effect of konjac glucomannan/carrageenan-based edible emulsion coatings with camellia oil on quality and shelf-life of chicken meat. *International Journal of Biological Macromolecules* 183: 331–339.

205 Kuai, L., Liu, F., Chiou, B.S. et al. (2021). Controlled release of antioxidants from active food packaging: a review. *Food Hydrocolloids* 120: 106992.

206 Becerril, R., Nerin, C., and Silva, F. (2020). Encapsulation systems for antimicrobial food packaging components: an update. *Molecules* 25: 1134.

207 Zhang, L., Yu, D., Regenstein, J.M. et al. (2021). A comprehensive review on natural bioactive films with controlled release characteristics and their applications in foods and pharmaceuticals. *Trends in Food Science & Technology* 112: 690–707.

208 Pascuta, M. and Vodnar, D. (2022). Nanocarriers for sustainable active packaging: an overview during and post COVID-19. *Coatings* 12.

209 Chu, Y., Cheng, W., Feng, X. et al. (2020). Fabrication, structure and properties of pullulan-based active films incorporated with ultrasound-assisted cinnamon essential oil nanoemulsions. *Food Packaging and Shelf Life* 25: 100547.

210 Yang, Y., Zheng, S., Liu, Q. et al. (2020). Fabrication and characterization of cinnamaldehyde loaded polysaccharide composite nanofiber film as potential antimicrobial packaging material. *Food Packaging and Shelf Life* 26: 100600.

211 Buendía-Moreno, L., Soto, S., Ros, M. et al. (2019). Innovative cardboard active packaging with a coating including encapsulated essential oils to extend cherry tomato shelf life. *LWT- Food Science and Technology* 116: 108584.

212 Chen, S., Brahma, S., Mackay, J. et al. (2020). The role of smart packaging system in food supply chain. *Journal of Food Science* 85: 517–525.

213 Chauhan, C., Dhir, A., Akram, M.U. et al. (2021). Food loss and waste in food supply chains. A systematic literature review and framework development approach. *Journal of Cleaner Production* 295: 126438.

214 Brennan, L., Langley, S., Verghese, K. et al. (2021). The role of packaging in fighting food waste: a systematised review of consumer perceptions of packaging. *Journal of Cleaner Production* 281: 125276.

215 United Nations (2017). *Work of the Statistical Commission pertaining to the 2030 Agenda for Sustainable Development*. United Nations.

216 Sivakumar, D. and Wall, M.M. (2013). Papaya fruit quality management during the postharvest supply chain. *Food Reviews International* 29: 24–48.

217 Wohner, B., Pauer, E., Heinrich, V. et al. (2019). Packaging-related food losses and waste: an overview of drivers and issues. *Sustainability* 11 (1).

218 Brancoli, P., Rousta, K., and Bolton, K. (2017). Life cycle assessment of supermarket food waste. *Resources, Conservation and Recycling* 118: 39–46.

219 Zhang, H., Hortal, M., Dobon, A. et al. (2015). The effect of active packaging on minimizing food losses: life cycle assessment (LCA) of essential oil component-enabled packaging for fresh beef. *Packaging Technology and Science* 28: 761–774.

220 European Nation (2004). Regulation (EC) No 1935/2004 of the European Parliament and of the Council of 27 October 2004 on materials and articles intended to come into contact with food and repealing Directives 80/590/EEC and 89/109/EEC in Official Journal of the European Union. European Nation.

221 European Nation (2009). Commission Regulation(EC) No 450/2009 of 29 May 2009 on active and intelligent materials and articles intended to come into contact with food, in Official Journal of the European Union. European Nation.

222 Dainelli, D., Gontard, N., Spyropoulos, D. et al. (2008). Active and intelligent food packaging: legal aspects and safety concerns. *Trends in Food Science & Technology* 19: S103–S112.

223 Sutaria, I. (2022). *An Overview of the Active Packaging Market Development (2023 to 2033).* Active Packaging Market. https://www.futuremarketinsights.com/reports/active-packaging-market#.

2

Additives in Active Food Packaging Systems

Debarshi Nath[1,2], *Amar K. Mohanty*[1,2], *Avik Mukherjee*[3], *Santosh Kumar*[3], *and Manjusri Misra*[1,2]

[1] *School of Engineering, Thornbrough Building, University of Guelph, Guelph, Ontario, Canada*
[2] *Department of Plant Agriculture, Bioproducts Discovery and Development Center, Crop Science Building, University of Guelph, Guelph, Ontario, Canada*
[3] *Department of Food Engineering and Technology, Central Institute of Technology Kokrajhar, Kokrajhar, Assam, India*

2.1 Introduction

Food contamination can occur while it is being harvested, processed, and distributed. Food products can be effectively shielded from external pollutants through the use of packaging, which can also stop deterioration due to chemical, physical, and biological changes during storage or food processing [1]. However, traditional packaging systems are unable to effectively regulate the reactions inside the food product [2]. The biggest issue with traditional packaging material is the use of non-biodegradable polymers for food packaging applications. The tremendous carbon footprint of such petrochemical-based synthetic polymers has caused havoc on the environment since packaging contributes to the highest amount of plastic waste. Packaging is the leading cause of litter worldwide, accounting for 146 million metric tons generated in 2015, 141 million metric tons of which remained unrecycled (96.6%). Packaging also has the shortest average working life of any industrial plastic industry. The operating life of single-use plastics, from use to disposal, can be as brief as a few minutes [3]. Current projections indicate that by 2050, 12 billion metric tons of plastic waste will be dumped in landfills or the natural environment, indicating a 4.9 billion metric tons increase from 2010 [4].

As such, there has been an increase in the use of biopolymers in food packaging and coating applications since petrochemical-based polymers are currently being banned by countries around the world. The performance of these biopolymeric materials as packaging materials is a concern since they exhibit poor light and moisture barriers as well as poor mechanical and thermal properties. Also, an increase in moisture permeation through the film results in microbial growth that rapidly deteriorates the packaged product. The existence of harmful microbes raises the likelihood of foodborne illnesses in humans, posing a public health hazard. The World Health Organization (WHO) estimated that in 2015, over

Smart Food Packaging Systems: Innovations and Technology Applications, First Edition.
Edited by Avik Mukherjee, Santosh Kumar, Manjusri Misra, and Amar K. Mohanty.
© 2025 John Wiley & Sons Ltd. Published 2025 by John Wiley & Sons Ltd.

600 million people were infected by foodborne diseases caused by microbial infections, with the death toll rising to approximately 420000 [5]. Another primary cause of food spoilage is oxidation, which causes a loss of essential fatty acids and vitamin degradation, eventually decreasing the nutritional content of the food. Despite extensive studies, the most common method for reducing oxidation still involves the direct administration of antioxidants to food products. Convenience and cost-effectiveness are the reasons behind this. Nevertheless, this strategy has two drawbacks: (i) antioxidants may lose or inhibit efficacy as a consequence of their interactions with food components during processing, and (ii) the food quality and taste gradually decline after the absorption of the active chemicals in the food [6].

Traditional food preservation methods, such as salting and applying heat, alter the food's taste, smell, color, and texture. Now more than ever, scientists are exploring new preservation methods due to a rising consumer demand for safer and more nutritious food options while retaining the original quality and taste of the food product. To aid in the preservation of quality and improve food safety, a new packaging approach known as active packaging (AP) has emerged because of advancements in materials science and engineering. AP, a relatively new innovation in comparison to the more common "inert packaging," uses components including antimicrobials, oxygen scavengers, and antioxidants. Utilizing both natural and artificial additives in the packaging system, this innovative approach to food packaging first emerged in the second part of the twentieth century [7]. The wide range of AP systems includes additives with a variety of active functions, including scavenging and absorbing properties (e.g., carbon dioxide, oxygen, moisture, ethylene, flavors, and UV light); releasing and emitting characteristics (e.g., antioxidants, carbon dioxide, ethanol, preservatives, and flavors); antimicrobial and quality retention properties; and component removal properties (catalyzing the removal of food components, such as cholesterol and lactose) [8]. These AP systems may be developed by introducing, coating, immobilizing, or altering the surface of the packing materials with active ingredients. They discharge or absorb chemicals into, or from, the packaged food products, or from the headspace surrounding the product [6]. By controlling the pace at which they are released, active compounds on the surface of foods can keep their concentration at an optimum level, thereby preventing the growth of microorganisms and/or the oxidation of the food [9]. Moreover, the addition of synthetic additives and preservatives has raised numerous concerns among consumers due to their adverse health effects on the human body and because they make packaging unrecyclable, resulting in excessive waste volumes. For example, the direct administration of synthetic antioxidants, such as butylated hydroxyanisole, in packaging systems improves the shelf life of food products by protecting against lipid oxidation [10]. However, butylated hydroxyanisole may have endocrine-disrupting effects in humans despite its usefulness in improving food quality and its widespread application in AP [11, 12]. Martillanes et al. [13] reported that a particularly effective method of limiting moisture from leaking in by the food product is the application of absorbent pads in food packaging. However, this packing technique has certain drawbacks since over time, unhygienic liquids become trapped in pads, generate foul odors, decay, and possibly grow foodborne pathogens.

A new packaging trend that has emerged in recent years was created to fill in the gaps in food protection by employing natural ingredients and guaranteeing optimum food safety during transportation. AP materials with antioxidants and nanomaterials integrated into the polymer matrix or on their surface have been the subject of substantial research in

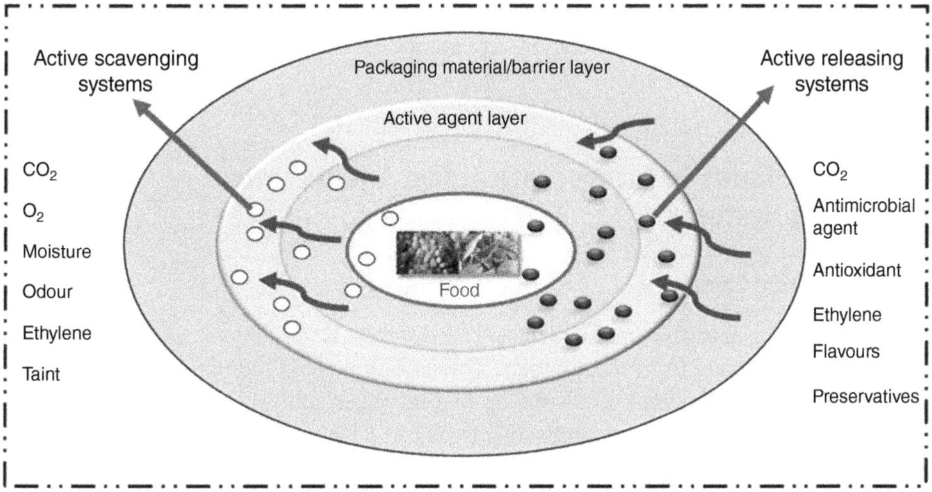

Figure 2.1 Schematic representation of the releasing systems and active scavenging mechanism adopted in food packaging. *Source:* Fadiji et al. [14]/MDPI/CC-BY-4.0.

recent years. Some of the additives that are incorporated into the polymer matrix that act as active agents include nanoparticles (NPs), essential oils (EOs), plant extracts, and antimicrobial peptides. For instance, the use of natural antioxidants in the AP material protects packaged meat against lipid oxidation. Antioxidants have the potential to interact with food items and their headspace, release antioxidants into the container, or absorb free radicals and other substances from the food or its surroundings. Additionally, AP made of natural antioxidants (EOs, polyphenols, etc.) is a less expensive option that may also reduces food safety hazards [10]. Figure 2.1 illustrates the active scavenging and releasing mechanisms used in modern food packaging.

Expert predictions state that the future of food packaging lies with the newest generations of AP. According to estimates, AP systems consist of 5% of the value of the packaging market – of which 35% is made up of active systems [15]. Despite having a modest percentage of the total value of package sales, there are signs that sales of AP could rise rapidly in the coming years. Comprehensive studies on AP have been done by several food research institutes and food enterprises across the world, and certain AP techniques have even been put into commercial use. Eden Agritech, a Thai firm, has created an edible covering called "Naturen" that is tailored to the preservation of freshly cut fruits and can increase their storage life by as much as three times. Similarly, Nature Fresh Farms, a California-based firm, has developed AP systems for cucumbers and Brussel sprouts and plans to expand their use. Several additional companies are working together on AP initiatives, such as the Avanaz-S project jointly carried out by Btsa Biotecnologías, Elpozo Alimentacion, Uvesa, Aplicadas, Monteloeder, Bandesur Alcala, Nurel, and SP Group [6].

Therefore, the objective of this chapter is to summarize the different types of additives commonly used in AP systems and provide a thorough overview of the latest advances in active food packaging. Moreover, the ecotoxicity of active ingredients commonly used in food packaging and their effect on organisms are also discussed. By analyzing the ecotoxicity of active ingredients, awareness of the ingredient's potential side effects is encouraged, resulting in the adoption of sustainable and eco-friendly alternatives. This will promote a deeper comprehension of the complex relationship between active agents and their ecotoxicological implications.

2.2 Techniques to Incorporate Active Agents Into Packaging Systems

2.2.1 Solution Casting

The direct casting approach has been widely used since it is the simplest and fastest method to develop films made from biopolymers. The casting process requires the use of a film-forming agent (biopolymers), a solvent, and a plasticizer/additive in order to develop the films. It is important to create a uniform, viscous film-forming solution with biopolymers that will be filtered, centrifuged, or subjected to other processes to remove insoluble particles and air bubbles before being dispersed on a flat surface and shaped to form the film matrix. Following this, the solvent is evaporated to close the void space between the polymer chains and promote interaction between them. The final step in this process is the removal of the film from the surface and storing them for application in food packaging. When using this technique, the film characteristics could differ slightly from one formulation to the other. Also, due to its high cost and production time, this technique is not yet employed for large-scale industrial manufacturing. In general, the enhancement in film properties was more noticeable in these cast films that used predetermined compositions. The application of the obtained films in food packaging demonstrated their potential as packaging films for assuring food safety, preserving food quality, and extending food shelf life [16].

2.2.2 Coating and Spraying

Coatings are often applied as liquids to the surfaces of food or other packaging materials. The substance is applied to the surface of the product by either soaking it in a solution or spraying it on and drying it. Although coatings are not themselves packaging, they can help improve food shelf life by limiting intrinsic factors and lowering barrier requirements. This coating can help keep food fresh for longer by preventing the growth of microorganisms and prolonging its shelf life. Several factors, including price, accessibility, functional features, and performance, influence the use of coatings in food applications. Parameters including structural matrix material, fabrication conditions, and additive concentration all play a role in determining these properties. The coating in direct contact with the food surface can affect gas permeability, and the antimicrobial components can migrate from the films and provide concentrated protection against ambient degrading elements [16, 17]. Figure 2.2 presents the casting and coating techniques used in AP.

2.2.3 Melt Blending

Commercial bioplastic packaging films are typically made using extrusion technology. Due to its shorter processing time and reduced energy usage, extrusion is commonly chosen over solution casting processes. The extrusion process consists of several steps, including

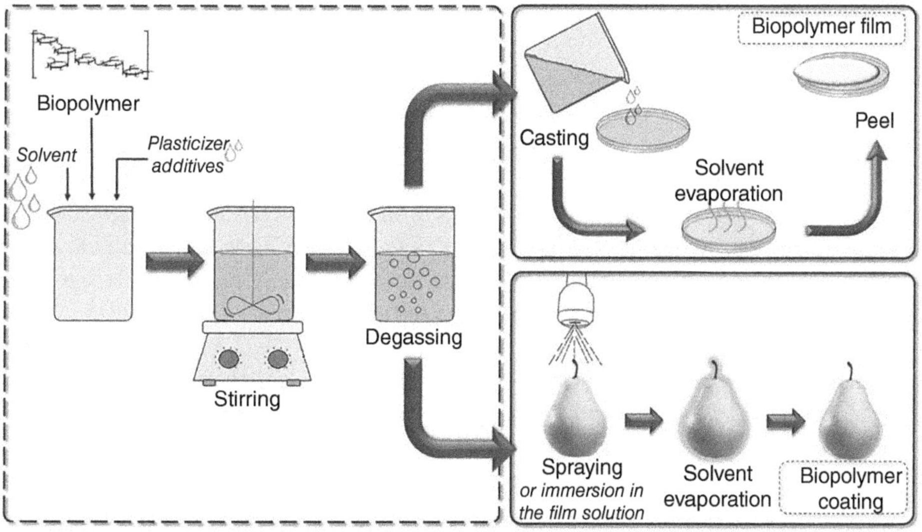

Figure 2.2 Simplified fabrication process for films and coatings using casting and spraying techniques, respectively. Casting and coating are the most common techniques used to develop active food packaging and coatings. *Source:* Marangoni Júnior et al. [16]/with permission of Elsevier.

formulation preparation using a variety of raw materials and their mixing, blending the mixture at high temperatures in an extruder to obtain strand-shaped extrudates of the composite, cutting the extrudates into pellets through the pelletizer, followed by drying of pellets in a hot-air oven. The pellets were then formed into sheets (using a cast film extruder and compression molding) or blown into a film by a blown film extruder. The extrusion procedure often results in films with satisfactory mechanical qualities and high thermal stability. The film is then rapidly cooled, removed from the die, and wrapped onto rolls for further use [17, 18].

2.3 Metal Oxide Nanoparticles (NPs) Used in Food Packaging

Typically, colloidal, solid particles with a diameter between 10 and 1000 nm are referred to as NPs. When particles are reduced to the nanoscale, their physical and chemical characteristics diverge significantly from those of particles with macro- or micro-scale sizes. NPs are being researched for their possible use in food packaging materials for their ability to enhance the mechanical properties, barrier qualities, and antibacterial activity of packaging materials, resulting in an improved shelf life of packaged products. Common nanomaterials used in food packaging are described below.

2.3.1 Zinc Oxide

Zinc Oxide (ZnO) exhibits a high surface-to-volume ratio, high thermal conductivity, and highly crystalline structure, which reinforces the mechanical properties of the matrix. Moreover, due to its size, ZnO has high moisture and UV barrier activity along with antibacterial properties. The Food and Drug Administration (FDA) classifies ZnO as a "Generally Recognized as Safe" (GRAS) substance because of its potential uses in the food and pharmaceutical industries [19]. All these properties of ZnO make it an ideal candidate for AP applications. ZnO NPs have been synthesized using a wide range of techniques, with parameter adjustments yielding a wide variety of ZnO particle morphologies. Examples include the sol–gel method of producing ZnO and silver (Ag)-doped ZnO NPs [20] and the effect of calcination temperature on the structure, morphology, and antibacterial activity of ZnO nanorods [21]. ZnO NPs were also synthesized using an ultrasonic technique, and their antimicrobial activity and the influence of their particle size were studied against *Escherichia coli* and *Staphylococcus aureus* growth [22]. In another study by [23], ZnO, magnesium oxide (MgO), and integrated ZnO-MgO NPs were synthesized using a chemical vapor synthesis technique, and their antibacterial efficacy against *E. coli* and *Bacillus subtilis* was investigated. In another study, the solvothermal technique yielded a variety of ZnO NP morphologies as well, where flower-shaped ZnO NPs were shown to be more effective against *E. coli* and *S. aureus* than rod- and sphere-shaped NPs [24].

ZnO has been integrated into coating materials and a variety of synthetic and biopolymers to create nanocomposites with active properties. Tamarind kernel powder-based packaging film reinforced with ZnO was developed successfully for packaging applications [25]. The developed films produced excellent moisture barrier properties with a maximum reduction of 48% and robust antimicrobial properties against *Bacillus cereus* and *E. coli*. Similarly, antifungal pullulan/collagen films incorporated with ZnO were developed by Bailore et al. [26]. ZnO NPs were able to enhance the nanocomposite film's rigidity and mechanical resistance along with inhibiting *Aspergillus niger* growth. Other biopolymer-based films were developed by incorporating ZnO into starch/kefiran [27], gelatin [28], and chitosan/cellulose acetate phthalate [29] matrices.

2.3.2 Titanium Dioxide

Titanium Dioxide (TiO_2) is a well-known metal oxide that is inexpensive, has great chemical stability, and is frequently employed in photocatalysis. TiO_2 has a biocidal impact on microorganisms due to its ability to oxidize the unsaturated poly phospholipid component of a microbial cell membrane. Several studies have shown that TiO_2, when exposed to sunlight or UV radiation, also forms reactive oxygen species (ROS) such as hydrogen peroxide, superoxide anions, and hydroxyl radicals, which directly attack microbial cell walls [30]. TiO_2 is one of the most adaptable materials and is utilized in different types of technology and food products. However, the European Food Safety Authority (EFSA) emphasized the necessity of more studies on TiO_2 safety in 2016 [19]. Due to the uncertainty around the potential genotoxicity of TiO_2 and the risk that it may accumulate in the body after consumption, the EFSA no longer views TiO_2 as safe when utilized as a food additive. However, the use of TiO_2 NPs in the food sector is not prohibited. The most popular way to produce TiO_2 is through the sol–gel process. Aerogels made of TiO_2 colloids created by the sol–gel technique and pectin have the potential to be used for food packaging [31]. Recent synthetic techniques include green synthesis of NPs, in which TiO_2 NPs are produced using plant extracts and exhibit effective antibacterial action against a wide range of bacteria [32, 33].

Different types of polymer matrices can be used in conjunction with TiO_2 to develop an AP system owing to its ability to enhance properties such as mechanical, barrier, thermal, and microbial growth inhibition, resulting in the extended shelf life of food products. Potato starch film, when incorporated with TiO_2, resulted in an excellent improvement in the light barrier properties and blocked ~90% of UV rays, which will effectively protect the packaged product from light-induced lipid oxidation. Moreover, the moisture barrier property improved by a maximum of 34% upon TiO_2 addition [34]. In another study, TiO_2 NPs were added into a sesame protein isolate matrix, and it was reported that the tensile strength of the composites increased while the water vapor transmission rate of the films increased due to the proper dispersion of TiO_2 NPs; additionally, 5% TiO_2 concentration could scavenge ~11% atmospheric O_2 over a six-hour period [35]. A similar reinforcing effect of TiO_2 was reported for agar/fish gelatin bilayer [36] and whey protein isolate [37] films.

2.3.3 Silicon Dioxide

Silica (SiO_2) and silicate are two of the most common forms of metalloid silicon, which is the most common solid element on Earth. Nano-SiO_2 is added to polymers to improve their durability and heat resistance. Additionally, nano-SiO_2 might be utilized in the production of non-stick coatings for containers, bottles, and bags [38]. In food packaging, most research has focused on using SiO_2 in conjunction with other nanomaterials, either as a synergistic antibacterial agent or as part of nanocomposites. Silver–silica nanocomposite antibacterial behavior, for example, was shown to be more effective than that of common materials like silver zeolite and silver nitrate [39]. Furthermore, Hetrick et al. found that about 99% of bacteria was eliminated when nitric oxide-releasing SiO_2 NPs were applied to biofilms of *Escherichia coli, Candida albicans, Staphylococcus epidermidis, Staphylococcus aureus,* and *Pseudomonas aeruginosa* [40]. The mechanical properties, like tensile strength and elongation at break of agar/sodium alginate nanocomposite film, improved by 65% and 60%, respectively, upon incorporating SiO_2. Furthermore, the water barrier property, thermal stability, and UV light barrier of the films were enhanced considerably [41]. Similarly, in another study, gallic acid-modified SiO_2, when incorporated with chitosan, improved the mechanical and barrier properties of the nanocomposite film. The films also exhibited robust antioxidant activity due to the presence of free amino groups of chitosan and the modification of SiO_2 [42]. Figure 2.3 depicts the numerous active ingredients used in food packaging.

2.4 Active Agents Derived from Organic Sources

2.4.1 Polyphenols

Polyphenols are present in all plant components, including the peel, leaves, pomace, stem, and roots. Due to their varied chemical structures and functional properties, they have found a widespread use in modern packaging. Plant sources of polyphenols include phenolic acid, flavonoids, polyphenolic amides, and lignans [43]. Plant extracts, which are high in polyphenol content, are added to foods to mask unpleasant colors and flavors or to

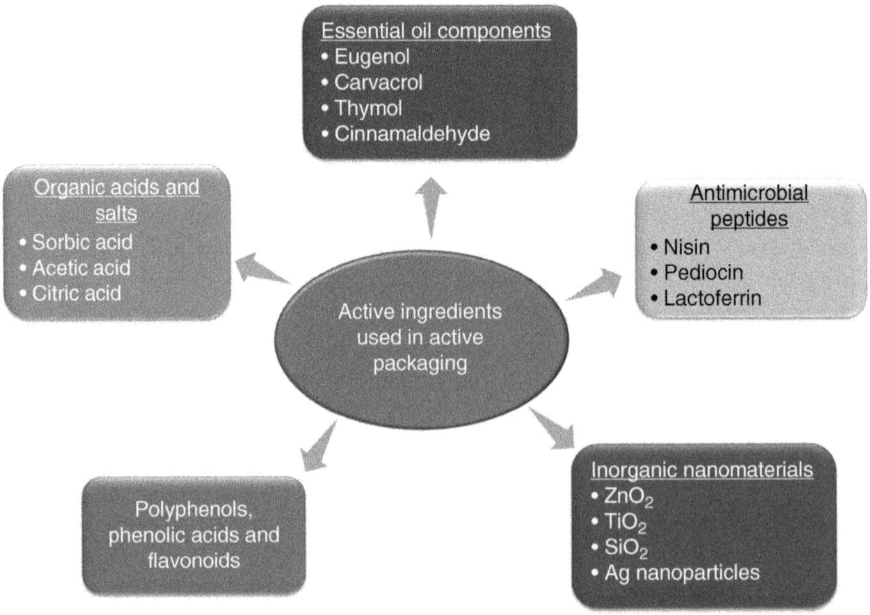

Figure 2.3 Different types of active agents used in food packaging.

enhance the food's desired quality, nutritional profile, and functional characteristics. Polyphenols can be used in a variety of edible films and coatings, including polysaccharide, lipid, or protein-based food packaging solutions [43, 44]. Phenolic compounds are used in biodegradable film and coatings, as antioxidants and antimicrobial agents, to enhance the shelf life of a packaged product and are used in either direct or indirect contact with the surface of the packing film.

2.4.2 Phenolic Acids

A subclass of phenolic compounds, known as phenolic acids, includes hydroxyphenyl acetic acids, hydroxybenzoic acids, and hydroxycinnamic acids, all of which have antibacterial and antioxidant properties [45]. According to Al Jitan et al. [46], both traditional and innovative techniques have been applied for phenolic acid extraction. The techniques include maceration, liquid-liquid extraction, Soxhlet extraction, microwave- and ultrasound-assisted extraction, supercritical extraction, and accelerated solvent extraction. Rashmi and Negi [47] reported in their study that hydroxycinnamic and hydroxybenzoic acids upon consumption improve the general well-being of the body because of their anti-inflammatory, antioxidant, antimutagenic, antimicrobial, antiplatelet, and hypoglycemic properties, which prevent cancer, stroke, and other cardiovascular diseases.

2.4.3 Flavonoids

Flavonoids are the most common type of polyphenol in the human diet and may be found in a wide variety of plant parts, such as fruits, vegetables, nuts, seeds, and leaves. All flavans have a double-ring structure; however, there are six different types of flavonoids distinguished by the pattern of hydroxylation, the degree of oxidation of the C-ring, and the moieties present in the 3 position. The different subclasses include flavanones present in citrus fruit; procyanidins and flavanols in red tea, cocoa, wine, and fruits; flavones and isoflavones in soy; anthocyanins in berries; and flavonols in fruits and vegetables [48]. Flavanols, commonly referred to as flavan-3-ols, are present as monomers and oligomeric procyanidins in a wide variety of plant sources. The foods that contain the most flavanols are cocoa, red wine, green tea, red grapes, berries, and pears. The constituents of flavanols are (+)-catechin and (−)-epicatechin, whereas procyanidins are oligomers of catechin and epicatechin. In contrast to other classes of flavonoids, which exist primarily in glucoside forms in plants, flavanols are typically present in the aglycone (which are free flavonoids or flavonoids without attached sugars) form as monomers, oligomers, or esterified with gallic acid to produce epicatechin gallate, and epigallocatechin gallate [48, 49]. Consumption of flavonoids is highly beneficial due to their ability to prevent a variety of diseases ranging from neurological to cardiovascular diseases. The antioxidant efficacy of certain flavonoids and extracts high in flavonoids has been demonstrated in several investigations, which has resulted in the use of these compounds in food packaging applications [43].

2.4.4 Tannins

Tannins are a class of water-soluble polyphenols with a wide variety of chemical structures and varying molecular weights. They are abundantly found in the bark and wood with some amount available in the fruits and leaves of herbaceous and woody plants [50, 51]. Plants rely on tannins for defense against a wide range of biological hazards, such as attacks from animals, insects, fungi, and bacteria and exposure to harmful light (UV rays and free radicals). Beverages, such as wine and green tea, which contain tannins, might help prevent numerous diseases. The distinctive chemical characteristics of tannin extract have the most recognizable mouthfeel for many food items and are used to enhance the quality of a wide range of food products. Commercially available tannins from wood extracts and castalagin from chestnuts exhibited antimicrobial properties against pathogens such as *Salmonella* typhimurium, *E. coli, S. aureus,* and *Vibrio* spp. [43, 52]. Thus, the antimicrobial, antioxidant, and UV absorption capabilities of tannins show the possibility for use in active food packaging.

2.4.5 Polyphenolic Amides

Functional substituents containing nitrogen may exist in certain polyphenols. Capsaicinoids in chili peppers [25] and avenanthramides in oats [26] are two such categories of polyphenolic amides that are noteworthy since they are present in commonly consumed food products. Capsaicinoids are found only in the fruits of plants of the genus Capsicum (chili peppers). They give the fruit its pungent flavor and exhibit possibly useful pharmacological effects [53]. The chemical structure of this compound consists of fatty acid bonded with an acid amide of vanillylamine [53]. Capsaicinoids, like capsaicin, which give chili peppers their heat, have also been proven to have potent antioxidant and anti-inflammatory effects, as well as the ability to modulate oxidative defense mechanisms of cells [54]. Although there are more than ten different structures, capsaicin and dihydrocapsaicin make up about 90% of all capsaicinoids. It is believed that the plant produces these compounds as a defense against fungi and herbivores [53].

Avenanthramides, or hydroxycinnamoyl anthranilate, are a subclass of low-molecular-weight polyphenols that are only found in oats [55]. Avenanthramides structurally consists of different cinnamic acids and various anthranilic acids [56]. They are antipathogenic and are produced by plants in response to any attack from fungi. Avenanthramides have been found to exhibit antioxidant properties, which are 10–30 times better than other phenolic antioxidants like caffeic acid and vanillin [57]. Other health benefits of avenanthramides include anticancer, anti-inflammatory, and vasodilation effects [55].

2.4.6 Other Polyphenols

2.4.6.1 Curcumin

Curcumin, also known as diferuloylmethane, is the primary polyphenol in the rhizome of the plant *Curcuma longa* (turmeric) [58]. Asian nations have traditionally employed *Curcuma longa* as a medicinal herb for numerous illnesses due to its antioxidant, anti-inflammatory, antimutagenic, antibacterial, and anticancer characteristics [59]. Curcumin's antioxidant and multifaceted therapeutic effects for chronic illnesses have been thoroughly documented. Curcumin's antioxidant properties has been traced back to its presence of ortho-methoxy groups [59, 60]. In its pure form, curcumin is a crystalline substance that is a vivid orange–yellow color [58]. Curcumin's limited water solubility reduces its bioavailability and prevents it from being easily administered orally. It is soluble in polar solvents, such as ethanol, and insoluble in water and hydrocarbon solvents. Curcumin has been shown to promote wound healing by influencing the inflammatory, proliferative, and maturation stages and can prevent respiratory and joint inflammation. As a result, curcumin might be presented as an option in the manufacturing of packaging systems and/or in biological applications (e.g., wound healing). Moreover, it is well accepted that a shift in pH may be used as a gauge of bacterial growth. Alterations in this component, as well as the generation of carbon dioxide and ammonia, may be detected by curcumin. Thus, several studies have been conducted to create curcumin-based active and intelligent packaging systems [59, 60].

2.4.6.2 Resveratrol

Resveratrol is a naturally produced phenolic compound from the family of phenolic compounds known as stilbenes. It exists in cis- or trans-geometrical isomers with the latter being the dominant isomer in plants [61]. Resveratrol is insoluble in water and soluble in ethanol and thus needs acidic conditions to maintain its activity. They are generally found in red wine, grape skin, and peanuts and have been associated with various health benefits by acting as an antioxidant, inhibiting platelet aggregation, promoting nitric oxide production, and enhancing high-density lipoprotein cholesterol [62]. Other health benefits of resveratrol include antiviral, cardioprotective, anticancer, anti-inflammatory, antiaging, and neuroprotection [63, 64]. Since resveratrol has been shown to have antioxidant properties, it has been proposed that it be utilized to reduce or prevent lipid

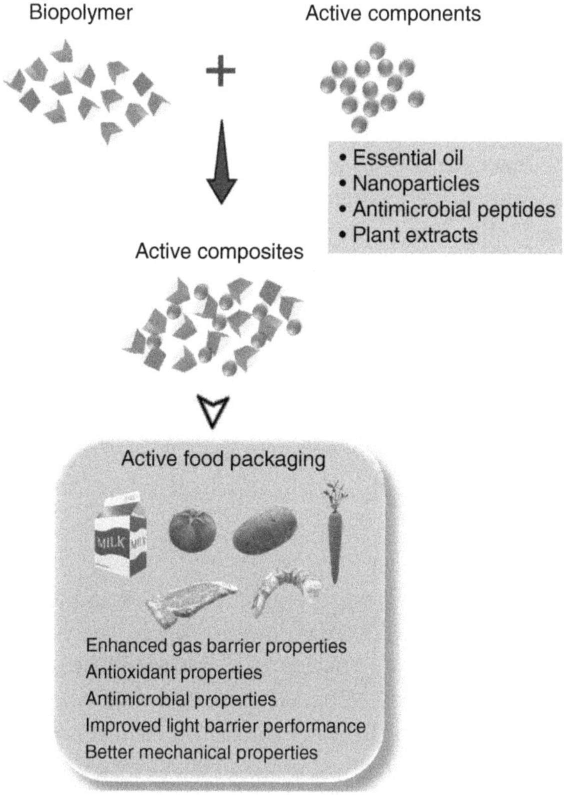

Figure 2.4 Fundamentals of AP: Development of biopolymer-based active composites and their roles in food preservation.

oxidation in pharmaceuticals and food products by preserving their nutritional value and increasing product longevity. Figure 2.4 illustrates the fundamentals of biopolymer-based AP systems.

2.5 Antimicrobial Peptides

Bacteriocins are a broad category of bacterially synthesized antibacterial peptides or proteins. Typically created by bacteria to compete with other microbes for resources in their environment, these substances have the power to suppress or kill closely related bacterial strains. Bacteriocins function in several ways, such as by forming pores in cell membranes, preventing the manufacture of cell walls, or preventing DNA replication. One crucial way that bacteriocins differ from conventional antibiotics is that they have a relatively small killing spectrum and are only harmful to bacteria that are closely linked to the strain that produces them [65]. Some bacteriocins, like nisin, pediocin, and lactoferrin, have been permitted to be used as food preservation agents and, when incorporated into active food packaging systems, prolong the storage life of a wide range of food products [66].

2.5.1 Nisin

Nisin, a bacteriocin (antimicrobial peptide), is a class I bacteriocin also known as lantibiotics. Lantibiotics are small peptides with uncommon amino acids like lanthionine, β-methyllanthionine, and different dehydrated amino acids [66, 67]. Nisin, which is obtained from a bacteria named *Lactococcus Lactis*, has been used as a food preservative agent for over 60 years, is regarded as the most important bacteriocin in food preservation, and is widely used in food packaging applications. Unlike other active agents, which lose their ability when subjected to different processing methods, nisin retains its active characteristics even after pasteurization or sterilization when absorbed into the food matrix, making it a very attractive active ingredient. Therefore, nisin has been extensively used in the production of antimicrobial packaging materials for increasing the shelf life of fish, meat, fruits, and vegetables [66]. Nisin A and nisin Z are two naturally occurring forms of nisin with similar antimicrobial spectra and structures, but they differ at amino acid position 27, where nisin A includes histidine and nisin Z has asparagine [68].

2.5.2 Pediocin

Pediocin is a heat-stable, Class II bacteriocin made from a peptide chain of 44 amino acids and is a basic and cationic peptide [69]. It is isolated from *Pediococcus acidilactici* and can also be used as a food preservative. The pediocin PA-1 structure is able to induce toxicity in sensitive cells because synthetically produced pediocin is bioactive and exhibits specific activity analogous to naturally produced pediocin. Other characteristic features of pediocin PA-1 are the absence of leucine, arginine, glutamine, or phenylalanine residues and the availability of three histidine and four lysine residues with one aspartate amino acid. Moreover, synthetic pediocin PA-1 also has two disulfide bonds [70]. Comparing it with the primary structures of other class II pediocin-like bacteriocins indicates that the pediocin polypeptide chain can be demarcated into two functional modules: an amphiphilic or hydrophobic C-terminal-α-helical domain and a conserved hydrophilic N-terminal-β-sheet [70]. Pediocin PA-1 exhibits a strong antimicrobial effect against a wide spectrum of Gram-positive bacteria, resulting in reduced food spoilage when incorporated into AP systems. Generally, Gram-negative bacteria are not affected due to the composition and structure of their outer membrane, which prevents pediocin PA-1 from accessing the cytoplasmic membrane [70]. Pediocin PA-1 is classified as GRAS, which has led to its use as an antimicrobial ingredient in food products. This peptide is used to extend the shelf life of food products, mainly ready-to-eat meat and canned products, exhibiting an inhibitory effect against *Listeria monocytogenes* [67].

2.6 Essential Oils

Essential oils (EOs) are volatile compounds that are obtained from different plant parts such as seeds, leaves, flowers, barks, fruits, woods, and whole plants using a variety of methods such as solvent extraction, steam distillation, and hydro-distillation. Naturally, plant features including growth stage, variety, geographic origin, plant portion used, age, season, and condition at harvest time all impact the content and quality of EOs. However, it is also influenced by the solvent, analytical conditions, and extraction technique. EOs have excellent solubility in ether, alcohol, and fixed oils but barely dissolves in water due to the density difference between the two. EOs are identified by their distinct aroma and are usually colorless and liquid at room temperature. EOs exhibit a wide range of biological properties such as antimicrobial, antioxidant, antidiabetic, antitumor, anti-inflammatory, insecticidal, and analgesic. Moreover, EOs and its components are made up of chemicals, such as phenolic compounds, that have antibacterial and antioxidant properties, thereby reducing or eliminating microbial growth and preventing lipid oxidation. As such, EOs and their components can be integrated with food packaging systems, which would be released during transportation/storage and would enhance the shelf life of the food product. Figure 2.5 shows the most common EOs used in active food packaging. The application of EO as a natural active agent in food packaging has been the subject of numerous research studies due to its capability to replace synthetic additives used in food preservation [10, 71].

EOs are extracted using a variety of techniques; however, the effectiveness of their extraction varies widely. Also, the quantity obtained is still very little (about 1%) and hence highly precious due to their limited abundance; only 10% of the plants consist of EOs and are called aromatic plants [72]. Hydro-distillation entails boiling a combination of water and plant materials in a closed vessel. The primary benefit of hydro-distillation is its ability to extract required material at temperatures below 100 °C, making it ideal for extracting from hydrophobic plants with a high boiling point. Steam distillation is another frequently employed extraction process. Steam from a steam generator is used to warm the plant matter. Boiling water is diverted so that it does not come into contact with the plant, while steam is permitted to run through it. The efficiency with which the EO in a plant is extracted depends on the amount of heat supplied by steam. Another widely used technique is the solvent extraction technique, where the plant material is mixed with solvents such as hexane, ethanol, ether, or acetone followed by mild heating, filtration, and solvent evaporation. The EO is then extracted by mixing the filtrate with alcohol before being distilled at a low temperature [10, 72]. According to their chemical composition, terpenes and hydrocarbon terpenes (isoprenes) make up the majority of EOs. The terpenes are further divided into hemiterpenes, monoterpenes, sesquiterpenes, diterpenes, etc. Most EOs – nearly 90%

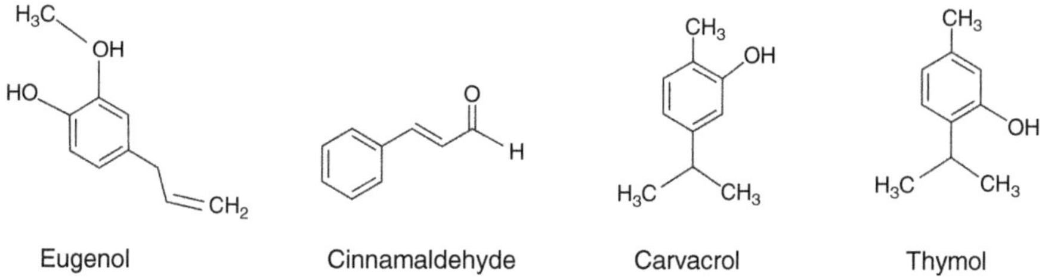

Figure 2.5 Chemical structure of different EOs components.

of them – are made of monoterpenes. Phenylpropanoids and their derivatives are another kind of oxygenated compound found in several EOs. They are mostly found in cinnamon bark, sassafras, clove, and vetiver [72].

2.6.1 Essential Oil Components

2.6.1.1 Cinnamaldehyde

Cinnamaldehyde (3-phenyl-2-propenal) is a major bioactive substance of the EO produced from the bark, leaves, and roots of the species *Cinnamomum* (~85%), which is around 100 different species. It is a yellow-colored viscous liquid isolated from cinnamon oil extraction created synthetically by condensation of benzaldehyde and acetaldehyde or by oxidizing cinnamyl alcohol [66]. Cinnamaldehyde is one of the numerous naturally occurring antibacterial agents, and it is a safe and natural bioactive molecule with extremely potent antibacterial action. Cinnamaldehyde exhibits antimicrobial properties against a wide spectrum of foodborne pathogens and is a natural additive for preserving food quality, making it ideal for application in food packaging. As such, it has been declared GRAS by the US FDA. Moreover, the Joint Food and Agriculture Organization (FAO)/WHO Expert Committee on Food Additives (JECFA) has approved its use for application as a safe and natural food additive [73, 74]. The most cited biological activities of cinnamaldehyde include antimicrobial, antioxidant, anticancer, antihyperglycemic, and other bioactivities [75, 76]. As a result, cinnamaldehyde has received considerable attention throughout several sectors, particularly the food and biomedical sectors. For example, cinnamaldehyde has been observed to increase antioxidant response in peripheral blood mononuclear cells in rheumatoid arthritis patients while decreasing proinflammatory cytokines TNF-α and IL-6 [77]. However, cinnamaldehyde's use is restricted by its low water solubility, susceptibility to light and air, and allergies to the human skin. One method for addressing the drawbacks is to create a sustained or controlled-release polymeric system for incorporating cinnamaldehyde [75].

2.6.1.2 Eugenol

Eugenol (4-allyl-2-methoxyphenol), a phenolic aromatic substance, is a major constituent (approximately 85%) of clove EO extracted from the buds and leaves of *Syzygium aromaticum* and plants of the genus *Cinnamomum* [78]. It is an allyl chain-substituted guaiacol and often appears as a yellowish-oily liquid. First isolated in 1929, eugenol is highly soluble in organic solvents and somewhat soluble in water. It was initially discovered in *Eugenia caryophyllata* as a volatile chemical. Eugenol can be easily made by the allylation of guaiacol with allyl chloride, and it could also be manufactured using biotechnological techniques like biotransformation using numerous microorganisms like *Pseudomonas sp.* HR199, *Streptomyces* sp., *E. coli*, *Amycolatop*, *Corynebacterium* sp., and *B. cereus* [79]. According to numerous reports, eugenol has biological features including antimicrobial, antioxidant, and anti-inflammatory properties. It typically functions as an anti-inflammatory and antioxidant at low concentrations, but at greater concentrations, it becomes a pro-oxidant and increases the production of tissue-damaging free radicals. It is also said to have antigenotoxic properties. For instance, it reduced the mutagenicity of *Salmonella typhimurium* caused by aflatoxin B and furylamide [80]. The FAO and the WHO both set a daily dietary limit for eugenol being 2.5 mg/kg body weight [16]. Additionally, eugenol has been deemed safe by the US-FDA and is neither mutagenic nor carcinogenic. As a result of all these functions, eugenol has been extensively used in the food, cosmetic, and pharmaceutical industries. Moreover, since eugenol is highly volatile and may lose its property during processing, it is best preserved by being encapsulated or incorporated into various materials such as films, emulsions, and nanostructured materials [66].

2.6.1.3 Carvacrol

Carvacrol, or 5-isopropyl-2-methyl phenol, is a phenolic monoterpenoid (isomeric with thymol) found in the EOs of thyme (*Thymus vulgaris*), wild bergamot (*Citrus aurantium*), and pepperwort (*Lepidium flavum*), but is most abundant in oregano (*Origanum vulgare*) [81]. Carvacrol smells and tastes like fresh marjoram and has an appealing tangy flavor. The robust antibacterial properties of carvacrol are essentially caused by the free hydroxyl group and its molecular structure. Carvacrol's antibacterial qualities have gained a lot of attention recently, but it also possesses a wide range of other pharmacological effects, including those of an antiplatelet, anti-inflammatory, antifungal, vasorelaxant, spasmolytic, antidepressant, anticancer, and anxiolytic agent [82, 83]. Multiple studies have demonstrated that carvacrol has potent cytotoxic, genotoxic, and proapoptotic effects against cancer cells, as well as affecting cell invasion via lowering the expression of matrix metalloprotease 2 and 9 [84]. Indeed, carvacrol has been tested with a wide variety of cancer cells, including the mouse B16 melanoma, the gastric carcinoma cells, and the chronic myeloid leukemia cells, the leiomyosarcoma cells, the Hep-2 human larynx carcinoma cells [81]. Carvacrol is a GRAS chemical that is widely used in the food industry as a flavoring agent [82]. Furthermore, the European Commission has added carvacrol to its list of chemical flavorings, and the FDA has given it approval for use in food flavoring [82]. As such, numerous research studies have been conducted in recent times using carvacrol as an additive in biodegradable polymer-based AP systems for enhancing the shelf life of packaged items.

2.6.1.4 Thymol

Thymol (2-isopropyl-5-methylphenol) is a naturally occurring monoterpenoid phenol crystalline that is colorless with a distinctive odor. It is an isomer of carvacrol having a different position of hydroxyl group and is present in the EO isolated from the *Lamiaceae* family, especially of the genera *Thymus Origanum, Ocimum, Thymbra, Satureja,* and *Monarda,* although it is mostly extracted from the plant *Thymus Vulgaris* [66, 85]. Biosynthesis of thymol starts with the aromatization of γ-terpinene to p-cymene, which is then hydroxylated to yield thymol. Genetic modification of microorganisms is also done to increase the quantity of thymol in the EO. For example, *Agrobacterium tumefaciens* was used to genetically modify the *Trachyspermum ammi* species in order to enhance the Ttps2 gene, which increased the quantity of thymol in the plant's EO [86]. Thymol demonstrates a variety of pharmacological effects, including antibacterial, free radical scavenging, antifungal, anti-inflammatory, antispasmodic, analgesic, antiseptic, and anticancer properties [87]. Thymol has been approved by the European Commission for use as a food flavoring agent due to

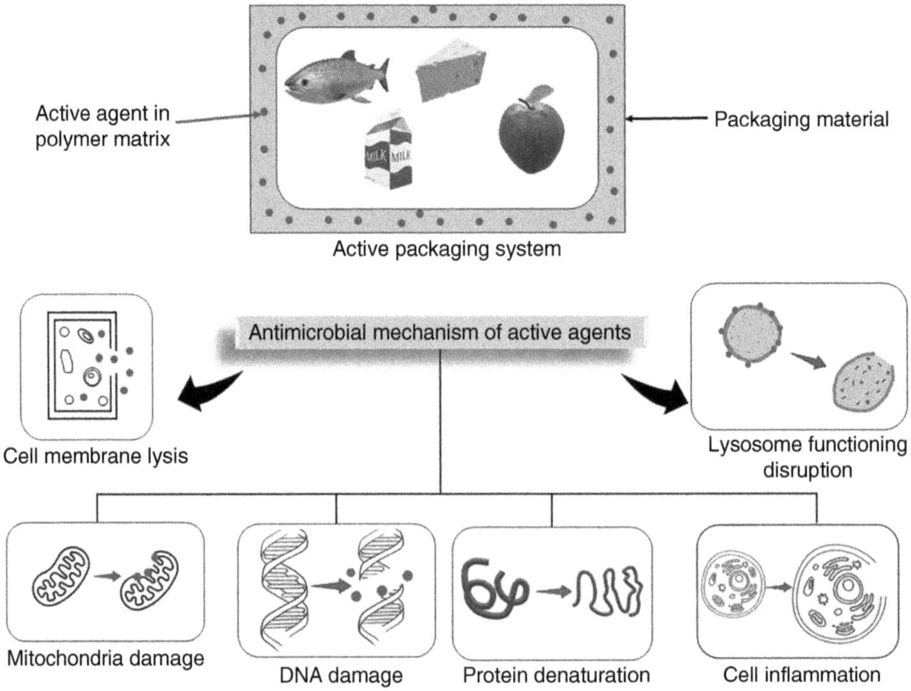

Figure 2.6 The antibacterial mechanism of natural antimicrobial chemicals in food-grade packaging.

health issues associated with its consumption, within acceptable limits. Moreover, the compound has been designated as GRAS or "approved food additives" by the US-FDA. Thymol has been the subject of much research in the food industry due to its robust antibacterial and antioxidant properties, which results in an enhanced shelf life of packaged food products [85]. Figure 2.6 describes the different mechanisms of microbial inhibition by active compounds.

2.7 Polymer Nanoparticles

Polymer NPs is an umbrella term encompassing all polymer NPs, including nanospheres and nanocapsules. These particles may encapsulate or immobilize a wide range of active substances but they are distinguished by their high surface area-to-volume ratio. Nanospheres are a type of matrix particle, meaning that their whole mass is composed of a solid and that molecules can either be adsorbed to the sphere's surface or contained within. The majority of these NPs have a spherical form, although nonspherical "nanospheres" have also been identified [21]. Nanocapsules are vesicular structures that contain the required components in a cavity formed by a liquid core (often water or oil) encapsulated by a polymeric shell [88]. For the preparation of polymer NPs, both naturally occurring and synthetic polymers have been used, although natural polymers are preferred over synthetic ones since the former is biodegradable and abundant in occurrence. Cellulose, starch, sodium alginate, chitin, polypeptides, albumin, and gelatin are all examples of common natural polymers used in the manufacture of polymer NPs. A variety of techniques, such as emulsion polymerization, solvent evaporation, and electrospinning, can be used to create polymer NPs. The size, shape, and stability of the NPs may all be influenced by the polymer used and the fabrication process.

Several advantages of using polymer NPs for the delivery of active ingredients in packaging systems include high drug encapsulation efficiency, greater intracellular uptake, and higher encapsulation stability of active ingredients. Polymer NPs may be tailored to deliver the active agents when incorporated into the packaging system, which will ultimately help in enhancing the shelf life of packaged products. Polymer NP's versatility, adaptability, and potential for multiple applications make them excellent agents for the controlled release of active ingredients [89]. For example, specific microorganisms, such as bacteria, fungi, and viruses, can be inhibited from growing by using antimicrobial polymer NPs. Numerous antimicrobial substances, including bacteriocins, organic acids, and EOs, can be included in these NPs. Moreover, antioxidants can also be loaded into the polymer NPs, which would scavenge the free radicals and prevent the oxidation of proteins and lipids, thereby improving their shelf life. The release rate of these agents can be controlled by changing the structure and composition of the NPs, as well as the surroundings of the packaging environment.

2.7.1 Chitosan

Chitosan is derived from chitin by its partial deacetylation and contains alternating units of $(1 \rightarrow 4)$ linked *N*-acetyl glucosamine and glucosamine units. It is a nitrogenous polysaccharide that is white, rigid, and inelastic. Chitosan has three functional groups, namely, amino groups and primary and secondary hydroxyl groups at C2, C3, and C6 locations. The chemical modification of chitosan can be done due to the hydroxyl groups, which attach side groups without effecting the properties of chitosan. Moreover, chitosan is nontoxic, biodegradable, and has antibacterial properties which makes it useful in a variety of applications including the fabrication of NPs for packaging applications [90]. Chitosan NPs were used to encapsulate polyphenol-rich pomegranate peel extract. The NPs were then incorporated into zein films to develop AP films. The chitosan NPs were able to inhibit the growth of *Listeria monocytogenes* on pork surfaces and extend the shelf life until day 10 of the study [91]. Similarly, quercetin-loaded chitosan NPs were developed by Roy et al. [92], which exhibited high antioxidant activity due to quercetin's exceptional radical scavenging ability. Moreover, the growth of microbes such as *E. coli* and *L. monocytogenes* was inhibited since quercetin binds to fatty acid chains in the lipid bilayer of bacteria, causing an aggregation that inhibits cell wall permeability and breakdown of bacterial membrane. Therefore, the effective release of active agents, like quercetin, can be achieved by the possible distribution of quercetin in a polymer matrix by the process of adsorption on hydrophilic NPs, such as chitosan NPs. Other active ingredients like curcumin [93], cinnamaldehyde [94], and tea polyphenol [95] exhibited robust antimicrobial and antioxidant properties due to the effective release of active ingredients from the chitosan matrix.

2.7.2 Alginate

Alginate is a biopolymer that is low-cost, abundant, biocompatible, mucoadhesive, non-toxic, and nonimmunogenic material that is obtained from bacteria and brown algae. Alginate is made up of β-D-mannuronic acid (M) and α-L-guluronic acid (G) residues linearly connected by 1,4-glycosidic bonds. The properties of alginate are affected by the content and sequencing of the M and G residues, which depend on the type of algae used. It is also possible to chemically modify alginate to change its characteristics [96]. Alginate NPs can be formed with the help of the gelation process, where divalent cations react with guluronic acid to form the "egg box" model, resulting in a gel structure. As a result of its gel formation ability, alginate is now used in a variety of biomedical and food packaging applications, including the encapsulation of active agents for sustained release and in AP for food [97]. Alginate can be used to encapsulate a variety of EOs. For example, oregano EO was encapsulated by chitosan-alginate NPs, and their application against microorganisms was evaluated. The NPs were able to inhibit the growth of both Gram-positive and Gram-negative pathogens, showing great promise for future use as food additives [98]. In another study, a nanostructured lipid carrier (NLC) and nanoemulsion of thymol-loaded alginate were developed by Khorrami et al. [99]. The NLC films released thymol into the air and model food matrices at a slower rate than nanoemulsion films. NLC films have extended antibacterial action due to their gradual release characteristics, unlike nanoemulsion-based films, which showed activity during their early stages. After adding NLC, the film's mechanical strength decreased, but it was still sufficient for food-packing applications. Similarly, nisin-loaded alginate-pluronic-chitosan NPs demonstrated an initial explosion release of nisin, which subsequently transformed into a gradual release. This resulted in the prolonged antimicrobial activity of nisin-loaded nanocomposite against *Micrococcus luteus, Pseudomonas aeruginosa, Salmonella enterica,* and *Enterobacter aerogenes* [100].

2.7.3 Gelatin

Gelatin is a protein biopolymer produced by acid- or base-catalyzed hydrolysis of collagen. Given that collagen, the most common protein in mammals, is used to make gelatin, the enzymatic breakdown of gelatin cannot result in any toxic by-products. Gelatin is a polyampholyte molecule, implying that it contains hydrophobic groups along with cation and anionic groups. Gelatin is triple helical in structure with repeating units of proline, glycine, and alanine, and this structure is denoted by $(Gly\text{-}X\text{-}Pro)_n$, where X is an amino acid, typically arginine, lysine, valine, and methionine. The three polypeptide chains that make up gelatin's distinctive triple helix shape account for the protein's exceptional stability. The fundamental structure of gelatin opens a wide range of possibilities for covalent drug attachment and chemical modification. These features along with the enormous potential for nano-sized delivery material enable gelatin NPs for various controlled-release applications in biomedical and food packaging applications [101]. Nanogels containing lemon peel extracts prepared from gelatin and guar gum in basil seed gum/casein films have been found to have applications for food packaging [102]. After the addition of the nanogels, the thermal stability, water vapor permeability, hydrophobicity, and mechanical properties improved. All nanogel-containing films exhibited potent anti-oxidative activity and showed a bactericidal effect against *E. coli* and *S. aureus*. Moreover, when tested for toxicity against

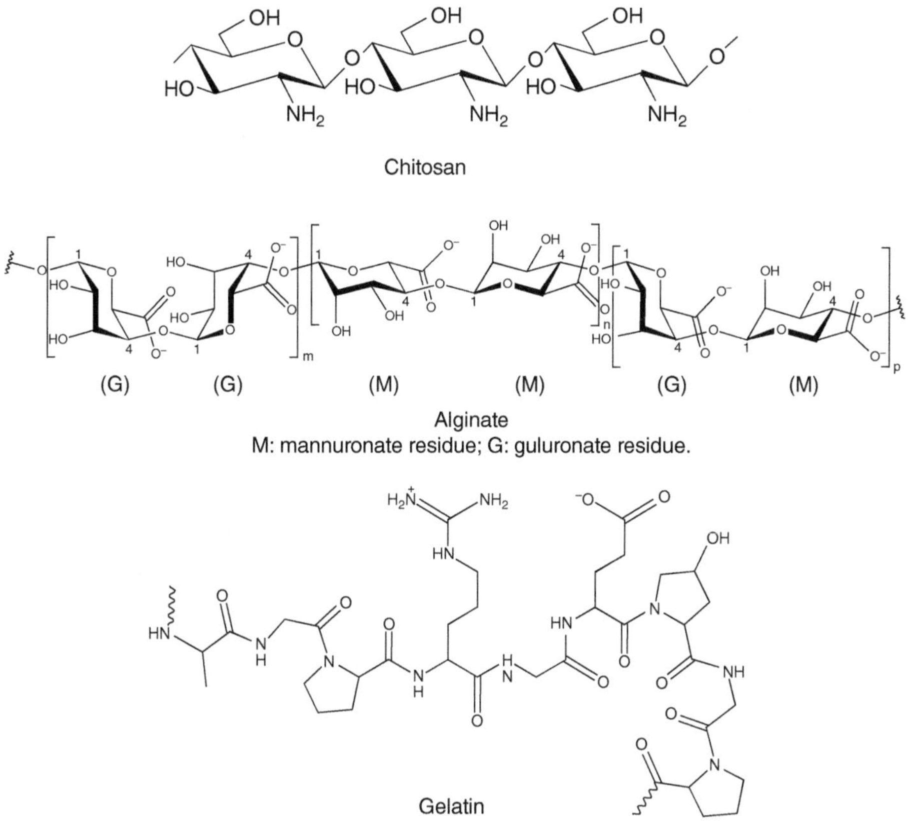

Figure 2.7 Structure of chitosan. *Source:* Krajewska [106]/MDPI/CC-BY-4.0. Alginate; Agulhon et al. [107]/American Chemical Society. Gelatin; Noor et al. [108]/MDPI/CC-BY-4.0.

an endothelial cell line for 72 hours, none of the films were harmful [102]. Similarly, the gelatin nanofiber membrane incorporated with eugenol NPs retarded the growth of *E. coli* and *S. aureus*. The eugenol NPs incorporated into the gelatin nanofiber membranes exhibited an improved and sustained release of active ingredients [103]. Electrospun gelatin nanofibers with clove oil-impregnated chitosan NPs were synthesized by ionic crosslinking and exhibited a bactericidal effect against *E. coli*; after eight hours of treatment, the nanomaterials achieved a 99.98% reduction in *E. coli* population. Upon use on cucumber, the flavor and color of the samples were maintained for four days [104]. Similar improvements in the shelf life of peeled giant freshwater prawns were demonstrated upon the application of *Mentha longifolia*-loaded electrospun carboxymethyl cellulose-gelatin nanofibrous films [105]. Figure 2.7 presents the chemical structure of chitosan, alginate, and gelatin.

2.8 Ecotoxicology

Biodegradable materials have the major benefit of quick biodegradation without significant ecotoxicity or negative environmental consequences after their use and disposal. However, the emergence of nanotechnology in this category of materials has brought up the problem of accidental pollution resulting from the biodegradation process. It is challenging to anticipate whether these nanomaterials will bio-accumulate or bio-concentrate in the food chain or act as an environmental contaminant since the distribution and fate of NPs when exposed to the ecosystem are still poorly known. Therefore, it is essential to properly assess the environmental risks associated with the NPs in the environment by conducting ecotoxicity tests. Recent ecotoxicity investigations have focused on nanomaterials that have previously been used in various industries (i.e., cosmetics, electrical materials, and pharmaceuticals), such as single-walled carbon nanotubes (SWCNTs), metal, and metal oxide NPs. More research is necessary to assess the ecotoxicity of bio-nanocomposites used in the food packaging sector.

In a study by Schiavo et al. [109], the authors compared the ecotoxicological impact of ZnO NPs on three frequently used marine organisms: the bioluminescent bacteria *Vibrio fischeri*, the microalgae *Dunaliella tertiolecta*, and the crustacean *Artemia salina*. ZnO NP toxicity was shown to be time-dependent and species-specific, with algae being the most sensitive followed by bacteria and crustaceans. ZnO NPs significantly reduced the vitality and body length of *A. salina* after a 14-day chronic exposure. Similarly, ZnO NPs at 1 mg/l concentration were shown to reduce the proliferation of four marine phytoplankton species (*Dunaliella tertiolecta, Thalassiosira pseudonana, Skeletonema marioni,* and *Isochrysis galbana*) by 50–75%. The release and absorption of free Zn ions are likely responsible for the toxicity of ZnO NPs [110]. In another study by Lin et al. [111], ZnO concentration at 2000 g/l was found to be hazardous to the seed germination of ryegrass and terminated root elongation of rape, radish, lettuce, ryegrass, corn, and cucumber. Moreover, the inhibition appeared during seed incubation instead of seed soaking.

TiO_2 is a commonly used material in sunscreen, paints, cosmetics, and food industries, among others. Therefore, identifying the ecotoxicity of TiO_2 is essential for evaluating its environmental impacts and formulating suitable risk management strategies. In the studies conducted on the ecotoxicity of TiO_2, the majority of toxicities against plants and algae were concentration- or time-dependent. In a study by Chen et al. [112], the ecotoxicity of TiO_2 NPs on *Chlamydomonas reinhardtii* (a green alga) was evaluated by examining the changes in its physiology due to the presence of the NP. The authors reported that the TiO_2 concentration of 1mg/L was harmful to the photosynthetic activity of the algae. Moreover, the *C. reinhardtii* cells began to experience damage as nano-TiO_2 concentrations rose. This damage included a decline in the number of chloroplasts, the degradation of various other organelles, the occurrence of plasmolysis, and the discovery of TiO_2 NPs inside the cell wall and membrane. The effect of TiO_2 on *Skeletonema costatum* and *Karenia brevis* was evaluated by Li et al. [113]. The authors found that the growth of the algae was inhibited due to the presence of the NPs. When exposed to TiO_2, the cell membrane of *K. brevis* was damaged, and the organelles within it were practically undetectable. In comparison to the control, *K. brevis* and *S. costatum* had significantly higher levels of malondialdehyde (MDA). Therefore, the growth of the selected samples was suppressed by the presence of nano-TiO_2 in algal suspensions. Internal ROS generation in the algal cells was found to be responsible for this oxidative stress. Similarly, the effect of TiO_2 and ZnO on the nematode *Caenorhabditis elegans* was conducted by Wang et al. [114]. Both the NPs were hazardous, preventing the nematode's growth, particularly its ability to reproduce. A similar negative impact on the growth and fertility of *Caenorhabditis elegans* was observed when exposed to cerium dioxide (CeO_2) and TiO_2 [115].

In a study by Hoecke et al. [116], the authors tried to assess if the toxicity of SiO_2 NPs is dependent on the surface area or mass of the NPs. The organism used in the study was algae *Pseudokirchneriella subcapitata*, and it was noted that indeed the surface area and particle size regulate the ecotoxicity of the SiO_2. Moreover, the SiO_2 NPs having sizes 12.5 and 27 nm were found to be toxic to the sample. In another study, Erdincmer et al. [117] looked at the hazardous effects of four nanomaterials (nano-HfO_2, nano-CeO_2, nano-SiO_2, and nano-Ta_2O_5) on three different types of organisms (the bioluminescent bacteria *Vibrio fischeri*, the green algae *Chlorella*, and the crustacean *Daphnia magna*). Ta_2O_5 was found to be most toxic to *V. fischeri*, while *Chlorella* was susceptible to all the nanomaterials because nanomaterials were readily captured by the algal cells, leading to significant inhibitions. Moreover, SiO_2 was found to be most toxic to *D. magna* (10.01 mg/l) when tested for 48 hours; similarly, SiO_2 was found to be most toxic to *Chlorella* (1.4 mg/l) after 72 hours. Figure 2.8 describes the antimicrobial mechanism by which NPs prevent microbial proliferations.

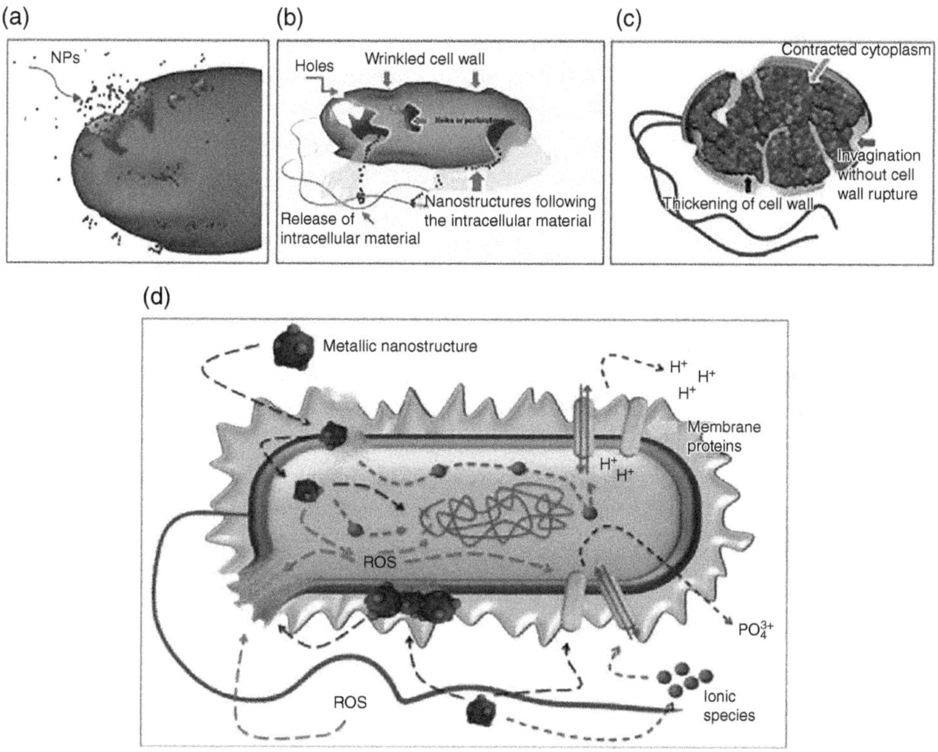

Figure 2.8 (a) Internalization and translocation of NPs into cells. NPs infiltrate cells via cell wall pores, pits, and other structural defects. (b) Schematic illustration of a collapsing cell demonstrating cytoplasmic extrusion and cell wall disintegration. (c) Significant differences in envelope composition (small invaginations and thickening of cell wall) and cytoplasmic extrusion are seen in this bacterial cell. (d) Possible mechanisms include cellular absorption of metal ions, intracellular depletion, DNA replication disruption, the release of metal ions, production of reactive oxygen species, buildup of NPs in the bacterial membrane, and NP dissolution. *Source:* Sirelkhatim et al. [118]/SpringerNature/CC-BY-4.0 and Diaz-Visurraga et al. [119].

When used properly, EOs are typically regarded to be safe and nontoxic; improper handling of them can have negative impacts on the environment and wildlife. Due to their complex chemical makeup and tendency to accumulate in the environment, EOs may have ecotoxic consequences. Some EOs have constituents that are harmful to aquatic life and can damage the aquatic ecology. As per Ferraz et al. [120], only 23% of the research focuses on the possible toxicity of plant-derived compounds on soil ecosystems, whereas a greater proportion of studies (76%) concentrate on the marine environment, notably in aquatic invertebrates (45%). While some extracts and EOs have reported to have no harmful impact on the target organisms or to have hazardous effects at elevated concentrations, other extracts and EOs have been documented to be toxic at concentrations less than the threshold established by international standards. Typically, extracts are less harmful than EOs, but the reverse can be true if the extract comes from plants that have been shown to produce harmful metabolites. Duringer et al. examined the effects of EOs derived from the leaves of *Juniperus occidentalis* and *Chamaecyparis lawsoniana* on the microalgae *R. subcapitata* to determine if the EOs were harmful [121]. It was found that the EOs from *J. occidentalis* was moderately poisonous to *R. subcapitata*, leading to a notable decrease in the algal cell population. However, the organism at the same trophic level as *R. subcapitata* would not experience any acute harmful effects from the leaching of EOs from *C. lawsoniana* into the surrounding ecosystem.

Researchers have also looked at the toxicity of plant extracts and EOs on several crustaceans, including *Daphnia magna, Daphnia pulex, Artemia salina,* and *Scapholeberis kingi.* Additionally, Ferraz et al. [120] reported that *D. magna* was the most-researched species among the publications that were retrieved in their analysis. International organizations like ASTM recommend using *Daphnia* genus species, specifically *D. magna*, to conduct ecotoxicity testing of chemicals and other compounds to evaluate the hazards of materials and chemicals to aquatic species [120]. *Ledum palustre* (wild rosemary) EO inhibited *D. manga* growth at 80 mg/l (mortality rate at 24 hours was 45%, and at 72 hours it was 47.5%) [122]. Similarly, EOs from *Peumus boldus,* an evergreen tree indigenous to Chile, with a concentration of 96.2 mg/l resulted in mortality of 46.2% and 66.2% after 24 hours and 48 hours of exposure to *D. magna,* respectively [123]. Similarly, in another study, acute toxicity to *D. magna* was also investigated using EOs from four species in the Myrtaceae family. The EC_{50} for immobilization of *D. magna* after 48 hours when exposed to EOs from *Melaleuca dissitiflora, Melaleuca linariiflora, Melaleuca quinquenervia,* and *Eucalyptus globulus* was 1.84, 8.94, 103.35, and 143.96 mg/l, respectively. It was also determined that these EOs persist in water, and after seven days, all the EO residues were below 22%, showing a high vulnerability to hydrolysis [124]. *Elaeagnus angustifolia* EO was evaluated for its toxicity on *A. salina.* Both the flower and leaf EOs were highly toxic, with the former having a 24-hour LC_{50} of 2.25 mg/l and the latter of 11.00 mg/l. The authors attribute this result to the EO's high ester component concentration, especially E-ethyl cinnamate [125].

Performing toxicity testing on fish larvae has been a common practice for quite some time. Regulations that prohibit using vertebrates as test subjects for toxicity have led to a rising restriction on this practice [120]. Therefore, to work around these restrictions, zebrafish (*Danio rerio*) embryotoxicity tests offer an alternative option; most of the present fish studies on the ecotoxicity of EO and plant extracts are done on zebrafish followed by rainbow trout (*Oncorhynchus mykiss*) and Nile tilapia (*Oreochromis niloticus*), among others. In another study by He et al. [126], the authors used the EO isolated from *Leonurus japonicus*, also known as motherwort, and studied its effect on *D. rerio*. The authors reported that the EO had a negative impact on the *D. rerio* embryo at a concentration of more than 6.25 mg/l. The EO reduced the heart rate of the embryos, and a significant reduction in egg hatching (77%) was seen at 6.25 and 12.5 mg. Dysplastic head and tail, incomplete heart development, and the absence of eyes were all issues for the tested embryos. The researchers concluded that *L. japonicus* EO was hazardous to zebrafish embryos, resulting in mortality and deformities, especially before the pharyngula stage, when the primary organs began to emerge. In some studies, the EO extracted from plants is deemed not harmful to the non-target specimen, while being fatal to another specimen. For example, low toxicity to *Gambusia affinis* was observed after 10 days of exposure to the EO of *Heracleum sprengelianum* leaf and 48 hours of exposure to the EO of *Zingiber nimmonii* rhizomes [127, 128], while the *Zingiber nimmonii* EO exhibited significant toxicity against disease-spreading vectors like *Aedes aegypti* (dengue vector), *Anopheles stephensi* (malaria vector), and *Culex quinquefasciatus* (lymphatic filariasis vector) [128].

Despite the widespread belief that products made from plants are "greener" and less hazardous alternatives to those made from chemicals, there is a shortage of empirical data to support this theory. As a result, it is crucial to expand the evaluation of these products to fully comprehend the effects they have on the ecosystem. As global laws regarding the ecological consequences of chemicals, chemical mixtures, and new products tighten, additional research will yield scientific evidence about the impact of EOs and other plants as potential new sources of bioactive substances.

2.9 Conclusion

Food packaging is an essential requirement for the preservation of packaged food items from different external factors like water vapor and oxygen permeability, pressure, mechanical shocks, heat, contaminants, and microbes, which tend to reduce the quality of the product. It promotes food safety and extends the shelf life of a packaged product. Due to the increase in consumer demands for safe products with no chemical or synthetic preservatives, AP is gaining a lot of interest because it is one of the recent emerging technologies that can enhance food quality, stability, and safety while lowering the presence of chemicals in the food material. It is also gaining consumer acceptability by incorporating active compounds like antimicrobials and antioxidants and has taken positive steps in solving ecological problems. This can be done by adding the active compounds into the biopolymer matrix during the film-forming process. Different types of active compounds used in developing AP films include metal oxides, polyphenols, antimicrobial peptides, and EOs, among others. These active agents have an impact on various functions of the biodegradable film, particularly by enhancing antimicrobial and antioxidant activity. The extent of this impact depends on the interaction and composition of the active agents with the polymer matrix. Moreover, the encapsulation of active ingredients within a polymer NP is another effective approach for developing AP materials due to the sustained release of compounds that prolong the activity of the films. The ecotoxicity of these active ingredients during biodegradation is another issue that cannot be ignored since it might present a significant threat to the ecosystem and different organisms during its life cycle. Studies on the ecotoxicity effect of active ingredients compel us to be vigilant while using active compounds since their excessive usage and leaching into the environment can be devastating to certain organisms. In conclusion, the utilization of active compounds obtained from natural sources presents a promising future for producing chemical-free, environmentally friendly, and minimally or non-harmful active food packaging material. However, more research is needed in this field to discover a vast array of active compounds from biobased sources, which are ecologically safe while exhibiting a more robust antioxidant and antimicrobial action toward the target pathogens.

CRediT Author Contributions

Debarshi Nath: Methodology, Investigation, Data curation, Formal analysis, Visualization, Writing – Original Draft Preparation

Amar K. Mohanty and Manjusri Misra: Project Conceptualization, Investigation, Methodology, Validation, Administration, Funding acquisition and Supervision, Writing, Review, and Editing

Avik Mukherjee and Santosh Kumar: Project Conceptualization, Writing, Review, and Editing; all authors contributed to the discussion, reviews, and approval of the manuscript for publication.

Conflicts of Interest

The authors declare that they have no known conflict of interest that could have appeared to influence the work reported in this paper.

Acknowledgments

The authors would like to thank the financial support of (i) the Ontario Agri-Food Innovation Alliance – Bioeconomy for Industrial Uses Research Program (Project Nos. 030699 and 030728); (ii) the Ontario Ministry of Agriculture, Food and Rural Affairs (OMAFRA) – University of Guelph Gryphon's Leading to the Accelerated Adoption of Innovative Research (LAAIR) Program (Project No. 030736); (iii) the Ontario Research Fund, Research Excellence Program; Round 11 (ORF-RE11) Ontario Ministry of Colleges and Universities (Project Nos. 056106 and 056184); (iv) the Natural Sciences and Engineering Research Council of Canada (NSERC), Canada Research Chair (CRC) program Project No. 460788; and the NSERC Discovery Grants Project No. 401716 to carry out this research. This study has also benefited from the facility funding to the Bioproducts Discovery and Development Center, University of Guelph, supported by FedDev Ontario, Canada; OMAFRA Canada; the Canada Foundation for Innovation (CFI); the Federal Post-Secondary Institutions Strategic Investment Fund (SIF), Canada; and the Bank of Montreal (BMO), Canada.

References

1 Nath, D., Santhosh, R., Pal, K., and Sarkar, P. (2022). Nanoclay-based active food packaging systems: a review. *Food Packaging and Shelf Life* 31: 100803. https://doi.org/10.1016/J.FPSL.2021.100803.
2 Mousavi Khaneghah, A., Hashemi, S.M.B., and Limbo, S. (2018). Antimicrobial agents and packaging systems in antimicrobial active food packaging: an overview of approaches and interactions. *Food and Bioproducts Processing* 111: 1–19. https://doi.org/10.1016/J.FBP.2018.05.001.
3 Rosenboom, J.G., Langer, R., and Traverso, G. (2022). Bioplastics for a circular economy. *Nature Reviews Materials* 7: 117–137. https://doi.org/10.1038/s41578-021-00407-8.
4 Cox, K.D., Covernton, G.A., Davies, H.L. et al. (2019). Human consumption of microplastics. *Environmental Science & Technology* 53: 7068–7074. https://doi.org/10.1021/ACS.EST.9B01517.
5 Nath, D., Santhosh, R., Ahmed, J., and Sarkar, P. (2022). Optical, mechanical, structural, and antimicrobial properties of tamarind kernel powder, halloysite, and cinnamaldehyde nanocomposite films. *Journal of Food Process Engineering* e14065. https://doi.org/10.1111/JFPE.14065.
6 Kuai, L., Liu, F., Sen, C.B. et al. (2021). Controlled release of antioxidants from active food packaging: a review. *Food Hydrocolloids* 120: 106992. https://doi.org/10.1016/J.FOODHYD.2021.106992.
7 Moeini, A., Germann, N., Malinconico, M., and Santagata, G. (2021). Formulation of secondary compounds as additives of biopolymer-based food packaging: a review. *Trends in Food Science and Technology* 114: 342–354. https://doi.org/10.1016/J.TIFS.2021.05.040.
8 Vilela, C., Kurek, M., Hayouka, Z. et al. (2018). A concise guide to active agents for active food packaging. *Trends in Food Science and Technology* 80: 212–222. https://doi.org/10.1016/J.TIFS.2018.08.006.
9 Gómez-Estaca, J., López-de-Dicastillo, C., Hernández-Muñoz, P. et al. (2014). Advances in antioxidant active food packaging. *Trends in Food Science and Technology* 35: 42–51. https://doi.org/10.1016/J.TIFS.2013.10.008.
10 Sharma, S., Barkauskaite, S., Jaiswal, A.K., and Jaiswal, S. (2021). Essential oils as additives in active food packaging. *Food Chemistry* 343: 128403. https://doi.org/10.1016/J.FOODCHEM.2020.128403.
11 Pop, A., Kiss, B., and Loghin, F. (2013). Endocrine disrupting effects of butylated hydroxyanisole (BHA – E320). *Clujul Medical* 86: 16.
12 Domínguez, R., Barba, F.J., Gómez, B. et al. (2018). Active packaging films with natural antioxidants to be used in meat industry: a review. *Food Research International* 113: 93–101. https://doi.org/10.1016/J.FOODRES.2018.06.073.
13 Martillanes, S., Rocha-Pimienta, J., Cabrera-Bañegil, M. et al. (2017). Application of phenolic compounds for food preservation: food additive and active packaging. *Phenolic Compounds – Biological Activity* https://doi.org/10.5772/66885.
14 Fadiji, T., Rashvand, M., Daramola, M.O., and Iwarere, S.A. (2023). A review on antimicrobial packaging for extending the shelf life of food. *Processes* 11: 590.
15 Wyrwa, J. and Barska, A. (2017). Innovations in the food packaging market: active packaging. *European Food Research and Technology* 243: 1681–1692. https://doi.org/10.1007/S00217-017-2878-2.

16 Marangoni Júnior, L., Vieira, R.P., and Anjos, C.A.R. (2020). Kefiran-based films: fundamental concepts, formulation strategies and properties. *Carbohydrate Polymers* 246: 116609. https://doi.org/10.1016/J.CARBPOL.2020.116609.

17 Kumar, S., Mukherjee, A., and Dutta, J. (2020). Chitosan based nanocomposite films and coatings: emerging antimicrobial food packaging alternatives. *Trends in Food Science and Technology* 97: 196–209. https://doi.org/10.1016/J.TIFS.2020.01.002.

18 Nath, D., Chetri, R., Santhosh, R., and Sarkar, P. (2022). Nanocomposites in food packaging. *Food, Medical, and Environmental Applications of Nanomaterials* 167–203. https://doi.org/10.1016/B978-0-12-822858-6.00007-8.

19 Nikolic, M.V., Vasiljevic, Z.Z., Auger, S., and Vidic, J. (2021). Metal oxide nanoparticles for safe active and intelligent food packaging. *Trends in Food Science and Technology* 116: 655–668. https://doi.org/10.1016/J.TIFS.2021.08.019.

20 Kaewklin, P., Siripatrawan, U., Suwanagul, A., and Lee, Y.S. (2018). Active packaging from chitosan-titanium dioxide nanocomposite film for prolonging storage life of tomato fruit. *International Journal of Biological Macromolecules* 112: 523–529. https://doi.org/10.1016/J.IJBIOMAC.2018.01.124.

21 Imani, S.M., Ladouceur, L., Marshall, T. et al. (2020). Antimicrobial nanomaterials and coatings: current mechanisms and future perspectives to control the spread of viruses including SARS-CoV-2. *ACS Nano* 14: 12341–12369. https://doi.org/10.1021/ACSNANO.0C05937.

22 Appierot, G., Lipovsky, A., Dror, R. et al. (2009). Enhanced antibacterial activity of nanocrystalline ZnO due to increased ROS-mediated cell injury. *Advanced Functional Materials* 19: 842–852. https://doi.org/10.1002/ADFM.200801081.

23 Vidic, J., Haque, F., Guigner, J.M. et al. (2014). Effects of water and cell culture media on the physicochemical properties of ZnMgO nanoparticles and their toxicity toward mammalian cells. *Langmuir* 30: 11366–11374. https://doi.org/10.1021/LA501479P.

24 Talebian, N., Amininezhad, S.M., and Doudi, M. (2013). Controllable synthesis of ZnO nanoparticles and their morphology-dependent antibacterial and optical properties. *Journal of Photochemistry and Photobiology. B* 120: 66–73. https://doi.org/10.1016/J.JPHOTOBIOL.2013.01.004.

25 Hoque, M., Sarkar, P., and Ahmed, J. (2022). Preparation and characterization of tamarind kernel powder/ZnO nanoparticle-based food packaging films. *Industrial Crops and Products* 178: 114670. https://doi.org/10.1016/J.INDCROP.2022.114670.

26 Bailore, N.N., Balladka, S.K., Doddapaneni, S.J.D.S., and Mudiyaru, M.S. (2021). Fabrication of environmentally compatible biopolymer films of pullulan/piscean collagen/ZnO nanocomposite and their antifungal activity. *Journal of Polymers and the Environment* 29: 1192–1201. https://doi.org/10.1007/S10924-020-01953-Y.

27 Shahabi-Ghahfarrokhi, I. and Babaei-Ghazvini, A. (2019). Using photo-modification to compatibilize nano-ZnO in development of starch-kefiran-ZnO green nanocomposite as food packaging material. *International Journal of Biological Macromolecules* 124: 922–930. https://doi.org/10.1016/J.IJBIOMAC.2018.11.241.

28 Shankar, S., Teng, X., Li, G., and Rhim, J.W. (2015). Preparation, characterization, and antimicrobial activity of gelatin/ZnO nanocomposite films. *Food Hydrocolloids* 45: 264–271. https://doi.org/10.1016/J.FOODHYD.2014.12.001.

29 Indumathi, M.P., Saral Sarojini, K., and Rajarajeswari, G.R. (2019). Antimicrobial and biodegradable chitosan/cellulose acetate phthalate/ZnO nano composite films with optimal oxygen permeability and hydrophobicity for extending the shelf life of black grape fruits. *International Journal of Biological Macromolecules* 132: 1112–1120. https://doi.org/10.1016/J.IJBIOMAC.2019.03.171.

30 Chaudhary, P., Fatima, F., and Kumar, A. (2020). Relevance of nanomaterials in food packaging and its advanced future prospects. *Journal of Inorganic and Organometallic Polymers and Materials* 30: 5180–5192. https://doi.org/10.1007/S10904-020-01674-8.

31 Nešić, A., Gordić, M., Davidović, S. et al. (2018). Pectin-based nanocomposite aerogels for potential insulated food packaging application. *Carbohydrate Polymers* 195: 128–135. https://doi.org/10.1016/J.CARBPOL.2018.04.076.

32 Subhapriya, S. and Gomathipriya, P. (2018). Green synthesis of titanium dioxide (TiO_2) nanoparticles by Trigonella foenum-graecum extract and its antimicrobial properties. *Microbial Pathogenesis* 116: 215–220. https://doi.org/10.1016/J.MICPATH.2018.01.027.

33 Thakur, B.K., Kumar, A., and Kumar, D. (2019). Green synthesis of titanium dioxide nanoparticles using *Azadirachta indica* leaf extract and evaluation of their antibacterial activity. *South African Journal of Botany* 124: 223–227. https://doi.org/10.1016/J.SAJB.2019.05.024.

34 Oleyaei, S.A., Zahedi, Y., Ghanbarzadeh, B., and Moayedi, A.A. (2016). Modification of physicochemical and thermal properties of starch films by incorporation of TiO_2 nanoparticles. *International Journal of Biological Macromolecules* 89: 256–264. https://doi.org/10.1016/J.IJBIOMAC.2016.04.078.

35 Fathi, N., Almasi, H., and Pirouzifard, M.K. (2019). Sesame protein isolate based bionanocomposite films incorporated with TiO_2 nanoparticles: study on morphological, physical and photocatalytic properties. *Polymer Testing* 77: 105919. https://doi.org/10.1016/J.POLYMERTESTING.2019.105919.

36 Vejdan, A., Ojagh, S.M., Adeli, A., and Abdollahi, M. (2016). Effect of TiO_2 nanoparticles on the physico-mechanical and ultraviolet light barrier properties of fish gelatin/agar bilayer film. *LWT-Food Science and Technology* 71: 88–95. https://doi.org/10.1016/J.LWT.2016.03.011.

37 Kadam, D.M., Thunga, M., Wang, S. et al. (2013). Preparation and characterization of whey protein isolate films reinforced with porous silica coated titania nanoparticles. *Journal of Food Engineering* 117: 133–140. https://doi.org/10.1016/J.JFOODENG.2013.01.046.

38 Garcia, C.V., Shin, G.H., and Kim, J.T. (2018). Metal oxide-based nanocomposites in food packaging: applications, migration, and regulations. *Trends in Food Science and Technology* 82: 21–31. https://doi.org/10.1016/J.TIFS.2018.09.021.

39 Egger, S., Lehmann, R.P., Height, M.J. et al. (2009). Antimicrobial properties of a novel silver-silica nanocomposite material. *Applied and Environmental Microbiology* 75: 2973–2976. https://doi.org/10.1128/AEM.01658-08.

40 Hetrick, E.M., Shin, J.H., Paul, H.S., and Schoenfisch, M.H. (2009). Anti-biofilm efficacy of nitric oxide-releasing silica nanoparticles. *Biomaterials* 30: 2782–2789. https://doi.org/10.1016/J.BIOMATERIALS.2009.01.052.

41 Hou, X., Xue, Z., Xia, Y. et al. (2019). Effect of SiO_2 nanoparticle on the physical and chemical properties of eco-friendly agar/sodium alginate nanocomposite film.

International Journal of Biological Macromolecules 125: 1289–1298. https://doi.org/10.1016/J.IJBIOMAC.2018.09.109.

42 Dong, W., Su, J., Chen, Y. et al. (2022). Characterization and antioxidant properties of chitosan film incorporated with modified silica nanoparticles as an active food packaging. *Food Chemistry* 373: 131414. https://doi.org/10.1016/J.FOODCHEM.2021.131414.

43 Deshmukh, R.K. and Gaikwad, K.K. (2022). Natural antimicrobial and antioxidant compounds for active food packaging applications. *Biomass Conversion and Biorefinery* 2022 (1): 1–22. https://doi.org/10.1007/S13399-022-02623-W.

44 Ali, A., Chen, Y., Liu, H. et al. (2019). Starch-based antimicrobial films functionalized by pomegranate peel. *International Journal of Biological Macromolecules* 129: 1120–1126. https://doi.org/10.1016/J.IJBIOMAC.2018.09.068.

45 Ordoñez, R., Atarés, L., and Chiralt, A. (2022). Biodegradable active materials containing phenolic acids for food packaging applications. *Comprehensive Reviews in Food Science and Food Safety* 21: 3910–3930. https://doi.org/10.1111/1541-4337.13011.

46 Al Jitan, S., Alkhoori, S.A., and Yousef, L.F. (2018). Phenolic acids from plants: extraction and application to human health. *Studies in Natural Products Chemistry* 58: 389–417. https://doi.org/10.1016/B978-0-444-64056-7.00013-1.

47 Rashmi, H.B. and Negi, P.S. (2020). Phenolic acids from vegetables: a review on processing stability and health benefits. *Food Research International* 136: 109298. https://doi.org/10.1016/J.FOODRES.2020.109298.

48 Hackman, R.M., Polagruto, J.A., Zhu, Q.Y. et al. (2008). Flavanols: digestion, absorption and bioactivity. *Phytochemistry Reviews* 7: 195–208. https://doi.org/10.1007/S11101-007-9070-4.

49 Hollman, P.C.H. (2001). Evidence for health benefits of plant phenols: local or systemic effects? *Journal of the Science of Food and Agriculture* 81: 842–852. https://doi.org/10.1002/JSFA.900.

50 Shirmohammadli, Y., Efhamisisi, D., and Pizzi, A. (2018). Tannins as a sustainable raw material for green chemistry: a review. *Industrial Crops and Products* 126: 316–332. https://doi.org/10.1016/J.INDCROP.2018.10.034.

51 Vera, M., Mella, C., García, Y. et al. (2023). Recent advances in tannin-containing food biopackaging. *Trends in Food Science and Technology* 133: 28–36. https://doi.org/10.1016/J.TIFS.2023.01.014.

52 Taguri, T., Tanaka, T., and Kouno, I. (2004). Antimicrobial activity of 10 different plant polyphenols against bacteria causing food-borne disease. *Biological & Pharmaceutical Bulletin* 27: 1965–1969. https://doi.org/10.1248/BPB.27.1965.

53 Whiting, S., Derbyshire, E., and Tiwari, B.K. (2012). Capsaicinoids and capsinoids. A potential role for weight management? A systematic review of the evidence. *Appetite* 59: 341–348. https://doi.org/10.1016/J.APPET.2012.05.015.

54 Tsao, R. (2010). Chemistry and biochemistry of dietary polyphenols. *Nutrients* 2: 1231. https://doi.org/10.3390/NU2121231.

55 Meydani, M. (2009). Potential health benefits of avenanthramides of oats. *Nutrition Reviews* 67: 731–735. https://doi.org/10.1111/J.1753-4887.2009.00256.X.

56 Beta, T. and Duodu, K.G. (2016). Bioactives: antioxidants. *Reference Module in Food Science* https://doi.org/10.1016/B978-0-08-100596-5.00110-4.

57 Dimberg, L.H., Theander, O., and Lingnert, H. (1993). Avenanthramides-a group of phenolic antioxidants in oats. *Cereal Chemistry* 70: 637.

58 Lestari, M.L.A.D. and Indrayanto, G. (2014). Curcumin. *Profiles of Drug Substances, Excipients and Related Methodology* 39: 113–204. https://doi.org/10.1016/B978-0-12-800173-8.00003-9.

59 Pulido-Moran, M., Moreno-Fernandez, J., Ramirez-Tortosa, C., and Ramirez-Tortosa, M.C. (2016). Curcumin and health. *Molecules* 21: 264. https://doi.org/10.3390/MOLECULES21030264.

60 Aliabbasi, N., Fathi, M., and Emam-Djomeh, Z. (2021). Curcumin: a promising bioactive agent for application in food packaging systems. *Journal of Environmental Chemical Engineering* 9: 105520. https://doi.org/10.1016/J.JECE.2021.105520.

61 Ma, D.S.L., Tan, L.T.H., Chan, K.G. et al. (2018). Resveratrol-potential antibacterial agent against foodborne pathogens. *Frontiers in Pharmacology* 9: 102. https://doi.org/10.3389/FPHAR.2018.00102.

62 Bhat, K.P.L., Kosmeder, J.W., and Pezzuto, J.M. (2004). Biological effects of resveratrol. *Antioxidants & Redox Signaling* 3: 1041–1064. https://doi.org/10.1089/152308601317203567.

63 Gülçin, I. (2010). Antioxidant properties of resveratrol: a structure–activity insight. *Innovative Food Science & Emerging Technologies* 11: 210–218. https://doi.org/10.1016/J.IFSET.2009.07.002.

64 Busolo, M.A. and Lagaron, J.M. (2015). Antioxidant polyethylene films based on a resveratrol containing clay of interest in food packaging applications. *Food Packaging and Shelf Life* 6: 30–41. https://doi.org/10.1016/J.FPSL.2015.08.004.

65 Riley, M.A. and Wertz, J.E. (2003). Bacteriocins: evolution, ecology, and application. *Annual Review of Microbiology* 56: 117–137. https://doi.org/10.1146/ANNUREV.MICRO.56.012302.161024.

66 Chakraborty, P., Nath, D., Hoque, M. et al. (2022). Biopolymer-based antimicrobial coatings for aquatic food products: a review. *Journal of Food Processing & Preservation* 46: e16465. https://doi.org/10.1111/JFPP.16465.

67 Santos, J.C.P., Sousa, R.C.S., Otoni, C.G. et al. (2018). Nisin and other antimicrobial peptides: production, mechanisms of action, and application in active food packaging. *Innovative Food Science & Emerging Technologies* 48: 179–194. https://doi.org/10.1016/J.IFSET.2018.06.008.

68 Sarkar, P., Bhunia, A., and Yao, Y. (2016). Nisin adsorption in colloidal systems formed with phytoglycogen octenyl succinate. *Food Biophysics* 11: 311–318. https://doi.org/10.1007/S11483-016-9436-5.

69 Ríos Colombo, N.S., Chalón, M.C., Navarro, S.A., and Bellomio, A. (2018). Pediocin-like bacteriocins: new perspectives on mechanism of action and immunity. *Current Genetics* 64: 345–351. https://doi.org/10.1007/S00294-017-0757-9.

70 Rodríguez, J.M., Martínez, M.I., and Kok, J. (2002). Pediocin PA-1, a wide-spectrum bacteriocin from lactic acid bacteria. *Critical Reviews in Food Science and Nutrition* 42: 91–121. https://doi.org/10.1080/10408690290825475.

71 Ribeiro-Santos, R., Andrade, M., de Melo, N.R., and Sanches-Silva, A. (2017). Use of essential oils in active food packaging: recent advances and future trends. *Trends in Food Science and Technology* 61: 132–140. https://doi.org/10.1016/J.TIFS.2016.11.021.

72 El, A.A., Miladi, K., Badri, W. et al. (2015). Essential oils: from extraction to encapsulation. *International Journal of Pharmaceutics* 483: 220–243. https://doi.org/10.1016/J.IJPHARM.2014.12.069.

73 Friedman, M. (2017). Chemistry, antimicrobial mechanisms, and antibiotic activities of cinnamaldehyde against pathogenic bacteria in animal feeds and human foods. *Journal of Agricultural and Food Chemistry* 65: 10406–10423. https://doi.org/10.1021/ACS.JAFC.7B04344.

74 Sun, J., Leng, X., Zang, J., and Zhao, G. (2022). Bio-based antibacterial food packaging films and coatings containing cinnamaldehyde: a review. *Critical Reviews in Food Science and Nutrition* https://doi.org/10.1080/10408398.2022.2105300.

75 Zhang, G., Li, T., Liu, J. et al. (2023). Cinnamaldehyde-contained polymers and their biomedical applications. *Polymers* 15: 1517. https://doi.org/10.3390/POLYM15061517.

76 Qureshi, K.A., Mohammed, S.A.A., Khan, O. et al. (2022). Cinnamaldehyde-based self-nanoemulsion (CA-SNEDDS) accelerates wound healing and exerts antimicrobial, antioxidant, and anti-inflammatory effects in rats' skin burn model. *Molecules* 27: 5225. https://doi.org/10.3390/MOLECULES27165225.

77 Mateen, S., Rehman, M.T., Shahzad, S. et al. (2019). Anti-oxidant and anti-inflammatory effects of cinnamaldehyde and eugenol on mononuclear cells of rheumatoid arthritis patients. *European Journal of Pharmacology* 852: 14–24. https://doi.org/10.1016/J.EJPHAR.2019.02.031.

78 Burt, S. (2004). Essential oils: their antibacterial properties and potential applications in foods – a review. *International Journal of Food Microbiology* 94: 223–253. https://doi.org/10.1016/J.IJFOODMICRO.2004.03.022.

79 Marchese, A., Barbieri, R., Coppo, E. et al. (2017). Antimicrobial activity of eugenol and essential oils containing eugenol: a mechanistic viewpoint. *Critical Reviews in Microbiology* 43: 668–689. https://doi.org/10.1080/1040841X.2017.1295225.

80 Huan, S.K.H., Wang, K.T., Yeh Der, S. et al. (2012). Antiproliferative and molecular mechanism of eugenol-induced apoptosis in cancer cells. *Molecules* 17: 6290–6304. https://doi.org/10.3390/MOLECULES17066290.

81 Sharifi-Rad, M., Varoni, E.M., Iriti, M. et al. (2018). Carvacrol and human health: a comprehensive review. *Phytotherapy Research* 32: 1675–1687. https://doi.org/10.1002/PTR.6103.

82 Friedman, M. (2014). Chemistry and multibeneficial bioactivities of carvacrol (4-isopropyl-2-methylphenol), a component of essential oils produced by aromatic plants and spices. *Journal of Agricultural and Food Chemistry* 62: 7652–7670. https://doi.org/10.1021/JF5023862.

83 Di, L.M.A., Cacciatore, S.I., Marinelli, L. et al. (2018). Carvacrol and its derivatives as antibacterial agents. *Phytochemistry Reviews* 17: 903–921. https://doi.org/10.1007/S11101-018-9569-X.

84 Fan, K., Li, X., Cao, Y. et al. (2015). Carvacrol inhibits proliferation and induces apoptosis in human colon cancer cells. *Anti-Cancer Drugs* 26: 813–823. https://doi.org/10.1097/CAD.0000000000000263.

85 Marchese, A., Orhan, I.E., Daglia, M. et al. (2016). Antibacterial and antifungal activities of thymol: a brief review of the literature. *Food Chemistry* 210: 402–414. https://doi.org/10.1016/J.FOODCHEM.2016.04.111.

86 Nomani, M., Sadat Noori, S.A., Tohidfar, M., and Ramshini, H. (2019). Overexpression of TPS2 gene to increase thymol content using *Agrobacterium tumefaciens*-mediated transformation in *Trachyspermum ammi* (Qom ecotype). *Industrial Crops and Products* 130: 63–70. https://doi.org/10.1016/J.INDCROP.2018.12.076.

87 Nagoor Meeran, M.F., Javed, H., Al, T.H. et al. (2017). Pharmacological properties and molecular mechanisms of thymol: prospects for its therapeutic potential and pharmaceutical development. *Frontiers in Pharmacology* 8: 380. https://doi.org/10.3389/FPHAR.2017.00380.

88 Rao, J.P. and Geckeler, K.E. (2011). Polymer nanoparticles: preparation techniques and size-control parameters. *Progress in Polymer Science* 36: 887–913. https://doi.org/10.1016/J.PROGPOLYMSCI.2011.01.001.

89 Lu, X.Y., Wu, D.C., Li, Z.J., and Chen, G.Q. (2011). Polymer nanoparticles. *Progress in Molecular Biology and Translational Science* 104: 299–323. https://doi.org/10.1016/B978-0-12-416020-0.00007-3.

90 Divya, K. and Jisha, M.S. (2017). Chitosan nanoparticles preparation and applications. *Environmental Chemistry Letters* 16: 101–112. https://doi.org/10.1007/S10311-017-0670-Y.

91 Cui, H., Surendhiran, D., Li, C., and Lin, L. (2020). Biodegradable zein active film containing chitosan nanoparticle encapsulated with pomegranate peel extract for food packaging. *Food Packaging and Shelf Life* 24: 100511. https://doi.org/10.1016/J.FPSL.2020.100511.

92 Roy, S. and Rhim, J.W. (2021). Fabrication of chitosan-based functional nanocomposite films: effect of quercetin-loaded chitosan nanoparticles. *Food Hydrocolloids* 121: 107065. https://doi.org/10.1016/J.FOODHYD.2021.107065.

93 Roy, S., Min, S.J., Biswas, D., and Rhim, J.W. (2023). Pullulan/chitosan-based functional film incorporated with curcumin-integrated chitosan nanoparticles. *Colloids and Surfaces A: Physicochemical and Engineering Aspects* 660: 130898. https://doi.org/10.1016/J.COLSURFA.2022.130898.

94 Hosseini, S.F., Ghaderi, J., and Gómez-Guillén, M.C. (2022). Tailoring physico-mechanical and antimicrobial/antioxidant properties of biopolymeric films by cinnamaldehyde-loaded chitosan nanoparticles and their application in packaging of fresh rainbow trout fillets. *Food Hydrocolloids* 124: 107249. https://doi.org/10.1016/J.FOODHYD.2021.107249.

95 Wang, Y., Zhang, R., Qin, W. et al. (2020). Physicochemical properties of gelatin films containing tea polyphenol-loaded chitosan nanoparticles generated by electrospray. *Materials and Design* 185: 108277. https://doi.org/10.1016/J.MATDES.2019.108277.

96 Paques, J.P., Van Der Linden, E., Van Rijn, C.J.M., and Sagis, L.M.C. (2014). Preparation methods of alginate nanoparticles. *Advances in Colloid and Interface Science* 209: 163–171. https://doi.org/10.1016/J.CIS.2014.03.009.

97 Choukaife, H., Doolaanea, A.A., and Alfatama, M. (2020). Alginate nanoformulation: influence of process and selected variables. *Pharmaceuticals* 13: 335. https://doi.org/10.3390/PH13110335.

98 Yoncheva, K., Benbassat, N., Zaharieva, M.M. et al. (2021). Improvement of the antimicrobial activity of oregano oil by encapsulation in chitosan – alginate nanoparticles. *Molecules* 26: 7017. https://doi.org/10.3390/MOLECULES26227017.

99 Karimi-Khorrami, N., Radi, M., Amiri, S. et al. (2022). Fabrication, characterization, and performance of antimicrobial alginate-based films containing thymol-loaded lipid nanoparticles: comparison of nanoemulsions and nanostructured lipid carriers.

International Journal of Biological Macromolecules 207: 801–812. https://doi.org/10.1016/J.IJBIOMAC.2022.03.149.

100 Bernela, M., Kaur, P., Chopra, M., and Thakur, R. (2014). Synthesis, characterization of nisin loaded alginate–chitosan–pluronic composite nanoparticles and evaluation against microbes. *LWT-Food Science and Technology* 59: 1093–1099. https://doi.org/10.1016/J.LWT.2014.05.061.

101 Sahoo, N., Sahoo, R.K., Biswas, N. et al. (2015). Recent advancement of gelatin nanoparticles in drug and vaccine delivery. *International Journal of Biological Macromolecules* 81: 317–331. https://doi.org/10.1016/J.IJBIOMAC.2015.08.006.

102 Mortazavi Moghadam, F.A., Khoshkalampour, A., Mortazavi Moghadam, F.A. et al. (2023). Preparation and physicochemical evaluation of casein/basil seed gum film integrated with guar gum/gelatin based nanogel containing lemon peel essential oil for active food packaging application. *International Journal of Biological Macromolecules* 224: 786–796. https://doi.org/10.1016/J.IJBIOMAC.2022.10.166.

103 Li, S., Shi, W., Wang, X. et al. (2022). The preparation and characterization of electrospun gelatin nanofibers containing chitosan/eugenol-sulfobutyl-β-cyclodextrin nanoparticles. *Colloids and Surfaces A: Physicochemical and Engineering Aspects* 648: 129109. https://doi.org/10.1016/J.COLSURFA.2022.129109.

104 Cui, H., Bai, M., Rashed, M.M.A., and Lin, L. (2018). The antibacterial activity of clove oil/chitosan nanoparticles embedded gelatin nanofibers against Escherichia coli O157:H7 biofilms on cucumber. *International Journal of Food Microbiology* 266: 69–78. https://doi.org/10.1016/J.IJFOODMICRO.2017.11.019.

105 Shahbazi, Y., Shavisi, N., Karami, N. et al. (2021). Electrospun carboxymethyl cellulose-gelatin nanofibrous films encapsulated with *Mentha longifolia* L. essential oil for active packaging of peeled giant freshwater prawn. *LWT* 152: 112322. https://doi.org/10.1016/J.LWT.2021.112322.

106 Krajewska, B. (2005). Membrane-based processes performed with use of chitin/chitosan materials. *Separation and Purification Technology* 41 (3): 305–312. https://doi.org/10.1016/j.seppur.2004.03.019.

107 Agulhon, P., Robitzer, M., David, L., and Quignard, F. (2012). Structural regime identification in ionotropic alginate gels: influence of the cation nature and alginate structure. *Biomacromolecules* 13: 215–220. https://doi.org/10.1021/bm201477g.

108 Noor, N.Q.I.M., Razali, R.S., Ismail, N.K. et al. (2021). Application of green technology in gelatin extraction: A review. *Processes* 9: 2227. https://doi.org/10.3390/pr9122227.

109 Schiavo, S., Oliviero, M., Li, J., and Manzo, S. (2018). Testing ZnO nanoparticle ecotoxicity: linking time variable exposure to effects on different marine model organisms. *Environmental Science and Pollution Research* 25: 4871–4880. https://doi.org/10.1007/S11356-017-0815-3.

110 Miller, R.J., Lenihan, H.S., Muller, E.B. et al. (2010). Impacts of metal oxide nanoparticles on marine phytoplankton. *Environmental Science & Technology* 44: 7329–7334. https://doi.org/10.1021/ES100247X.

111 Lin, D. and Xing, B. (2007). Phytotoxicity of nanoparticles: inhibition of seed germination and root growth. *Environmental Pollution* 150: 243–250. https://doi.org/10.1016/J.ENVPOL.2007.01.016.

112 Chen, L., Zhou, L., Liu, Y. et al. (2012). Toxicological effects of nanometer titanium dioxide (nano-TiO_2) on Chlamydomonas reinhardtii. *Ecotoxicology and Environmental Safety* 84: 155–162. https://doi.org/10.1016/J.ECOENV.2012.07.019.

113 Li, F., Liang, Z., Zheng, X. et al. (2015). Toxicity of nano-TiO_2 on algae and the site of reactive oxygen species production. *Aquatic Toxicology* 158: 1–13. https://doi.org/10.1016/J.AQUATOX.2014.10.014.

114 Wang, H., Wick, R.L., and Xing, B. (2009). Toxicity of nanoparticulate and bulk ZnO, $Al2O_3$ and TiO_2 to the nematode *Caenorhabditis elegans*. *Environmental Pollution* 157: 1171–1177. https://doi.org/10.1016/J.ENVPOL.2008.11.004.

115 Roh, J.Y., Park, Y.K., Park, K., and Choi, J. (2010). Ecotoxicological investigation of CeO_2 and TiO_2 nanoparticles on the soil nematode *Caenorhabditis elegans* using gene expression, growth, fertility, and survival as endpoints. *Environmental Toxicology and Pharmacology* 29: 167–172. https://doi.org/10.1016/J.ETAP.2009.12.003.

116 Van Hoecke, K., De Schamphelaere, K.A.C., Van Der Meeren, P. et al. (2008). Ecotoxicity of silica nanoparticles to the green alga pseudokirchneriella subcapitata: importance of surface area. *Environmental Toxicology and Chemistry* 27: 1948–1957. https://doi.org/10.1897/07-634.1.

117 Erdinçmer, N. and Sponza, D.T. (2020). Evaluation of the effects of nano-metal oxide (nano-SiO_2, nano-HfO_2, nano-CeO_2, nano-Ta_2O_5) on microtox, algae and *Daphnia magna*. *Asian Basic and Applied Research Journal* 2: 76–81.

118 Sirelkhatim, A., Mahmud, S., Seeni, A. et al. (2015). Review on zinc oxide nanoparticles: antibacterial activity and toxicity mechanism. *Nanomicro Letters* 7: 219–242. https://doi.org/10.1007/s40820-015-0040-x.

119 Díaz-Visurraga, J., Gutiérrez, C., Von Plessing, C., and García, A. (2011). Metal nanostructures as antibacterial agents. In: *Science Against Microbial Pathogens: Communicating Current Research and Technological Advances*, Vol. 1, No. 51, 210–218.

120 Ferraz, C.A., Pastorinho, M.R., Palmeira-de-Oliveira, A., and Sousa, A.C.A. (2022). Ecotoxicity of plant extracts and essential oils: a review. *Environmental Pollution* 292: 118319. https://doi.org/10.1016/J.ENVPOL.2021.118319.

121 Duringer, J.M., Swan, L.R., Walker, D.B., and Craig, A.M. (2010). Acute aquatic toxicity of western juniper (*Juniperus occidentalis*) foliage and Port Orford cedar (*Chamaecyparis lawsoniana*) heartwood oils. *Environmental Monitoring and Assessment* 170: 585–598. https://doi.org/10.1007/S10661-009-1259-0.

122 Benelli, G., Pavela, R., Cianfaglione, K. et al. (2020). Ascaridole-rich essential oil from marsh rosemary (*Ledum palustre*) growing in Poland exerts insecticidal activity on mosquitoes, moths and flies without serious effects on non-target organisms and human cells. *Food and Chemical Toxicology* 138: 111184. https://doi.org/10.1016/J.FCT.2020.111184.

123 Pavela, R., Benelli, G., Petrelli, R. et al. (2019). Exploring the insecticidal potential of boldo (*Peumus boldus*) essential oil: toxicity to pests and vectors and non-target impact on the microcrustacean *Daphnia magna*. *Molecules* 24: 879. https://doi.org/10.3390/MOLECULES24050879.

124 Park, H.M., Kim, J., Chang, K.S. et al. (2011). Larvicidal activity of myrtaceae essential oils and their components against *Aedes aegypti*, acute toxicity on daphnia magna, and aqueous residue. *Journal of Medical Entomology* 48: 405–410. https://doi.org/10.1603/ME10108.

125 Torbati, M., Asnaashari, S., and Heshmati, A.F. (2016). Essential oil from flowers and leaves of *Elaeagnus angustifolia* (Elaeagnaceae): composition, radical scavenging and general toxicity activities. *Advanced Pharmaceutical Bulletin* 6: 163–169. https://doi.org/10.15171/apb.2016.023.

126 He, Y.L., Shi, J.Y., Peng, C. et al. (2018). Angiogenic effect of motherwort (*Leonurus japonicus*) alkaloids and toxicity of motherwort essential oil on zebrafish embryos. *Fitoterapia* 128: 36–42. https://doi.org/10.1016/J.FITOTE.2018.05.002.

127 Govindarajan, M. and Benelli, G. (2016). Eco-friendly larvicides from Indian plants: effectiveness of lavandulyl acetate and bicyclogermacrene on malaria, dengue and Japanese encephalitis mosquito vectors. *Ecotoxicology and Environmental Safety* 133: 395–402. https://doi.org/10.1016/J.ECOENV.2016.07.035.

128 Govindarajan, M., Rajeswary, M., Arivoli, S. et al. (2016). Larvicidal and repellent potential of *Zingiber nimmonii* (J. Graham) Dalzell (Zingiberaceae) essential oil: an eco-friendly tool against malaria, dengue, and lymphatic filariasis mosquito vectors? *Parasitology Research* 115: 1807–1816. https://doi.org/10.1007/S00436-016-4920-X.

3

Antimicrobial Food Packaging Systems

Jigyasa[1], *Jaspreet Kaur Rajput*[2], *and Smriti Dogra*[2]

[1] *Department of Chemistry, Maharishi Markandeshwar University, Mullana, Ambala, Haryana, India*
[2] *Department of Chemistry, Dr B R Ambedkar National Institute of Technology, Jalandhar, Punjab, India*

3.1 Introduction

Food rotting is the process of contamination of food that leads to loss of color, texture, and nutritional content as well as stimulates the growth of pathogenic bacteria, degrading the product's quality and making it inedible. When food is exposed to the environment during processing and packaging, contamination can happen [1]. Food losses are a significant financial worry for the food business as well as foodborne outbreaks pose a serious risk to the public health. One of the main causes of food loss may be attributed to the cross-contamination of spoilage bacteria on raw materials and processed goods [2]. The addition of antimicrobials into food can lower the danger of tampering and the likelihood of cross-contamination of food products with pathogenic and rotting microorganisms. The fabrication of packaging materials with natural antibacterial agents is becoming increasingly popular. The growth and survival of numerous microbes can be efficiently prevented by adding synthetic antimicrobial agents directly to meals. However, consumers prefer less processed and preservative-free food items with an extended shell life. An additional and ultimate barrier that may restrict the growth of foodborne infections is provided by antimicrobial packaging [3]. Conventional food packaging seeks to increase the product's shelf life, maintain its quality, and guarantee its safety. Antimicrobial substances that are designed to prevent food contamination are currently gaining a lot of scientific interest worldwide, which results in significant research on them [4]. The primary methods of food preservation used today are chemical preservatives, which, however, provide further advantages by reducing post-process contamination and achieving higher inhibitory effect against bacteria. There are various active food packaging ideas, such as oxygen scavengers, moisture absorbers, UV barriers, and other processes supplying antioxidants and flavors [5]. In an innovative approach known as antimicrobial packaging system, an antimicrobial agent is incorporated into the packaging system through various technologies. Figure 3.1 depicts the various benefits offered by antimicrobial food packaging system.

Smart Food Packaging Systems: Innovations and Technology Applications, First Edition.
Edited by Avik Mukherjee, Santosh Kumar, Manjusri Misra, and Amar K. Mohanty.
© 2025 John Wiley & Sons Ltd. Published 2025 by John Wiley & Sons Ltd.

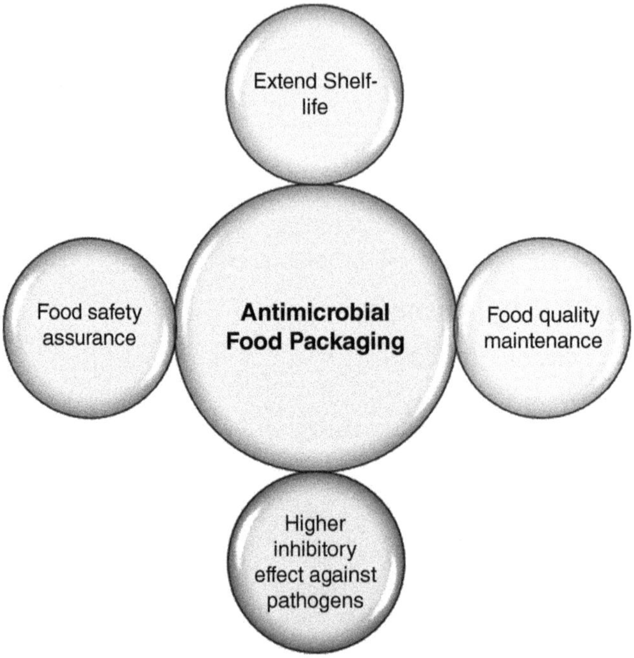

Figure 3.1 Various benefits of antimicrobial food packaging.

3.2 Foodborne Pathogens and Their Control

Foodborne pathogens are basically categorized as foodborne bacteria, foodborne viruses, and foodborne parasites. The diseased condition caused by them mainly affects the intestinal system, which can later worsen and affect other systems of the body. Tables 3.1, 3.2, and 3.3 shown in this section discuss the various causative foodborne bacteria, foodborne viruses, and foodborne parasites, respectively, that cause human illness and their control.

3.3 Antimicrobial Active Agents

The antimicrobial active agents utilized in food packaging may have their origins from plants, animals, microbes, or synthetic. They inhibit microbial growth through various mechanisms, as shown in Figure 3.2.

3.3.1 Antimicrobials from Plant Origin

Plants have always been advantageous in providing biologically active compounds that serve as sources of useful drugs [6]. The secondary metabolites found in plants, such as alkaloids, flavonoids, terpenoids, essential oils, sterols, glycosides, and tannins, possess

Table 3.1 Foodborne bacterial illnesses and their control.

Causative organism	Disease/illness	Incubation period	Foods involved	Control of spread
Arizona hinshawii	Arizona infection – nausea, headache, fever, and weakness	2–46 hours	Meat products, ice cream, and bakery products	Chilling of food done in a controlled manner; sanitization of storage equipment
Bacillus cereus	Gastroenteritis – severe diarrhea and abdominal cramps	8–16 hours	Precooked starchy foods like cereals, rice, pasta, meat, and sauces	Cook food above 65°C kills its spores, cool cooked food quickly, and maintain personal hygiene
Brochothrix thermosphacta			Meat and meat products, frozen vegetables, and dairy products are stored aerobically or in a modified atmosphere	Atmospheric cold plasma for storage of food
Campylobacter jejuni and *Campylobacter coli*	Food poisoning – abdominal pain, diarrhea, and vomiting	2–5 days	Raw milk, poultry products, fresh meat, pork sausages, and beef	The species is sensitive to heat and freezing; hence, adequate cooking can prevent the spread, along with good food handling practices
Clostridium botulinum	Botulism	12–36 hours	Preserved meats and fish, inadequately home-canned foods	Heat processing of canned food, boiling of suspected food for at least 15 minutes, good sanitation, rejection of swollen canned food
Clostridium perfringens	Gastroenteritis	8–24 hours	Meat and poultry products' spores are found in soil and other natural environment	Good personal hygiene, rapid cooling of cooked meat, and reheating leftover foods
Corynebacterium diphtheriae	Diphtheria – fever, swelling of the neck, skin lesions, breathing difficulty, and even death	2–5 days	Smear-ripened cheese, raw grated beetroot, and poultry products	Good hygienic practices in slaughterhouses to prevent cross-contamination from infected animals and antibiotic treatment for infected animals

(*Continued*)

Table 3.1 (Continued)

Causative organism	Disease/illness	Incubation period	Foods involved	Control of spread
Cronobacter sakazakii	Meningitis, seizures, and sepsis in premature and full-term infants	6–8 hours	Powdered infant milk powder, dried meats, legumes, herbal tea, dried nuts, flours, and spices	Adequate care taken during preparation of formula milk
Escherichia coli	Enteropathogenic *E. coli* infection – infantile diarrheal disease and traveler's diarrhea	8–24 hours	Inadequate preservation of cooked foods like cheese and fish	Refrigeration of food, personal hygiene, and adequate cooking time
Listeria monocytogenes	Abdominal cramps, headache, nausea, vomiting, fever, diarrhea	24 hours	Infected cattle shed the organism in the milk, and manure containing this species can contaminate the crop	Pasteurization of food can reduce bacteria count
Plesiomonas shigelloides	Gastroenteritis – diarrhea of various degrees of severity	48 hours	Seafood – fish, oysters, and carbs and contaminated water supply	Avoid eating raw shellfish, clean and sanitize food handling place
Salmonella choleraesuis, java, derby, infantis	Salmonellosis – abdominal pain, vomiting, dehydration, fever, diarrhea, headache	12–36 hours	Uncooked fresh meat and meat products, milk and milk products, eggs	Good hygiene of food handler, cooking at appropriate temperature, pasteurization of milk, refrigeration of food
Salmonella typhi	Typhoid fever	7–28 days	Foods contaminated with feces and urine of infected person	Washing the food articles before use, appropriate cooking temperature, and pasteurization of milk
Shigella sonnei, S. dysenteriae, S. boydii	Shigellosis – abdominal pain, fever, chills, watery stools, nausea, and headache	1–7 days	Milk, moist foods like salad and seafood	Thorough cooking of food at optimum temperature, utilization of treated water in food preparation, control of flies

Table 3.1 (Continued)

Causative organism	Disease/illness	Incubation period	Foods involved	Control of spread
Staphylococcus aureus	Gastroenteritis or inflammation of intestinal lining	2–4 hours	Bakery goods, meat, fish, milk, and milk products	Pasteurization of milk, personal hygiene of food employees
Streptococcus pyogenes	Local infections on skin that may later develop into sepsis, high fever, tonsillitis, in severe cases, inflammatory disorder of respiratory tract	1–3 days	Milk and milk products, eggs, and steamed seafood	Chill food rapidly after cooking, pasteurization of food
Vibrio parahaemolyticus	Gastroenteritis – diarrhea, mild fever, chills, vomiting, nausea, and headache	24–48 hours	Uncooked sea foods – oysters, shrimp, and crabs	Proper cooking of food, avoid using sea water for washing of foods
Yersinia enterocolitica, Y. pseudotuberculosis	Yersiniosis – severe abdominal pain and diarrhea	24–36 hours	Beef, pork, milk, turkey, tofu, and seafood	Pasteurization of food is highly effective in treating the organism

Source: Centres for Disease Control and Prevention/U.S. Department of Health & Human Services/Public domain.

antimicrobial properties. These phytochemicals show antimicrobial activity through: (i) complexation with microbial cell wall; (ii) inactivation of enzymes; (iii) substrate deprivation; (iv) metal ion complexation; (v) inhibits the release of autocoids and prostaglandins; (vi) destroy nucleic acids; and (vii) deform amino acids after protein-lipid peroxidation [7]. Some popular antimicrobials from plant origin are clove, ginger, cinnamon, onion, nutmeg, turmeric, neem, etc. Majority of plant-based antimicrobials are generally recognized as safe (GRAS) by the U.S. Food and Drug Administration. Hence, this plant-based antimicrobial food packaging is widely accepted by consumers. The procedure followed in manufacturing food packaging comprises applying the packaging material or format a coating or active liquid containing the phytochemical. The common methods adopted for extraction of the phytochemicals include cold pressing, soaking in alcohol, and steam distillation [8]. The packaging polymers commonly utilized are polypropylene (PP) or polyethylene/ethylene vinyl alcohol copolymer. These plant-based packagings have been found to be effective against a wide range of antimicrobials such as *Candida albicans, Rhodotorula glutinis, Schizosaccharomyces pombe, Escherichia coli* O157:H7, *Bacillus subtilis, Staphylococcus aureu, and Pseudomonas aeruginosa.*

Table 3.2 Foodborne virus illnesses and their control.

Causative organism	Disease/illness	Incubation period	Foods involved	Control of spread
Astrovirus	Gastroenteritis – stomach ache, loss of appetite, watery diarrhea for 2-4 days, vomiting, abdominal pain	3–4 days	Feco-orally transmitted and has been found in river water and shellfish	Proper handwashing and hygiene practices while handling food
Enteric adenovirus	Acute Gastroenteritis	2 days–2 weeks	Feco-orally transmitted through food and water	Proper handwashing and hygiene practices while handling food
Hepatitis A	Fatigue, nausea, stomach pain, joint pain, and jaundice	15–50 days (mean: 28 days)	Contaminated food or drink	Proper and hygienic handling of food throughout the food chain
Hepatitis E	Fever, fatigue, loss of appetite, vomiting, abdominal pain, and jaundice	15–60 days (mean: 40 days)	Fecally contaminated food, raw or undercooked pork, venison, wild boar meat, or shellfish	Avoiding raw pork and venison and using sanitized water for cooking
Norwalk-like viruses	Gastroenteritis – nausea, vomiting, abdominal cramps, and diarrhea	12–48 hours	It spread through fecally contaminated food	Consumption of food after proper cooking, adequate level of sanitation at hotels, cruises, and restaurants, exclusion of ill food handlers for 48–72 hours after resolution of illness
Norovirus	Gastroenteritis – nausea, vomiting, diarrhea, myalgias, abdominal pain, headache, and fever	12–48 hours	Contaminated food or water and aerosols created during vomiting and fomites	Good manufacturing practices at food preparation site and general hygiene of food handlers
Rotavirus	Watery diarrhea, vomiting, and low-grade fever	1–2 days	Fecal-oral route by contaminated hands, food, objects, and water	Good manufacturing practices at food preparation site and general hygiene of food handlers

Source: Centres for Disease Control and Prevention/U.S. Department of Health & Human Services/Public domain.

Table 3.3 Foodborne parasitic illnesses and their control.

Causative organism	Disease/illness	Incubation period	Foods involved	Control of spread
Anisakis spp.	Anisakiasis – abdominal pain, nausea, vomiting, allergic reactions on skin	Several days	Raw or undercooked fish or squid	Thorough cooking of fish or squid (internal temperature of at least 65 °C)
Capillaria philippinensis	Intestinal capillariasis – severe diarrhea, anorexia, vomiting, and weight loss	About 2 weeks	Raw or undercooked infected shrimp or fish	Thorough cooking of sea food at optimum temperature
Clonorchis sinensis	Liver disease – infects bile duct and gall bladder, decreases gallbladder contractions, and feeds on bile	Approx. 21 days	Raw or undercooked fish	Cooking meat at appropriate temperature to kill the oocysts
Cryptosporidium parvum	Cryptosporidiosis – diarrheal illness may be accompanied by fever or fatigue	2–10 days (average 7 days)	Thick-walled oocysts contaminate water and food (raw milk, uncooked meat),	Use treated water for cleaning food, wash hands thoroughly before cooking and eating, thoroughly cook meat, and drink pasteurized milk
Cyclospora cayetanensis	Cyclosporiasis – gastrointestinal disease characterized by nausea, vomiting, and low-grade fever	1 week	Oocyst (from feces), contaminated food (leafy vegetables and fruits), and water	Refrigerate cut, peeled, or cooked fruits and vegetables, wash fruits and vegetables under running water thoroughly, cutting boards for meat should be washed with soap and hot water regularly
Echinococcus spp.	Cystic echinococcosis – cysts in nervous system, kidney, lungs, and liver that may later rupture causing severe anaphylactic reactions Alveolar echinococcosis – parasitic tumors in the liver, lung, and brain	Many years	Embryonated eggs, contaminated food, or water	General sanitation while handling food, washing food thoroughly with running water before cooking, and avoiding consumption of food and water contaminated by fecal matter from dogs

(*Continued*)

Table 3.3 (Continued)

Causative organism	Disease/illness	Incubation period	Foods involved	Control of spread
Diphyllobothrium	Diphyllobothriasis – intestinal obstruction and gall bladder disease	4–6 weeks	Raw or undercooked fish	Freezing or adequately cooking fish will kill the parasite
Endoamoeba histolytica	Amoebic dysentery – diarrheal illness, or extraintestinal infection, lysis of intestinal cells	1–3 weeks	ingestion of mature cysts through fecal-contaminated water or food	Cleaning the fruits and vegetables before use, disposal of human feces away from agricultural land
Giardia duodenalis	Giardiasis – gastrointestinal disease associated with malabsorption of fat, disaccharidase deficiencies, protein loss, and weight loss	3–35 days but may be longer	Ingestion of contaminated water and food	Proper disposal of wastewater and feces, avoiding eating uncooked foods, cleaning, and disinfecting cleaning area
Opisthorchiidae	Hepatology can finally lead to obstructive jaundice and bile cancer	3–4 weeks	Undercooked or raw fish	Cook fish adequately (to an internal temperature of ~63 °C), storing the fish at −20 °C or below for at least 24 hours
Taenia spp.	Taeniasis – digestive problems, diarrhea, abdominal pain, loss of appetite	10–14 weeks	Consumption of vegetation/raw or undercooked meat contaminated with eggs or gravid proglottid of Taenia spp.	Cooking meat at a temperature higher than 140 °F, sanitizing water for cooking food, and general hygiene of food handlers
Toxoplasma gondii	Toxoplasmosis – the parasite can migrate to brain and eyes causing destruction of cells, focal chorioretinitis may result in impaired vision and large necrotic abscesses	5–23 days	Undercooked meat of infected animals, consuming food or water contaminated with cat feces	Cooking of meat (at least 71 °C), freezing meat for several days at sub-zero (0 °F) temperatures before cooking, avoiding drinking unpasteurized milk, and practicing proper management of cats

Table 3.3 (Continued)

Causative organism	Disease/illness	Incubation period	Foods involved	Control of spread
Trichinella spp.	Trichinellosis – gastrointestinal infection characterized by abdominal pain, nausea, vomiting, headache, and fatigue	1–2 weeks	Ingestion of undercooked meat containing encysted larvae of trichinella	General hygiene while handling meat, Freeze pork less than 6 in. thick for 20 days at −15 °C, cleaning meat grinders thoroughly
Trypanosoma cruzi	Human African Trypanosomiasis (sleeping sickness) – hemorrhages and extravascular hemolysis may be seen in some organs, behavioral changes, confusion, and sensory disturbances	7–14 days	The disease is transmitted by infected tsetse flies' bite and consumption of food/ beverages contaminated with feces/urine of infected blood-sucking triatomine bugs	Good hygiene practices in food preparation, transportation, and storage

Source: Centres for Disease Control and Prevention/U.S. Department of Health & Human Services/Public domain.

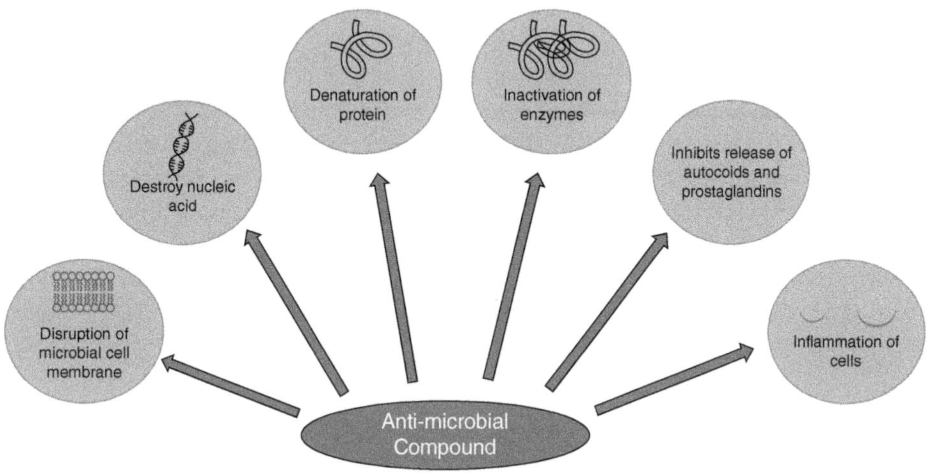

Figure 3.2 Mechanism of action of antimicrobial compounds.

3.3.2 Antimicrobials from Animal Origin

3.3.2.1 Chitosan

Chitosan is a nontoxic, unbranched, and renewable biopolymer obtained from chitin. Chitin constitutes exoskeleton of insects, fungi, and certain marine creatures like crabs, lobsters, shrimps, etc. [9]. Its deacetylation in alkaline medium results in the generation of chitosan [10]. Chitosan has a wide range of applications in the fields of pharmaceuticals, cosmetics, biomedical, agriculture, food, and water industries as antimicrobial, antitumor, anticoagulant, anti-inflammatory, bio fungicide, and artificial skin crafts [11]. Its monomer unit comprises hydroxyl and amine groups as main functional groups [12]. This amino group during methylation of chitosan converts to quaternary amino group ($-N^+(CH_3)_3$), thus resulting in generation of positively charged polymer backbone. It can be easily processed in forms like gel, films, fibers, and even beads, resulting in its wider application. It possesses certain incredible properties that crown it as an ideal food packaging material: antimicrobial action, good oxygen/moisture transmission coefficient, biodegradability and nontoxic.

Studies have shown that antimicrobial activity of chitosan is dependent on the nature of chitosan polymer (molecular weight, degree of deacetylation, chain length, charge distribution, and water retention values), solvent medium (ionic strength, pH, and micro-organism (Gram-positive/Gram-negative bacteria) [13]. Its antimicrobial action is a result of multiple detrimental actions that could involve its interaction with (i) outer cell components, (ii) cytoplasmic membrane, and (iii) cytoplasmic constituents [11]. Its biocide action could be extracellular (high molecular weight chitosan) or intracellular (low molecular weight chitosan) depending on its molecular weight. In extracellular mode of action, the chitosan polymer, with positively charged amino group backbone, electrostatically interacts with negatively charged microbial cell structures [14]. In gram-positive bacteria, peptidoglycans and teichoic acid layer, whereas in gram-negative bacteria, lipopolysaccharide attached with phosphorylated groups constitute the polyanionic layer. The studies conducted by Helander et al. [15] and No et al. [16] clearly indicated that the electrostatic interaction between positively charged chitosan and negatively charged microbial cell envelopes is responsible for biocide action. The binding of chitosan to the microbial cell envelope leads to disruption of normal ionic flow across membrane, alterations in cell surface, leakage of small intracellular molecules, and membrane depolarization. All these factors combined lead to loss of cell integrity and functionality [17]. Other authors proposed that chitosan molecules due to their chelating nature can complex with Ca^{2+} in cell membrane, causing destabilization of cell walls [18]. In intracellular mode of action, shown by low molecular weight of chitosan, it passes through the cell wall and disrupts the normal functioning of the cell. It can influence transcription in DNA, translation in RNA and protein, and even affect mitochondrial activity [19]. It can chelate with essential trace metal ions in cells like Ni, Mn, Zn, Fe, and Cu), causing impaired microbial growth [17].

3.3.2.2 Defensin

Defensins are naturally occurring cysteine-rich, cationic antimicrobial peptides with potent antimicrobial activity [20]. They are found in animals, plants, and fungi. In addition to antimicrobial activity, they also show anti-inflammatory and wound-repair activity [21]. Its antimicrobial activity has been known for centuries. But recently, more research has been driven toward immobilizing this antimicrobial peptide on surfaces (like biofilms) to enhance its lifespan and stability. The leading reasons behind its antimicrobial actions are: (i) interaction of defensin with lipid bilayer of bacterial cell membrane causing generation of pores/channels that finally leads to leakage of cell components; and (ii) invasion of peptides inside cytoplasm causing cellular inactivation through overproduction of reactive oxygen species (ROS) [22]. A study conducted by Shwaiki et al. [23] demonstrated that defensin derived from barley endosperm showed inhibitory action against food spoilage yeasts without causing adverse effects (proven by hemolytic and cytotoxic assays). This group of researchers successfully applied their findings to the preservation of different beverages, like apple juice. Similar studies were also conducted by Kraszewska et al. [24] in which defensin's antibacterial activity against beverage-spoiling bacteria was studied. Defensin has also shown promising anti-fungal activity by delaying the fungal growth in dough by two days [25].

3.3.2.3 Lactoferrin

Lactoferrin is an antimicrobial peptide isolated by gastric pepsin cleavage of bovine lactoferrin and human lactoferrin [26]. It is an active component of the defense system of colostrum, milk, and mucosal surfaces as well as nutritional component of milk [27]. It shows antimicrobial activity against wide spectrum of bacterial strains. Its high affinity toward iron is believed to play a crucial role in its bactericidal action [28]. It removes iron from bacteria, thus inhibiting bacterial growth [29]. However, studies show the existence of other mechanisms also like: (i) direct interaction between cationic lactoferrin and anionic lipid A, a component of bacterial cell membrane, thus altering the membrane permeability; (ii) displacement of divalent cations like Mg^{2+}/Ca^{2+} from lipopolysaccharide through chelation by lactoferrin; and (iii) degrading various secretion proteins produced by bacteria during invasion into cell, thus blocking bacterial uptake into host cells [30, 31]. This peptide finds wide usage in food industry as a biopreservative for inhibition of the yeast *Dekkera bruxellesis* in wines, active packaging on cheese to prevent growth of Pseudomonas, and control of *Escherichia coli* O157: H7 and *Pseudomonas fluorescens* in the chicken fillets [32].

3.3.2.4 Lactoperoxidase

Lactoperoxidase is an animal-origin antimicrobial heme protein found in bovine milk, saliva, tears, and colostrum of animals and humans [33, 34]. It has been well explored as food biopreservative for preservation of dairy products [35, 36], fish [37], and beverages [38]. It forms a strong antimicrobial system in combination with H_2O_2 and SCN^- or halogen. In this system, H_2O_2 plays a crucial role in initiation step, wherein it converts lactoperoxidase (Fe^{3+}) from the ground state into the so-called Compound I state (Fe^{2+}), and SCN^- or halogen acts as electron donor. Lactoperoxidase shows antimicrobial action against both foodborne bacteria, fungi, and viruses [33]. Its well-established utilization as food preservative is to enhance shelf life and functionality of milk during storage. Its bactericidal activity involves: (i) blockage of glycolytic pathways; (ii) oxidative destruction of essential molecules like oxidation of sulphydryls to sulphenyl derivatives that later inhibit bacteria; and (iii) interference in cytopathic effects [39].

3.3.2.5 Lysozyme

Lysozyme is an antimicrobial hydrophilic protein that shows prominent bactericidal action against foodborne gram-positive bacteria [40]. It is an animal-origin enzyme that is found in tears, saliva, mucus, human milk, and egg whites [41]. Its bactericidal action involves attack on the major bacterial cell wall component, peptidoglycan. It basically hydrolyzes β-1,4-linkage between acetylmuramic acid and *N*-acetylglucosamine of peptidoglycan, thus leading to cell lysis [42, 43]. Its usage as a food biopreservative [44] is well established and is commercially utilized for preservation of fruits, vegetables, infant food, seafood, meat, and wine [42]. The researchers have successfully immobilized lysozyme as active-packaging material into polymers like poly (ethylene terephthalate), polyvinyl alcohol [45], pullulan [46], chitosan, zein film [47], gelatin [48], and cellulose. To achieve this immobilization techniques such as covalent bonding, adsorption, and encapsulation have been utilized.

3.3.2.6 Pleurocidin

Pleurocidin is natural antimicrobial peptide isolated from mucosal skin tissues of winter flounder (*Pleuronectes americanus*) [49, 50]. It exhibits some very essential properties that signify its utilization as a biopreservative in food: heat stability, broad-spectrum antimicrobial activity, low cytotoxicity, activity in wide spectrum of pH, and low hemolysis [51, 52]. It shows distinctive antimicrobial activity against *V. parahaemolyticus, L. monocytogenes, E. coli, C. albicans, S. cerevisiae*, and *Penicillium expansum* [53]. Its antimicrobial action involves its interaction with neutral and acidic anionic bacterial phospholipid membranes. This interaction results in strong membrane translocation due to pore formation that finally leads to loss of bacterial cell integrity [49]. Studies have also shown that it interacts not only with bacterial membrane but also targets their DNA [54]. Wang et al. [55] developed pleurocidin-poly (vinyl alcohol) electrospun antimicrobial nanofibers. The incorporation of poly (vinyl alcohol) was meant to enhance the stability of pleurocidin in food system. The antimicrobial activity of electrospun nanofiber was studied against *Escherichia coli* and *Staphylococcus aureus*. This pleurocidin antimicrobial fiber was successfully applied for preservation of apple cider.

3.3.2.7 Protamine

Protamine is an arginine-rich polycationic antimicrobial peptide that is extracted from sperm cells of vertebrates like salmon fish [56, 57]. Its bactericidal activity has been studied against a range of foodborne microbes like *Escherichia coli, Staphylococcus aureus, Listeria monocytogenes, and Clostridium botulinum* [58–60]. Aspedon and Groisman [61] have put forward various evidence to explain the mechanism involved in antimicrobial action of protamine. They proposed that a higher delta psi value of electrical membrane potential causes increased bacterial cell lysis by protamine. Protamine affects the cell's nutrient uptake, especially proline and leucine amino acids. This action later affects the energy transduction system of cell by reducing the cellular ATP content.

3.3.3 Antimicrobials from Microbial Origin

3.3.3.1 Bacteriophages

Since the enhancement in cases of antibiotic-resistant bacterial strains, the need for an efficient antimicrobial has increased. The bacteriophages, a class of viruses, were utilized against such bacterial strains. They can kill bacteria without causing negative side effects on host cells. The utilization of bacteriophage's phage therapy dates to 1919, and its demand boomed during the second World War. Some common bacteriophages are caudovirals, podoviruses, siphoviruses, and myoviruses. They initiate their "cell lytic" approach through their tail fibers, which bind to bacterial cell surface. They later inject their DNA into the bacterial cell, which causes re-programming in bacterial cells that results in generation of multiple bacteriophages within bacterial cell. These newly formed bacteriophages burst open the bacterial cell causing the cell death. This phage therapy for microbes finds lots of utilization in food processing and food production sectors.

3.3.3.2 Natamycin

Natamycin is a natural anti-fungal compound, a polyene macrolide, which is derived from *Streptomyces natalensis* and *Streptococcus lactis* (actinomycetes). It is highly effective against yeast and molds at concentration levels between <1 and 10 ppm but ineffective against bacteria and viruses [62, 63]. It is a white, creamy powder that is insoluble in water and mineral oils. It is thermally unstable, hence added post-food processing step that requires high temperature. However, its encapsulation within cyclodextrin complex enhances its thermal stability and solubility without compromising its anti-fungal property [64]. It is normally sprayed or applied directly over cheese and sausages [65]. It offers lots of benefits as a good preservative material, like its applicability in a wide pH range, minimum concentration requirement, and low rate of migration into food [64]. It has been utilized for preservation of fruit juices [66, 67], fruits [68–70], poultry, dairy [71–73], and bakery products [74, 75]. Its mode of action toward molds and yeast is believed to involve interaction between natamycin's polyene chain and ergosterol in fungi. Ergosterol is major component of fungi's cell membrane, which plays a crucial role in fungal growth and development [76]. It protects lipids against oxidation in liposomes and is responsible for yeast's resistance to tert-butyl hydroperoxide [77]. Natamycin forms a stable polyene-sterol complex with ergosterol. This complexation results in enhancement of cell permeability, thus leading to loss of essential metal ions and small peptides, which causes yeast and fungus death [62].

3.3.3.3 Nisin

Nisin is an antimicrobial peptide produced by Lactococcus lactis. It shows antimicrobial activity against *L. monocytogenes*, *Bacillus cereus*, *Lactobacillus plantarum*, *Micrococcus luteus*, *L. acidophilus*, and *Enterococcus faecalis*. It has been well-explored food preservative in dairy products, meat, canned food, and vegetables. Its antimicrobial action is believed to occur when nisin interacts with anionic lipid present on cytoplasmic membrane of microbes. This interaction results in generation of pores that cause leakage of various essential cellular components like amino acids and ATP, thus finally leading to cell collapse. It possesses properties like excellent water solubility, no negative immune response by human body, low molecular weight, and high heat-tolerance. Hence, it is considered as safe food preservative.

3.3.3.4 Reuterin

Reuterin is an antimicrobial compound with formula $HOCH_2CH_2CHO$ (hydroxy and aldehyde as functional groups), produced by *Lactobacillus reuteri* (inhabitant of human intestine) [78]. For several years, its antimicrobial mode of action remained unclear. Lately, researchers have done extensive research and proposed various mechanisms to explain its biocide action. It was proposed that its biocidal mechanism is due to combined or solo effect of following actions: (i) generation of ROS that leads to oxidative stress in microbial cells [78, 79]; (ii) inhibition in activity of bacterial ribonucleotide reductase (enzyme required for conversion of ribonucleotide to deoxyribonucleotide) [80]; and (iii) oxidative damage to free sulfhydryl groups that constitute reduced glutathione and various enzymes [81]. Its antimicrobial action has been studied against various microbes like *E. coli*, Bacteroides spp., Bifidobacterium spp. [80], *Clostridioides difficile* [79], *Campylobacter* spp. [82], etc. It possesses certain characteristic features that idealize its use as food preservative: antimicrobial, action in wide pH range, water-soluble, and resistance toward proteolytic and lipolytic enzymes [80]. It has been successfully applied as antimicrobial preservative in dairy products (cheese, milk, and yogurt) [83, 84], meat and fish products (fillets, ham, and chicken carcasses) [85, 86], and protective fruit coatings (strawberries) [87].

3.3.4 Antimicrobials Nanomaterials

This combination of nanotechnology and food packaging gives rise to the term "Nano-packaging." Nanomaterials have been used in antimicrobial films as antibiotic transporters as well as growth inhibitors. Numerous studies demonstrate the biological efficacy of inorganic nanoparticles (NPs), which have potent antibacterial properties at low doses [88]. Food processing facilities have employed antimicrobial nanocoating on food preparation surfaces for sanitary reasons and contamination control [89]. The use of polymeric nanoparticles in food packaging to prevent or treat microbial infections can be categorized into two approaches: either the production of nanoparticles using polymers that are naturally antimicrobial or the use of polymeric nanoparticles that contain and release sustained and therapeutic amounts of the antimicrobial agent. Among the most researched polymers as antimicrobial nanoparticles for food packaging are chitosan, poly (lactide-co glycoside), sodium alginate, and pluronic [90]. The metallic nanoparticles have also shown strong suppression of bacteria, fungi, and protozoa even at low concentrations [91]. The common nanomaterials utilized in food packaging are ZnO NPs [54], Ag NPs [92], MgO NPs [93], TiO_2 NPs [94], and CuO NPs [95]. These nanoparticles show their antimicrobial action through: (i) generation of ROS; (ii) increased redox ability through fast donation and acceptance of electrons, which later destroy the microbial cell components; and (iii) positively charged nanoparticles bind with negatively charged bacterial cell components causing flocculation.

3.4 Technologies of Incorporation of Antimicrobials in Food Packaging System

3.4.1 Ionic or Covalent Bonding Between Antimicrobial and Food Packaging System

This method is generally applied to those antimicrobials that cannot withstand high temperature involved in extrusion process [96]. Chitin, chitosan, and polymer matrix-based food packaging film serve as ideal sites for binding of antimicrobials. They contain functional groups like —COOH, —NH_2, and —OH, which participate in condensation reactions resulting in covalent bond formation. In case of adsorption of antimicrobials on packaging material, ionic linkages participate, e.g., lysozyme (antimicrobial), can get adsorbed on neutral chitin or cationic chitosan [97].

3.4.2 Antimicrobial Sachets in Packaging System

It is an innovative approach wherein the sachets comprising active-releasing compounds or active scavenging systems are added to packaging systems to enhance the shelf life of food by improving its quality and integrity. These sachets usually comprise oxygen scavengers, CO_2 scavengers, moisture absorbers, ethanol emitters, ethylene scavengers, odor removers, aroma releasers, and antioxidant releasers [96]. Bahmid et al. [98] developed a packaging system wherein the sachet released allyl isothiocyanate (AITC) vapor (antimicrobial compound) via sinigrin–myrosinase reaction in situ. This method of placing the carrier/emitting sachets into the packaging system allows a controlled release of compounds into the system. The common sachet materials used for release of active compounds are low-density polyethylene, nonwoven fabrics, paper/ethylene vinyl acetate laminates, paper/PP laminates, filter paper, cellulose bags, etc. [2].

3.4.3 Antimicrobial-Coated Films

In this technique, antimicrobials are coated on the surface of a film via coating techniques. The film substrate is made from plastics (low-density polyethylene, polyethylene terephthalate, PP, and polyvinyl chloride) or biopolymers (cellulose, chitin, chitosan, and polylactic acid) [99]. The coating techniques to obtain an effective antimicrobial-coated packaging film include flexography coating, gravure roll coating, electrospinning, spin coating, solution casting, dip coating, ex-situ coating, sol–gel process, layer-by-layer coating, bar coating, spray coating, and knife coating [100]. Antimicrobial films from plastics and bio-based origins have been well explored in food packaging such as chitosan/carboxymethylcellulose (CMC)/ZnO [101], polylactic acid/ZnO NPs [102], chitosan/gelatin/starch/nanocellulose [103], and cellulose/lignin NPs [104].

3.5 Release of Antimicrobials from Food Packaging

An efficient antimicrobial food packaging material must have a dual capability, i.e., to retain as well as release antimicrobials at controlled level. This innovative practice of slow release of antimicrobial from packaging materials is considered better than its direct addition to food matrix. Although immediate reduction in microbe population is achieved using direct addition technique, studies have shown that for total inhibition against a microbe a much higher concentration of antimicrobial is required, and moreover, this method cannot affect the survivor microbial population as they will continue to grow after the depletion of antimicrobials. The antimicrobial food packaging allows certain benefits over direct addition techniques, i.e., continuous antimicrobial effect on the food during extended exposure while not affecting the sensorial characteristics of food [105]. As discussed in previous section, encapsulation of antimicrobial in polymer matrix is a common approach followed for designing active antimicrobial packaging. As per studies conducted by various researchers, the release profile of antimicrobial depends on following factors: the properties of both antimicrobial component and the matrix as well as the environmental conditions. These properties of antimicrobial component include: (i) difference in solubility of antimicrobial component in matrix and food; (ii) its amount within the matrix; and (iii) localization of antimicrobial component within matrix. A higher release is observed initially when the amount of antimicrobial is high in packaging film, which later becomes saturated. The antimicrobial components, which are positioned superficially on matrix, are released in a burst initially, followed by a sustained release later. The properties of polymer matrix that govern the release profile are: (i) swelling behavior of polymer matrix and (ii) glass transition point of polymer. Wang et al. [55] studied the release profile of antimicrobial pleurocidin and observed that initially there was a burst release of pleurocidin through the packaging film at all temperatures (37, 25, and 4 °C). However, at higher temperatures, an enhanced release was observed. They observed that, in addition to temperature, the swelling behavior of polymer matrix (polyvinyl alcohol) plays a crucial role in release profile. The interaction of polymer matrix with release medium (food) causes its swelling, thus inducing fast diffusion of antimicrobial into the food matrix. A similar swelling behavior was observed by Hari and Nair [106]. They encapsulated streptomycin in starch nanocrystals, which were further entrapped into chitosan matrix. They observed a controlled streptomycin release profile, which was dependent on swelling behavior of polymer matrix. Temperature response studies were conducted [107]. They observed that by switching the glass transition temperature (T_g) of a polymer, the release of antimicrobial can be regulated. Octanedione (an antimicrobial) was entrapped within polymer (Eudragit Rs100 and poly(methyl methacrylate)). They observed that this polymer matrix was turned "on" at 37 °C and "off" at 25 °C for the release of antimicrobial compound. The temperature range of 34.8–36.5 °C was examined as the wet T_g range wherein maximum release of antimicrobial was observed. Thus, the choice of polymer with specific glass transition points enables release of antimicrobials at the temperature at which microbial growth accelerates. The environmental factors that control the release profile are: (a) temperature; (b) pH; and (c) moisture. As the moisture increases in the environment, the porosity of matrix wall varies resulting in enhanced release of antimicrobial component [108]. The effect of pH on release profile of antimicrobial was observed by Keawchaoon and Yoksan [109], wherein an enhanced release of carvacrol from chitosan nanoparticles was observed in acidic medium than in alkaline or neutral medium. Recently, Li et al. [110] fabricated a pH-sensitive nanocomposite coating and applied it to enhance shelf life of wheat noodles. The antimicrobial compound cinnamaldehyde was loaded into the halloysite nanotubes (HNTs), and the ends of the tube were capped with oppositely charged polyelectrolytes. The pH of wheat noodles gradually reduces as their shelf life increases, owing to increased microbial growth. This nanocomposite coating showed an enhanced release of antimicrobial at lower pH, whereas at higher pH, the release was restrained.

3.6 Applications of Antimicrobial Active Packaging

3.6.1 Fruits and Vegetables

The fruits and vegetables may get contaminated by microorganisms during transportation, storage, or processing. During transportation, mechanical damage or extreme temperature conditions may increase growth of microorganisms. Damp and moist storage places favor the growth of microorganisms. During processing, improper food handling practices like contaminated benches, trays, and equipment; contaminated water for washing; unhygienic food handlers, etc. cause contamination of fruits and vegetables. Inclusion of such contaminated fruits and vegetables in juices later produces poor-quality juices. The antimicrobial coatings enhance shelf life of a fruit/vegetable by: (i) lowering the ascorbic acid content; (ii) declining the browning rate and fruit decay; (iii) lowering the content of malondialdehyde (toxic product) in fresh-cut fruits/vegetables; and (iv) delayed rate of respiration. Chitosan, as discussed in an earlier section, is an animal-based antimicrobial. Its potential as an efficient antimicrobial has been well utilized to fabricate active antimicrobial packaging for fruits and vegetables. Chitosan can act as antimicrobial packaging either as a solo nanoparticle coating or as a matrix for incorporation of other antimicrobial compounds. Pilon et al. [111] synthesized a chitosan nanoparticle of sizes 110 and 300 nm and applied it as a coating on Gale apple slices. A comparative microbiological analysis clearly revealed that chitosan-based nanoparticles functioned as an effective coating against molds in comparison to conventional chitosan coating. Chitosan serves as an efficient polymer matrix for generating blended active-packaging materials. These blended materials have enhanced synergistic effects. This approach has been utilized by various researchers, like Yang et al. [112], who blended antimicrobial phenolic compounds from blueberry leaf extract with chitosan matrix. The 2% (w/v) chitosan coating having 12% blueberry extract and with modified atmosphere packaging (MAP) (O_2 + CO_2) was successfully applied to extend shelf life of fresh blueberries for around 40 days postharvest. A similar blending approach was also utilized by Takala et al. [113], who prepared a trilayer matrix film of polycaprolactone/methylcellulose/polycaprolactone and incorporated organic acid extracts of rosmarinic acid and Asian and Italian essential oils. The film was able to control growth of various foodborne bacteria at temperature of 4 °C for up to 12 days. An edible coating comprising CMC and chitosan was fabricated by Arnon et al. [114] as substitute for synthetic waxes. This edible coating was applied over citrus fruits like mandarins, oranges, and grapes. The performance of this coating was in comparison with commercially used synthetic waxes in terms of maintaining acidity level and total soluble solids. On similar grounds, banana flour/chitosan composite film was fabricated by Martínez-Abad et al. [115] with antimicrobial property against *Staphylococcus aureus*. This composite film was modified into sealable bags for storage of fresh-cut vegetables. Nabifarkhani et al. [116] incorporated chitosan matrix with nanocellulose fiber (1%) and thyme oil (1%). This coating was able to effectively enhance shelf life of sweet cherry when stored at 1 °C for five weeks.

The nanomaterials as discussed in Section 3.3.4 offer promising antimicrobial properties. They are usually dispersed in suitable matrices to fabricate coating materials of desired properties. Fernández et al. [117] dispersed AgNPs into cellulose fibers, and this coating was able to enhance shelf life of fresh-cut melon for 10 days at 4 °C. Ding et al. [118] dispersed ZnO and AgNPs into low-density polyethylene film and applied it to enhance shelf life of lotus root for 12 days at 4 °C. Plant-based antimicrobials have also been excessively utilized to fabricate active antimicrobial packaging. Zhu et al. [119] synthesized edible films comprising carvacrol and cinnamaldehyde against *Salmonella* Newport. The film showed promising antimicrobial results when tested with organic leafy green vegetables of iceberg such as romaine lettuce, mature spinach, and baby spinach. These plant-based essential oils are also dispersed in suitable matrices such that the so-obtained matrix has enhanced antimicrobial properties. A few examples under these categories are: polypropylene/ethylene vinyl alcohol (PP/EvOH) with oregano essential oil and citral [105]; oregano essential oil incorporated in poly (lactic acid)–cellulose nanocrystals for mixed vegetables against *Listeria monocytogenes* [120]; Garlic oil in β-cyclodextrin for tomato slices against yeast and molds [121]; blueberry leaf extract in chitosan matrix for blueberry fruit against various bacterial and fungal strains [112]; lemongrass, oregano oil, and vanillin incorporated in apple puree-alginate edible coatings for fresh-cut "Fuji" apples [122].

3.6.2 Bakery and Other Food Products

Bakery products are an essential component of our everyday food habits. These broadly include breads, cakes, pastries, pasta, biscuits, breakfast cereals, and macaroni. The dough prepared for manufacturing these products is highly susceptible to microbial growth. It can get infested by microbes at any stage of food processing, which includes: (i) contaminated grains for manufacturing of products; (ii) contamination of freshly knitted dough from working platforms and food handlers; and (iii) slicing/wrapping of finished products. Among the various microbes, molds are the major spoilage-causing organisms. The bakery products have generally low water activity (a_W); however, a little moistening will favor mold growth, which can cause serious problems. A study was conducted by Legan and Voysey [123] during 1982–1983 on commercially produced breads. These were stored at 22 °C and results were analyzed for months. He concluded that molds such as Penicillium, Aspergillus, and Cladosporium were common microbial infections in bread. The mold spores are generally killed during baking process. However, mold species from air, equipment, and food handlers can be cultivated during storage of raw materials. Next to molds, bacteria, and yeast are the microorganisms that can be involved in spoilage of bakery products. If the moistening level in these products is high, it will allow for the growth of yeast and bacteria. Ropiness of bread is caused by bacteria *Bacillus subtilis*, which leads to yellow or brown spots on bread. Red bread is caused by bacteria *Serratia marcescens,* which is characterized by appearance of blood-red drops on bread. Cakes and pastries usually get infested by microbes because their toppings or fillings are susceptible to microbial growth when not stored in an adequate storage condition. Other dry bakery products, like biscuits, get contaminated during the drying/cooling step in food manufacturing. Antimicrobial active packaging is a trending practice that is being considered to prevent the spoilage of bakery products. The active packaging usually comprises antimicrobial compounds incorporated within biodegradable/nonbiodegradable polymer films. Essential oils, sorbates, and propionates are the antimicrobial compounds that are usually utilized to enhance shelf life of bakery products. Biodegradable cellulose acetate films containing 0%, 2%, and 4% sodium propionate were prepared by Soares et al. [124]. The film was placed between bread slices packed in low-density polythene (LDPE) packages. It enhanced the shelf life of bread slices by 15 days at room temperature. On a similar line but with enhanced shelf life, Silveira et al. [125] synthesized a cellulose-based biodegradable polymer matrix with an average thickness of 25–70 μM and containing 7 and 3% sorbic acid (antimicrobial agent), respectively. The film was applied to pastry dough. They studied the migration index (σ) of sorbic acid from the film for 40 days at storage temperature of 8 ± 1 °C. The antimicrobial activity of the fil was studied against yeast, fungi, *Staphylococcus* spp., and *E. coli*. The film with 25 μM thickness and 7% sorbic acid performed better than others in terms of slow and gradual release of sorbic acid into pastry dough. Lopes et al. [126] also utilized a cellulose-derivative polymer as a film to incorporate cinnamaldehyde (antimicrobial). The film was applied to wrap bread and pastries. The film was able to reduce the concentration of yeast, mold, and *Staphylococcus* spp. by 2–3 log cycles. Taking further the benefits of biodegradable cellulose polymers, Otoni et al. [127] fabricated an edible film from methylcellulose and nanoemulsions of clove and oregano essential oils. They utilized Tween 80 and ultrasonication to prepare stable nanoemulsions. Herein, for the first time, the effect of the droplet size on the quality of the film provided an interesting result. The introduction of nanodroplets had a positive plasticizing effect on methylcellulose film by reducing the elastic modulus and increasing the elongation at break of the film. The film effectively enhanced shelf life of bread for 15 days. Gutiérrez et. al. have contributed immensely toward the development of efficient antimicrobial packaging films. In the year 2009, Gutiérrez et al. [128] reported the fabrication of cinnamon essential oil-infused PP film for packaging of bakery products. The antimicrobial study was conducted on 54 independent samples of bakery products over a 28-days period. The sensory evaluation results indicated that the packaging can effectively enhance the shelf life from 3 to 10 days within good safety limits. The same research group carried this study forward, and in the year 2011 Gutiérrez et al. [129] published an article wherein a comparative study was made between effectiveness of MAP and active packaging toward antimicrobial protection of bakery products. The active packaging comprised self-adhesive label containing cinnamon essential oil. The MAP comprised 60% CO_2 and 40% N_2. The statistical analysis revealed that the active packaging performed better than MAP in terms of antimicrobial activity and did not alter the sensory properties of the food article. On similar line, Passarinho et al. [130] fabricated sachets fitted with oregano essential oil diffused polymeric resin. The sachets with varied concentrations of oregano essential oil were placed inside sliced bread packages and incubated for 15 days at room temperature. The antimicrobial active components thymol and carvacrol present in oregano essential oil effectively controlled the growth of yeast, mold, coliform, and Salmonella spp. while enhancing the shelf life of bread for 15 days. A biodegradable colloidal carbohydrate pectin containing cinnamon and clove essential oils was drawn into thin film by Sachdeva et al. [131]. An oil-in-water emulsion was prepared of essential oil and Tween 80. The microbes in test were bread spoilers *Aspergillus* and *Penicillium*.

3.7 Conclusion and Future Perspective

The introduction of antimicrobial compounds in the packaging system can efficiently limit the growth of germs by lengthening the lag period and reducing the growth rate, respectively. The antimicrobials can be obtained from various sources, both natural and synthetic. Both these sources offer their respective pros and cons. The main concern while designing antimicrobial food packaging is to ensure that it maintains the integrity of food while enhancing its shelf life. Antimicrobials can be introduced in food packaging systems through various means; however, surface coating is considered promising as compared to other methods. The bio-nanocomposites are trending materials utilized in food packaging. They are biodegradable hence have attracted the interest of researchers. The nanomaterials utilized in such composites enhance the antimicrobial, thermal, and mechanical properties of the packaging material. However, still the requirement of "ideal food packaging system" is a long road to travel. The issues of toxicity and migration of antimicrobial agents into food are a matter of concern that need to be addressed.

References

1 Malhotra, B., Keshwani, A., and Kharkwal, H. (2015). Antimicrobial food packaging: potential and pitfalls. *Frontiers in Microbiology* 6: 611. https://doi.org/10.3389/fmicb.2015.00611.

2 Otoni, C.G., Espitia, P.J.P., Avena-Bustillos, R.J., and McHugh, T.H. (2016). Trends in antimicrobial food packaging systems: Emitting sachets and absorbent pads. *Food Research International* 83: 60–73. https://doi.org/10.1016/j.foodres.2016.02.018.

3 Irkin, R. and Esmer, O.K. (2015). Novel food packaging systems with natural antimicrobial agents. *Journal of Food Science and Technology* 52 (10): 6095–6111. https://doi.org/10.1007/s13197-015-1780-9.

4 Yam, K.L. and Lee, D.S. (2012). 1 – Emerging food packaging technologies: an overview. In: *Emerging Food Packaging Technologies* (ed. K.L. Yam and D.S. Lee), 1–9. Woodhead Publishing. https://doi.org/10.1533/9780857095664.1.

5 Ahvenainen, R. and Hurme, E. (1997). Active and smart packaging for meeting consumer demands for quality and safety. *Food Additives and Contaminants* 14 (6–7): 753–763. https://api.semanticscholar.org/CorpusID:44996378.

6 Imran, M., Khan, S., Arif, D., and M. (2017). Evaluation of antimicrobial properties of some indian medicinal plants against MDR pathogenic bacteria. *International Journal of Research in Pharmaceutical Sciences* 8: 522–533.

7 Solomakos, N., Govaris, A., Koidis, P., and Botsoglou, N. (2008). The antimicrobial effect of thyme essential oil, nisin and their combination against *Escherichia coli* O157:H7 in minced beef during refrigerated storage. *Meat Science* 80 (2): 159–166. https://doi.org/10.1016/j.meatsci.2007.11.014.

8 Mustapha, A., Jai, J., Hamidon, F.S. et al. (2017). Antimicrobial agents from Malaysian plants and their potential use in food packaging material: review. *Chemical Engineering Research Bulletin* 19: 57. https://doi.org/10.3329/cerb.v19i0.33797.

9 Aranaz, I., Alcántara, A.R., Civera, M.C. et al. (2021). Chitosan: an overview of its properties and applications. *Polymers (Basel)* 13 (19): 3256–3256.

10 Mondal, A., Dhar, A.K., Banerjee, S. et al. (2022). Chapter 2 – Antimicrobial uses of chitosan. In: *Chitosan in Biomedical Applications* (ed. M.S. Hasnain, S. Beg, and A.K. Nayak), 13–36. Academic Press. https://doi.org/10.1016/B978-0-12-821058-1.00009-5.

11 Raafat, D. and Sahl, H.G. (2009). Chitosan and its antimicrobial potential – a critical literature survey. *Microbial Biotechnology* 2 (2): 186–201.

12 Goy, R.C., Morais, S.T.B., and Assis, O.B.G. (2016). Evaluation of the antimicrobial activity of chitosan and its quaternized derivative on *E. coli* and *S. aureus* growth. *Revista Brasileira de Farmacognosia* 26 (1): 122–127. https://doi.org/10.1016/j.bjp.2015.09.010.

13 Oh, H.I., Kim, Y.J., Chang, E.J., and Kim, J.Y. (2001). Antimicrobial characteristics of chitosans against food spoilage microorganisms in liquid media and mayonnaise. *Bioscience, Biotechnology, and Biochemistry* 65 (11): 2378–2383. https://doi.org/10.1271/bbb.65.2378.

14 Savard, T., Beaulieu, C., Boucher, I., and Champagne, C.P. (2002). Antimicrobial action of hydrolyzed chitosan against spoilage yeasts and lactic acid bacteria of fermented vegetables. *Journal of Food Protection* 65 (5): 828–833. https://doi.org/10.4315/0362-028X-65.5.828.

15 Helander, I.M., Nurmiaho-Lassila, E.-L., Ahvenainen, R. et al. (2001). Chitosan disrupts the barrier properties of the outer membrane of Gram-negative bacteria. *International Journal of Food Microbiology* 71 (2): 235–244. https://doi.org/10.1016/S0168-1605(01)00609-2.

16 No, H.K., Park, N.Y., Lee, S.H. et al. (2002). Antibacterial activities of chitosans and chitosan oligomers with different molecular weights on spoilage bacteria isolated from Tofu. *Journal of Food Science* 67 (4): 1511–1514. https://doi.org/10.1111/j.1365-2621.2002.tb10314.x.

17 Bellich, B., D'Agostino, I., Semeraro, S. et al. (2016). "The good, the bad and the ugly" of chitosans. *Marine Drugs* 14 (5): https://doi.org/10.3390/md14050099.

18 Young, D.H., Kohle, H., and Kauss, H. (1982). Effect of chitosan on membrane permeability of suspension-cultured *Glycine max* and *Phaseolus vulgaris* cells. *Plant Physiology* 70 (5): 1449–1454.

19 Ke, C.-L., Deng, F.-S., Chuang, C.-Y., and Lin, C.-H. (2021). Antimicrobial actions and applications of chitosan. *Polymers* 13 (6): https://doi.org/10.3390/polym13060904.

20 Divakarla, S.K., Das, T., Chatterjee, C. et al. (2022). Antimicrobial and anti-inflammatory gallium–defensin surface coatings for implantable devices. *ACS Applied Materials & Interfaces 14* (7): 9685–9696. https://doi.org/10.1021/acsami.1c19579.

21 Hiemstra, P.S. (2006). Defensins. In: *Encyclopedia of Respiratory Medicine* (ed. G.J. Laurent and S.D. Shapiro), 7–10. Academic Press. https://doi.org/https://doi.org/10.1016/B0-12-370879-6/00110-1.

22 Bruggeman, M., Ijakipour, H., and Stamboulis, A. (2019). Defensin-like peptides and their antimicrobial activity in free-form and immobilized on material surfaces. In: *Peptide Synthesis* (ed. J.T. Varkey). IntechOpen. https://doi.org/10.5772/intechopen.85508.

23 Shwaiki, L.N., Sahin, A.W., and Arendt, E.K. (2021). Study on the inhibitory activity of a synthetic defensin derived from barley endosperm against common food spoilage yeast. *Molecules* 26 (1): 165.

24 Kraszewska, J., Beckett, M.C., James, T.C., and Bond, U. (2016). Comparative analysis of the antimicrobial activities of plant defensin-like and ultrashort peptides against food-spoiling bacteria. *Applied and Environmental Microbiology* 82 (14): 4288–4298.

25 Thery, T. and Arendt, E.K. (2018). Antifungal activity of synthetic cowpea defensin Cp-thionin II and its application in dough. *Food Microbiology* 73: 111–121. https://doi.org/10.1016/j.fm.2018.01.006.

26 Tomita, M., Takase, M., Wakabayashi, H., and Bellamy, W. (1994). Antimicrobial peptides of lactoferrin. *Advances in Experimental Medicine and Biology* 357: 209–218.

27 Özer, B. (2014). Natural antimicrobial systems|lactoperoxidase and lactoferrin. In: *Encyclopedia of Food Microbiology*, 2e (ed. C.A. Batt and M.L. Tortorello), 930–935. Academic Press. https://doi.org/https://doi.org/10.1016/B978-0-12-384730-0.00241-X.

28 Wang, B., Timilsena, Y.P., Blanch, E., and Adhikari, B. (2019). Lactoferrin: structure, function, denaturation and digestion. *Critical Reviews in Food Science and Nutrition* 59 (4): 580–596. https://doi.org/10.1080/10408398.2017.1381583.

29 Niaz, B., Saeed, F., Ahmed, A. et al. (2019). Lactoferrin (LF): a natural antimicrobial protein. *International Journal of Food Properties* 22 (1): 1626–1641. https://doi.org/10.1080/10942912.2019.1666137.

30 Bellamy, W., Takase, M., Wakabayashi, H. et al. (1992). Antibacterial spectrum of lactoferricin B, a potent bactericidal peptide derived from the N-terminal region of bovine lactoferrin. *The Journal of Applied Bacteriology* 73 (6): 472–479.

31 Jenssen, H. and Hancock, R.E.W. (2009). Antimicrobial properties of lactoferrin. *Biochimie* 91 (1): 19–29. https://doi.org/10.1016/j.biochi.2008.05.015.

32 Duran, A. and Kahve, H. (2017). The use of lactoferrin in food industry. *Academic Journal of Science* 7: 89–94.

33 Davidson, P.M. and Zivanovic, S. (2003). 2 – The use of natural antimicrobials. In: *Food Preservation Techniques* (ed. P. Zeuthen and L. Bøgh-Sørensen), 5–30. Woodhead Publishing https://doi.org/10.1533/9781855737143.1.5.

34 Zhang, W. and Rhim, J.-W. (2022). Functional edible films/coatings integrated with lactoperoxidase and lysozyme and their application in food preservation. *Food Control* 133: 108670. https://doi.org/10.1016/j.foodcont.2021.108670.

35 Earnshaw, R.G., Banks, J.G., Defrise, D., and Francotte, C. (1989). The preservation of cottage cheese by an activated lactoperoxidase system. *Food Microbiology* 6 (4): 285–288. https://doi.org/10.1016/S0740-0020(89)80010-3.

36 Haddadin, M.S., Ibrahim, S.A., and Robinson, R.K. (1996). Preservation of raw milk by activation of the natural lactoperoxidase systems. *Food Control* 7 (3): 149–152. https://doi.org/10.1016/0956-7135(96)00023-0.

37 Jooyandeh, H., Aberoumand, A., and Nasehi, B. (2011). Application of lactoperoxidase system in fish and food products: a review.

38 Popper, L. and Knorr, D. (1997). Inactivation of yeast and filamentous fungi by the lactoperoxidase-hydrogen peroxide-thiocyanate-system. *Food/Nahrung* 41 (1): 29–33. https://doi.org/10.1002/food.19970410108.

39 Seifu, E., Buys, E.M., and Donkin, E.F. (2005). Significance of the lactoperoxidase system in the dairy industry and its potential applications: a review. *Trends in Food Science & Technology* 16 (4): 137–154. https://doi.org/10.1016/j.tifs.2004.11.002.

40 Hughey, V.L. and Johnson, E.A. (1987). Antimicrobial activity of lysozyme against bacteria involved in food spoilage and food-borne disease. *Applied and Environmental Microbiology* 53 (9): 2165–2170.

41 Corradini, C., Alfieri, I., Cavazza, A. et al. (2013). Antimicrobial films containing lysozyme for active packaging obtained by sol–gel technique. *Journal of Food Engineering* 119 (3): 580–587. https://doi.org/10.1016/j.jfoodeng.2013.05.046.

42 Silvetti, T., Morandi, S., Hintersteiner, M., and Brasca, M. (2017). Chapter 22 – Use of hen egg white lysozyme in the food industry. In: *Egg Innovations and Strategies for Improvements* (ed. P.Y. Hester), 233–242. Academic Press. https://doi.org/10.1016/B978-0-12-800879-9.00022-6.

43 Syngai, G.G. and Ahmed, G. (2019). Chapter 11 – Lysozyme: a natural antimicrobial enzyme of interest in food applications. In: *Enzymes in Food Biotechnology* (ed. M. Kuddus), 169–179. Academic Press. https://doi.org/10.1016/B978-0-12-813280-7.00011-6.

44 Proctor, V.A., Cunningham, F.E., and Fung, D.Y.C. (1988). The chemistry of lysozyme and its use as a food preservative and a pharmaceutical. *C R C Critical Reviews in Food Science and Nutrition* 26 (4): 359–395. https://doi.org/10.1080/10408398809527473.

45 Mastromatteo, M., Lecce, L., de Vietro, N. et al. (2011). Plasma deposition processes from acrylic/methane on natural fibres to control the kinetic release of lysozyme from PVOH

monolayer film. *Journal of Food Engineering* 104 (3): 373–379. https://doi.org/10.1016/j.jfoodeng.2010.12.032.

46 Silva, N.H.C.S., Vilela, C., Almeida, A. et al. (2018). Pullulan-based nanocomposite films for functional food packaging: exploiting lysozyme nanofibers as antibacterial and antioxidant reinforcing additives. *Food Hydrocolloids* 77: 921–930. https://doi.org/10.1016/j.foodhyd.2017.11.039.

47 Mecitoğlu, Ç., Yemenicioğlu, A., Arslanoğlu, A. et al. (2006). Incorporation of partially purified hen egg white lysozyme into zein films for antimicrobial food packaging. *Food Research International* 39 (1): 12–21. https://doi.org/10.1016/j.foodres.2005.05.007.

48 Pattarasiriroj, K., Kaewprachu, P., and Rawdkuen, S. (2020). Properties of rice flour-gelatine-nanoclay film with catechin-lysozyme and its use for pork belly wrapping. *Food Hydrocolloids* 107: 105951. https://doi.org/10.1016/j.foodhyd.2020.105951.

49 Mogoşanu, G.D., Grumezescu, A.M., Bejenaru, C., and Bejenaru, L.E. (2017). 11 – Natural products used for food preservation. In: *Food Preservation* (ed. A.M. Grumezescu), 365–411. Academic Press. https://doi.org/10.1016/B978-0-12-804303-5.00011-0.

50 Yusuf, M. (2018). Chapter 12 – Natural antimicrobial agents for food biopreservation. In: *Food Packaging and Preservation* (ed. A.M. Grumezescu and A.M. Holban), 409–438. Academic Press. https://doi.org/10.1016/B978-0-12-811516-9.00012-9.

51 Tao, R., Tong, Z., Lin, Y. et al. (2011). Antimicrobial and antibiofilm activity of pleurocidin against cariogenic microorganisms. *Peptides* 32 (8): 1748–1754. https://doi.org/10.1016/j.peptides.2011.06.008.

52 Villalobos-Delgado, L.H., Nevárez-Moorillon, G., and v, Caro, I., Quinto, E. J., & Mateo, J. (2019). 4 – Natural antimicrobial agents to improve foods shelf life. In: *Food Quality and Shelf Life* (ed. C.M. Galanakis), 125–157. Academic Press. https://doi.org/10.1016/B978-0-12-817190-5.00004-5.

53 Burrowes, O.J., Hadjicharalambous, C., Diamond, G., and Lee, T.-C. (2004). Evaluation of antimicrobial spectrum and cytotoxic activity of pleurocidin for food applications. *Journal of Food Science* 69 (3): FMS66–FMS71. https://doi.org/10.1111/j.1365-2621.2004.tb13373.x.

54 Ko, S.J., Kang, N.H., Kim, M.K. et al. (2019). Antibacterial and anti-biofilm activity, and mechanism of action of pleurocidin against drug resistant *Staphylococcus aureus*. *Microbial Pathogenesis* 127: 70–78. https://doi.org/10.1016/j.micpath.2018.11.052.

55 Wang, X., Yue, T., and Lee, T. (2015). Development of pleurocidin-poly(vinyl alcohol) electrospun antimicrobial nanofibers to retain antimicrobial activity in food system application. *Food Control* 54: 150–157. https://doi.org/10.1016/j.foodcont.2015.02.001.

56 Fujiki, M., Abe, K., Hayakawa, T. et al. (2019). Antimicrobial activity of protamine-loaded calcium phosphates against oral bacteria. *Materials* 12 (17): https://doi.org/10.3390/ma12172816.

57 Kim, Y.-H., Kim, S.M.O.O., and Lee, S.I.Y. (2015). Antimicrobial activity of protamine against oral microorganisms. *Biocontrol Science* 20 (4): 275–280. https://doi.org/10.4265/bio.20.275.

58 Hansen, L.T. and Gill, T.A. (2000). Solubility and antimicrobial efficacy of protamine on *Listeria monocytogenes* and *Escherichia coli* as influenced by pH. *Journal of Applied Microbiology* 88 (6): 1049–1055. https://doi.org/10.1046/j.1365-2672.2000.01074.x.

59 Johansen, C., Verheul, A., Gram, L. et al. (1997). Protamine-induced permeabilization of cell envelopes of gram-positive and gram-negative bacteria. *Applied and Environmental Microbiology* 63 (3): 1155–1159.

60 Truelstrup Hansen, L., Austin, J.W., and Gill, T.A. (2001). Antibacterial effect of protamine in combination with EDTA and refrigeration. *International Journal of Food Microbiology* 66 (3): 149–161. https://doi.org/10.1016/S0168-1605(01)00428-7.

61 Aspedon, A. and Groisman, E.A. (1996). The antibacterial action of protamine: evidence for disruption of cytoplasmic membrane energization in *Salmonella typhimurium*. *Microbiology* 142 (12): 3389–3397. https://doi.org/10.1099/13500872-142-12-3389.

62 Meena, M., Prajapati, P., Ravichandran, C., and Sehrawat, R. (2021). Natamycin: a natural preservative for food applications – a review. *Food Science and Biotechnology* 30 (12): 1481–1496. https://doi.org/10.1007/s10068-021-00981-1.

63 Thomas, L. and v, & Delves-Broughton, J. (2003). NATAMYCIN. In: *Encyclopedia of Food Sciences and Nutrition*, 2e (ed. B. Caballero), 4109–4115. Academic Press. https://doi.org/10.1016/B0-12-227055-X/01380-8.

64 Delves-Broughton, J. (2014). Preservatives|permitted preservatives – natamycin. In: *Encyclopedia of Food Microbiology*, 2e (ed. C.A. Batt and M.L. Tortorello), 87–91. Academic Press. https://doi.org/10.1016/B978-0-12-384730-0.00269-X.

65 Lantano, C., Alfieri, I., Cavazza, A. et al. (2014). Natamycin based sol–gel antimicrobial coatings on polylactic acid films for food packaging. *Food Chemistry* 165: 342–347. https://doi.org/10.1016/j.foodchem.2014.05.066.

66 Alimoradi, V., Afshar Mogaddam, M.R., Farajzadeh, M.A. et al. (2022). Surfactant-assisted salting-out homogenous liquid–liquid extraction based on deep eutectic solvents using central composite design; application in the extraction of natamycin from fruit juices before its determination by HPLC-UV. *Microchemical Journal* 179: 107504. https://doi.org/10.1016/j.microc.2022.107504.

67 Karaman, K., Sagdic, O., and Yilmaz, M.T. (2020). Potential of natamycin to control growth of Zygosaccharomyces spp. in apple juice during storage. *International Journal of Food Microbiology* 332: 108771. https://doi.org/10.1016/j.ijfoodmicro.2020.108771.

68 Pan, H., Zhong, C., Xia, L. et al. (2022). Antifungal activity of natamycin against kiwifruit soft rot caused by Botryosphaeria dothidea and potential mechanisms. *Scientia Horticulturae* 305: 111344. https://doi.org/10.1016/j.scienta.2022.111344.

69 Xu, X., Peng, X., Huan, C. et al. (2023). Development of natamycin-loaded zein-casein composite nanoparticles by a pH-driven method and application to postharvest fungal control on peach against Monilinia fructicola. *Food Chemistry* 404: 134659. https://doi.org/10.1016/j.foodchem.2022.134659.

70 Zhang, C., Gong, H., and Liu, Y. (2022). Effects of postharvest coating using chitosan combined with natamycin on physicochemical and microbial properties of sweet cherry during cold storage. *International Journal of Biological Macromolecules* 214: 1–9. https://doi.org/10.1016/j.ijbiomac.2022.06.057.

71 Delves-Broughton, J., Thomas, L.V., and Williams, G.C. (2006). Natamycin as an antimycotic preservative on cheese and fermented sausages. *Food Australia* 58: 19–21.

72 Şanlı, S., Kılıçarslan, S., and Şanlı, N. (2022). Evaluation of natamycin in commercial dairy products by a green capillary zone electrophoresis method and confirmation with a liquid chromatography-mass spectrometry. *Food Bioscience* 50: 102114. https://doi.org/10.1016/j.fbio.2022.102114.

73 Sara, A.E., Ibrahim, M.A., Abdou, M.M., and Hamdi, A. (2014). The role of natamycin fortification to extend shelf life of plain yoghurt. *Benha Veterinary Medical Journal* 27: 140–149.

74 Delves-Broughton, J., Steenson, L., Dorko, C. et al. (2010). 12 – Use of natamycin as a preservative on the surface of baked goods: a case study. In: *Case Studies in Novel Food Processing Technologies* (ed. C.J. Doona, K. Kustin, and F.E. Feeherry), 303–320. Woodhead Publishing. https://doi.org/10.1533/9780857090713.3.303.

75 Tsiraki, M.I., El-Obeid, T., Yehia, H.M. et al. (2018). Effects of chitosan and natamycin on vacuum-packaged phyllo: a pastry product. *Journal of Food Protection* 81 (12): 1982–1987. https://doi.org/10.4315/0362-028X.JFP-18-236.

76 Hu, D., Gao, H., and Yao, X. (2020). 1.18 – Biosynthesis of triterpenoid natural products. In: *Comprehensive Natural Products III* (ed. H.-W. (Ben) Liu and T.P. Begley), 577–612. Elsevier. https://doi.org/10.1016/B978-0-12-409547-2.14678-5.

77 Dupont, S., Fleurat-Lessard, P., Cruz, R.G. et al. (2021). Antioxidant properties of ergosterol and its role in yeast resistance to oxidation. *Antioxidants* 10 (7): https://doi.org/10.3390/antiox10071024.

78 Schaefer, L., Auchtung, T.A., Hermans, K.E. et al. (2010). The antimicrobial compound reuterin (3-hydroxypropionaldehyde) induces oxidative stress via interaction with thiol groups. *Microbiology* 156 (6): 1589–1599. https://doi.org/10.1099/mic.0.035642-0.

79 Engevik, M.A., Danhof, H.A., Shrestha, R. et al. (2020). Reuterin disrupts Clostridioides difficile metabolism and pathogenicity through reactive oxygen species generation. *Gut Microbes* 12 (1): 1795388. https://doi.org/10.1080/19490976.2020.1795388.

80 Cleusix, V., Lacroix, C., Vollenweider, S. et al. (2007). Inhibitory activity spectrum of reuterin produced by Lactobacillus reuteri against intestinal bacteria. *BMC Microbiology* 7 (1): 101. https://doi.org/10.1186/1471-2180-7-101.

81 Sun, M.-C., Hu, Z.-Y., Li, D.-D. et al. (2022). Application of the reuterin system as food preservative or health-promoting agent: a critical review. *Foods* 11 (24): https://doi.org/10.3390/foods11244000.

82 Asare, P.T., Zurfluh, K., Greppi, A. et al. (2020). Reuterin demonstrates potent antimicrobial activity against a broad panel of human and poultry meat *Campylobacter* spp. isolates. *Microorganisms* 8 (1): https://doi.org/10.3390/microorganisms8010078.

83 Stevens, M., Vollenweider, S., Lacroix, C., & Zurich, E. T. H. (2011). 5 – The potential of reuterin produced by Lactobacillus reuteri as a broad spectrum preservative in food. In C. Lacroix (Ed.), *Protective Cultures, Antimicrobial Metabolites and Bacteriophages for Food and Beverage Biopreservation* (pp. 129–160). Woodhead Publishing. https://doi.org/10.1533/9780857090522.1.129

84 Vimont, A., Fernandez, B., Ahmed, G. et al. (2019). Quantitative antifungal activity of reuterin against food isolates of yeasts and molds and its potential application in yogurt. *International Journal of Food Microbiology* 289: 182–188. 10.1016/j.ijfoodmicro.2018.09.005.

85 El-Ziney, M.G., van den Tempel, T., Debevere, J., and Jakobsen, M. (1999). Application of reuterin produced by Lactobacillus reuteri 12002 for meat decontamination and preservation. *Journal of Food Protection* 62 (3): 257–261. https://doi.org/10.4315/0362-028x-62.3.257.

86 Mei, J., Ma, X., and Xie, J. (2019). Review on natural preservatives for extending fish shelf life. *Foods* 8 (10): https://doi.org/10.3390/foods8100490.

87 Hernández-Carrillo, J.G., Orta-Zavalza, E., González-Rodríguez, S.E. et al. (2021). Evaluation of the effectivity of reuterin in pectin edible coatings to extend the shelf-life of strawberries during cold storage. *Food Packaging and Shelf Life* 30: 100760. https://doi.org/10.1016/j.fpsl.2021.100760.

88 Wang, L., Hu, C., and Shao, L. (2017). The antimicrobial activity of nanoparticles: present situation and prospects for the future. *International Journal of Nanomedicine* 12: 1227–1249. https://doi.org/10.2147/IJN.S121956.

89 Anvar, A.A., Ahari, H., and Ataee, M. (2021). Antimicrobial properties of food nanopackaging: a new focus on foodborne pathogens. *Frontiers in Microbiology* 12: 690706.

90 Stanić, V. and Tanasković, S.B. (2020). Chapter 11 – Antibacterial activity of metal oxide nanoparticles. In: *Nanotoxicity* (ed. S. Rajendran, A. Mukherjee, T.A. Nguyen, et al.), 241–274. Elsevier. https://doi.org/10.1016/B978-0-12-819943-5.00011-7.

91 Raghunath, A. and Perumal, E. (2017). Metal oxide nanoparticles as antimicrobial agents: a promise for the future. *International Journal of Antimicrobial Agents* 49 (2): 137–152. https://doi.org/10.1016/j.ijantimicag.2016.11.011.

92 Mathew, S., Snigdha, S., Mathew, J., and Radhakrishnan, E.K. (2019). Biodegradable and active nanocomposite pouches reinforced with silver nanoparticles for improved packaging of chicken sausages. *Food Packaging and Shelf Life* 19: 155–166. https://doi.org/10.1016/j.fpsl.2018.12.009.

93 De Silva, R.T., Mantilaka, M.M.M.G.P.G., Ratnayake, S.P. et al. (2017). Nano-MgO reinforced chitosan nanocomposites for high performance packaging applications with improved mechanical, thermal and barrier properties. *Carbohydrate Polymers* 157: 739–747. https://doi.org/10.1016/j.carbpol.2016.10.038.

94 Kaewklin, P., Siripatrawan, U., Suwanagul, A., and Lee, Y.S. (2018). Active packaging from chitosan-titanium dioxide nanocomposite film for prolonging storage life of tomato fruit. *International Journal of Biological Macromolecules* 112: 523–529. https://doi.org/10.1016/j.ijbiomac.2018.01.124.

95 Zhang, W., Roy, S., and Rhim, J.-W. (2023). Copper-based nanoparticles for biopolymer-based functional films in food packaging applications. *Comprehensive Reviews in Food Science and Food Safety* 22 (3): 1933–1952. https://doi.org/10.1111/1541-4337.13136.

96 Contreras, C.B., Charles, G., Toselli, R., and Strumia, M.C. (2017). Antimicrobial active packaging. Bio packaging (1st edition), 9781315152349.

97 Anastas, P.T., Rodriguez, A., de Winter, T.M. et al. (2021). A review of immobilization techniques to improve the stability and bioactivity of lysozyme. *Green Chemistry Letters and Reviews* 14 (2): 302–338. https://doi.org/10.1080/17518253.2021.1890840.

98 Bahmid, N.A., Pepping, L., Dekker, M. et al. (2020). Using particle size and fat content to control the release of Allyl isothiocyanate from ground mustard seeds for its application in antimicrobial packaging. *Food Chemistry* 308: 125573. https://doi.org/10.1016/j.foodchem.2019.125573.

99 Al-Tayyar, N.A., Youssef, A.M., and Al-hindi, R. (2020). Antimicrobial food packaging based on sustainable Bio-based materials for reducing foodborne pathogens: a review. *Food Chemistry* 310: 125915. https://doi.org/10.1016/j.foodchem.2019.125915.

100 Fu, Y. and Dudley, E.G. (2021). Antimicrobial-coated films as food packaging: a review. *Comprehensive Reviews in Food Science and Food Safety* 20 (4): 3404–3437. https://doi.org/10.1111/1541-4337.12769.

101 Youssef, A.M., EL-Sayed, S.M., EL-Sayed, H.S. et al. (2016). Enhancement of Egyptian soft white cheese shelf life using a novel chitosan/carboxymethyl cellulose/zinc oxide bionanocomposite film. *Carbohydrate Polymers* 151: 9–19. https://doi.org/10.1016/j.carbpol.2016.05.023.

102 Zhang, H., Hortal, M., Jordá-Beneyto, M. et al. (2017). ZnO-PLA nanocomposite coated paper for antimicrobial packaging application. *LWT* 78: 250–257. https://doi.org/10.1016/j.lwt.2016.12.024.

103 Noorbakhsh-Soltani, S.M., Zerafat, M.M., and Sabbaghi, S. (2018). A comparative study of gelatin and starch-based nano-composite films modified by nano-cellulose and chitosan for food packaging applications. *Carbohydrate Polymers* 189: 48–55. https://doi.org/10.1016/j.carbpol.2018.02.012.

104 Yang, W., Fortunati, E., Dominici, F. et al. (2016). Synergic effect of cellulose and lignin nanostructures in PLA based systems for food antibacterial packaging. *European Polymer Journal* 79: 1–12. https://doi.org/10.1016/j.eurpolymj.2016.04.003.

105 Muriel-Galet, V., Cerisuelo, J.P., López-Carballo, G. et al. (2012). Development of antimicrobial films for microbiological control of packaged salad. *International Journal of Food Microbiology* 157 (2): 195–201. https://doi.org/10.1016/j.ijfoodmicro.2012.05.002.

106 Hari, N. and Nair, A.J. (2016). Development and characterization of chitosan-based antimicrobial films incorporated with streptomycin loaded starch nanoparticles. *New Horizons in Translational Medicine* 3 (1): 22–29. https://doi.org/10.1016/j.nhtm.2016.04.002.

107 Pan, F., Amarjargal, A., Altenried, S. et al. (2021). Bioresponsive hybrid nanofibers enable controlled drug delivery through glass transition switching at physiological temperature. *ACS Applied Bio Materials* 4 (5): 4271–4279. https://doi.org/10.1021/acsabm.1c00099.

108 Boostani, S. and Jafari, S.M. (2020). Chapter two – Controlled release of nanoencapsulated food ingredients. In: *Release and Bioavailability of Nanoencapsulated Food Ingredients*, vol. 5 (ed. S.M. Jafari), 27–78. Academic Press. https://doi.org/10.1016/B978-0-12-815665-0.00002-3.

109 Keawchaoon, L. and Yoksan, R. (2011). Preparation, characterization and in vitro release study of carvacrol-loaded chitosan nanoparticles. *Colloids and Surfaces B: Biointerfaces* 84 (1): 163–171. https://doi.org/10.1016/j.colsurfb.2010.12.031.

110 Li, Q., Ren, T., and Perkins, P. (2022). The development and application of nanocomposites with pH-sensitive "gates" to control the release of active agents: extending the shelf-life of fresh wheat noodles. *Food Control* 132: 108563. https://doi.org/10.1016/j.foodcont.2021.108563.

111 Pilon, L., Spricigo, P.C., Miranda, M. et al. (2015). Chitosan nanoparticle coatings reduce microbial growth on fresh-cut apples while not affecting quality attributes. *International Journal of Food Science and Technology* 50: 440–448.

112 Yang, G., Yue, J., Gong, X. et al. (2014). Blueberry leaf extracts incorporated chitosan coatings for preserving postharvest quality of fresh blueberries. *Postharvest Biology and Technology* 92: 46–53. https://doi.org/10.1016/j.postharvbio.2014.01.018.

113 Takala, P.N., Salmieri, S., Boumail, A. et al. (2013). Antimicrobial effect and physicochemical properties of bioactive trilayer polycaprolactone/methylcellulose-based films on the growth of foodborne pathogens and total microbiota in fresh broccoli. *Journal of Food Engineering* 116 (3): 648–655. https://doi.org/10.1016/j.jfoodeng.2013.01.005.

114 Arnon, H., Zaitsev, Y., Porat, R., and Poverenov, E. (2014). Effects of carboxymethyl cellulose and chitosan bilayer edible coating on postharvest quality of citrus fruit. *Postharvest Biology and Technology* 87: 21–26. https://doi.org/10.1016/j.postharvbio.2013.08.007.

115 Martínez-Abad, A., Ocio, M.J., Lagarón, J.M., and Sánchez, G. (2013). Evaluation of silver-infused polylactide films for inactivation of *Salmonella* and feline calicivirus in vitro and on fresh-cut vegetables. *International Journal of Food Microbiology* 162 (1): 89–94. https://doi.org/10.1016/j.ijfoodmicro.2012.12.024.

116 Nabifarkhani, N., Sharifani, M., Daraei Garmakhany, A. et al. (2015). Effect of nano-composite and thyme oil (*Tymus vulgaris* L.) coating on fruit quality of sweet cherry (Takdaneh Cv) during storage period. *Food Science & Nutrition* 3 (4): 349–354. https://doi.org/10.1002/fsn3.226.

117 Fernández, A., Picouet, P., and Lloret, E. (2010). Cellulose-silver nanoparticle hybrid materials to control spoilage-related microflora in absorbent pads located in trays of fresh-cut melon. *International Journal of Food Microbiology* 142 (1): 222–228. https://doi.org/10.1016/j.ijfoodmicro.2010.07.001.

118 Ding, Y., Yun, J., Li, X. et al. (2012). Evaluation of nano-packing on the shelf life of fresh-cut lotus root (NelumbonuciferaGaerth). In: *Advances in Technology and Management* (ed. H. Kim), 775–780. Berlin, Heidelberg: Springer.

119 Zhu, L., Olsen, C., McHugh, T. et al. (2014). Apple, carrot, and hibiscus edible films containing the plant antimicrobials carvacrol and cinnamaldehyde inactivate salmonella newport on organic leafy greens in sealed plastic bags. *Journal of Food Science* 79 (1): M61–M66. https://doi.org/10.1111/1750-3841.12318.

120 Salmieri, S., Islam, F., Khan, R.A. et al. (2014). Antimicrobial nanocomposite films made of poly(lactic acid)–cellulose nanocrystals (PLA–CNC) in food applications – part B: effect of oregano essential oil release on the inactivation of *Listeria monocytogenes* in mixed vegetables. *Cellulose* 21 (6): 4271–4285. https://doi.org/10.1007/s10570-014-0406-0.

121 Ayala-Zavala, J.F. and González-Aguilar, G.A. (2010). Optimizing the use of garlic oil as antimicrobial agent on fresh-cut tomato through a controlled release system. *Journal of Food Science* 75 (7): M398–M405. https://doi.org/10.1111/j.1750-3841.2010.01723.x.

122 Rojas-Graü, M.A., Raybaudi-Massilia, R.M., Soliva-Fortuny, R.C. et al. (2007). Apple puree-alginate edible coating as carrier of antimicrobial agents to prolong shelf-life of fresh-cut apples. *Postharvest Biology and Technology* 45 (2): 254–264. https://doi.org/10.1016/j.postharvbio.2007.01.017.

123 Legan, J.D. and Voysey, P.A. (1991). Yeast spoilage of bakery products and ingredients. *Journal of Applied Bacteriology* 70 (5): 361–371. https://doi.org/10.1111/j.1365-2672.1991.tb02950.x.

124 Soares, N.F.F., Rutishauser, D.M., Melo, N. et al. (2002). Inhibition of microbial growth in bread through active packaging. *Packaging Technology and Science* 15 (3): 129–132. https://doi.org/10.1002/pts.576.

125 Silveira, M.F.A., Soares, N.F.F., Geraldine, R.M. et al. (2007). Antimicrobial efficiency and sorbic acid migration from active films into pastry dough. *Packaging Technology and Science* 20 (4): 287–292. https://doi.org/10.1002/pts.757.

126 Lopes, F.A., de Fátima Ferreira Soares, N., de Cássia Pires Lopes, C. et al. (2014). Conservation of bakery products through cinnamaldehyde antimicrobial films. *Packaging Technology and Science* 27 (4): 293–302. https://doi.org/10.1002/pts.2033.

127 Otoni, C.G., Pontes, S.F.O., Medeiros, E.A.A., and de Soares, N.F.F. (2014). Edible films from methylcellulose and nanoemulsions of clove bud (*Syzygium aromaticum*) and Oregano (*Origanum vulgare*) essential oils as shelf life extenders for sliced bread. *Journal of Agricultural and Food Chemistry* 62 (22): 5214–5219. https://doi.org/10.1021/jf501055f.

128 Gutiérrez, L., Sánchez, C., Batlle, R., and Nerín, C. (2009). New antimicrobial active package for bakery products. *Trends in Food Science & Technology* 20 (2): 92–99. https://doi.org/10.1016/j.tifs.2008.11.003.

129 Gutiérrez, L., Batlle, R., Andújar, S. et al. (2011). Evaluation of antimicrobial active packaging to increase shelf life of gluten-free sliced bread. *Packaging Technology and Science* 24 (8): 485–494. https://doi.org/10.1002/pts.956.

130 Passarinho, A.T.P., Dias, N.F., Camilloto, G.P. et al. (2014). Sliced bread preservation through oregano essential oil-containing sachet. *Journal of Food Process Engineering* 37 (1): 53–62. https://doi.org/10.1111/jfpe.12059.

131 Sachdeva, A., Vashist, S., Chopra, R., and Puri, D. (2017). Antimicrobial activity of active packaging film to prevent bread spoilage. *International Journal of Food Sciences and Nutrition* 2: 29–37.

4

Oxygen Scavengers in Active Food Packaging Systems

Arihant Ahuja, Dimple Singh, Varsha Singh, Anamika Singh, Satish Kumar Bachala, and Vibhore Kumar Rastogi

Department of Paper Technology, Indian Institute of Technology Roorkee, Roorkee, Uttarakhand, India

4.1 Introduction

Food packaging describes creating a protective shield for food against external factors such as moisture, oxygen, dust, UV radiation, chemicals, or microbes [1]. A common term for such packaging is passive packaging [2]. Recently food packaging technologies are developing active packaging with emitters or absorbers to enhance the quality of food products packed by further increasing the shelf life [3]. Oxygen scavengers, ethylene scavengers, carbon dioxide absorbers or emitters, antioxidant releasers, moisture scavengers, and antimicrobial packaging are part of active packaging. Oxygen scavengers are used if the food packed is affected by oxygen. Oxygen is a highly reactive molecule that can cause various chemical reactions in food products, leading to the degradation of quality, nutritional value, and sensory attributes. These reactions include the oxidation of unsaturated fats and oils, browning reactions, vitamin degradation, flavor loss, and growth of aerobic bacteria [4], which lead to the deterioration of food shelf life, flavor, color, and smell of the food. High oxygen barrier packaging is employed to avoid the interaction of oxygen with the product. However, some products release oxygen during storage, which causes oxidation and microbial growth. In this case, active packaging is employed, which, contrary to conventional packaging, interacts with the contents and the environment [5]. Active packaging uses active components (artificial or natural), which, when incorporated into the package, interact with the foods to increase the shelf life, as shown in Figure 4.1 [6].

Oxygen scavengers are the best-known and most widely used active packaging to remove or scavenge oxygen from the package headspace. The shelf life of food can be extended with the help of oxygen scavengers by reducing the oxygen level to less than 0.1 vol.% [4], thereby preventing the oxidation of delicate substances such as fats and oils. This technique is frequently utilized in the food and beverage sector, where packaging free of oxygen is necessary for the products. Examples of food products that may benefit from using oxygen scavengers include snack foods, processed meats, bakery products, and pet food [7–9].

Smart Food Packaging Systems: Innovations and Technology Applications, First Edition.
Edited by Avik Mukherjee, Santosh Kumar, Manjusri Misra, and Amar K. Mohanty.
© 2025 John Wiley & Sons Ltd. Published 2025 by John Wiley & Sons Ltd.

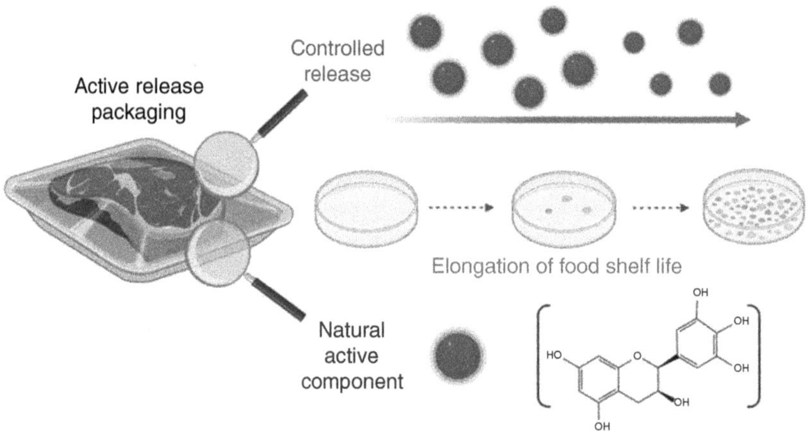

Figure 4.1 Interaction of active compound with packed food. *Source:* Westlake et al. [6]/American Chemical Society/CC-BY-4.0.

This chapter discusses the role of oxygen scavengers and different types of scavengers based on metal, bio/polymers, enzymes, and microorganisms. Moreover, the reaction mechanism of different scavengers is addressed briefly. Furthermore, the incorporation of techniques in the primary package and its usage for fruits, vegetables, dairy products, muscle food, and bakery products have also been widely discussed.

4.2 Oxygen Scavengers

Oxygen scavengers work by removing or reducing the presence of oxygen in food packaging through a chemical reaction that consumes the oxygen in the package headspace [10]. Figure 4.2 shows various properties of incorporating oxygen scavengers in food packaging, including delaying fruit and vegetable ripening, preventing mold development or growth on food products, extending food shelf life, reducing headspace oxygen concentration to avoid food oxidation, and preventing early food degradation. In food packaging, oxygen scavengers can be integrated into various packaging materials, including films, pouches, and bottles [7], providing greater flexibility and cost-effectiveness compared to high oxygen barrier packaging, which is a limiting factor in packaging materials. They can be combined with other components, such as moisture absorbers, antimicrobial agents, and ethylene absorbers, to provide a comprehensive solution for maintaining product quality and safety [11]. The use of active food packaging systems not only extends shelf life and maintains product quality but also reduces food waste and improves sustainability by minimizing the need for preservatives and additives. Therefore, integrating oxygen scavengers in active food packaging systems offers a promising solution for improving packaged food product's safety, quality, and sustainability. Compared to high oxygen barrier packaging, oxygen scavengers have several advantages. High oxygen barrier packaging can block oxygen from entering the package, but residual oxygen can remain inside the package and cause oxidative degradation of sensitive compounds. In contrast, oxygen scavengers

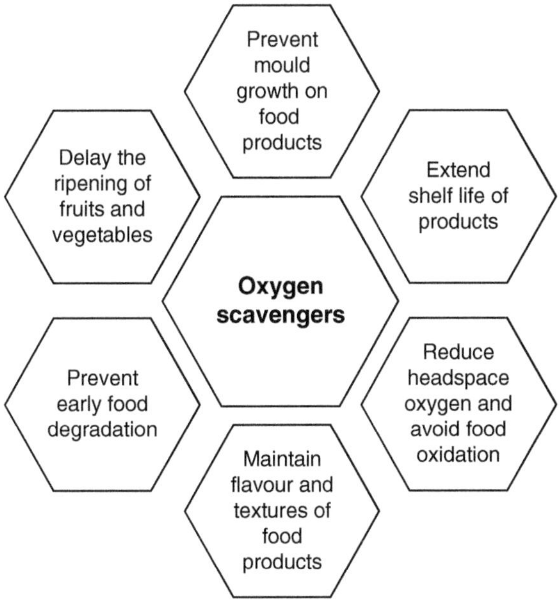

Figure 4.2 Oxygen scavengers and their role in maintaining food quality.

Table 4.1 Commercially available oxygen scavengers and their active ingredient.

Oxygen scavenger	Active ingredient	References
Ageless OMAC	Iron powder	[13, 14]
Bioka	Enzymes	[15]
Grace Darex	Ascorbic acid	[15]
Cryovac OS1000	Polybutadiene	[16]
ColorMatrix Amosorb	Polybutadiene	[17]

actively consume residual oxygen in the package environment, maintaining product quality and extending shelf life [12]. Overall, the use of oxygen scavengers in packaging is a highly effective and efficient approach to ensure the preservation of food and beverage products.

Table 4.1 shows commercially available oxygen scavengers, such as Ageless OMAC, Bioka, Grace Darex, and Cryovac OS1000, have been developed to cater to specific food applications. Of these, Ageless OMAC is a metal-based scavenger that uses iron or iron oxide as the active ingredient [13, 14], while enzyme-based scavengers, such as Bioka, utilize enzymes to catalyze the reaction that consumes oxygen [15]. Grace Darex utilizes ascorbic acid [15], a form of vitamin C, as the active ingredient to consume oxygen. Amosorb and Cryovac OS1000 is a polymer-based scavenger that utilizes polybutadiene to react with oxygen [16, 17]. Various types of oxygen scavengers can be categorized based on their chemical composition and mechanism of action. One type is metal-based scavenging systems which utilize iron, zinc, palladium, and titanium as the active component. Bio-based scavenging systems use organic compounds such as gallic acid, pyrogallol, ascorbic acid, alpha-tocopherol, leghemoglobin, and polyunsaturated fatty acids to remove oxygen. Enzyme-based scavenging systems employ enzymes such as glucose oxidase, oxalate oxidase, ethanol oxidase, and laccase to break down oxygen. Polymer-based scavenging systems utilize unsaturated polymer rubbers such as polybutadiene and polyisoprenes to remove oxygen. Finally, microorganism-based scavenging systems make use of microorganisms that consume oxygen.

4.2.1 Metal-Based Scavenging System

Metal-based oxygen scavenging systems are a type of oxygen scavenger that utilizes metal particles or compounds to consume oxygen in the package environment [4]. The metal particles or compounds are typically embedded in a polymer matrix or incorporated into an absorbent pad, sachet, or film inside the package. These scavengers function through a reaction, where the metal catalyst reacts with oxygen in the package environment to form metal oxides or hydroxides. This process occurs through the transfer of electrons from the metal catalyst to the oxygen molecule, resulting in the formation of stable metal oxide or hydroxide species and the removal of oxygen from the package environment. The reaction kinetics of metal-based oxygen scavengers depend on various factors, including the metal catalyst's concentration and distribution, oxygen availability in the package environment, and temperature and humidity conditions. Most metal-based oxygen scavengers involve a reaction between the metal catalyst and oxygen. The principle of oxidation in most metal cases is based on oxidation in the presence of Lewis acids and moisture [4]. Two sequential reactions can describe this process. First, the metal catalyst (M) in the scavenger reacts with water (H_2O) to produce metal hydroxide (MOH) and hydrogen gas (H_2) (Eq. 4.1). The metal hydroxide species then reacts with oxygen (O_2) in the package environment to form metal oxide (MO) and water (H_2O) (Eq. 4.2).

$$\mathrm{M} + \mathrm{H_2O} \rightarrow \mathrm{MOH} + \mathrm{H_2} \tag{4.1}$$

$$2\,\mathrm{MOH} + \frac{1}{2}\mathrm{O_2} \rightarrow 2\mathrm{MO} + \mathrm{H_2O} \tag{4.2}$$

Iron-based scavengers are one of the most used oxygen scavengers in food packaging applications. Iron can be used in sachets inside the packaging alongside the food or incorporated into polymeric films. The latter technique is more efficient as the iron particles are homogenously spread throughout the packaging, providing maximum area coverage inside the package. However, the oxygen scavenging properties of iron-based scavengers also depend upon the polymer matrix, as the oxygen permeability of different polymers varies [4]. Recently, nano-iron instead of iron has been explored to increase oxygen scavenging activity. The oxygen scavenging rate of nano-iron is nine times higher than that of conventional iron powders [18]. This nano-iron also works in the presence of moisture and anhydrous conditions. The first major commercial metal-based oxygen scavenger was an iron-based oxygen scavenger introduced in Japan in 1977 by Mitsubishi Gas and Chemical Company. It was also made available in the United States during the same year and was

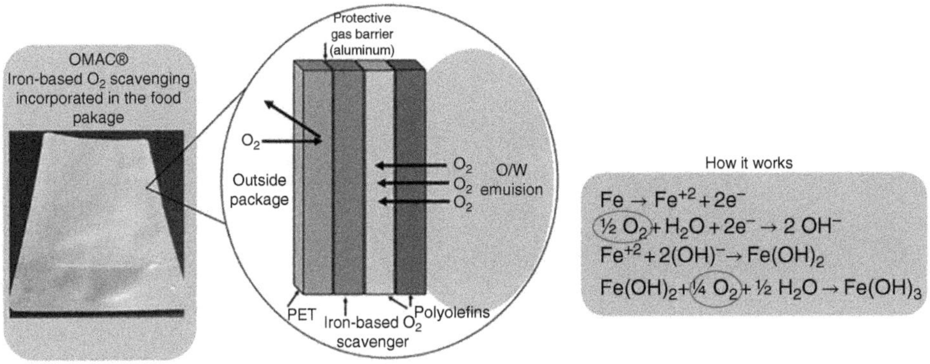

Figure 4.3 Iron-based Ageless OMAC oxygen scavenger and its reaction mechanism. *Source:* Reproduced from Johnson et al. [13]/with permission of Elsevier.

sold by Ageless OMAC [19]. The mechanism of Ageless OMAC is based on oxidizing iron to iron oxide, as shown in Figure 4.3. In this system, the iron-based oxygen scavenger is packed between the layers of packaging. Therefore, it is compatible with liquid products as the active compound is not in direct contact with the product.

Similar to iron, other metals such as zinc also possess oxygen scavenging activity. It also reduces oxygen from the headspace, as in the case of iron, by converting it to metal oxides. The nanostructure particles of zinc have also shown good oxygen scavenging capacity. The advantage of using nanostructure zinc particles is that, after consuming the oxygen from the atmosphere and turning it into zinc oxide nanoparticles, they act as antimicrobial agents as nano ZnO has inherent antimicrobial properties [20]. Zinc with ascorbic acid extended the shelf life of bakery products such as bun and bread [21]. The moisture requirement of iron or zinc-based scavengers makes them ineffective in moisture-sensitive applications. Palladium is an interesting metal-based oxygen scavenger that does not work by oxidizing itself but by hydrolysis of the hydrogen present in the atmosphere to form water. Thus, the headspace of the package must contain hydrogen. Due to the flammability of hydrogen, it can only be set within the range of 5–7% with nitrogen. Therefore, hydrogen restriction allows scavenging a maximum of 2.5 vol% oxygen if palladium is coated onto the packaging [22].

In the presence of UV-A irradiation, titania (or titanium dioxide) starts an oxidative mineralization process, thus mediating the destruction of many organic compounds. Ethyl cellulose incorporated titania with a reducing agent triethanolamine produces a flexible film (Figure 4.4a). When exposed to UV irradiation, these films resulted in oxidation of ethyl cellulose and triethanolamine, thus reducing the headspace oxygen. Increasing the UV irradiation time resulted in the luminescence of the polymer-coated glass. The maximum luminescence was reached at 14 h, corresponding to less than 1% oxygen (Figure 4.4b). UV irradiation time was observed to be proportional to the oxygen scavenging activity. This system can scavenge oxygen at 0.017 cm^3 O_2/h-cm^2 over one day [23]. Titania is a large-band semiconductor that requires UV light to start the semiconductor photocatalysis reaction. The mechanism of the titania-based scavenger is based on the oxidative mineralization process. The photo-mineralization of ethyl cellulose polymer and the reducing agent starts

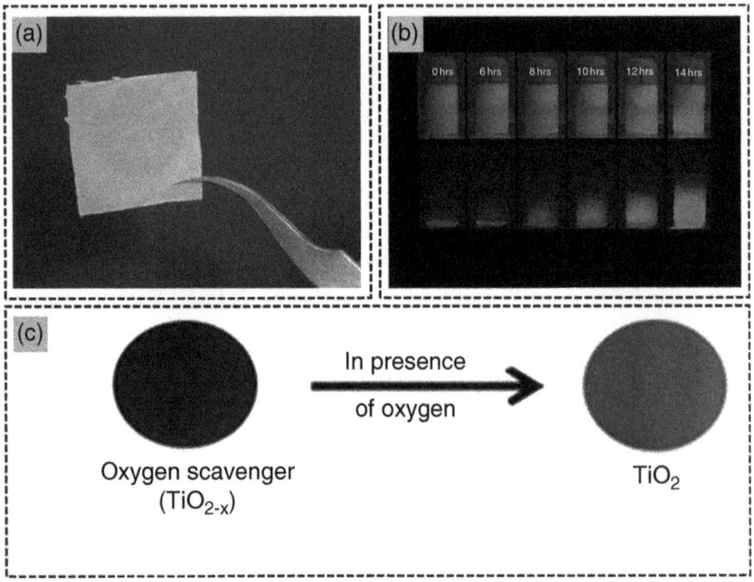

Figure 4.4 (a) Titania-based flexible active film. (b) UV irradiation time reflecting the titania-based oxygen scavenging in glass vials. *Source:* Reproduced with permission from Mills et al. [23]/Elsevier. (c) Color change of Titan based nanotube. *Source:* Reproduced from Tulsyan et al. [24]/with permission of John Wiley & Sons.

with the help of UV light, which reduces oxygen from the ambient environment (Eq. 4.3). The requirement of UV exposure for the oxygen scavenging activity makes it difficult for practical application. However, titania-based nanotubes showed oxygen scavenging without UV light, and oxygen is reacted and consumed by the titania nanotubes themselves. More interestingly, the titania-based nanotubes change color when they consume oxygen, changing color from dark blue to yellowish as they react with oxygen (Figure 4.4c) [24].

$$\text{Organic} + O_2 \xrightarrow[\text{TiO}_2]{h\nu \geq 3.2\,\text{eV}} CO_2 + H_2O + \text{mineral acids} \tag{4.3}$$

4.2.2 Bio-Based Scavenging System

Bio-based scavenging systems are naturally occurring substances that remove oxygen from the package. These are generally organic substances such as phenolics (gallic acid, pyrogallol), vitamins (ascorbic acid, alpha-tocopherol), and proteins (leghemoglobin). The bio-based system's primary advantage over metal-based scavengers is that they are environment-friendly, less costly, and nontoxic. Phenolic acids, such as gallic acid, activate upon contact with moisture; hence they are used as oxygen scavengers for foods with high moisture content. Langowski and Wanner first introduced gallic acid as an oxygen scavenger for packaging applications [25]. Galic acid, under alkaline circumstances, absorbs a significant amount of oxygen. When combined with sodium carbonate, gallic acid can scavenge oxygen up to 447 mg/g in 25 days when used in powder form. When blended inside the polyethylene matrix, the gallic acid has an oxygen scavenging capacity of 430 mg/g in 25 days [26]. The mechanism of gallic acid to scavenge oxygen is based on auto-oxidation in the presence of alkaline pH. Gallic acid with alkaline media transfers its four acidic protons to an acceptor base. Further, it oxidizes to form gallate radicals and produces hydrogen peroxide (Eq. 4.4, 4.5). Simultaneously, the gallate radical forms a gallic acid dimer (Eq. 4.6), which can further oxidize itself to quinone of gallic dimer and produce hydrogen peroxide (Eq. 4.7). And finally, the quinone of gallic dimer can oxidize to form open ring product of dimer (Eq. 4.8) [27].

$$\text{Gallic acid} + O_2 \rightarrow \text{Gallic}\bullet + HO_2\bullet \tag{4.4}$$

$$HO_2\bullet + \text{Gallic acid} \rightarrow \text{Gallic}\bullet + H_2O_2 \tag{4.5}$$

$$2\,\text{Gallic}\bullet \rightarrow \text{Gallic acid dimer} \tag{4.6}$$

$$\text{Gallic acid dimer} + O_2 \rightarrow \text{Quinone of dimer} + 2H_2O_2 \tag{4.7}$$

$$\text{Quinone of dimer} + 2O_2 \rightarrow \text{Open ring product of dimer} \tag{4.8}$$

Similarly, pyrogallol, a byproduct of gallic acid when decomposed at higher temperatures, can be used as an oxygen scavenger in food packaging with sodium carbonate and can scavenge 6.48 ml of oxygen daily for every gram of oxygen scavenger material in the package [28]. Ascorbic acid, commonly known as Vitamin C, is oxidized as oxygen scavenging material to form dehydroascorbic acid. Ascorbic acid, when incorporated in the whey protein films at 0.2 M concentration, had an oxygen scavenging capacity of 35.6 ± 2.3 ml O_2/g dry film [29]. The oxidation process is slow and thus can be accelerated with the help of metals as a catalyst for increasing the oxidation reaction of ascorbic acid. The metal-catalyzed ascorbic acid oxidation mechanism follows the formation of metal-ascorbate complex compounds and ascorbate monoanionic, which scavenge oxygen from the headspace rapidly. The fully protonated ascorbic acid (AH_2) is converted to ascorbate monoanionic (HA^-) (Eq. 4.9), which is converted to metal-ascorbate complex (MHA) (Eq. 4.10). MHA consumes oxygen and oxidizes to form an intermediate complex (MHA[O_2]) (Eqs. 4.11 and 4.12). This intermediate complex is then converted to a semiquinone form of ascorbic acid ($HA^{\bullet -}$) (Eq. 4.13). The semiquinone form is finally converted to dehydroascorbic acid (A) with the help of a metal catalyst (Eq. 4.14). Simultaneously, the semiquinone form of ascorbic acid also reacts with oxygen to form dehydroascorbic acid (A) and produces free radicals ($O_2^{\bullet -}$) (Eq. 4.15) [4].

$$AH_2 \leftrightarrow HA^- + H^+ \tag{4.9}$$

$$HA^- + M^{n+} \leftrightarrow MHA^{(n-1)+} \tag{4.10}$$

$$MHA^{(n-1)+} + O_2 \leftrightarrow MHA(O_2)^{(n-1)+} \tag{4.11}$$

$$MHA(O_2)^{(n-1)+} \rightarrow MHA^{\bullet}(O_2^{\bullet})^{(n-1)+} \tag{4.12}$$

$$MHA^{\bullet}(O_2^{\bullet})^{(n-1)+} \rightarrow HA^{\bullet -} + M^{n+} + O_2^{\bullet} \tag{4.13}$$

$$HA^{\bullet -} + M^{n+} \rightarrow A + M^{(n-1)+} \tag{4.14}$$

$$HA^{\bullet -} + O_2 \rightarrow A + O_2^{\bullet -} \tag{4.15}$$

Metal catalysts are avoided in foods due to the incompatibility of food metal detectors and microwave ovens. To overcome these disadvantages, activated carbon-ascorbic acid was recently investigated. In the presence of water, the oxygen scavenging capacity was 2.2 times higher than the commercialized oxygen scavengers and can be used to package meat products [30]. Grace Darex Packaging Technologies developed an ascorbic acid/sulfite-based oxygen scavenger commercially sold by the name Darex. This system can be placed inside the bottle's crown, metal or plastic closures. It consumes oxygen and converts the ascorbate to dehydroascorbic acid and sulfite to sulfate, potentially increasing the shelf life of beer by protecting it from oxidation [15].

α-tocopherol is another oxygen scavenger that reacts with oxygen to produce a stable free radical. This stable free radicals' reactivity is lower than oxygen, so there are fewer chances of food spoilage. α-tocopherol requires the presence of transition metal to increase the scavenging. The metal catalyst transforms the oxygen into reactive oxygen species, and finally, the tocopherols scavenge reactive oxygen species. The transition metal catalyst transforms the oxygen into singlet electron oxygen (Eq. 4.16). α-tocopherol can react with this singlet electron oxygen and produce tocopherol hydroperoxydienone, tocopherylquinone, and quinine epoxide (Eq. 4.17). Simultaneously, singlet electron oxygen can be transformed into reactive oxygen species (Eq. 4.17), which takes electrons from α-tocopherol to scavenge the reactive oxygen species (Eq. 4.18). The transition metal catalyst with α-tocopherol has an oxygen scavenging rate of 0.11 ml O_2/g-day and an oxygen scavenging capacity of 6.72 ml O_2/g [31].

$$\mathrm{Fe(II)} + \mathrm{O_2} \rightarrow \mathrm{Fe(III)} + \mathrm{O_2^-} \bullet \tag{4.16}$$

$$\text{Tocopherol} + \mathrm{O_2^-} \bullet \rightarrow \text{Tochopherol} \bullet + \mathrm{HO_2} \bullet \tag{4.17}$$

$$\text{Tocopherol} \bullet + \mathrm{HO_2} \bullet \rightarrow \text{Tochopherol} \bullet + \mathrm{H_2O_2} \tag{4.18}$$

Leghemoglobin obtained from soybean roots has shown oxygen scavenging activity as it binds with oxygen and creates anaerobic conditions [7]. However, it has not been used in packaging, so it can potentially be employed as a bio-based oxygen scavenging system for application in food packaging. Polyunsaturated fatty acids such as oleic, linoleic, and linoleic acids are reactive species that can also work as oxygen scavenger oxygen for dry foods. The main advantage of these fatty acids is that they do not require water for oxidation; therefore, these can be used for dry food products. The presence of unsaturation, i.e. double bonds in their structure, makes them susceptible to oxidation. When these fatty acids are exposed to oxygen, the double bonds go through oxidation reactions and use the oxygen in the surrounding environment. The fatty acids are commonly mixed with a carrier oil, such as soybean, sesame, or cottonseed oil, to create a solid oxygen scavenger composition for a sachet [32]. Generally, metal catalysts are compounded with these fatty acids to accelerate the oxidation of fatty acids. However, during the catalytic autoxidation of the fatty acids, they can generate aldehydes producing a foul smell inside the package [33].

4.2.3 Enzyme-Based Scavenging System

Enzymes-based oxygen scavengers are a particular category of a nontoxic and environmentally friendly method to scavenge oxygen from the package headspace. Enzymes are a special type of biological protein that accelerates a chemical reaction in case of oxygen scavenging;

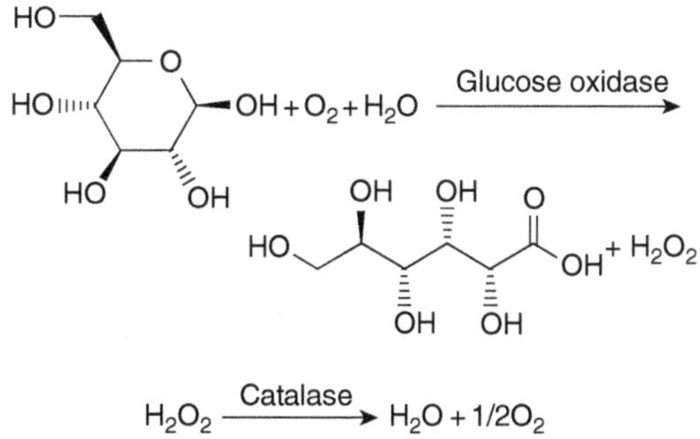

Figure 4.5 Reaction mechanism of oxygen scavenging by glucose oxidase and catalase. *Source:* Reproduced from Shi et al. [36]/with permission of American Chemical Society.

an enzyme surface is for reacting with a substrate to remove oxygen [34]. The packaged products can either include enzymes directly, or an isolated sachet can contain them separately. The most common enzyme-based oxygen scavengers are glucose oxidase, ethanol oxidase, oxalate oxidase, and laccase. Glucose oxidase is used as an oxygen scavenging along with another enzyme, "catalase." These two enzymes are frequently used in food packaging for scavenging oxygen [35]. For oxygen scavenging, the glucose oxidase enzyme needs glucose to act as a substrate either in packaging or mixed in the product. The glucose oxidase works by the reaction mechanism of oxidoreductase in the presence of moisture (Figure 4.5). It transfers hydrogen from the glucose substrate to headspace oxygen and thus forms glucono–delta–lactone and hydrogen peroxide. This hydrogen peroxide is broken down into water with the help of another enzyme-catalyzed [36, 37]. Glucose oxidase potential usage is removing oxygen from bottled beer and wine. It is possible to immobilize catalase and glucose oxidase as coatings on various polymers. However, the food must be in direct contact with the surface containing the active enzyme to start the redox reaction, which can further decrease the effectiveness of enzyme-based films used for food packaging [34]. Bioka is a commercially available enzyme-based oxygen scavenger film that utilizes glucose oxidase/catalase. It is asserted that the reactive powder's ingredients and the reaction products are safe for consumer consumption and the environment. It eradicates oxygen from the package's headspace in 12–48 hours at 20 °C. Moreover, it scavenges oxygen at low temperatures also; it can scavenge oxygen completely in 24–96 hours at 2–6 °C. However, due to the heat sensitivity, it cannot be used above 60 °C [33].

Oxalate oxidase is another enzyme used to scavenge oxygen from packaged food. Oxalate oxidase belongs to *cupin* superfamily a, which are germin-like protein. It is usually used for the analysis of oxalic acid in clinical samples. Oxalate oxidase works in a similar manner to that of glucose oxidase. Oxalate oxidase can catalyze the oxidation reaction of oxalic acid to produce hydrogen peroxide and carbon dioxide (Eq. 4.19). This way, oxalate oxidase can scavenge the headspace oxygen from the package. Simultaneously, catalase can be used to remove the hydrogen peroxide (Eq. 4.20, 4.21), as seen in the case of the glucose oxidase/catalase enzyme [38].

$$HOOC-COO^- + H^+ + O_2 \rightarrow 2CO_2 + H_2O_2 \quad (4.19)$$

$$H_2O_2 \rightarrow H_2O + \frac{1}{2}O_2 \quad (4.20)$$

$$HOOC-COO^- + H^+ + \frac{1}{2}O_2 \rightarrow 2CO_2 + H_2O \quad (4.21)$$

Alcohol oxidase or ethanol oxidase enzyme can also be used as an oxygen scavenger. The sources of ethanol oxidase are fungi, bacteria, plants, yeast, and mollusks. Ethanol oxidase catalyzes ethanol's oxidation and converts it into aldehydes with additional byproduct hydrogen peroxide, thus removing the headspace oxygen from the system (Eq. 4.22). Although the working of ethanol oxidase enzyme does not require adding water, a lot of ethanol is needed to absorb ample oxygen from the package, which could result in a foul smell. Furthermore, the aldehyde is produced during this reaction which could result in a yogurt-like odor in food [39]. Therefore, ethanol oxidase is not widely used in industries.

$$CH_3CH_2OH + O_2 \rightarrow CH_3CHO + H_2O_2 \tag{4.22}$$

Laccases are multicopper oxidase enzymes found in plants, fungi, insects, and bacteria and are also used for scavenging oxygen. These enzymes do not produce hydrogen peroxide in contrast to other enzyme systems. Its working mechanism is oxidizing phenolic hydroxyls to phenoxy radicals [40]. Therefore, compounds with phenolic hydroxyls should be present in combination with laccases. Lignin (alkali lignin, hydrolytic lignin, organosolv lignin) and its derivates (lignosulfonates and fractionated lignosulfonates) have been used as phenolic hydroxyl compounds with laccases to scavenge oxygen [40–42]. The advantage of a laccase-based enzyme is that the reactivity of these enzymes is relatively high at 4 °C. Thus, they can also be used as oxygen scavengers for refrigerated products [43]. These systems utilize enzymes to scavenge oxygen, and are susceptible to pH, temperature, salt concentration, and water activity. They cannot be used for food products with low moisture content because they require water for activation [44]. Although they have numerous packaging advantages, the major disadvantage is that water is required; thus, they cannot be used for low-moisture food products [35]. Moreover, the enzymatic approach is typically more expensive than other systems [34].

4.2.4 Polymer-Based Scavenging System

Polymer-based oxygen scavengers can be directly blended with the packaging material or used as a standalone material. The most used polymer-based oxygen scavengers are rubbers. These can be either synthetic (polybutadiene) or natural (Polyisoprene) rubbers. These polymers may require a transition metal catalyst, an initiator, or an activator [45]. The catalyst and the initiator are used to accelerate the oxidation reaction. In contrast, the activator activates the reaction, so the oxidation reaction does not start before product packaging. Synthetic rubbers such as polybutadiene, a type of unsaturated hydrocarbon, can scavenge oxygen. These unsaturated hydrocarbons have a very high capacity for oxygen scavenging [4]. In the presence of metal catalysts such as cobalt neo decanoate, the polybutadiene can absorb 15 wt.% of the oxygen in seven days [46]. The mechanism reaction of polybutadiene-based scavengers follows the formation of free radicals and then the scavenging of radicals. The cobalt catalyst converts the oxygen to free radicals (Eq. 4.23–4.25). The RH in Eq. (4.23) denotes the polymeric molecule that is susceptible to oxidation. Moreover, the unsaturated double bonds are also sources of oxygen consumption as they can form epoxides. Potentially, the conversion of aldehydes to carboxylic acid can also be the reason for scavenging oxygen [46]. Therefore, the scavenging rate of oxygen increases with metal catalyst loading. Amosorb is a commercially available oxygen absorber using polybutadiene mixed with polyester to scavenge oxygen [17].

$$O_2 + RH \rightarrow ROOH \quad (4.23)$$

$$Co^{2+} + ROOH \rightarrow Co^{3+} + OH^- + RO\bullet \quad (4.24)$$

$$Co^{3+} + ROOH \rightarrow Co^{2+} + H^+ + ROO\bullet \quad (4.25)$$

The presence of a metal catalyst with polybutadiene makes it toxic; therefore, these transition metals should be avoided. Another interesting way is to photo-oxidize the polybutadiene in the presence of UV irradiation. five hours of UV irradiated polybutadiene/polyethylene composite can absorb up to 15% of the oxygen (4.4 ml O_2/g film) in 6 days. The UV irradiation helped in the photooxidation of double bonds in polybutadiene and the formation of epoxides, carbonyl groups, and hydroxyl groups. The UV exposure accelerated the oxidation of vinyl bonds in the polybutadiene and thus increased the oxygen absorption capacity. The oxidation of double bonds leads to the formation of epoxides, carbonyl groups, and hydroxyl groups [47]. However, these films required a lot of UV exposure time, which might not be practically feasible. The requirement for low-cost and low-toxic catalysts led to the usage of titanium dioxide. Titanium dioxide can also be used to catalyze oxygen scavenging. Polylactic acid containing 13 wt.% hydroxy-terminated polybutadiene had oxygen absorption of 28 ml/g in 30 days [48]. Cryovac Division of Sealed Air developed a polybutadiene-based oxygen scavenger, which uses light as a triggering mechanism (Figure 4.6a) and can be used to package food products (Figure 4.6b).

Apart from Polybutadienes, and synthetic rubbers, other unsaturated polymers, such as polyisoprene, can be used as oxygen scavengers. Polyisoprene is a natural rubber available in two isomers: *cis*-polyisoprene and *trans*-polyisoprene. *Cis*-polyisoprene is obtained from *Hevea brasiliensis* and is commonly known as natural rubber latex. When utilizing the *cis*-polyisoprene in its form in the presence of a metal catalyst and a

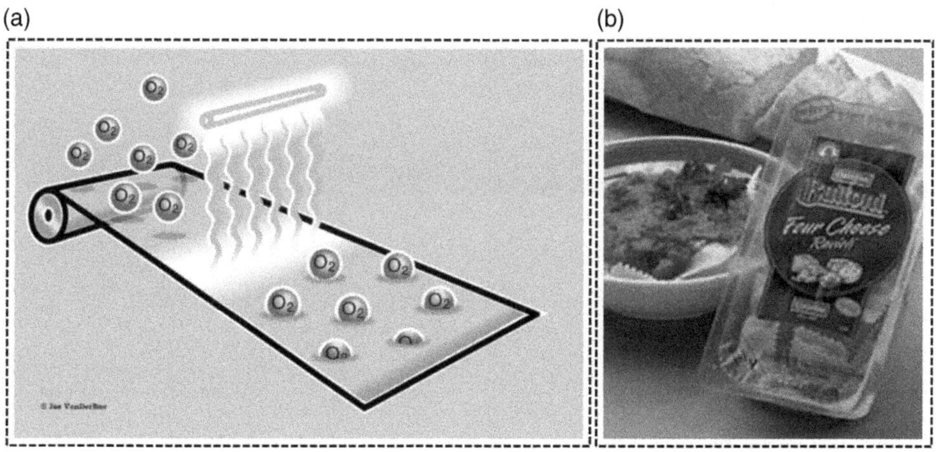

Figure 4.6 Cryovac Oxygen scavenging films (a) being activated by light-activated. *Source:* Cruz et al. [32]/IntechOpen/CC BY 3.0. (b) Used for packaging cheese-containing foods. *Source:* Reproduced with permission from Cooksey [16]/John Wiley & Sons.

photo-initiator (Figure 4.7a), it can remove the atmospheric oxygen (20.9%) within 60 days at 25 °C. At 45 °C, oxygen from the headspace can be removed entirely within 20 days. However, scavenging cannot occur at lower temperatures (5 °C). In the case of the *cis*-polyisoprene oxygen scavenging mechanism, acetophenone is added to the polymer to activate in the presence of UV light, followed by the generation of triplet oxygen (free radicals). These triplet oxygens are then converted to singlet oxygen with transition metals catalysts such as manganese chloride. Simultaneously, the singlet oxygen is reacted with *cis*-polyisoprene favoring the α-scission (Figure 4.7b) [49]. Similarly, *trans*-polyisoprene also acts as a polymer-based oxygen scavenger. *Trans*-polyisoprene blended with LDPE (20 wt.%) needs a transition metal catalyst such as cobalt chloride. The cobalt chloride will produce free radicals, as seen in the case of *cis*-polyisoprene.

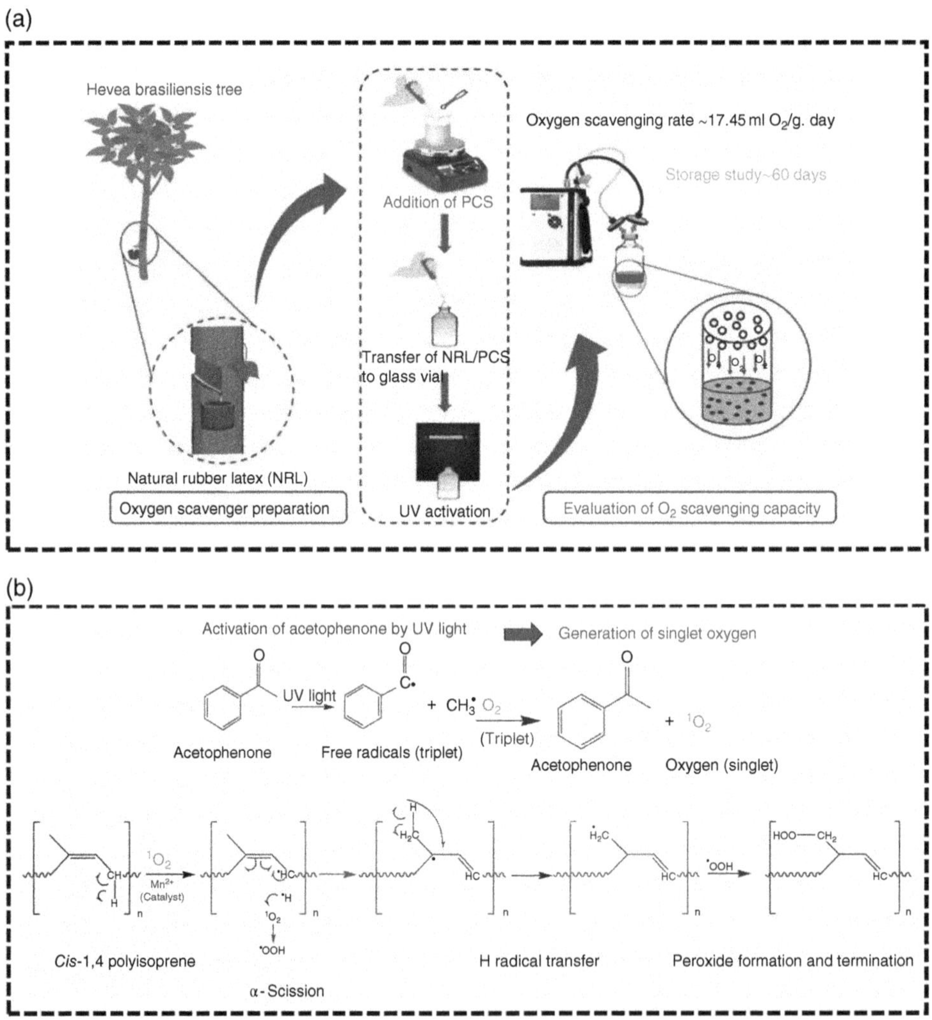

Figure 4.7 (a) Composition of *cis*-isoprene based oxygen scavenger. (b) Reaction mechanism of oxygen scavenging activity of *cis*-polyisoprene. *Source:* Reproduced from Ramakanth et al. [49]/with permission of Elsevier.

This system can consume oxygen from 21.1% to 15.53% in eight days with a scavenging capacity of 16.72 ml O_2/g. These films can be used to maintain the qualitative properties of roasted peanuts [45] and beef jerky by reducing the aerobic microbial growth as the oxygen concentration in jerky beef headspace was lowered with time [50].

4.2.5 Microorganism-Based Oxygen Scavenger System

Microorganisms that take up oxygen from the environment can be utilized as oxygen scavenging systems by entrapping in a polymeric matrix [4]. These biological oxygen scavengers are effective in preserving food items, safe to use, consumer-friendly, and affordable [32]. Microorganisms such as bacteria (*Pseudomonas putida*, *Bacillus amyloliquefaciens*, *Kocuria varians*), yeast (*Saccharomyces cerevisiae*, *Pichia subpelliculosa*) and fungi (*Aspergillus niger*) can be used as oxygen scavenger system [4, 51]. For uniformity of these microorganisms in the package, the immobilization can be done on the interior surface of a container and even on surfaces of a closure. Yeast can be used as a microbial oxygen scavenger by immobilization in wax and can be activated with moisturizer or ascorbic acid so that it respires and consumes oxygen. The advantage of immobilized yeast is that if the closed container requires to be pasteurized, the yeast is still viable enough to extract oxygen. If entrapped in wax, yeast can scavenge oxygen from beer bottles from 1.8 to 1.2 ppm in 19 days [4]. In particular, the oxygen uptake rate of *Saccharomyces cerevisiae* yeast is about 1.14 μl/min/mg when immobilized in calcium-alginate gel [5]. Immobilized *Saccharomyces cerevisiae*, if coated onto the bottle closure, can be used for beer packaging as, during respiration, it will consume oxygen while producing carbon dioxide and alcohol. Therefore, this type of oxygen scavenger can be added to alcoholic beverages, such as beer, without producing determinate changes [52]. Apart from yeast, some bacteria, such as *Aspergillus niger*, can also be immobilized in the calcium-alginate gel. It has an oxygen uptake rate of 0.65 $\mu l\,min^{-1}\,mg^{-1}$ [51].

Apart from immobilization in solid wax, the microorganism can also be compounded in flexible films such as hydroxyethyl cellulose and polyvinyl alcohol. Yeast (*Pichia subpelliculosa*) and bacteria (*Kocuria varians*) have been entrapped in these films for oxygen scavenging. Polyvinyl alcohol has higher macromolecular mobility than hydroxyethyl cellulose; therefore, the entrapment of microorganisms is much better. Therefore, polyvinyl alcohol shows better oxygen scavenging [53]. To prolong the viability of the microorganisms in the films, the manufacturing cycle of films, their storage, and the rehydration, i.e. the contact of films with the foodstuff, must be controlled [32].

The limitation of entrapped microorganism systems is the heat liability of integrated biological compounds during plastic containers such as PET bottles manufacturing at high temperatures and pressure. This restricts their use in food packaging because currently available biological systems are mixed into water-soluble polymers, gels, or low-melting wax or paraffin. For these cases, bacterial endospores are more appropriate for direct incorporation into a plastic material because these spores are typically recognized to have high resistance to dry heat and pressure. Bacterial endospores such as bacillus amyloliquefaciens are heat resistant and can be used as an active component in poly(ethylene terephthalate, 1,4-cyclohexane dimethanol (PETG) at 210 °C (Figure 4.8). The oxygen scavenging capacity for this system is 0.10 ± 0.02 ml O_2/g-day. PETG significantly

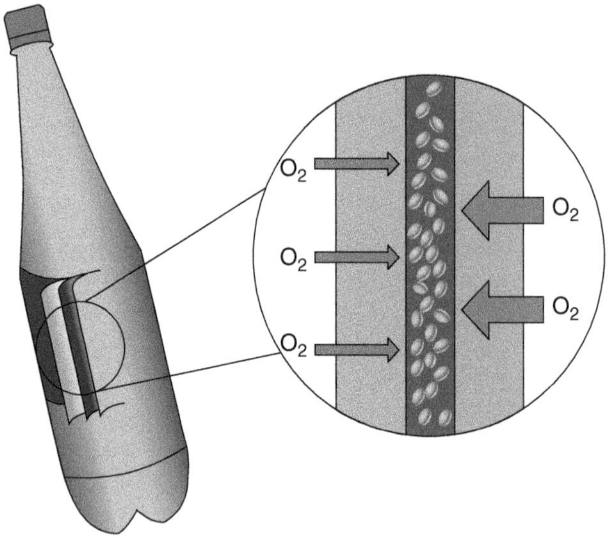

Figure 4.8 PETG-based bottle containing bacterial spores. *Source:* Reproduced from Anthierens et al. [54]/with permission of Elsevier.

absorbs more moisture at 30 °C, making it a better carrier material for spore rehydration and germination. Therefore, spores germinate when water is absorbed from the food into the PETG matrix. This method presents some significant advantages over the existing chemical solutions, including being a biologically based system and having the ability to deal with problems with polymer compatibility and recyclability. In addition, the ability of the new system to incorporate the active ingredient directly into the polymer matrix eliminates the need for coatings or sachets, which is a clear advantage over previously described bio-catalytic systems [54].

4.3 Loading of Oxygen Scavengers in Active Packaging

The oxygen scavengers can be loaded into the packaging in two different ways, (i) sachets, where the active compound is not in contact with the primary package but kept inside the package beside the food products, and (ii) to incorporate active compounds in the primary package, which can be done by blending, multilayer, coating, or immobilization, as shown in Figure 4.9.

4.3.1 Sachets or Labels

In most cases, sachets are the most frequently used forms of loading technique of oxygen scavengers in active food packaging technology. Metal-based oxygen scavengers such as iron are widely used and loaded into the sachets placed inside the primary packaging. These sachets are made up of materials highly permeable to oxygen so that oxygen can interact with the metal powder packed inside the sachets. Sachets containing metal powder can lower the oxygen level below 0.01% and therefore, are used in various foods such as meat, cheese, bakery products, nuts, and chips [11, 37, 55]. Labels incorporated with oxygen scavengers are another form of the self-adhesive oxygen-absorbing system used in

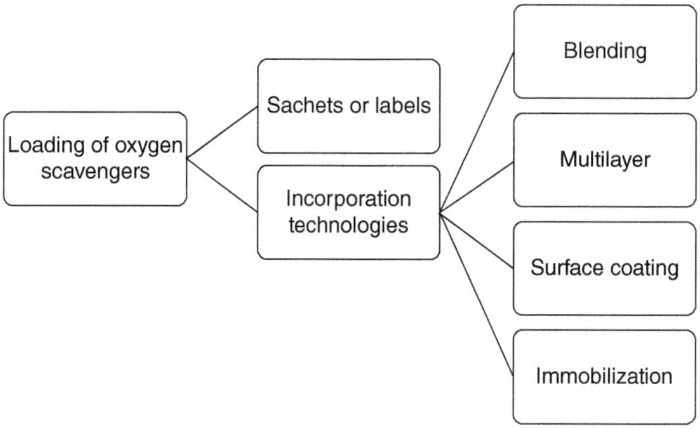

Figure 4.9 Loading techniques of oxygen scavenger in packaging.

active packaging technology. These labels adhere to the inside wall of the packaging to minimize the oxygen level present in the packaging system [31, 55]. This type of oxygen scavenging technology is used successfully in packaging various food products such as fish, cooked and cured meat, poultry, bakery, and pizzas [56]. The sachets and labels are affordable, but they also have several demerits, including accidental release and ingestion by the consumer resulting in health concerns. Moreover, these sachets and labels are unsuitable for liquid products [31, 55].

4.3.2 Incorporation Techniques

Incorporating active compounds into the packaging material is an alternative loading technique that can be used even for liquid products [57]. Adding oxygen scavenging compounds directly into packaging material instead of sachets solves many problems. For example, in tightly packed food products such as cheese packets, sachets do not help to remove oxygen because this tight packaging inhibits the functionality of sachets, whereas incorporating oxygen-absorbing compounds in this packaging is more effective. This technique can be used in bottles, containers, and closures [57]. There are different incorporation methods developed according to the physiochemical properties of both scavengers and polymer matrix, which include direct blending of oxygen scavengers to the polymer film or matrix, implementation of multilayer structure, spreading of oxygen scavengers as a thin coating layer, and immobilization or fixing of oxygen scavengers onto the inner side of the packaging.

4.3.2.1 Coatings

Coatings are usually applied in the liquid state and are usually nonsupporting layers applied on the surface of the substrate to impart some technical superiority (Figure 4.10a). Food packaging surfaces are coated with oxygen scavengers to reduce the oxygen from the headspace, which can be done by coating a solution and evaporating the solvent [17]. The active agents work either by contacting food materials or without contact with food items by producing a vapor phase that transfers oxygen scavenging properties to the outer

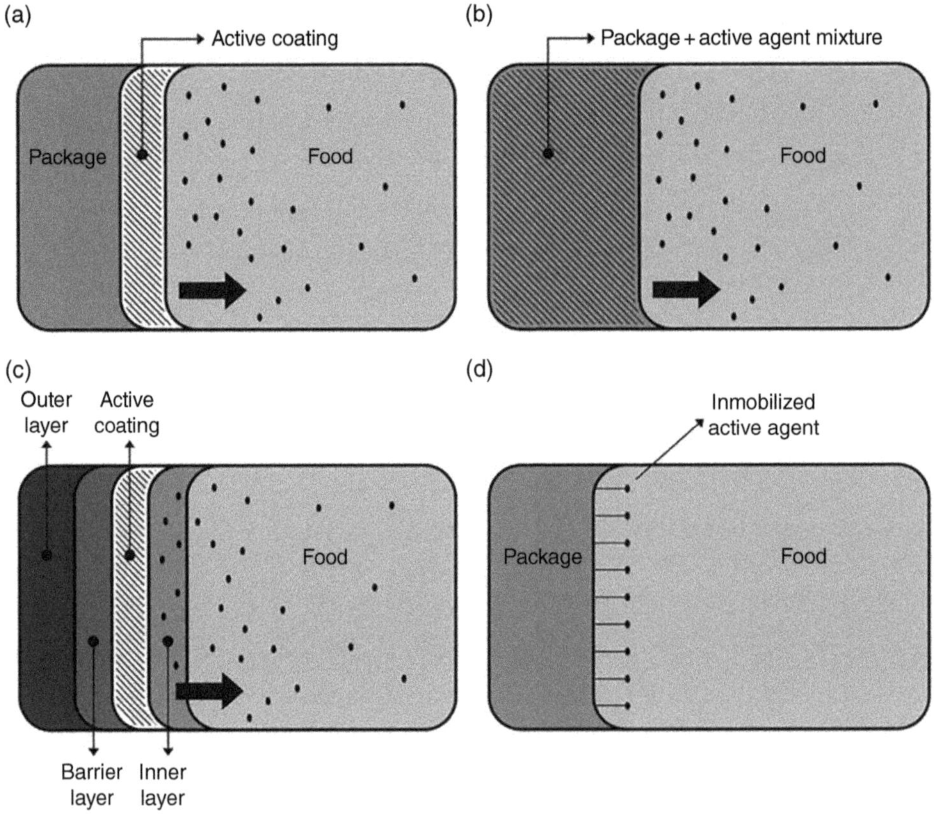

Figure 4.10 Different routes of incorporating oxygen scavenger in primary package. (a) Active coating. (b) Blending in polymer. (c) Multilayer structure. (d) Immobilization of enzyme. *Source:* Reproduced from María Roberta [34]/with permission of Elsevier.

surface [58]. Oxalate oxidase or glucose oxidase trapped in coating based on the polymeric material are examples that can be used as oxygen scavengers in active food packaging systems [38, 59, 60] and help to evaluate the effectiveness of oxygen depleting property of a coating. Coating techniques reduce the thickness or surface area of the coating, which allows the use of a less amount of oxygen scavenger. However, good compatibility of the active phase should be maintained between the solvent and polymer matrix. To avoid delamination of the coating layer from the substrate surface, there must be a good affinity between the coating layer and the substrate surface [35].

4.3.2.2 Blending

Blending oxygen scavenging material in the polymer matrix is one of the efficient ways to use oxygen scavengers (Figure 4.10b). One of the simple ways to add oxygen scavengers into a barrier layer is by directly blending it with a polymer matrix [61]. This technique is suitable for developing the layer either by extrusion [58, 62–64] or by casting method [29, 65, 66]. An incomplete blending of these polymers results in the formation of particles, usually spherical within the dispersing matrix, whose size varies depending on mass ratios, rheology, and compounding conditions [67]. Therefore, extrusion blends oxygen scavenging polymers with insufficient structural properties [35]. Literature in the past mentioned numerous examples of different polymers containing polybutadiene-based oxygen scavengers blended with polyethylene terephthalate focusing on rigid applications for wine and juice [68–72]. The blending of oxygen scavenger seems to be the best incorporation technique as it is mixed with polymer matrix efficiently; however, it can have an adverse effect on the mechanical properties of the packaging materials. Oxygen scavengers, when blended with packaging, can affect the structural behavior of the package as the oxygen scavenger is modified with time. The scavenging reaction can affect the active system's rheological behavior and structural modification. The oxygen scavenger's compatibility with the polymer matrix and its thermal stability is essential for selecting the extrusion configuration and process conditions [73]. Moreover, the scavenging performance and functional properties are affected by these parameters.

4.3.2.3 Multilayer Structure

Multilayer structures depict flexible and integrated systems of different layers, each with different functions, which cannot be achieved by a single-layer system (Figure 4.10c). These multilayer systems can be used to further extend the shelf life of food and, at the same time, maintain user-centric properties of packaging such as printability, mechanical properties, sealability, and transparency. In a multilayer system, an oxygen layer is inserted between two inert layers [58]; this system overcomes some issues but is costlier. Multilayer iron-based packaging film has been commercialized and sandwiched between multiple inert layers of packaging [74]. Similarly, gallic acid-based oxygen scavengers have been extruded and laminated in a polymer matrix to develop an oxygen scavenging multilayer package [25]. This method is not widely used as this method of loading oxygen scavenger is costly, and the manufacturing process is complex [58].

4.3.2.4 Immobilization

Immobilization involves the fixation of oxygen scavengers on the packaging surface; this technique is commonly applied to enzyme-based oxygen scavengers (Figure 4.10d). Enzyme immobilization is attaching enzymes to the surface or a matrix. It can be achieved by either physical adsorption, chemical method, entrapment, or using both methods to acquire full activity. One way to immobilize the enzyme is through chemical immobilization, which involves a covalent bond between the enzyme and the polymer that usually occurs by chemical activation of the package surface and protein nucleophilic group's attachment of the enzyme [35]. The immobilized enzymes can be activated by UV irradiation [75]. In this case, the immobilization technique is irreversible, and the enzyme's movement to food is intended. In some exceptional cases, crosslinkers such as glyoxal, formaldehyde, glutaraldehyde, or transglutaminase attach enzymes to the packaging's surface. Weak secondary forces, such as hydrogen bonds, van der Waals forces, and ionic or hydrophobic interaction, can also be utilized to adsorb the enzymes onto the packaging surfaces. Alternatively, enzymes can be encapsulated in droplets to immobilize and coat on package. Moreover, this technique is used for the development of antimicrobial applications based on enzymes [35].

4.4 Applications of Oxygen Scavengers in Active Packaging

The presence of oxygen in packaging can lead to a reduction in the nutritional value of food products, cause fat rancidity, discoloration, and promote bacteria or mold growth [76]. The presence of residual oxygen in the headspace of packaging also causes degradation in proteins [77]. If present inside the package, oxygen scavengers can scavenge the residual oxygen. They thus can be used for various applications such as (i) Fruits and vegetables, (ii) Dairy products, (iii) Muscle foods, and (iv) Bakery foods.

4.4.1 Fruits and Vegetables

Fruits and vegetables provide human nutrition through protein, vitamins, dietary fibers, fatty acids, and antioxidants [78]. The presence of oxygen has some adverse effects on fruits and vegetables; it results in the loss of Vitamin C or ascorbic acid, speeding up the respiration of fresh fruits and vegetables. It also leads to enzymatic or nonenzymatic phenolic browning of fruits and vegetables [79]. For various fresh fruits and vegetables, oxygen scavenging technology is progressing and is essential because it results in delaying discoloration and ripening and lowering the respiration rate. It also helps prevent the growth of spoilage microorganisms and reduces off-odor [19]. Oxygen scavengers are used worldwide to maintain the overall quality of slightly processed vegetables and fruits and lengthen their shelf life [80]. Oxygen scavengers are effective for fresh-cut vegetables and fruits [79]. Table 4.2 shows different fruits and vegetables and the oxygen scavengers used to package these food products. Oxygen scavengers are proven to be advantageous for various fruits and vegetables such as apples, litchi, peeled garlic, Leafy greens, strawberries, and blueberries. These oxygen scavengers are beneficial in extending shelf life and preserving the color, flavor, and texture of various fruits and vegetables.

4.4.2 Dairy Products

Dairy products such as yogurt, butter, cheese, and milk are the principal sources of different essential nutrients [56]. When dairy products such as dry powdered milk, yogurt, cheese, butter, and milk are stored and packed in packaging film, they should not have any contact with oxygen; otherwise, it leads to oxidation and spoilage of the given dairy product. Oxygen

Table 4.2 Different oxygen scavengers in fruits and vegetables with their benefits.

Food	Oxygen scavenger	Benefits	Reference
Apple	Amosorb DFC 4020	Increases shelf life, preserves color and flavor	[81]
Litchi	Pyrogallol	Pyrogallol coating extends the shelf life of litchi fruit	[82]
Peeled garlic	Pyrogallol	Prevents browning and oxidation, maintains texture and flavor	[83]
Strawberry	Iron-based	Increases antioxidant activity and extends shelf life	[84]

Table 4.3 Different oxygen scavengers in dairy products with their benefits.

Food	Oxygen scavenger	Benefits	Reference
Milk	Glucose oxidase in small packets	Prevents oxidation, extends shelf life	[85]
Cheese	Iron-based	Inhibits spoilage organisms and extends shelf	[86]
Yogurt	Iron and ascorbic acid-based	Inhibits bacterial growth and extends shelf life	[87]
Butter	Iron-based	Prevents oxidation, extends shelf life	[88]
Ice cream	Glucose oxidase in small packets	Inhibits spoilage organisms and extends shelf	[85]

scavenging technology helps in protecting dairy products against oxygen's adverse effects, prolongs their shelf life, and reduces wastage [77]. Different dairy products such as milk, cheese, yogurt, butter, and ice cream, along with different types of oxygen scavengers used and their benefits, are mentioned as under in Table 4.3. For milk and ice creams, glucose oxidase in small packets is used to prevent oxidation, resulting in prolonged shelf life. For cheese, butter, and yogurt iron-based oxygen scavengers are used to inhibit spoilage bacteria growth.

4.4.3 Muscle Foods

Meat and poultry are generally packaged in a vacuum; however, oxygen might enter the package during the supply chain, resulting in the growth of microbes, discoloration, and loss of nutritional value. It also results in the development of off-odor and off-flavor. All these factors cause deterioration in overall food quality, shelf life, and safety [56]. Therefore, oxygen scavengers are preferred for muscle foods to maintain flavor, odor, and appearance. It also helps stabilize the product's composition and helps delay the spoilage caused by microbial growth [89]. Table 4.4 represents various muscle foods such as meat, fish, beef, pork, and ham, the respective oxygen scavengers used in their packaging, and the benefits. For meat products, sodium tripolyphosphate is used as an oxygen scavenger to improve meat tenderness and lengthen shelf life. For fish, beef, and pork, commercial oxygen scavengers are used to maintain their color and help in extending shelf life. Similarly, palladium-based oxygen scavengers are used for ham packaging to extend shelf life.

Table 4.4 Different oxygen scavengers in muscle foods with their benefits.

Food	Oxygen scavenger	Benefits	Reference
Meat	Iron-based	Improves meat tenderness and extends shelf life	[90]
Fish	Iron-based	Reduces oxygen concentration and shelf life	[91]
Beef	Polyisoprene	Maintain oxidative stability of beef	[50]
Ham	Palladium-based	Preventing discoloration of ham	[92]

Table 4.5 Different oxygen scavengers in bakery and other food products with their benefits.

Food	Oxygen scavenger	Benefits	Reference
Bun and bread	Zinc, iron, or ascorbic acid	Prevents mold growth and extends shelf life	[21]
Pastry	Palladium-based	Prevents mold growth and extends shelf life	[95]
Nuts	Iron-based	Prevents oxidation and extends shelf life	[18]
Peanuts	Ascorbate	Prevents oxidation and extends shelf life	[96]
Dried fruits	Oxygen scavengers dispersed polypropylene	Prevents oxidation and extends shelf life	[97]

4.4.4 Bakery and Other Food Products

The presence of oxygen can lead to narrowing the shelf life of various bakery items by three factors: the aerobic microorganism's growth that can cause deterioration, the chemical effects of oxygen, and insect pest attacks. These factors can cause odor, flavor, and color changes if present alone or with one another. Hence affecting the overall quality of food. However, oxidation is the leading cause of degradation in bakery products [93]. Oxygen scavengers are used as they help to extend their shelf life, preserve their freshness, and prevent spoilage [56]. If in other food products packaging, such as dried fruits, roasted peanuts, nuts, and snack foods, oxygen is present, it can cause oxidation of lipids, growth of microbes, browning of enzymes, and nutritional loss, which results in declining the shelf life of such products [94]. Different bakery and other food products such as a bun, bread, pastry, nuts, peanuts, and dried fruits, along with respective oxygen scavengers and their benefits, are depicted in Table 4.5; for bread and buns, Zinc, Iron, or ascorbic acids are used as oxygen scavengers to prevent the growth of molds and lengthen the shelf life. Palladium-based and iron-based oxygen scavengers are used for pastries and nuts to enhance the quality by prolonging the shelf life. For peanuts, sodium ascorbate is used as an oxygen scavenger to lengthen the shelf life by preventing oxidation. For dry fruits, oxygen scavengers dispersed polypropylene has proven to extend the shelf life.

4.5 Conclusion and Future Perspective

Oxygen scavengers are crucial for preserving the quality and extending the shelf life of packaged food products by reducing the adverse effects of oxygen. While metal-based oxygen scavengers are currently the most widely commercialized form, their limitations, such as toxicity and unsuitability for liquid foods, have spurred the development of bio-based and polymer-based alternatives. Bio-based and polymer-based oxygen scavengers offer advantages in terms of nontoxicity and ease of blending or coating on packaging materials. At the same time, they require transition metal catalysts for optimal performance. Enzyme and microbial-based oxygen scavengers, although less investigated, are commercially challenging due to production, formulation, storage, and regulatory considerations, and they may hold potential for future advancements. The future of oxygen scavengers in food packaging may involve reducing or eliminating transition metal-based catalysts, which are currently added to many oxygen scavenger formulations. Research efforts may focus on finding alternative catalysts or catalyst-free systems to address concerns related to toxicity and environmental impact associated with transition metal-based catalysts. This could pave the way for future sustainable and eco-friendly oxygen scavenger technologies, aligning with the growing demand for safer and more environmentally conscious packaging solutions in the food industry.

Acknowledgments

Author Arihant Ahuja would like to thank the Indian Institute of Technology Roorkee and Prime Minister Research Fellowship for providing online resources and support during this chapter writing.

References

1 Kumar, H., Ahuja, A., Kadam, A.A. et al. (2023). Antioxidant film based on chitosan and tulsi essential oil for food packaging. *Food and Bioprocess Technology* 16 (2): 342–355. https://doi.org/10.1007/S11947-022-02938-6.

2 Yildirim, S., Röcker, B., Pettersen, M.K. et al. (2018). Active packaging applications for food. *Comprehensive Reviews in Food Science and Food Safety* 165–199. https://doi.org/10.1111/1541-4337.12322.

3 Sharma, P., Ahuja, A., Dilsad Izrayeel, A.M. et al. (2022). Physicochemical and thermal characterization of poly(3-hydroxybutyrate-co-4-hydroxybutyrate) films incorporating thyme essential oil for active packaging of white bread. *Food Control* 133: 108688. https://doi.org/10.1016/j.foodcont.2021.108688.

4 Dey, A. and Neogi, S. (2019). Oxygen scavengers for food packaging applications: A review. *Trends in Food Science & Technology* 90: 26–34. https://doi.org/10.1016/j.tifs.2019.05.013.

5 Pereira de Abreu, D.A., Cruz, J.M., and Paseiro Losada, P. (2012). Active and intelligent packaging for the food industry. *Food Reviews International* 28 (2): 146–187. https://doi.org/10.1080/87559129.2011.595022.

6 Westlake, J.R., Tran, M.W., Jiang, Y. et al. (2022). Biodegradable active packaging with controlled release: principles, progress, and prospects. *ACS Food Science & Technology* 2 (8): 1166–1183. https://doi.org/10.1021/acsfoodscitech.2c00070.

7 Gupta, P. (2023). Role of oxygen absorbers in food as packaging material, their characterization and applications. *Journal of Food Science and Technology* 1: 1–11. https://doi.org/10.1007/s13197-023-05681-8.

8 Ladikos, D. and Lougovois, V. (1990). Lipid oxidation in muscle foods: a review. *Food Chemistry* 35 (4): 295–314. https://doi.org/10.1016/0308-8146(90)90019-Z.

9 Robards, K., Kerr, A.F., and Patsalides, E. (1988). Rancidity and its measurement in edible oils and snack foods. A review. *The Analyst* 113 (2): 213. https://doi.org/10.1039/an9881300213.

10 Cherpinski, A., Szewczyk, P., Gruszczyński, A. et al. (2019). Oxygen-scavenging multilayered biopapers containing palladium nanoparticles obtained by the electrospinning coating technique. *Nanomaterials* 9 (2): 262. https://doi.org/10.3390/nano9020262.

11 Ozdemir, M. and Floros, J.D. (2004). Active food packaging technologies. *Critical Reviews in Food Science and Nutrition* 44 (3): 185–193. https://doi.org/10.1080/10408690490441578.

12 Rooney, M.L. (2005). Oxygen-scavenging packaging. In: *Innovations in Food Packaging*, 123–137. Elsevier https://doi.org/10.1016/B978-012311632-1/50040-1.

13 Johnson, D.R., Inchingolo, R., and Decker, E.A. (2018). The ability of oxygen scavenging packaging to inhibit vitamin degradation and lipid oxidation in fish oil-in-water emulsions. *Innovative Food Science & Emerging Technologies* 47: 467–475. https://doi.org/10.1016/j.ifset.2018.04.021.

14 Sanchez, S.A., Nunberg, S., Cnossen, K., and Eckelman, M.J. (2023). Life cycle assessment of anoxic treatments for cultural heritage preservation. *Resources, Conservation and Recycling* 190: 106825. https://doi.org/10.1016/j.resconrec.2022.106825.

15 Brody, A.L. (2014) Micro-oxygen: the next giant step in packaged food preservation. https://www.ift.org/~/media/food technology/pdf/2014/03/0314_col_packaging.pdf (accessed 29 March 2023).

16 Cooksey, K. (2010). Oxygen scavenging packaging systems, encyclopedia of polymer science and technology, Hoboken, NJ, USA. https://onlinelibrary.wiley.com/doi/full/10.1002/0471440264.pst570 (accessed 29 March 2023).

17 Demicheva, M. (2015). Novel oxygen scavenger systems for functional coatings. Arcada – Nylands svenska yrkeshögskola http://www.theseus.fi/handle/10024/87205 (accessed 28 March 2023).

18 Mu, H., Gao, H., Chen, H. et al. (2013). A nanosised oxygen scavenger: preparation and antioxidant application to roasted sunflower seeds and walnuts. *Food Chemistry* 136 (1): 245–250. https://doi.org/10.1016/j.foodchem.2012.07.121.

19 Cichello, S.A. (2015). Oxygen absorbers in food preservation: a review. *Journal of Food Science and Technology* 52 (4): 1889–1895. https://doi.org/10.1007/s13197-014-1265-2.

20 Calderon, V. and S., Gomes, B., Ferreira, P.J. and Carvalho, S. (2017). Zinc nanostructures for oxygen scavenging. *Nanoscale* 9 (16): 5254–5262. https://doi.org/10.1039/C7NR01367A.

21 Matche, R.S., Sreekumar, R.K., and Raj, B. (2011). Modification of linear low-density polyethylene film using oxygen scavengers for its application in storage of bun and bread. *Journal of Applied Polymer Science* 122 (1): 55–63. https://doi.org/10.1002/app.33718.

22 Yildirim, S., Röcker, B., Rüegg, N., and Lohwasser, W. (2015). Development of palladium-based oxygen scavenger: optimization of substrate and palladium layer thickness. *Packaging Technology and Science* 28 (8): 710–718. https://doi.org/10.1002/pts.2134.

23 Mills, A., Doyle, G., Peiro, A.M., and Durrant, J. (2006). Demonstration of a novel, flexible, photocatalytic oxygen-scavenging polymer film. *Journal of Photochemistry and Photobiology A: Chemistry* 177 (2–3): 328–331. https://doi.org/10.1016/j.jphotochem.2005.06.001.

24 Tulsyan, G., Richter, C., and Diaz, C.A. (2017). Oxygen scavengers based on titanium oxide nanotubes for packaging applications. *Packaging Technology and Science* 30 (6): 251–256. https://doi.org/10.1002/pts.2296.

25 Pant, A., Sängerlaub, S., and Müller, K. (2017). Gallic acid as an oxygen scavenger in bio-based multilayer packaging films. *Materials* 10 (5): 489. https://doi.org/10.3390/ma10050489.

26 Pant, A.F., Dorn, J., and Reinelt, M. (2018). Effect of temperature and relative humidity on the reaction kinetics of an oxygen scavenger based on gallic acid. *Frontiers in Chemistry* 6: https://doi.org/10.3389/fchem.2018.00587.

27 Ahn, B.J., Gaikwad, K.K., and Lee, Y.S. (2016). Characterization and properties of LDPE film with gallic-acid-based oxygen scavenging system useful as a functional packaging material. *Journal of Applied Polymer Science* 133 (43): https://doi.org/10.1002/app.44138.

28 Gaikwad, K.K., Singh, S., and Lee, Y.S. (2017). A pyrogallol-coated modified LDPE film as an oxygen scavenging film for active packaging materials. *Progress in Organic Coatings* 111: 186–195. https://doi.org/10.1016/j.porgcoat.2017.05.016.

29 Janjarasskul, T., Tananuwong, K., and Krochta, J.M. (2011). Whey protein film with oxygen scavenging function by incorporation of ascorbic acid. *Journal of Food Science* 76 (9): E561–E568. https://doi.org/10.1111/j.1750-3841.2011.02409.x.

30 Lee, J.-S., Chang, Y., Lee, E.-S. et al. (2018). Ascorbic acid-based oxygen scavenger in active food packaging system for raw meatloaf. *Journal of Food Science* 83 (3): 682–688. https://doi.org/10.1111/1750-3841.14061.

31 Byun, Y., Darby, D., Cooksey, K. et al. (2011). Development of oxygen scavenging system containing a natural free radical scavenger and a transition metal. *Food Chemistry* 124 (2): 615–619. https://doi.org/10.1016/j.foodchem.2010.06.084.

32 Cruz, R.S., Camilloto, G.P., and dos Santos Pires, A.C. (2012). Oxygen scavengers: an approach on food preservation. In: *Structure and Function of Food Engineering* (ed. A.A. Eissa). IntechOpen. https://doi.org/10.5772/48453.

33 Vermeiren, L., Heirlings, L., Devlieghere, F., and Debevere, J. (2003). Oxygen, ethylene and other scavengers. In: *Novel Food Packaging Techniques* (ed. R. Ahvenainen), 22–49. Elsevier. https://doi.org/10.1533/9781855737020.1.22.

34 María Roberta, A. (2020). *Oxygen Scavenging Films and Coating of Biopolymers for Food Application*. Elsevier. https://doi.org/10.1016/B978-0-12-818134-8.00022-5.

35 Apicella, A. and Incarnato, L. (2019). Oxygen scavengers in food packaging. In: *Reference Module in Food Science*. Elsevier. https://doi.org/10.1016/B978-0-08-100596-5.22942-9.

36 Shi, X., Lim, J., and Ha, T. (2010). Acidification of the oxygen scavenging system in single-molecule fluorescence studies: in situ sensing with a ratiometric dual-emission probe. *Analytical Chemistry* 82 (14): 6132–6138. https://doi.org/10.1021/ac1008749.

37 Vermeiren, L., Devlieghere, F., van Beest, M. et al. (1999). Developments in the active packaging of foods. *Trends in Food Science & Technology* 10 (3): 77–86. https://doi.org/10.1016/S0924-2244(99)00032-1.

38 Winestrand, S., Johansson, K., Järnström, L., and Jönsson, L.J. (2013). Co-immobilization of oxalate oxidase and catalase in films for scavenging of oxygen or oxalic acid. *Biochemical Engineering Journal* 72: 96–101. https://doi.org/10.1016/j.bej.2013.01.006.

39 Labuza, T.P. and Breene, W.M. (1989). Applications of "active packaging" for improvement of shelf-life and nutritional quality of fresh and extended shelf-life foods 1. *Journal of food processing and preservation* 13 (1): 1–69.

40 Johansson, K., Gillgren, T., Winestrand, S. et al. (2014). Comparison of lignin derivatives as substrates for laccase-catalyzed scavenging of oxygen in coatings and films. *Journal of Biological Engineering* 8 (1): 1. https://doi.org/10.1186/1754-1611-8-1.

41 Johansson, K., Winestrand, S., Johansson, C. et al. (2012). Oxygen-scavenging coatings and films based on lignosulfonates and laccase. *Journal of Biotechnology* 161 (1): 14–18. https://doi.org/10.1016/j.jbiotec.2012.06.004.

42 Winestrand, S., Järnström, L., and Jönsson, L.J. (2021). Fractionated lignosulfonates for laccase-catalyzed oxygen-scavenging films and coatings. *Molecules* 26 (20): 6322. https://doi.org/10.3390/molecules26206322.

43 Chatterjee, R., Johansson, K., Järnström, L., and Jönsson, L.J. (2011). Evaluation of the potential of fungal and plant laccases for active-packaging applications. *Journal of Agricultural and Food Chemistry* 59 (10): 5390–5395. https://doi.org/10.1021/jf103811g.

44 González, A. and Alvarez Igarzabal, C.I. (2013). Soy protein – poly(lactic acid) bilayer films as biodegradable material for active food packaging. *Food Hydrocolloids* 33 (2): 289–296. https://doi.org/10.1016/j.foodhyd.2013.03.010.

45 Gaikwad, K.K., Singh, S., Negi, Y.S., and Lee, Y.S. (2020). The effect of trans-polyisoprene/LDPE based active films on oxidative stability in roasted peanuts. *Journal of Food Measurement and Characterization* 14 (4): 1857–1864. https://doi.org/10.1007/s11694-020-00433-0.

46 Li, H., Tung, K.K., Paul, D.R. et al. (2012). Characterization of oxygen scavenging films based on 1,4-polybutadiene. *Industrial & Engineering Chemistry Research* 51 (21): 7138–7145. https://doi.org/10.1021/ie201905j.

47 Juan-Polo, A., Maestre Pérez, S.E., Monedero Prieto, M. et al. (2021). Double-function oxygen scavenger and aromatic food packaging films based on LDPE/polybutadiene and peanut aroma. *Polymers* 13 (8): 1310. https://doi.org/10.3390/polym13081310.

48 Kordjazi, Z. and Ajji, A. (2021). Development of TiO_2 catalyzed HTPB based oxygen scavenging films for food packaging applications. *Food Control* 121: 107639. https://doi.org/10.1016/j.foodcont.2020.107639.

49 Ramakanth, D., Akhila, K., Gaikwad, K.K., and Maji, P.K. (2022). UV-activated oxygen scavenging system based on natural rubber latex from *Hevea brasiliensis* for active packaging applications. *Industrial Crops and Products* 178: 114658. https://doi.org/10.1016/j.indcrop.2022.114658.

50 Gaikwad, K.K., Singh, S., Shin, J., and Lee, Y.S. (2020). Novel polyisoprene based UV-activated oxygen scavenging films and their applications in packaging of beef jerky. *LWT* 117: 108643. https://doi.org/10.1016/j.lwt.2019.108643.

51 Gosmann, B. and Rehm, H.J. (1986). Oxygen uptake of microorganisms entrapped in Ca-alginate. *Applied Microbiology and Biotechnology* 23 (3–4): 163–167. https://doi.org/10.1007/BF00261907.

52 Edens, L., Farin, F., Ligtvoet, A.F. and Van der Platt, J.B. (1991) Dry yeast immobilized in wax or paraffin for scavenging oxygen. US Patent. https://patents.google.com/patent/US5106633A/en (accessed March 2023).

53 Altieri, C., Sinigaglia, M., Corbo, M. et al. (2004). Use of entrapped microorganisms as biological oxygen scavengers in food packaging applications. *LWT- Food Science and Technology* 37 (1): 9–15. https://doi.org/10.1016/S0023-6438(03)00115-4.

54 Anthierens, T., Ragaert, P., Verbrugghe, S. et al. (2011). Use of endospore-forming bacteria as an active oxygen scavenger in plastic packaging materials. *Innovative Food Science & Emerging Technologies* 12 (4): 594–599. https://doi.org/10.1016/j.ifset.2011.06.008.

55 Tian, F., Decker, E.A., and Goddard, J.M. (2013). Controlling lipid oxidation of food by active packaging technologies. *Food & Function* 4 (5): 669–680. https://doi.org/10.1039/C3FO30360H.

56 Hogan, S.A. and Kerry, J.P. (2008). Smart packaging of meat and poultry products. In: *Smart Packaging Technologies for Fast Moving Consumer Goods* (ed. J. Kerry and P. Butler), 33–59. Wiley. https://doi.org/10.1002/9780470753699.

57 Nwakaudu, A.A., Nwakaudu, M.S., Owuamanam, C.I., and Iheaturu, N.C. (2015). The use of natural antioxidant active polymer packaging for food preservation: a review. *Applied Signals Reports* 2 (4): 38–50.

58 Gaikwad, K.K., Singh, S., and Youn, · and Lee, S. (2018). Oxygen scavenging films in food packaging. *Environmental Chemistry Letters* 16: 523–538. https://doi.org/10.1007/s10311-018-0705-z.

59 Gohil, R.M. and Wysock, W.A. (2014). Designing efficient oxygen scavenging coating formulations for food packaging applications. *Packaging Technology and Science* 27 (8): 609–623. https://doi.org/10.1002/PTS.2053.

60 Nestorson, A., Neoh, K.G., Kang, E.T. et al. (2008). Enzyme immobilization in latex dispersion coatings for active food packaging. *Packaging Technology and Science* 21 (4): 193–205. https://doi.org/10.1002/pts.796.

61 Ferrari, M.C., Carranza, S., Bonnecaze, R.T. et al. (2009). Modeling of oxygen scavenging for improved barrier behavior: blend films. *Journal of Membrane Science* 329 (1–2): 183–192. https://doi.org/10.1016/j.memsci.2008.12.030.

62 Byun, Y., Kim, Y.T., and Whiteside, S. (2010). Characterization of an antioxidant polylactic acid (PLA) film prepared with α-tocopherol, BHT and polyethylene glycol using film cast extruder. *Journal of Food Engineering* 100 (2): 239–244. https://doi.org/10.1016/j.jfoodeng.2010.04.005.

63 Jamshidian, M., Arab Tehrany, E., Cleymand, F. et al. (2012). Effects of synthetic phenolic antioxidants on physical, structural, mechanical and barrier properties of poly lactic acid film. *Carbohydrate Polymers* 87 (2): 1763–1773. https://doi.org/10.1016/j.carbpol.2011.09.089.

64 Marcos, B., Sárraga, C., Castellari, M. et al. (2014). Development of biodegradable films with antioxidant properties based on polyesters containing α-tocopherol and olive leaf extract for food packaging applications. *Food Packaging and Shelf Life* 1 (2): 140–150. https://doi.org/10.1016/j.fpsl.2014.04.002.

65 Byun, Y., Bae, H.J., and Whiteside, S. (2012). Active warm-water fish gelatin film containing oxygen scavenging system. *Food Hydrocolloids* 27 (1): 250–255. https://doi.org/10.1016/j.foodhyd.2011.06.010.

66 Fortunati, E., Luzi, F., Dugo, L. et al. (2016). Effect of hydroxytyrosol methyl carbonate on the thermal, migration and antioxidant properties of PVA-based films for active food packaging. *Polymer International* 65 (8): 872–882. https://doi.org/10.1002/pi.5090.

67 Apicella, A., Scarfato, P., Di Maio, L., and Incarnato, L. (2018). Transport properties of multilayer active PET films with different layers configuration. *Reactive and Functional Polymers* 127: 29–37. https://doi.org/10.1016/j.reactfunctpolym.2018.03.015.

68 Bacigalupi, C., Lemaistre, M.H., Boutroy, N. et al. (2013). Changes in nutritional and sensory properties of orange juice packed in PET bottles: an experimental and modelling approach. *Food Chemistry* 141 (4): 3827–3836. https://doi.org/10.1016/j.foodchem.2013.06.076.

69 Dombre, C., Guillard, V., and Chalier, P. (2015). Protection of methionol against oxidation by oxygen scavenger: an experimental and modelling approach in wine model solution. *Food Packaging and Shelf Life* 3: 76–87. https://doi.org/10.1016/j.fpsl.2015.01.002.

70 Di Felice, R., Cazzola, D., Cobror, S., and Oriani, L. (2008). Oxygen permeation in PET bottles with passive and active walls. *Packaging Technology and Science* 21 (7): 405–415. https://doi.org/10.1002/pts.820.

71 Ros-Chumillas, M., Belissario, Y., Iguaz, A., and López, A. (2007). Quality and shelf life of orange juice aseptically packaged in PET bottles. *Journal of Food Engineering* 79 (1): 234–242. https://doi.org/10.1016/j.jfoodeng.2006.01.048.

72 Sängerlaub, S. and Müller, K. (2017). Long-time performance of bottles made of PET blended with various concentrations of oxygen scavenger additive stored at different temperatures. *Packaging Technology and Science* 30 (1–2): 45–58. https://doi.org/10.1002/pts.2283.

73 Galdi, M.R. and Incarnato, L. (2011). Influence of composition on structure and barrier properties of active PET films for food packaging applications. *Packaging Technology and Science* 24 (2): 89–102. https://doi.org/10.1002/pts.917.

74 Shin, Y., Shin, J., and Lee, Y.S. (2011). Preparation and characterization of multilayer film incorporating oxygen scavenger. *Macromolecular Research* 19 (9): 869–875. https://doi.org/10.1007/s13233-011-0912-y.

75 Goddard, J.M. and Hotchkiss, J.H. (2007). Polymer surface modification for the attachment of bioactive compounds. *Progress in Polymer Science* 32 (7): 698–725. https://doi.org/10.1016/j.progpolymsci.2007.04.002.

76 Gaikwad, K.K. and Lee, Y.S. (2016). Novel natural phenolic compound-based oxygen scavenging system for active packaging applications. *Journal of Food Measurement and Characterization* 10 (3): 533–538. https://doi.org/10.1007/s11694-016-9332-1.

77 Foltynowicz, Z. (2020). Oxygen scavengers applications in the dairy industry. *Journal of Dairy Research & Technology* 3 (1): 1–6. https://doi.org/10.24966/DRT-9315/100016.

78 Deshwal, G.K., Tiwari, S., Panjagari, N.R., and Masud, S. (2021). Active packaging of fruits and vegetables: quality preservation and shelf-life enhancement. In: *Packaging and Storage of Fruits and Vegetables* (ed. T. Alam), 109–131. New York: Apple Academic Press. https://doi.org/10.1201/9781003161165-6.

79 Bodbodak, S. and Rafiee, Z. (2016). Recent trends in active packaging in fruits and vegetables. In: *Eco-Friendly Technology for Postharvest Produce Quality* (ed. M.W. Siddiqui), 77–125. Elsevier. http://doi.org/10.1016/B978-0-12-804313-4.00003-7.

80 Wilson, M.D., Stanley, R.A., Eyles, A., and Ross, T. (2019). Innovative processes and technologies for modified atmosphere packaging of fresh and fresh-cut fruits and vegetables. *Critical Reviews in Food Science and Nutrition* 59 (3): 411–422. https://doi.org/10.1080/10408398.2017.1375892.

81 Di Maio, L., Scarfato, P., Galdi, M.R., and Incarnato, L. (2015). Development and oxygen scavenging performance of three-layer active PET films for food packaging. *Journal of Applied Polymer Science* 132 (7): https://doi.org/10.1002/app.41465.

82 Jing, G., Huang, H., Yang, B. et al. (2013). Effect of pyrogallol on the physiology and biochemistry of litchi fruit during storage. *Chemistry Central Journal* 7 (1): 19. https://doi.org/10.1186/1752-153X-7-19.

83 Singh, S., Gaikwad, K.K., and Lee, Y.S. (2019). Development and application of a pyrogallic acid-based oxygen scavenging packaging system for shelf life extension of peeled garlic. *Scientia Horticulturae* 256: 108548. https://doi.org/10.1016/j.scienta.2019.108548.

84 Aday, M.S. and Caner, C. (2013). The shelf life extension of fresh strawberries using an oxygen absorber in the biobased package. *LWT- Food Science and Technology* 52 (2): 102–109. https://doi.org/10.1016/j.lwt.2012.06.006.

85 Meyer, R.I. and Jokay, L. (1960). The effect of an oxygen scavenger packet, desiccant in-package system on the stability of dry whole milk and dry ice cream mix. *Journal of Dairy Science* 43 (6): 844.

86 Patel, J., Al-Ghamdi, S., Zhang, H. et al. (2019). Determining shelf life of ready-to-eat macaroni and cheese in high barrier and oxygen scavenger packaging sterilized via microwave-assisted thermal sterilization. *Food and Bioprocess Technology* 12 (9): 1516–1526. https://doi.org/10.1007/s11947-019-02310-1.

87 Norouzbeigi, S., Vahid-Dastjerdi, L., Yekta, R. et al. (2021). Effects of using different O_2 scavengers on the qualitative attributes of bifidus yogurt during refrigerated storage. *Food Research International* 140: 109953. https://doi.org/10.1016/j.foodres.2020.109953.

88 Otero-Pazos, P., Sendón, R., Martínez, I. et al. (2018). Evaluation of oxygen absorber system effectiveness in butter containers. *CyTA Journal of Food* 16 (1): 205–212. https://doi.org/10.1080/19476337.2017.1379561.

89 Rooney, M.L. (2005). Introduction to active food packaging technologies. In: *Innovations in Food Packaging* (ed. J.H. Han), 63–79. Elsevier. https://doi.org/10.1016/B978-012311632-1/50037-1.

90 Chounou, N., Chouliara, E., Mexis, S.F. et al. (2013). Shelf life extension of ground meat stored at 4 °C using chitosan and an oxygen absorber. *International Journal of Food Science & Technology* 48 (1): 89–95. https://doi.org/10.1111/j.1365-2621.2012.03162.x.

91 Mohan, C.O., Ravishankar, C.N., and Srinivasagopal, T.K. (2008). Effect of O_2 scavenger on the shelf-life of catfish (*Pangasius sutchi*) steaks during chilled storage. *Journal of the Science of Food and Agriculture* 88 (3): 442–448. https://doi.org/10.1002/jsfa.3105.

92 Hutter, S., Rüegg, N., and Yildirim, S. (2016). Use of palladium based oxygen scavenger to prevent discoloration of ham. *Food Packaging and Shelf Life* 8: 56–62. https://doi.org/10.1016/j.fpsl.2016.02.004.

93 Berenzon, S. and Saguy, I.S. (1998). Oxygen absorbers for extension of crackers shelf-life. *LWT- Food Science and Technology* 31 (1): 1–5. https://doi.org/10.1006/fstl.1997.0286.

94 Gaikwad, K.K., Deshmukh, R.K., and Lee, Y.S. (2022). Natural phenolic compound coated oxygen scavenging active polyolefin film for preserving quality of fish cake. *Biomass Conversion and Biorefinery* 1: 3. https://doi.org/10.1007/s13399-022-02678-9.

95 Rüegg, N., Röcker, B., and Yildirim, S. (2022). Application of palladium-based oxygen scavenger to extend the mould free shelf life of bakery products. *Food Packaging and Shelf Life* 31: 100771. https://doi.org/10.1016/j.fpsl.2021.100771.

96 Modaresi, A.S. and Niazmand, R. (2021). Characterization of oxygen scavenger film based on sodium ascorbate: extending the shelf life of peanuts. *Food and Bioprocess Technology* 14 (6): 1184–1193. https://doi.org/10.1007/s11947-021-02631-0.

97 Miranda, G., Berna, A., and Mulet, A. (2019). Dried-fruit storage: an analysis of package headspace atmosphere changes. *Foods* 8 (2): 56. https://doi.org/10.3390/foods8020056.

5

Carbon Dioxide (CO_2) Scavengers and Emitters in Food Packaging

Narender Raju Panjagari and Deepak Kumar

Dairy Technology Division, Food Packaging Laboratory, ICAR-National Dairy Research Institute, Karnal, Haryana, India

5.1 Introduction

Packaging plays an essential role in food preservation by protecting food products from external factors that can cause spoilage, contamination, or deterioration. Its primary functions in food preservation include barriers to external elements such as oxygen, moisture, light, microorganisms, and physical protection. Further, it provides tamper evidence, portion control, convenience, and serves as a means of communication. Effective food preservation packaging solutions are designed based on the specific requirement of the food product, considering factors such as its composition, moisture content or water activity, sensitivity to oxygen and light, and intended shelf life. Different packaging materials and technologies are chosen to optimize the preservation of different types of foods, from fresh produce to shelf-stable canned goods. Active food packaging (AP) refers to a type of packaging technology that goes beyond the traditional role of containing and protecting food products. It actively interacts with packaged food to enhance its shelf life, maintain freshness, and improve safety. This interaction is achieved through various mechanisms and components integrated into the packaging materials or systems [1].

Broadly speaking, AP can be categorized into nonmigratory systems, where scavengers are used to eliminate undesirable elements from the packaging environment without deliberate migration, and active migratory systems, which involve emitters that permit controlled migration of beneficial substances into the packaging environment, positively affecting the food product [2, 3]. As per the European Union Commission Regulation [4], active materials and articles are defined as "materials and articles that are intended to preserve or enhance packaged food quality." Active materials are specifically developed to incorporate components that intentionally absorb or release substances from or into the packaged food or from its surrounding environment [5]. An active agent may be placed inside the package or could be an integral part of a package film/layer. An illustration of the working principle of active packaging systems is shown in Figure 5.1. Some of the key aspects of active packaging are moisture absorption, oxygen scavenging, carbon dioxide release or absorption, ethylene removal, and

Smart Food Packaging Systems: Innovations and Technology Applications, First Edition.
Edited by Avik Mukherjee, Santosh Kumar, Manjusri Misra, and Amar K. Mohanty.
© 2025 John Wiley & Sons Ltd. Published 2025 by John Wiley & Sons Ltd.

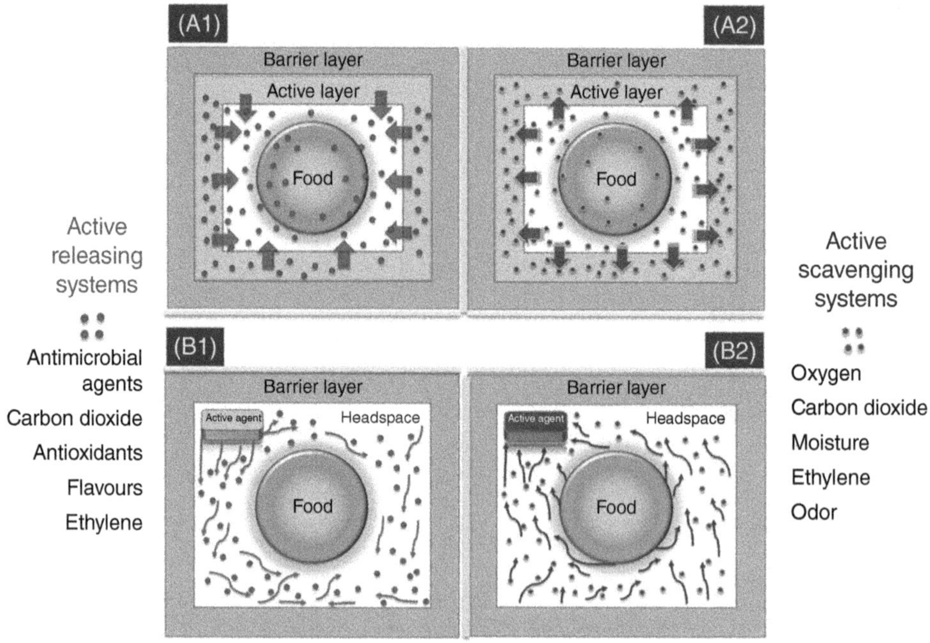

Figure 5.1 An illustration of the working principles of active packaging systems (A1 and B1: Active releasing systems; A2 and B2: Active absorbing systems).

odor removal. Active packaging plays a vital role in enhancing food safety, decreasing food waste, and ensuring that consumers receive products with optimal quality and freshness.

AP is used in various food industry sectors, including fresh produce, meats, dairy products, baked goods, and ready-to-eat meals. The number of studies on the utilization of active packaging in food has gradually increased over the years. This can be evidenced by the fact that when we searched Elsevier's Science Direct, one of the leading online scientific and technical content publishing companies, on May 5, 2023, with "active food packaging" as keywords, we got 438 994 publication results. According to one of the market intelligence reports, in the year 2021, the active and intelligent packaging market had a value of USD 18.84 billion, and it is calculated to exhibit a compound annual growth rate (CAGR) of 6.64% from 2022 to 2027 [6]. Among the AP systems, carbon dioxide (CO_2) emitters and scavengers are components or materials used to regulate the concentration of CO_2 within the packaging environment to enhance the shelf life of food products. CO_2-based scavengers/absorbers are becoming more popular due to their antibacterial and antioxidant characteristics. Both CO_2 scavengers and emitters help optimize the packaging environment to better preserve the safety and quality of food products [7].

Although CO_2 serves as a beneficial gas in the modified atmosphere packaging (MAP) of foods, excessive CO_2 buildup in a package may damage the product's quality and/or package integrity, particularly in the case of CO_2-releasing items such as fresh produce and fermented foods. In such instances, food quality and integrity are ensured by the addition of CO_2 scavengers to packaging material. Higher concentration of CO_2 sufficiently prevents bacterial attack in foods, preserving their freshness and extending their

Table 5.1 Commercially available CO_2 scavengers and emitters.

Commercial/trade name	Manufacturer
CO_2 scavengers	
Evert-fresh	Evert-fresh type G, USA
ATCO CO_2 absorber	Emco Fresh Technologies Ltd, UK
Freshock, Ageless E	Mitsubishi Gas Chemical Co. Inc.
Lipmen (G-Type)	Lipmen Company Ltd., Korea
Ever-fresh green bags	Evert-fresh Corp., USA
CO_2 emitters	
VitalonÒ G	Toagosei Chemical Co. (Japan)
ATCO CO_2 emitter	Emco Fresh Technologies Ltd, UK
AgelessÒ (E-Type and GE-Type)	Mitsubishi Gas Chemical Co. (Japan)
LipmenÒ CO_2 absorber	EMCO (UK)
San DryÒ CO_2 absorber	Hengsan Vietnam Co., Vietnam
FreshPax Type MÒ	Multisorb Technologies (USA)

Source: Gaikwad and Lee [11], Vermeiren et al. [12], Lee et al. [13], and Vilela et al. [14].

shelf life [8, 9]. The most prominent CO_2 absorbers are calcium hydroxide, activated charcoal, zeolite, and magnesium oxide [10]. Commercially a good number of CO_2 scavengers and emitters are available for food packaging applications (Table 5.1). Considering the aforementioned context, the objective of this chapter is to examine the practical applications and mechanisms of action of CO_2 scavengers and emitters used in food packaging.

5.2 CO_2 Scavengers and Emitters

5.2.1 CO_2 Scavengers

CO_2 is produced in most of the horticultural produce due to respiration while in some foods as a result of degradation reactions. To prevent food spoilage and/or package deterioration, it is necessary to eliminate the generated CO_2 from the package [13, 15, 16]. Freshly ground or roasted coffees, which produce large amounts of carbon dioxide, are particularly suitable for the usage of carbon dioxide scavengers as coffee can retain up to 15 atm of dissolved CO_2. The reason to produce CO_2 is Strecker's degradation reaction. On one hand, leaving freshly roasted or ground coffees unpackaged is not advisable because they will collect air and moisture and lose the desired aroma quality. By contrast, in the case of coffee which is hermetically sealed within a pack shortly after it has been roasted, CO_2 releases will build up over time and this may lead to an increased risk of burst. Fresh produce consumes oxygen (O_2) and produces CO_2. Fruits and vegetables may suffer physiological damage because of too much CO_2 accumulation. Huge amounts of CO_2 are produced during the fermentation of unpasteurized vegetables by lactic acid bacteria (LAB), which expands the package or increases the pressure inside [17]. Therefore, in these kinds of packing situations, it is desirable to carefully remove CO_2 from the package headspace. Two solutions have been suggested to address this issue. The first option is to utilize packaging with innovative valves designed to enable the escape of generated/excess CO_2. The second choice involves using either a carbon dioxide scavenger or a combined system that can scavenge both CO_2 and oxygen (O_2). In general, scavengers are fabricated to remove undesirable elements that are present in the packaging environment and do not directly migrate from the package or packaging system to the product. Instead, they only improve the packaging conditions, leading to an extended product shelf life. The double-action scavenger forms namely sachets and labels, capable of scavenging both O_2 and CO_2 are widely used for canned and foil-wrapped coffees [18].

5.2.2 CO_2 Emitters

Carbon dioxide (CO_2) helps in keeping food products fresh and prevents the development of odor-related degradation reactions [19]. High CO_2 levels (60–80%) generate a partial vacuum in the product package, inhibiting microbial growth and extending shelf life. In general, for poultry, fish, fresh meat, fish, and non-climacteric fruits to efficiently inhibit the proliferation of microorganisms, elevated CO_2 concentrations are favored. This is achieved through cytoplasmic pH changes, inhibition of bacterial enzymes as well as alterations in the cell membranes of bacteria ([14, 20]). Hence, the product shelf life can be increased by including a CO_2-generating system or by adding CO_2-generating materials within the package or its design. However, to maintain the desired concentration, CO_2 must be generated continuously as most plastic films have higher CO_2 permeability [21]. Foods such as beef and poultry should have high carbon dioxide levels (10–80%) to prevent microbial development on the surface and lengthen shelf life [22]. High CO_2 levels, on the other hand, may lead to the occurrence of undesirable anaerobic glycolysis in fruits and alterations in product flavor [23, 24]. Flexible packaging may collapse because of the removal of O_2 or the dissolution of CO_2 in the product, so an O_2 scavenger and CO_2 generator could be combined to prevent package collapse brought on by O_2 absorption [9, 13]. When O_2 is removed from a container, a partial vacuum is created, which might cause flexible packaging to collapse. A partial vacuum is also created when food contained in the package is added with a combination of different gases containing CO_2 because the CO_2 easily dissolves into perishable and high water-containing products including dairy products, meat, and seafood [25, 26]. A higher CO_2 concentration in the packaging's headspace is necessary for proper CO_2 dissolution into the food products, enabling its antimicrobial effects. Under these circumstances, it is preferable to insert active agents to simultaneously release CO_2, which consumes O_2. Therefore, CO_2 emitters offer benefits and are essential for maintaining the required CO_2 content in the packaging, to make sure the effectiveness of the MAP system.

5.3 Mechanistic Principles of CO_2 Scavenger

In a comprehensive review of CO_2 absorbers for food packaging, Lee [16] described that due to the impact of environmental CO_2 on climate change, the focus of recent innovative CO_2 scavenging technologies has been on mitigating global warming, and hence, limited direct applications in AP were observed. Irrespective of the technologies adopted, nonhazardous chemical, and physical adsorption principles are required for the removal of CO_2 from food containers. CO_2-scavenging materials may be placed in a pouch or bag, inserted into a food package, or produced in the form of a sheet/coating. Combined physical and chemical absorbents may be included in the formulas containing additives or catalysts. The mechanisms of CO_2 absorption by these chemical and physical agents are given further in the chapter.

5.3.1 Chemical Absorbers

Different salts and alkaline solutions can scavenge CO_2 from the headspace of packaged food and among them, $Ca(OH)_2$ is the most widely used CO_2 scavenger in food packaging. It is considered safe for accidental food contact and the reaction requires no additional reactants to absorb CO_2 [27]. It produces calcium carbonate ($CaCO_3$) and water as harmless byproducts (Eq. 5.1). Among the alkaline salts, Na_2CO_3 can react with CO_2 under humid conditions to produce $NaHCO_3$ (Eq. 5.2). Hence, Na_2CO_3 is used with high-moisture food or humid environment due to the requirement of H_2O molecules for this reaction to occur [28]. While, CaO (calcium oxide) is often used in perishable produce transportation vehicles for the absorption of large volumes of CO_2 absorption. Salts such as $Mg(OH)_2$ and MgO have not been successfully utilized in food packaging due to their low efficiency in reactions and/or uncommon use as food additives. Among the amino acid salt solutions, sodium glycinate ($C_2H_4NNaO_2$) reacts with CO_2 and removes it (Eq. 5.3). However, sodium glycinate being a liquid phase, is difficult to use with food packaging systems. Wang et al. [29] successfully incorporated sodium glycinate into an agar film, to provide solid support and to enable moisture absorption for the CO_2-scavenging reaction to happen. A sizable number of patent applications have claimed the use of different combinations of chemical agents as CO_2 absorbers for food applications such as dough products; vacuum-packaged (VP) coffee; raw bread and coffee beans; roasted coffee beans; and coffee, tea, and juices [13].

$$\mathrm{Ca(OH)_2 + CO_2 \rightarrow CaCO_3 + H_2} \tag{5.1}$$

$$\mathrm{Na_2CO_3 + CO_2 + H_2O \rightarrow NaCO_3} \tag{5.2}$$

$$\mathrm{NaCOOCH_2NH_2 + CO_2 + H_2O \rightarrow NaCOOCH_2NH_3^+ + HCO_3^-} \tag{5.3}$$

The thermodynamic properties of chemical CO_2 absorbers and their CO_2-absorbing reactions in terms of maximum CO_2-absorbing capacity, the change in enthalpy (ΔH^o), the change in entropy (ΔS^o), and the change in Gibbs free energy (ΔG^o) are presented in Table 5.2. Negative ΔG^o and ΔH^o values depict the tendency toward spontaneity under standard conditions and the exothermic characteristics of the reactions, respectively. It can be seen from the table that calcium hydroxide has the highest CO_2-absorbing ability. The attributes of the food product must be considered when choosing an appropriate CO_2 absorber. To achieve optimization, it is crucial to consider the absorber's capacity, and rate of CO_2 absorption. The CO_2 scavenger sachet/pouch film's moisture permeability can be altered to achieve the desired rate of CO_2 absorption.

Table 5.2 Thermodynamic properties of some CO_2-absorbing chemical reactions and physical CO_2 absorption at 25 °C.

CO_2 absorbers	Maximum Absorption capacity ($mol\,g^{-1}$)	Change in enthalpy, ΔH^o ($KJ\,mol^{-1}$)	Change in entropy, ΔS^o ($J\,mol^{-1}\,K^{-1}$)	Change in Gibbs free energy, ΔG^o ($KJ\,mol^{-1}$)
Chemical reactions				
Na_2CO_3	9.43×10^{-3}	−91.6	−215.3	−26.1
$Ca(OH)_2$	1.35×10^{-2}	−114.7	−135.5	−74.4
CaO	1.78×10^{-2}	−179.2	−160.1	−131.4
MgO	2.48×10^{-2}	−95.7	−175.0	−48.4
$Mg(OH)_2$	1.72×10^{-2}	−63.6	−141.3	−21.4
$NaCOOCH_2NH_2$	1.03×10^{-2}	–	–	−72.5
Physical absorbers				
Activated carbon	2.83×10^{-3}	−15.0	24.4	−7.65
Zeolite 4A	3.26×10^{-3}	−12.9	18.5	−7.30
Zeolite 13X	4.22×10^{-3}	−11.3	17.6	−7.93

Source: Lee [16] and Hauchhum and Mahanta [30].

5.3.2 Physical Absorbers

Physical absorption is another common technique that is used for the scavenging of CO_2 gas. In contrast to chemical CO_2 absorbers, the physical adsorption process is reversible, with the equilibrium shifting either backward or forward direction based on the environmental conditions. Further, it is an exothermic process ($-\Delta H^o$) occurring at low temperatures (Table 5.1). The surface area, pore volume, as well as size distribution play crucial roles in CO_2 adsorption [16]. At a constant temperature, the quantity of adsorbed CO_2 rises with increasing partial pressure of CO_2, which is best described by Henry's law or various physical adsorption isotherm models [31]. However, the presence of moisture in a food package may have an impact on CO_2 adsorption or may cause it to compete with moisture adsorption. In general, it has been reported that the presence of a minute quantity of moisture can enhance physical CO_2 sorption on the adsorbent while on the contrary, a large amount of moisture can inhibit it [32].

Zeolite possesses crystalline void cages-like structures comprising open channels and cavities of varying molecular structures, sizes, and shapes, which aid in the absorption of CO_2 molecules. Zeolite exhibits 3D structures comprised of tetrahedral primary units composed of $(AlO_4)^{4-}$ and $(SiO_4)^{4-}$. Due to the arrangement of this primary unit, different open channels and voids are formed which help in absorbing CO_2 gas molecules. Different types of zeolites are available for gas adsorption such as type 4A, type 5A, and type 13X. Type 4A and type 13X are characterized by the presence of sodium as a major cation while type 5A contains either calcium or magnesium. The aperture sizes of types 4A, 5A, and 13X are 4, 5, and 10 Å, respectively [16]. The surface area, pore size, and bulk density of zeolites are 400–700 $m^2\,g^{-1}$, 0.4–1.0 nm, and 470–720 $kg\,m^{-3}$, respectively. The CO_2 absorption capacity of zeolite is 3.26–4.22 $\times 10^{-3}\,mol\,g^{-1}$ based on dry state condition at 1 bar and 15 °C. The performance of zeolites is hindered in the presence of water vapor. In terms of CO_2 absorption capacity, zeolite surpasses activated carbon and retains CO_2 adsorption equilibrium when protected from water vapor. A plastic film with high CO_2 transmission and low water vapor permeability can control the zeolite sachet's CO_2 absorption rate [33]. However, their stability is reduced at high temperatures due to thermal destruction, making them unsuitable for high-temperature CO_2 extraction processes. The addition of alkali metals or alkaline earth to the adsorbent's surface creates basic sites that have a strong affinity for acidic gas molecules such as carbon dioxide [34]. Vatalis et al. [35] explored the possible application of Greek natural zeolites for the collection and storage of carbon dioxide. Activated carbon has an inherent porous structure, contributing to its considerable surface area (500–2500 $m^2\,g^{-1}$). The pore size (10–30 Å) and bulk density (250–600 $kg\,m^{-3}$) of activated carbon are larger than zeolite, resulting in a higher CO_2 absorption capacity compared to zeolite. Additionally, activated carbon is not influenced by the presence of moisture due to its nonpolar nature [36].

5.4 Mechanism of CO_2 Emitters

CO_2 can dissolve in both the lipid and aqueous phases of food, leading to the formation of H_2CO_3 (carbonic acid) and subsequent lowering of the pH of the food product. The solubility of CO_2 in food is described by Henry's law (Eq. 5.4).

$$C_{CO_2} = H_{CO_2} P_{CO_2} \tag{5.4}$$

where C_{CO_2}, P_{CO_2}, and H_{CO_2} are CO_2 content dissolved in food ($mol\,kg^{-1}$), partial pressure of CO_2 gas in the headspace (atm), and Henry's law constant taken as solubility ($mol\,kg^{-1}\,atm^{-1}$), respectively[13]. Foods with higher H_{CO_2} would have a higher tendency to dissolve CO_2 in their matrix, which may cause the collapse of packages such as O_2 scavenging in a can cause pressure decrease, and CO_2 dissolution from CO_2 enriched headspace in MAP products in semirigid plastic trays may collapse or create a partial vacuum. In these situations, the production of CO_2 gas in an amount equal to the gas that has disappeared can aid in maintaining the integrity and aesthetics of the package. Hence, without a continuous CO_2 refill, the package may experience deformation due to the formation of under pressure; an issue that can be addressed with a CO_2 emitter. A CO_2 emitter compensates for the loss of CO_2 gas, ensuring the package maintains its integrity and reduces physical stress on moist meat products. The mechanisms of CO_2 emitters used in AP are described further in the chapter.

5.4.1 CO_2 Emitters

The first CO_2 emitter formulation used in packaging applications involved ferrous carbonate. Under moist conditions, upon oxygen consumption, Fe_2CO_3 (ferrous carbonate) generates CO_2 as shown in Eq. 5.5. Ferrous carbonate was utilized to overcome the reduction in pressure resulting from oxygen depletion in oxidative food packaging. However, since Fe_2CO_3 is very unstable under ambient conditions, metal halide, and other catalysts are incorporated with it to control the reaction rate of CO_2 production and achieve a balance between CO_2 emission and oxygen absorption [17].

$$4FeCO_3 + 6H_2O + O_2 \rightarrow 4Fe(OH)_3 + 4CO_2 \tag{5.5}$$

In another CO_2 generation system, under damp conditions, sodium bicarbonate ($NaHCO_3$) or sodium ascorbate ($C_6H_7O_6Na$) can react with organic acid to produce CO_2 (Eq. 5.6). Norway research group reported application of a CO_2 emitter by the use of sodium bicarbonate (1–17 mg) and citric acid in the presence of moisture [13]. It is well-known that promoting an aqueous environment for the ingredients to dissolve is crucial to starting the CO_2-releasing reaction. Due to the requirement for an aqueous environment, all of these emitters are typically in the shape of a pad or sachet and are always in contact with a liquid absorber as depicted in Eqs. 5.5 and 5.6.

$$3NaHCO_3 + C_3H_5OH(COOH)_3 \rightarrow C_3H_4OH(COONa)_3 + 3CO_2 + 3H_2O \tag{5.6}$$

5.4.2 Active Valves for CO_2 Emission or Removal to Environment

In an innovative method, freshly roasted or ground coffees, which produce significant quantities of CO_2, a unique device that releases excess CO_2 through a one-way valve was reported by Lee et al. [13]. The active valve principle is based on keeping the package pressure equilibrated with atmospheric pressure. When excess pressure is built into the package due to accumulated CO_2, the valve gets opened and removes the excess pressure till it gets equilibrated with the outside atmosphere (Figure 5.2).

When excess pressure is released, the valve automatically gets closed. The main components of active valves are the sealing layer (membrane/silicone oil), body, and cap (Figure 5.2). They should be attached to the package wall typically with heat sealing. Small holes are present on the body, cap, and surface of packaging material (polyethylene or polypropylene). The active valve's opening and closing mechanism relies on the elastic action of a membrane primarily composed of polyester film or rubber adhered to the body. Upon reaching a specific internal overpressure threshold (200–900 Pa), the system initiates the detachment of the membrane from the oil, allowing gas to pass through perforations from within the package to the external environment. To enable the release of internal gas, the internal overpressure needs to surpass the adhesion force that holds the membrane onto

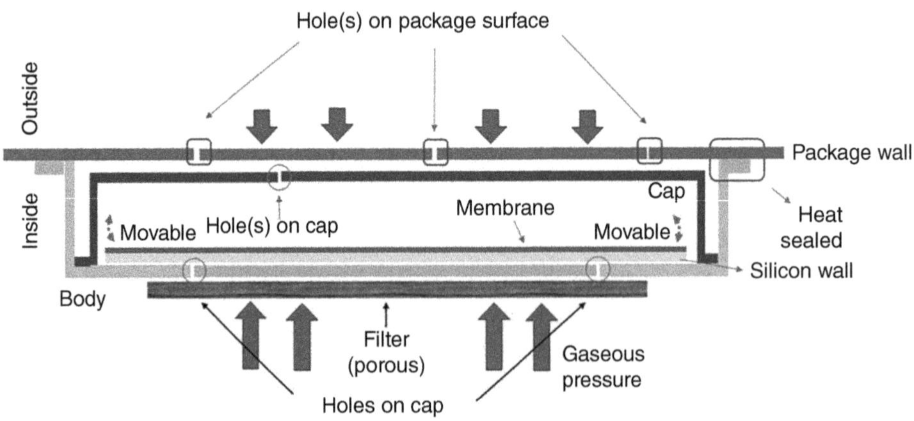

Figure 5.2 One-way active valve with typical design. *Source:* Lee et al. [13]/with permission of John Wiley & Sons.

the surface of the body through the oil [13]. The membrane returns to its original location while remaining connected to the oil when the pressure disparity diminishes. This kind of system is already being used commercially in Italy, Korea, Taiwan, Switzerland, and China (Jeong 2018) for roasted coffee, fermented foods, and bread dough.

5.5 Incorporation of CO_2 Scavengers and Emitters in Active Food Packaging

Incorporating CO_2-absorbing agents, into food packaging can be achieved through various integration or formulation methods. Absorbents in powder form can be encased in flexible or rigid packaging such as a film/bag or sachet [16]. The binder can be utilized to transform the absorber materials into bead-like or pellet-like forms. Adsorbent-filled sachets are frequently used as food-active packaging because they are efficient and practical for gas control. The emitters are commonly available in the form of pads or sachets. When the absorbent pad captures any leaking liquid from the product, the chemicals inside the sachet react with the liquid, resulting in CO_2 release. But sachets can potentially compromise food safety owing to accidental breaks; moreover, they are moisture-sensitive and need extra packing procedures [20]. Studies on the integration of an adsorbent directly into packing films have been done to address the shortcomings of sachets. By combining and melting linear low-density polyethylene (LLDPE) and a carbon dioxide adsorbent, packaging films with the ability to absorb carbon dioxide were created.

5.6 CO_2 Emitters in Active Food Packaging

5.6.1 Antimicrobial Applications

The evidence of CO_2's ability to inhibit microbial growth is abundant. The content of CO_2 dissolved in the food product and its solubility rate in both aqueous and lipid phases have a significant impact on the antimicrobial effect. The solubility of carbon dioxide varies among different food products and is influenced by specific characteristics, such as composition, pH, surface area, and as well as temperature reduction [37]. The partial pressure of a gas is directly proportional to the antimicrobial property of CO_2 gas [20]. The bactericidal impact of CO_2 relies on several variables, such as the level of pressure, time duration, CO_2 concentration, and temperature of the environment [26]. CO_2, through the reduction of relative O_2 levels and antibacterial action, exerts inhibitory effects on a diverse range of aerobic bacteria and fungi. In active packaging systems, CO_2 emitters are employed to extend the lag phase and generation time of microorganisms during the logarithmic growth phase in perishables [38, 39]. CO_2 reacts with H_2O, resulting in the formation of H_2CO_3, which subsequently undergoes deprotonation. The function of membrane proteins and the condition of the cell membrane are both hampered by CO_2 and HCO_3 [40]. To effectively reduce surface microflora when preserving meat-based products, it is advisable to use higher concentrations of CO_2 (10–80%) [12, 41]. However, the impact of CO_2 on microorganisms varies: moderate (10%) to high (20%) can hinder the growth of aerobic bacteria, while it promotes the growth of LAB. Furthermore, Yingyuad et al. [42] reported that an elevated CO_2 concentration (above 50%) inhibits the growth of *Listeria monocytogenes, C. botulinum*, and *C. perfringens*. According to Mohan et al. [43], elevated CO_2 levels have positive effects on reducing the growth of *Pseudomonas* spp., which otherwise leads to discoloration and off-flavor in fish. The impact of CO_2 on bacterial growth is multifaceted, and researchers have identified four pathways associated with it as mentioned as under:

- Modifications in the functioning of the cell membrane can affect the process of nutrient absorption and uptake
- Enzyme activity is being hampered
- By penetrating the cell membranes, it alters the pH
- By changing the physicochemical properties of the protein

The combination of these actions resulted in a bacteriostatic effect, extending the growth period and lag phase of microorganisms. When stored at room temperature, elevated CO_2 levels within the packaging inhibit the growth of *Escherichia coli, Clostridium sporogenes, Vibrio* spp., *Bacillus cereus, and Staphylococcus aureus* [44]. The effect of combined action of O_2 scavenger plus CO_2 emitter in improving the storage stability of lamb loins (*Longissimus lumborum*) stored at $1 \pm 1\,°C$ was studied by Trindade et al. [45] and microbial counts were determined. It was reported that according to the Brazilian legislation, the standards for vacuum-packed meat, not aged, are 10^4 colony forming units (CFU) per gram sample for thermotolerant coliform, 3×10^3 CFU g^{-1} for coagulase positive *Staphylococci* and absence of *Salmonella* in 25 g of sample. The study revealed that the microbial counts of VP lamb loin meat with a dual-function sachet (O_2 scavenger plus CO_2 emitter) remained within the acceptable limits established by regulatory standards. The effect of different levels of CO_2 concentrations (30%, 60%, and 100%) on the storage stability of chicken fillets was studied by Holck et al. [40]. In this study, they used a CO_2 emitter comprising $NaHCO_3$ and citric acid added to a liquid absorber. It was reported that including a CO_2 emitter avoided package collapse, significantly reduced drip loss, and slowed down microbial growth. This effect was observed when CO_2 gas concentration in the headspace gas mixture increased from 60 to 100% due to the CO_2 emitter. The effect of VP, MAP, and MAP with CO_2 emitter on the quality of fresh meat (reindeer) stored at 4 °C for 21 days was studied by Pettersen et al. [46] who reported that the MAP (CO_2:N_2 of 60 : 40) with CO_2 emitter samples resulted in enhanced microbial stability compared to VP and MAP alone as CO_2 emitter increased partial pressure of CO_2. Further, it was reported that maintenance of gas-to-product (G/P) ratio of 1 : 1 was too low in maintaining the quality of the meat.

In a study conducted by Hansen et al. [47], the CO_2 emitter comprising $NaHCO_3$ and citric acid added to a liquid absorber and activated by sterile water was reported to enhance the shelf life of MAP cod from 14 to 21 days. It was reported that as expected in the CO_2 emitter containing samples, the *Carnobacterium* spp. was found to be dominating while no change in the *Photobacterium* spp. In another study, Hansen et al. [19] investigated the effect of a CO_2 emitter pad on the storage life of VP or MAP (G/P ratio: 1.6/1) cod loins (*Gadus morhua*) and reported that the initial freshness of the cod loins was better preserved by adding a CO_2 emitter in both VP and MAP. It was reported that the shelf life of MAP samples with added CO_2 emitter was 13 days compared to 9 days of VP samples with added CO_2 emitter and MAP samples containing no CO_2 emitter. Further, it was reported that CO_2 emitter did not alter the dominant spoilage index organism (*Photobacterium phosphoreum*) of MAP fish.

5.6.2 Antioxidant Applications

CO_2 emitters can indeed act as antioxidants in AP. The concept behind this is to use the controlled release of CO_2 within the packaging environment to help preserve the quality of the food and freshness. CO_2 can help retard oil/fat oxidation, which is one of the primary causes of spoilage of food. Research studies have been conducted to explore the effectiveness of CO_2 emitters in AP [48]. These studies typically assess the impact of CO_2 on food quality, shelf-life extension, and the inhibition of oxidation processes. However, the effectiveness of CO_2 emitters as antioxidants can vary depending on several factors. It is essential to consider the specific requirements of the food product and conduct thorough testing to determine the optimal approach for preservation. Sachets and labels with dual functionality, designed to scavenge O_2 and emit CO_2 in equal amounts, are commercially available in the market. These systems serve the purpose of removing O_2 and producing CO_2 simultaneously. They prove particularly useful for products where not only maintaining package volume is crucial but also appearance, especially under partial vacuum conditions. Commercially available sachets (Ageless™ type GE; FreshPax® type M) with this dual-action feature operate through self-reactions using either Fe_2CO_3 or a combination of ascorbic acid and $NaHCO_3$.

5.7 Application of CO_2 Scavenging and Emitting Systems in Active Packaging

5.7.1 Fruits and Vegetables

Fresh horticultural produce predominantly produces CO_2 as a result of respiration and if this CO_2 is not properly managed and allowed to accumulate to elevated levels within the packaging, it can exert detrimental effects on the biological processes of the fresh produce. Hence, carefully managed CO_2 removal can aid in preserving product quality. The elevated CO_2 concentration in the headspace of packaged produce reduces the growth of aerobic microorganisms and diminishes the rates of respiration and senescence processes [49]. Fresh produce is said to have a low respiration rate and less microbial development on its surface at CO_2 partial pressures above 0.1 atm [38]. The impact of CO_2 on food can be either advantageous or detrimental, depending on the specific properties of the food in question. Therefore, it becomes necessary to employ an optimal CO_2 concentration to achieve the desired outcomes. Nugraha et al. [50] suggested that utilizing a CO_2 scavenger could potentially prevent or delay the enzymatic browning of pears when packed in plastic bags with 6–8% O_2 concentration. It is important to note that pears are susceptible to CO_2 damage when exposed to concentrations exceeding 2%. A MAP system with a CO_2 absorber, maintaining O_2 content around 9% and CO_2 concentration between 1% and 4%, proved to be effective in reducing the decay of high CO_2-sensitive mushrooms. Lee [16] demonstrated that utilizing a CO_2 scavenger to maintain a CO_2 content of 0.4% within the package effectively prevented or delayed chilling injury in eggplants stored at 4 °C for up to five days. Wang et al. [29] conducted a study on Shiitake mushrooms, employing an agar-based label incorporated with Na_2CO_3 as a CO_2 absorber. The results indicated reduced color change, improved texture, enhanced flavor, and a decreased bacterial count in the mushrooms.

In kimchi, a traditional fermented Korean food, pasteurization is not a viable option considering its sensory attributes, as the fermentation process persists, leading to the simultaneous production of CO_2 [51]. Scavengers can be advantageous in this context, as they assist in mitigating the issue of bloating or potential bursting of packages due to the accumulation of CO_2. The carbon dioxide generated during the fermentation of kimchi can readily deteriorate the packaging material, leading to volume expansion and breakage. To address these concerns arising from carbon dioxide, adsorbents have been developed. Lee et al. [17] developed food packaging material comprising $Ca(OH)_2$ and a porous medium with CO_2-adsorptive functionality. The incorporation of calcium hydroxide-impregnated adsorption powder and porous media in the packaging materials led to enhanced mechanical qualities and exhibited significant chemical and physical reactions, resulting in noteworthy adsorptive power and reusability. The created adsorbents were integrated into low-density polyethylene (LDPE) to produce kimchi packaging materials. The adsorption packaging showed no impact on the ripening of the kimchi. For kimchi packaging, the optimal CO_2 concentration fell within the range of 40–55%, and the proposed adsorption packaging materials proved effective in regulating the carbon dioxide concentration. A novel one-way valve was devised in a study conducted by Jeong [52] to mitigate package breakage caused by CO_2 gas accumulation in kimchi. The implementation of this device effectively decreased the pressure buildup in the headspace due to CO_2 without compromising the sensory attributes of kimchi. Lee et al. [53] introduced a novel food packaging film that incorporated $Ca(OH)_2$ as a CO_2 adsorbent, utilizing a fatty acid solution to prevent adsorbent aggregation in the film. Stearic acid acted as a surfactant, leading to reduced aggregates in the films, resulting in improved mechanical, optical, and barrier properties. This enhancement in mechanical and physical characteristics allowed the films to demonstrate effective CO_2 adsorption capability without interfering with the ripening process of kimchi when employed as packaging materials. Aday et al. [54] applied two CO_2 (EMCO®) absorbers and one O_2 absorber during packaging of strawberries and reported that CO_2 scavengers prolonged the storage stability of strawberry fruit by delaying the respiration rate, thereby preserving its quality, and retaining firmness.

5.7.2 Dairy Products

Milk and milk products have a limited shelf life as they possess high water activity (a_w) and a pH conducive to microbial growth. Most dairy products are prone to rapid deterioration concerning flavor, texture, and color alterations due to these factors. Fermented dairy products such as yogurt, kumiss, kefir, and cheese generate CO_2 during storage. Although excessive CO_2 buildup can adversely affect quality and package integrity, a reasonably elevated level of CO_2 production proves beneficial in imparting an appealing flavor or maintaining the best quality in certain instances [16, 55]. Some fermented foods taste better when there is a somewhat low CO_2 concentration because it encourages the growth of

LAB. [56]. CO_2 removal from dairy products is not common, as certain products such as carbonated yogurt drinks benefit from a specific level of carbonation. However, CO_2 scavengers can be utilized when the need arises to decrease elevated CO_2 levels in dairy products. CO_2 scavenger/emitter can be used in a variety of dairy products such as whole milk powder, dairy whitener, and cheese in creating a modified atmosphere (MA) environment inside the package with the help of different levels of CO_2. Gill and Tan [57] reported that as the temperature drops, the inhibitory effect of CO_2 concentration increases. Rajagopal et al. [58] examined the effects of storing raw milk at 5, 6.1, 10, and 20 °C under CO_2 pressures of 68–689 kPa on the local microbiota and reported that combined low CO_2 pressure and refrigeration temperatures may increase the microbial quality and safety of raw milk and helps in transporting raw or concentrated milk over long distances.

Whole milk powder is shelf-stable for six months only primarily due to oxidative changes in the fat. Oxidative rancidity can be reduced by lowering the O_2 content of the package. To address this concern, Day [59] presented two commonly employed techniques. The first option involves incorporating specially designed one-way valves in the packaging to enable the escape of carbon dioxide. The second option is to utilize a system with carbon dioxide scavengers and dual-action oxygen and CO_2 systems. CO_2 concentration in whole milk powder can be maintained by applying CO_2 emitter in the packaging material. MAP has been primarily used for cheese packaging in dairy industry. However, due to the varying compositions and shelf life within the cheese category, the packaging requirements of each type of cheese need to be considered carefully and separately. Another thing to consider is that some cheeses produce CO_2, while others do not. The ideal gas mixture to be employed depends on the properties of the cheese when it is packaged in an MA [60]. Controlling the CO_2 levels is crucial because it has been shown that for some cheeses, high CO_2 levels can give the cheese an unpleasant flavor. Additionally, if the CO_2 levels are too high without N_2 in the package, the product may absorb the CO_2, which could cause the pack to collapse. So, CO_2 emitter can be used in maintaining CO_2 concentration in the MAP of cheese.

5.7.3 Muscle Foods

CO_2 generators find extensive use in meat preservation due to their inhibitory effects against various aerobic bacteria and fungi. Among the three gases commonly employed in MAP systems, namely oxygen, CO_2, and nitrogen, CO_2 stands out as the sole gas with a direct antimicrobial impact. The inhibitory action of CO_2 affects different microorganisms differently, as moderate (10%) to high (20%) levels of CO_2 can hinder *Pseudomonas* spp, while LAB may be stimulated by CO_2. However, certain pathogens are least affected by CO_2 levels below 50%. Nevertheless, a potential concern lies in inhibiting spoilage-causing microorganisms, as this may give the impression that the food product is safe for consumption while it could contain a significant number of pathogens that have proliferated in the absence of competition from natural microflora [42]. In animal food preservation, high levels of CO_2 are typically preferred as they effectively increase the shelf life by inhibiting the growth of microorganisms on the surface. Pettersen et al. [46] reported the impact of various packing techniques on fresh reindeer meat. The research revealed that meat packed in an MA containing CO_2 and N_2 in the ratio of 60 : 40 both with or without a CO_2 emitter, exhibited an extended shelf life of 21 days meat. The CO_2 levels in the MA packages (G/P ratio of 1 : 1) with an emitter experienced an initial drop to approximately 45% on the first day but later rose and reached 65% to 70% over the storage period.

Hansen et al. [19] reported that VP cod fillets containing a CO_2 emitter resulted in an extension of shelf life by two days, resulting in a total shelf life of nine days. The use of CO_2 emitters in active MAP has been investigated for extending the shelf life of chicken fillets [40], ready-to-eat meat products, cod loin, and lamb [19, 45, 61]. The shelf life of chicken meat is commonly extended with MAP with significant levels of CO_2. However, during storage at lower temperatures, CO_2 dissolves in the meat, causing the packaging to collapse and squeezing the meat, leading to significant drip loss. Holck et al. [40] reported that the use of CO_2 emitter sachets (100% CO_2) and MAP resulted in reduced drip loss in chicken fillets with shelf life extended by seven days. In MAP, CO_2 at levels ranging from 10% to 80% is commonly used on the surfaces of perishable foods such as cheese, poultry, and meat or moisture and/or O_2-sensitive bakery goods as CO_2 acts both as a bacteriostatic and fungistatic agent. The lag phase of the CO_2-sensitive microorganisms could be extended and the rate of growth of *Pseudomonas* could be suppressed by CO_2 emitters [62]. However, it is important to note that a high concentration of CO_2 can promote the growth of yeast and LAB. Moreover, CO_2 presence has little to no impact on anaerobic food pathogens. Hence, where there is a significant risk of foodborne illness, the use of high CO_2 is not recommended.

5.7.4 Bakery and Other Food Products

Baked goods are a significant component of a well-balanced diet, and a diverse range of these items can be discovered on the shelves of grocery stores. However, bakery products are vulnerable to physical, chemical, and microbiological deterioration, just like other processed foods. Growth of fungus is the main issue that affects bakery foods' shelf life [63]. High CO_2 concentrations are frequently used to restrict fungal growth and increase shelf life [64, 65]. However, CO_2's high solubility can result in flavor alterations and package collapse; however, these issues can be readily avoided by using N_2 [66]. Active packaging components namely moisture absorbers, antioxidants, antimicrobials, and ethanol emitters have been reported to effectively enhance the shelf life of bakery products [67]. However, CO_2 emitters of absorbers applications are limited.

5.8 Conclusion

Due to lifestyle changes and concurrent increasing demand for safe and quality foods, there are continuous developments in food packaging technologies too. Among the different food packaging technologies, the type of packaging system that actively interacts with the food with the aim of maintaining freshness, and enhancing safety is called active packaging. This interaction is achieved through various mechanisms and components integrated into the packaging materials or systems. CO_2 scavengers and emitters are important components of active packaging systems. By actively removing excess CO_2 from the food packaging environment, CO_2 scavengers prevent the accumulation of this gas, which can lead to product deterioration, spoilage, and loss of nutritional value. CO_2 emitters retard ripening, inhibit microbial growth, and preserve the quality of fruits, vegetables, and meats. Furthermore, the use of CO_2 emitters in MAP systems either alone or in combination with O_2 absorbers improves the acceptability of packaged foods. The most prominent CO_2 absorbers are calcium hydroxide, activated charcoal, zeolite, and magnesium oxide while ferrous carbonate was utilized as a CO_2 emitter. When selecting an appropriate CO_2 absorber, one must consider the rate of production of CO_2, the level of CO_2 required, and packaging variables. CO_2 emitters or scavengers have been largely employed in the packing of meat, coffee, and fruits and vegetables. However, more research and development are required for applications in bakery and other products.

References

1 Ahmed, M.W., Haque, M.A., Mohibbullah, M. et al. (2022). A review on active packaging for quality and safety of foods: current trends, applications, prospects and challenges. *Food Packaging and Shelf Life* 33: 100913.

2 Dainelli, D., Gontard, N., Spyropoulos, D. et al. (2008). Active and intelligent food packaging: legal aspects and safety concerns. *Trends in Food Science & Technology* 19 (Suppl. 1): S103–S112.

3 Robertson, G. (2013). *Food Packaging: Principles and Practices*. Boca Raton: CRC Press.

4 EC (2009). Active and intelligent materials and articles intended to come into contact with food. European Union Commission Regulation No. 450/2009. Document 32009R0450.

5 Brody, A.L. (2001). What's active in active packaging. *Food Technology* 55 (9): 104–106.

6 Modor Intelligence (2023). Active and intelligent packaging market size and share analysis-growth trends and forecasts. https://www.mordorintelligence.com/industry-reports/active-and-intelligent-packaging-market-industry (accessed 15 May 2023).

7 Wyrwa, J. and Barska, A. (2017). Innovations in the food packaging market: active packaging. *European Food Research and Technology* 243: 1681–1692.

8 Cutter, C.N. (2002). Microbial control by packaging: A review. *Critical Reviews in Food Science and Nutrition* 42 (2): 151–161.

9 Puligundla, P., Jung, J., and Ko, S. (2012). Carbon dioxide sensors for intelligent food packaging applications. *Food Control 25* (1): 328–333.

10 Bodbodak, S. and Rafiee, Z. (2016). Recent trends in active packaging in fruits and vegetables. In: *Eco-Friendly Technology for Postharvest Produce Quality* (ed. M.W. Siddiqui), 77–125. London: Academic Press.

11 Lee, Y.S. (2017). Current scenario of gas scavenging systems used in active packaging – a review. *Journal of Packaging Science and Technology*. 23 (2): 109–117.

12 Vermeiren, L., Devlieghere, F., van Beest, M. et al. (1999). Developments in the active packaging of foods. *Trends in Food Science and Technology* 10 (3): 77–86.

13 Lee, D.S., Wang, H.J., Jaisan, C., and An, D.S. (2022). Active food packaging to control carbon dioxide. *Packaging Technology and Science* 35 (3): 213–227.

14 Vilela, C., Kurek, M., Hayouka, Z. et al. (2018). A concise guide to active agents for active food packaging. *Trends in Food Science and Technology* 80: 212–222.

15 Floros, J.D., Dock, L.L., and Han, J.H. (1997). Active packaging technologies and applications. *Food Cosmetics and Drug Packaging* 20 (1): 10–17.

16 Lee, D.S. (2016). Carbon dioxide absorbers for food packaging applications. *Trends in Food Science and Technology* 57: 146–155.

17 Lee, D.S., Yam, K.L., and Piergiovanni, L. (2019). *Food Packaging Science and Technology* Special Indian edition. Boca Raton, FL: CRC Press.

18 Day, B.P.F. and Potter, L. (2011). Active packaging. In: *Food and Beverage Packaging Technology* (ed. R. Coles and M. Kirwan). West Sussex: Wiley-Blackwell.

19 Hansen, A.Å., Moen, B., Rødbotten, M. et al. (2016). Effect of vacuum or modified atmosphere packaging (MAP) in combination with a CO_2 emitter on quality parameters of cod loins (*Gadus morhua*). *Food Packaging and Shelf Life* 9: 29–37.

20 Yildirim, S., Röcker, B., Pettersen, M.K. et al. (2018). Active packaging applications for food. *Comprehensive Reviews in Food Science and Food Safety* 17 (1): 165–199.

21 Mangaraj, S., Goswami, T.K., and Mahajan, P.V. (2009). Applications of plastic films for modified atmosphere packaging of fruits and vegetables: a review. *Food Engineering Reviews* 1: 133–158.

22 Mortazavi, S.M.H., Kaur, M., Farahnaky, A. et al. (2023). The pathogenic and spoilage bacteria associated with red meat and application of different approaches of high CO_2 packaging to extend product shelf-life. *Critical Reviews in Food Science and Nutrition* 63 (12): 1733–1754.

23 Ali, M., Batool, S., Khalid, N. et al. (2023). Alcoholic off-flavor disorders in fresh fruits. *Journal of Food Biochemistry* 2023: 1–16.

24 Floros, J.D., Nielsen, P.V., and Farkas, J.K. (2000). Advances in modified atmosphere and active packaging with applications in the dairy industry. *Bulletin of the International Dairy Federation* 346: 22–28.

25 Mohan, C.O.A. and Ravishankar, C.N. (2019). Active and intelligent packaging systems – application in seafood. *World Journal of Aquaculture Research and Development* 1: 10–15.

26 Singh, P., Wani, A.A., Karim, A.A., and Langowski, H.C. (2012). The use of carbon dioxide in the processing and packaging of milk and dairy products: a review. *International Journal of Dairy Technology* 65 (2): 161–177.

27 Barnes, K., Sinclair, R., and Watson, D. (ed.) (2006). *Chemical Migration and Food Contact Materials*. England: Woodhead Publishing.

28 Shin, D.H., Cheigh, H.S., and Lee, D.S. (2002). The use of Na_2CO_3-based CO_2 absorbent systems to alleviate pressure buildup and volume expansion of kimchi packages. *Journal of Food Engineering* 53 (3): 229–235.

29 Wang, H.J., An, D.S., Rhim, J.-W., and Lee, D.S. (2015). A multi-functional biofilm used as an active insert in modified atmosphere packaging for fresh produce. *Packaging Technology and Science* 28 (12): 999–1010.

30 Hauchhum, L. and Mahanta, P. (2014). Carbon dioxide adsorption on zeolites and activated carbon by pressure swing adsorption in a fixed bed. *International Journal of Energy and Environmental Engineering* 5: 349–356.

31 Yang, R.T. (1997). *Gas Separation by Adsorption Processes*. Stoneham: Butterworth.

32 Marx, D., Joss, L., Hefti, M. et al. (2013). The role of water in adsorption-based CO_2 capture systems. *Energy Procedia* 37: 107–114.

33 Lee, D.S., Shin, D.H., Lee, D.U. et al. (2001). The use of physical carbon dioxide absorbents to control pressure buildup and volume expansion of kimchi packages. *Journal of Food Engineering 48* (2): 183–188.

34 Hinkov, I., Lamari, F., Langlois, P. et al. (2016). Carbon dioxide capture by adsorption. *Journal of Chemical Technology And Metallurgy* 51 (6): 609–626.

35 Vatalis, K.I., Laaksonen, A., Charalampides, G., and Benetis, N.P. (2012). Intermediate technologies towards low-carbon economy. The Greek zeolite CCS outlook into the EU commitments. *Renewable and Sustainable Energy Reviews* 16 (5): 3391–3400.

36 Xu, D., Xiao, P., Zhang, J. et al. (2013). Effects of water vapour on CO_2 capture with vacuum swing adsorption using activated carbon. *Chemical Engineering Journal* 230: 64–72.

37 Devlieghere, F. and Debevere, J. (2000). Influence of dissolved carbon dioxide on the growth of spoilage bacteria. *LWT- Food Science and Technology* 33 (8): 531–537.

38 Lopez-Rubio, A., Almenar, E., Hernandez-Muñoz, P. et al. (2004). Overview of active polymer-based packaging technologies for food applications. *Food Reviews International* 20 (4): 357–387.

39 Suppakul, P., Miltz, J., Sonneveld, K., and Bigger, S.W. (2003). Active packaging technologies with an emphasis on antimicrobial packaging and its applications. *Journal of Food Science* 68 (2): 408–420.

40 Holck, A.L., Pettersen, M.K., Moen, M.H., and Sørheim, O. (2014). Prolonged shelf life and reduced drip loss of chicken filets by the use of carbon dioxide emitters and modified atmosphere packaging. *Journal of Food Protection* 77 (7): 1133–1141.

41 Biji, K.B., Ravishankar, C.N., and Mohan, C.O. (2015). Smart packaging systems for food applications: a review. *Journal of Food Science and Technology* 52 (10): 6125–6135.

42 Yingyuad, S., Ruamsin, S., Reekprkhon, D. et al. (2006). Effect of chitosan coating and vacuum packaging on the quality of refrigerated grilled pork. *Packaging Technology and Science: An International Journal* 19 (3): 149–157.

43 Mohan, C.O., Ravishankar, C.N., and Gopal, T.K. (2010). Active packaging of fishery products – a review. *Fishery Technology* 47 (1): 1–18.

44 Scannell, A.G.M., Hill, C., Ross, R.P. et al. (2000). Development of bioactive food packaging materials using immobilized bacteriocins Lacticin and Nisaplin®. *International Journal of Food Microbiology* 60 (2–3): 241–249.

45 Trindade, M.A., Villanueva, N.D.M., Antunes, C.V., and Freire, M.T.D.A. (2013). Active packaged lamb with oxygen scavenger/carbon dioxide emitter sachet: physical-chemical and microbiological stability during refrigerated storage. *Brazilian Journal of Food Technology* 16: 216–225.

46 Pettersen, M.K., Hansen, A.A., and Mielnik, M. (2014). Effect of different packaging methods on quality and shelf life of fresh Reindeer meat. *Packaging Technology and Science* 27 (12): 987–997.

47 Hansen, A.Å., Mørkøre, T., Rudi, K. et al. (2007). Quality changes during refrigerated storage of MA-packaged pre-rigor fillets of farmed Atlantic cod (*Gadus morhua* L.) using traditional MAP, CO_2 emitter, and vacuum. *Journal of Food Science* 72 (9): M423–M430.

48 Vasile, C. and Baican, M. (2021). Progresses in food packaging, food quality, and safety – controlled-release antioxidant and/or antimicrobial packaging. *Molecules* 26 (5): 1263.

49 Chakraverty, A. and Singh, R.P. (2016). *Postharvest Technology and Food Process Engineering*. Boca Raton: CRC Press.

50 Nugraha, B., Bintoro, N., and Murayama, H. (2015). Influence of CO_2 and C_2H_4 adsorbents to the symptoms of internal browning on the packaged 'Silver Bell' Pear (*Pyrus communis* L.). *Agriculture and Agricultural Science Procedia* 3: 127–131.

51 Jung, J.Y., Lee, S.H., and Jeon, C.O. (2014). Kimchi microflora: history, current status, and perspectives for industrial kimchi production. *Applied Microbiology and Biotechnology* 98: 2385–2393.

52 Jeong, S., Cho, C.H., Lee, H.G. et al. (2018). Development and application of multi-function valve to solve major problems of expansion and off-odor leakage in the packaging of Kimchi. *Korean Journal of Packaging Science and Technology* 24 (3): 113–120.

53 Lee, H.G., Cho, C.H., Kim, H.K., and Yoo, S. (2020). Improved physical and mechanical properties of food packaging films containing calcium hydroxide as a CO_2 adsorbent by stearic acid addition. *Food Packaging and Shelf Life* 26: 100558.

54 Aday, M.S., Caner, C., and Rahvali, F. (2011). Effect of oxygen and carbon dioxide absorbers on strawberry quality. *Postharvest Biology and Technology* 62 (2): 179–187.

55 Jansson, S.E.A., Edsman, C.J., Gedde, U.W., and Hedenqvist, M.S. (2001). Packaging materials for fermented milk: effects of material crystallinity and polarity on food quality. *Packaging Technology and Science: An International Journal* 14 (3): 119–127.

56 Caplice, E. and Fitzgerald, G.F. (1999). Food fermentations: role of microorganisms in food production and preservation. *International Journal of Food Microbiology* 50 (1-2): 131–149.

57 Gill, C.O. and Tan, K.H. (1980). Effect of carbon dioxide on growth of meat spoilage bacteria. *Applied and Environmental Microbiology* 39 (2): 317–319.

58 Rajagopal, M., Werner, B.G., and Hotchkiss, J.H. (2005). Low pressure CO_2 storage of raw milk: microbiological effects. *Journal of Dairy Science* 88 (9): 3130–3138.

59 Day, B.P.F. (2003). Active packaging. In: *Food Packaging Technologies* (ed. R. Coles, D. McDowell, and M. Kirwan), 282–302. Boca Raton: CRC Press.

60 Gonzalez-Fandos, E., Sanz, S., and Olarte, C. (2000). Microbiological, physicochemical and sensory characteristics of Cameros cheese packaged under modified atmospheres. *Food Microbiology* 17 (4): 407–414.

61 Chen, J. and Brody, A.L. (2013). Use of active packaging structures to control the microbial quality of a ready-to-eat meat product. *Food Control* 30 (1): 306–310.

62 Chaix, E., Guillaume, C., and Guillard, V. (2014). Oxygen and carbon dioxide solubility and diffusivity in solid food matrices: a review of past and current knowledge. *Comprehensive Reviews in Food Science and Food Safety* 13 (3): 261–286.

63 Smith, J.P., Daifas, D.P., El-Khoury, W. et al. (2004). Shelf life and safety concerns of bakery products – a review. *Critical Reviews in Food Science and Nutrition* 44 (1): 19–55.

64 Hempel, A.W., O'Sullivan, M.G., Papkovsky, D.B., and Kerry, J.P. (2013). Use of smart packaging technologies for monitoring and extending the shelf-life quality of modified atmosphere packaged (MAP) bread: application of intelligent oxygen sensors and active ethanol emitters. *European Food Research and Technology* 237: 117–124.

65 Kotsianis, I.S., Giannou, V., and Tzia, C. (2002). Production and packaging of bakery products using MAP technology. *Trends in Food Science and Technology* 13 (9–10): 319–324.

66 Rodriguez-Aguilera, R. and Oliveira, J.C. (2009). Review of design engineering methods and applications of active and modified atmosphere packaging systems. *Food Engineering Reviews* 1: 66–83.

67 Qian, M., Liu, D., Zhang, X. et al. (2021). A review of active packaging in bakery products: applications and future trends. *Trends in Food Science and Technology* 114: 459–471.

6

Ethylene Scavengers for Active Packaging of Fresh Horticultural Produce

Gurbuz Gunes

Department of Food Engineering, Faculty of Chemical and Metallurgical Engineering, Istanbul Technical University, Istanbul, Turkey

6.1 Introduction

Fresh horticultural produce includes fresh fruits and vegetable products for human consumption. Fresh produce products constitute an important part of our diet. They are rich in vitamins, minerals, dietary fibers, and bioactive constituents that are important for consumer health. Fresh fruits and vegetables have living cells that are metabolically active during postharvest storage until their fresh consumption. The energy required for these metabolic activities is generated by cellular respiration. Cellular respiration takes place in mitochondria in a series of complicated biochemical reactions that utilize a reserve of carbon sources, several enzymes, and molecular oxygen, resulting in the emission of CO_2 and H_2O. There is a close association between the rate of respiration and the storage life of fresh produce: the higher the respiration rates, the lower the storage life. Fresh produce is classified as climacteric or non-climacteric based on its postharvest ripening and ethylene metabolism. A climacteric fresh produce (mainly fruits) can ripen and synthesize a substantial amount of ethylene during postharvest life. Climacteric fruits are usually harvested at the mature but unripe stage so that they can resist mechanical damage during postharvest handling operations and have a longer storage life. This also enables long-distance shipping to distant markets. These commodities are ripened during storage before being served to the final consumers. Fully ripe fruits are extremely sensitive to mechanical damage during postharvest handling. Reductions in firmness and acidity, increase in sugar level, and color change are typical changes observed during ripening. Controlling the ripening process is a tool for extension of storage life. Synthesis of ethylene initiates and accelerates the ripening process. Ethylene has autocatalytic activity in its synthesis, and this is associated with accelerated ripening. Therefore, control of ethylene synthesis and its action on fresh produce is an important strategy for the extension of storage life. On the other hand, exposure of climacteric fresh fruits to

Smart Food Packaging Systems: Innovations and Technology Applications, First Edition.
Edited by Avik Mukherjee, Santosh Kumar, Manjusri Misra, and Amar K. Mohanty.
© 2025 John Wiley & Sons Ltd. Published 2025 by John Wiley & Sons Ltd.

external ethylene in a controlled environment (artificial accelerated ripening) is used in the ripening of fruits that have been harvested at an unripe stage before marketing.

Non-climacteric fresh produce must be harvested at the horticultural maturity that is optimum for consumption or further processing since it cannot ripen after harvest. These commodities normally do not produce substantial ethylene during postharvest unless mechanically injured, but they are sensitive to external ethylene exposure. Thus, the control of ethylene in the surrounding environment of non-climacteric fruits is also important for quality maintenance and shelf-life extension. Minimally processed fresh produce (fresh-cut produce products) is an emerging class of products that also contain living cells. The process involves peeling, size reductions, and some nonthermal preservation treatments so that the resulting products are ready-to-eat or ready to cook while retaining their freshness. Because the tissues are damaged/wounded during minimal processing, respiration and ethylene production increase, and thus they become even more sensitive to ethylene exposure, regardless of whether they are climacteric or non-climacteric. Therefore, it is important to minimize wound-induced and/or external ethylene for fresh-cut products during storage life.

Packaging is an inevitable part of fresh produce products. Active packaging involves the inclusion of some active ingredients, such as antioxidants, antimicrobials, gas emitters, or absorbers, that are added to the packaging material or headspace of the package to preserve the quality of foods. Ethylene scavengers can be included in active packaging to eliminate the accumulation of ethylene in the headspace of fresh produce packages. Ethylene scavengers can be applied in a sachet that is added to the package headspace, or they can be impregnated in the packaging material that is to be used in fresh produce packaging. They would control ethylene-induced quality degradations and contribute to the extension of the shelf life of the product. Conscious reviews on ethylene scavengers and packaging applications have been published [1–5]. This chapter discusses the use of ethylene scavengers in the active packaging of fresh produce with up-to-date information.

6.2 Biosynthesis Pathway and Effects of Ethylene on Ripening of Fruits and Vegetables

6.2.1 Ethylene Biosynthesis

Ethylene is a simple gaseous molecule naturally synthesized by fresh produce. It regulates, at nanomolar concentrations, a lot of biochemical changes in fresh produce, such as fruit ripening and senescence, seed germination, adventitious root formation, leaf and flower senescence, and responses to stress. It has a strong impact on postharvest quality maintenance of especially climacteric fruits. The ethylene biosynthetic pathway was first proposed by Yang in the 1950s. The details of the biosynthesis are described by the Yang cycle or methionine cycle as shown in Figure 6.1 [6]. The precursor of ethylene is 1-aminocyclopropane-1-carboxylic acid (ACC). L-Methionine is first converted to *S*-adenosyl-methionine (SAM) by SAM synthase enzyme. The resultant SAM is then transformed to 5-methyltioadenosine and ACC by the ACC synthase enzyme. The 5-methyltioadenosine is further utilized in the regeneration of

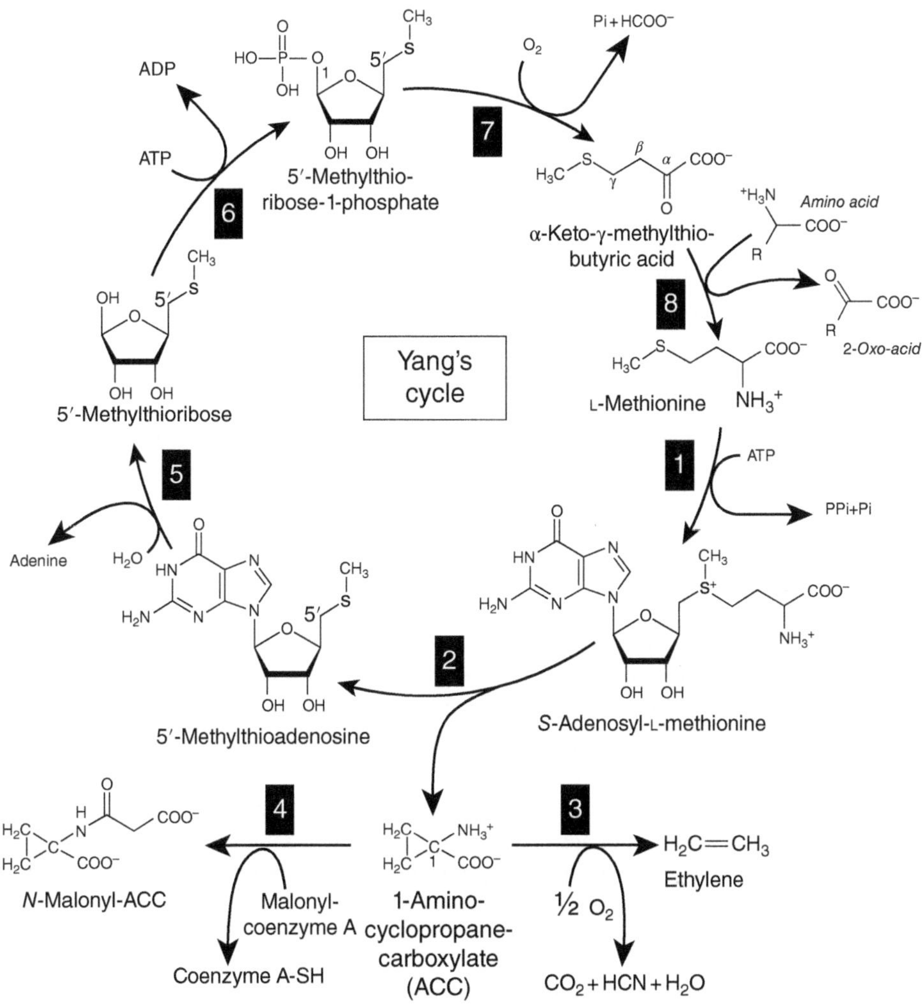

Figure 6.1 Ethylene biosynthesis in plant cells as illustrated by Yang's cycle. *Source:* Reproduced from Botton et al. [6]/with permission of Elsevier.

L-methionine in the Yang cycle. The final product, ethylene, is synthesized from the oxidation of ACC-by-ACC oxidase in the presence of molecular oxygen. Thus, ethylene synthesis is an O_2-dependent reaction and is affected by the concentration of O_2 in the environment surrounding the product. The final step of the reaction also produces CO_2, HCN, and H_2O. Ethylene can increase its biosynthesis through upregulation of ACC synthase genes, and this effect is known as autocatalytic effect. Ethylene is perceived by a collection of receptors present in the endoplasmic reticulum, and this perception results in activation of various genes that take part in various physiological reactions, including its biosynthesis. Chemical characteristics of ethylene, such as being able to freely pass the membrane, get solubilized, and be freely diffused in the cellular environment, facilitate its perception through binding its receptors.

6.2.2 Effects of Ethylene on Ripening

Some fruits, such as tomatoes, apples, pears, plums, peaches, strawberries, cherries, and grapes, are consumed at the ripe stage while others, such as cucumber, green peas, green beans, and zucchini, are consumed at the unripe stage. Ripening occupies the stage between the later stage of growth and development and the early stage of senescence. Ripening processes involve many biochemical reactions, resulting in a lot of changes in composition, color, texture, and overall sensory attributes that determine the eating quality of fruits that are consumed after ripening. Ethylene is known as the primary actor (a hormone) in regulating fruit ripening and general plant senescence. All the ripening-related changes in fresh produce such as loss of chlorophyll and color change, formation of aroma and flavor compounds, loss of firmness, and ethylene biosynthesis are stimulated by ethylene during the pre- and postharvest periods. Various stress conditions, such as mechanical damage, wilting, and pathogen, attack can also enhance ethylene production and action. The effect of ethylene on ripening is especially important in climacteric fruits. Climacteric fruits are distinguished from non-climacteric ones by a sharp increase in respiration and ethylene production rate at the onset of ripening. Climacteric fruits can ripen after harvest, while this is not the case with non-climacteric ones. Some common examples of climacteric and non-climacteric fruits are listed in Table 6.1. Ethylene regulates all the ripening-related phenomena in the climacteric fruits, while its role in non-climacteric ripening is minimal. On the other hand, it is known that there are some ethylene-independent ripening pathways in climacteric fruits and some ethylene-dependent ripening pathways in non-climacteric fruits [7]. It must be emphasized that ethylene has a significant impact on the qualities of both climacteric and non-climacteric fruits in the postharvest period.

Ethylene produced by fresh produce (especially the climacteric ones) can accumulate in the storage environment and act as an external source of ethylene. Ethylene can also be released from other sources, such as pyrolysis, of hydrocarbon fuels and microbial

Table 6.1 Some common examples of climacteric and non-climacteric fruits.

Climacteric fruits	Non-climacteric fruits
Apples	Berries
Pears	Cherries
Peaches	Watermelon
Tomatoes	Grapes
Plums	Citrus fruits
Apricot	Pineapple
Banana	Pepper
Melon (cantaloupe, honeydew)	Olive
Mango	Cranberry
Papaya	Eggplant

activities. During the postharvest period, exposure of fruits and vegetables to ethylene can have significant impacts on their quality. These impacts may be desirable in some cases while undesirable in other cases. Ethylene binds to specific receptors producing signals that are transduced to initiate several biochemical reactions that include ripening metabolisms. The typical effect of ethylene exposure on climacteric fruits is the acceleration of ripening and related quality changes. This is considered undesirable because it shortens storage life unless it is intentionally utilized to ripen fruits before marketing. Exposure of non-climacteric ripe fruits to ethylene during postharvest would have detrimental effects such as increased susceptibility to pathogens, chilling injury, and other stress conditions. Ethylene has autocatalytic effects on its biosynthesis, so exposure of fresh commodities to ethylene accelerates ethylene biosynthesis, and thus the ethylene actions are also accelerated [8]. Therefore, a common strategy to extend the storage life of fresh produce is to avoid their exposure to external or internal ethylene during postharvest storage. It is recommended to keep the ethylene concentration in the storage atmosphere below $0.015\,\mu l\,l^{-1}$ [9].

The control of ethylene action on fresh produce can be achieved by methods targeting biosynthesis, receptor level, or minimization of ethylene concentration in the environment. The use of some inhibitors such as aminoethoxyvinylglycine (AVG) and aminooxyacetic acid (AOA) for ACC synthase can be considered for the reduction of ethylene precursor (ACC) concentration [10]. Another approach could be the use of ACC oxidase inhibitors like Co^{2+}, α-aminoisobutyric acid (AIB), ethanol, and acetaldehyde. Controlled or modified atmospheres with reduced O_2 and elevated CO_2 can also decrease ethylene production. Treatments with naturally occurring polyamines such as spermine, spermidine, and putrescine reduce ethylene biosynthesis by interacting with SAM, an ethylene precursor. The impact of ethylene can also be controlled at the receptor levels by various treatments such as 1-MCP, silver thiosulfate, and irradiated diazocyclopentadiene [10]. Among these, 1-MCP has the most potential and is commercially applicable as it is safe to use in foods and is very effective at low concentrations [11]. Another approach is to prevent ethylene accumulation in the atmosphere surrounding fresh produce. This can be achieved by using various ethylene absorbers/scavengers and ethylene scrubbers in the packages and storage areas.

6.3 Types of Ethylene Scavengers

As described earlier, ethylene accelerates its biosynthesis, known as autocatalytic effects. Thus, the removal of ethylene from the environment where the fresh produce is present is very important for inhibiting ethylene action. Ethylene scavengers of various types are utilized to do this during the storage, transportation, and packaging of fresh produce (Table 6.2). Some scavengers are involved in the oxidation of ethylene, while others are simply adsorbing/absorbing it, making ethylene unavailable for its action. Some compounds, on the other hand, do not interact with ethylene but prevent its action through binding and locking ethylene receptors in the cell. Specific compounds in each of these groups are briefly discussed below.

Table 6.2 Classification of ethylene control systems.

Oxidizers	Adsorbers	Receptor blockers
Ethylene control systems		
Potassium permanganate	Activated carbon	1-Mehylcyclopropene
Sodium permanganate	Zeolites	
Titanium dioxide	Clays	
Novel oxidizers (ZSM-5/WO_3-Pt-nanorod, Pt-loaded mesoporous silica)	Metal–organic frameworks	

6.3.1 Potassium and Sodium Permanganate

Potassium permanganate ($KMnO_4$) is a well-known ethylene scavenger that oxidizes ethylene to CO_2 and H_2O as well as some carboxylic acids and alcohols [12, 13]. The purple color of $KMnO_4$ changes to brown during oxidation (MnO_2^- is reduced to MnO_2), providing a visual indication of its activity. An efficient contact of $KMnO_4$ with ethylene is required for effective scavenging of ethylene. This can be facilitated by adsorbing $KMnO_4$ onto various carriers such as celite, alumina, vermiculite, zeolite, perlite, and clay. The performances of these carriers are positively correlated with their bulk density or porosity. The design of devices containing $KMnO_4$ carriers is also important, as the efficient contact of the ethylene-containing atmosphere with these scrubbers is the key to their performance. $KMnO_4$ is toxic and thus must not be in contact with foods. This is the main limitation of the use of $KMnO_4$, especially in packaging applications. Thus, it is utilized in some other materials, like sachets, filters, etc. In a sachet system, $KMnO_4$ on a carrier like alumina is placed in a small package that is highly permeable to ethylene. The rate of ethylene oxidation and the total oxidation capacity are the two critical characteristics that should be considered in practical use. The oxidation rate of a sachet determines how fast the ethylene is removed from the environment, while the total oxidation capacity is the maximum amount of ethylene that can be oxidized by a sachet. These properties depend on the type, characteristics, and amount of carrier in the sachet, as well as the amount of $KMnO_4$ adsorbed onto it. Sodium permanganate ($NaMnO_4$) in an aqueous solution can also be impregnated in a carrier, in a similar way to $KMnO_4$, and used to scavenge ethylene through oxidation. It was found that $NaMnO_4$ has better efficiency in ethylene removal than $KMnO_4$ due to its higher solubility in water, and thus higher amount in a carrier is possible [14]. However, its commercial use is limited due to the higher cost of its production [1].

6.3.2 Titanium Oxide

Titanium oxide (TiO_2) catalyzes photooxidation of ethylene to CO_2 and H_2O. Ultraviolet (UV) light around 380 nm wavelength together with O_2 and H_2O is required for TiO_2 to oxidize ethylene. Among the three polymorphic structures, anatase has the most photocatalytic activity while brookite has the least activity [15]. Hussain et al. [16] developed optimized titanium nanoparticles with higher anatase content and found that they had higher photocatalytic degradation of ethylene than Degussa P25 material. Reduction of particle size, increase of surface area, improved dispersion, and doping with metal ions and nonmetals are the factors affecting the capacity of TiO_2 for oxidizing ethylene [17, 18]. TiO_2 is relatively cheap and approved as a food additive by the U.S. FDA; thus, it can be used in food contact systems [18]. However, the required UV light exposure limits its practical use in packaging applications. The potential degradative effects of TiO_2 on volatile flavor compounds can also be a concern.

6.3.3 Novel Oxidizers

Platinum catalyst supported on mesoporous silica with enhanced hydrophobicity was shown to oxidize ethylene effectively [19]. Similarly, a nanocomposite containing Pt-loaded WO_3 nanorods on ZSM-5 was developed by Kim et al. [20]. This catalyst was shown to be activated by visible light, which can be an advantage over UV-light-activated TiO_2-based catalysts. Na et al. [21] developed another platinum-containing catalyst, Pt-loaded mesoporous silica, that was coated on ceramic scaffolds. The authors found that this material effectively decomposed ethylene with the following proposed mechanism: the platinum particles adsorb ethylene and O_2, and the double bond of ethylene is attacked by atomic oxygen, forming formaldehyde that is eventually converted to water and CO_2.

6.3.4 Activated Carbon

Activated carbon is a material with a high surface area that can remove ethylene by adsorption. Materials with low inorganic matter such as wood, nut shells, fruit pulps, stalks, and coal are good sources of activated carbon. The surface area, chemistry, porosity, temperature, and humidity affect ethylene adsorption onto activated carbon. Depending on the methods and the raw materials activated, carbons with surface area in the range of 300–5000 $m^2 g^{-1}$ in granular, powdered, or fiber forms can be obtained [22]. Granular forms are shown to adsorb more ethylene than the other forms. Activation with hydrogen, thermal treatment, and palladium impregnation increase the adsorption capacity of activated carbon [23].

6.3.5 Zeolite

Zeolites are volcanic crystalline aluminosilicates with microporous structures and negative charges. The porous structure with interconnected cages and channels enables the absorption of volatiles such as ethylene. More than a hundred types of zeolites, either natural or synthetic, are available. Zeolites have been integrated into different materials, such as filters and packaging films, or contained in sachets for ethylene adsorption. The relatively low ethylene adsorption capacity of zeolites can be improved by modifying their surface by cationic materials. Modification of zeolites with sodium, potassium, rubidium, and cesium caused a significant increase in their ethylene adsorption [24–26]. Palladium is another material that enhances the ethylene-scavenging capacity of adsorbers like zeolite and activated carbon when it is impregnated onto them [23, 27, 28].

6.3.6 Clays

Clays are also aluminosilicates. Their crystal structure is tetrahedral or octahedral in the form of a layer [1]. These layers have different ions dominated: Si^{4+} in tetrahedral and Mg^{2+} in octahedral [2]. Halloysite, bentonite, sepiolite, and kaolinite are some examples of clays, and their ethylene adsorption potentials are varied depending on microstructure, temperature, and activation methods like alkali, acid, drying, etc. [22, 29–31]. Among these clays, halloysite nanotubes show great potential in active packaging systems with their good dispersing properties in polymers, GRAS status, economy, and suitability of encapsulation [2]. Clays are usually contained in sachets made of ethylene-permeable materials, but their integration into the packaging materials is also of interest.

6.3.7 Metal–Organic Frameworks

A synthetic porous structure called metal–organic frameworks (MOFs) is formed when groups of metal ions bind to an organic molecule. The MOFs have an extremely high surface area and can be synthesized with different microstructures using different organic and inorganic building blocks [32]. Thus, their ethylene adsorption capacities are higher than the other adsorbers mentioned above under ambient relative humidity or dry conditions. On the other hand, ethylene absorption by MOFs is low under high relative humidity as water gets absorbed rather than ethylene. This could be a major limitation since fresh produce is usually stored at high relative humidity. Synthesis of MOFs with lower hydrophilicity would reduce this disadvantage.

6.3.8 Methylcyclopropene

1-methylcyclopropene (1-MCP) is a compound having a very high affinity for ethylene receptors in plant cells [11]. Thus, it does not physically interact with ethylene to remove it from the environment, but it inhibits ethylene perception by the cells. 1-MCP binds ethylene receptors irreversibly, resulting in fewer receptors available for ethylene. It is effective at extremely low concentrations in the range of $2.5\,nl\,l^{-1}$–$1\,\mu l\,l^{-1}$ [11]. Various commercial powdered formulas based on α-cyclodextrin are available, and they are mixed with water to release 1-MCP. Inhibition of the binding of ethylene to their receptor would not only control all ethylene-induced physiological changes but also the autocatalytic ethylene production.

6.4 Principles and Mechanisms of Ethylene Scavenging

Ethylene scavenging can be achieved by either chemical or physical mechanisms. The chemical mechanism involves the conversion of ethylene to other compounds through some chemical reactions. One very well-known chemical scavenging is the oxidation of ethylene by $KMnO_4$ to CO_2 and H_2O. The double bond in the ethylene is attacked by $KMnO_4$ in the oxidation reaction. The purple color of $KMnO_4$ transforms to a brown color due to MnO_2 in the reaction. As mentioned in the earlier section, $KMnO_4$ is the most commercially utilized ethylene scavenger in the industry. However, $KMnO_4$ must not be in contact with the food due to its toxicity. Thus, it is usually impregnated into some carriers, like

SiO_2, nanoparticles, and contained in a highly permeable sachet. Another chemical scavenging is through titanium oxide (TiO_2), which exhibits a photocatalytic oxidation of ethylene to CO_2 and H_2O upon exposure to UV light around 380 nm [16]. The UV light excites electrons on the surface of TiO_2 generating free radicals in the presence of O_2 and water [17]. Although the exact mechanism is not known, these free radicals are thought to take part in the oxidation of ethylene. The requirement of the special UV wavelength, deactivation of TiO_2, and low efficiency are drawbacks for the practical application of this system, but chemical modification of the catalyst and improvements in the design of the system can reduce these drawbacks. For example, the use of metal ions (vanadium, chromium, manganese, iron, etc.) or nonmetals (carbon, nitrogen, etc.) together with TiO_2 has been shown to oxidize ethylene with visible light [17].

Ethylene can also be scavenged physically through its adsorption by some materials, such as activated carbon and zeolite. Ethylene can be adsorbed onto the adsorbers by two mechanisms. One of them is a physical binding through van der Waals forces, and the other one is chemical adsorption through a chemical bond at the active adsorption centers on the adsorbers. The rate and capacity of adsorption depend on many factors including ethylene concentration, surface area to volume ratio of adsorber, pore structure, surface chemistry of adsorber, temperature, and pressure. The chemistry of the adsorption centers would have a key role in ethylene adsorption. For instance, the adsorption centers in sepiolites have oxygen ions, water molecules, Mg ions, and groups of SiOH in their microstructure, through which ethylene may interact [33]. The activation methods (acid, alkali, H_2, salt, impregnation with palladium, heat treatment, etc.) applied to the adsorbers would modify the surface chemistry and have a significant impact on their ethylene adsorption [34]. For example, the alkaline treatment of halloysite and acid treatment of sepiolites increased their ethylene-scavenging capacity [22, 29]. It is thought that ethylene binds to the metals on the surface of MOFs through an electrostatic interaction. The particle size of an adsorber is a key factor affecting its ethylene-scavenging activity. Lower particle size would increase the surface area and thus the contact with the ethylene, resulting in higher adsorption [34, 35]. Ethylene-scavenging capacity is generally increased with increased temperature, but considering a narrow range of storage temperature of the packaged fresh produce, its impact would be limited. The package headspace of fresh produce would have high relative humidity and contain O_2 and CO_2 in addition to ethylene. Thus, the impacts of these environments on the ethylene-scavenging activity of particular adsorbers must be considered. Apart from chemical and physical scavenging as described above, ethylene can also be controlled at the receptor level. 1-MCP does not remove ethylene from the environment, but it controls its action by occupying the ethylene receptors [36]. 1-MCP has a very high affinity for ethylene receptors and competes with ethylene to bind the receptors. This makes the cells less responsive to ethylene present in the cell and environment.

6.5 Advances and Applications of Ethylene-Scavenging Systems

Beyond the chemical degradation of ethylene by $KMnO_4$, photocatalytic degradation of ethylene is another means of eliminating ethylene from the storage and/or package atmosphere. Titanium oxide (TiO_2) nanoparticles have been utilized in the photocatalytic degradation of ethylene. Ethylene is degraded by UV light in the presence of TiO_2 nanoparticles [16]. The TiO_2 nanoparticles act as a photocatalyst without which the ethylene is not degraded. The presence of humidity and O_2 is also required, which are thought to be involved in the generation of free radicals that react with ethylene. The higher surface area of the TiO_2 photocatalyst also has positive effects on ethylene degradation. The incorporation of ethylene scavengers into the packaging material rather than their use in sachets has also been a great subject of interest. Active packaging systems that have ethylene scavengers are emerging. Zeolite, clays, activated carbons, TiO_2, and $KMnO_4$ impregnated in SiO_2 have been incorporated into polymeric films in various studies. The active packaging films showed ethylene-scavenging activity, but in some instances, they had adverse effects on the gas barrier and optical and mechanical barrier properties. Incorporating ethylene scavengers into a packaging material can be done by coating it, mixing it with the polymer melt in the extrusion process, or solvent casting. A successful incorporation of halloysite nanotubes into polyethylene film has been made, and the resultant active film performed well in ethylene scavenging and packaging of fresh produce such as tomato and banana [31]. The ethylene-scavenging capacity of halloysite nanotubes was significantly improved by alkaline treatment [22]. Active packaging films with ethylene scavengers have generally a lower capacity of ethylene absorption due to the lower quantity of scavengers that can be added to the polymer as compared to the sachet system. Incorporation of scavengers into polymeric films homogeneously without adversely affecting film properties remains the major challenge. The sachet system can be in different forms, such as porous or highly permeable bags, blankets, and slabs that can be utilized in package headspace or storage and transport vehicles. The capacity of the sachet system can be adjusted to the desired levels, and they can be protected from ethylene in the ambient conditions before their use. The development of sachet material that is highly permeable to ethylene but effectively separates its content from the packaged product is needed for improved performance.

Another approach in ethylene scavenging is to target ethylene receptors. Ethylene must bind to its receptor to show its physiological effects. Blocking of ethylene receptors can be achieved by 1-MCP, AVG, and silver thiosulfate. Among these, 1-MCP has captured significant interest, and its commercial utilization has increased. However, 1-MCP and the other receptor blockers are mainly applied to fruits before packaging and storage; their use in active packaging systems has not been well established yet [37]. Active packaging systems with 1-MCP released to the headspace would have great potential. Effective impregnation of polymeric packaging films with 1-MCP was shown to be a challenge; the solution casting approach was not successful, while hot-pressing polymer with 1-MCP showed some promise [38]. The authors showed that low-temperature hot-pressed films released a significant level of 1-MCP that would have an impact on packaged produce. 1-MCP contained in paper or sachets can also be included in packages of fresh produce. Lee et al. [39] showed that 1-mcp loaded with silica in sachets of PVA released 1-MCP under high humidity conditions, and thus it can have the potential to be used in active packaging of fresh produce. There is also interest in storing 1-MCP in MOFs and using them in fresh produce packages so that 1-MCP can be released back into the headspace of the package [40, 41]. However, sufficient release of 1-MCP adsorbed by the MOFs in the package is a challenge to be resolved. Controlled release systems integrated with packages are a growing interest with high potential [40]. Some specific applications of active packaging with ethylene control systems on various fresh produce are briefly discussed in the following section.

6.5.1 Banana

Banana is a climacteric fruit that is highly sensitive to ethylene. The fruit is harvested at the mature green stage for longer storage life. Postharvest exposure to endogenous or external ethylene stimulates the ripening process and thus shortens their shelf life. Therefore, inhibition of their ethylene production and avoidance of their exposure to ethylene are desired for banana storage. Active packaging with ethylene scavenging is a tool that can be used in this perspective. Several research results have confirmed the efficiency of this approach in the preservation of the fruit. Nano-$KMnO_4$ treated with silane and zinc stearate was incorporated into LDPE film at up to 7% by weight and tested on bananas by Khosravi et al. [42]. They found that the nanocomposite film delayed color change and weight loss and thus extended the shelf life of banana as compared with a neat LDPE film. Hassan et al. [43] used sachets containing $KMnO_4$ adsorbed onto vermiculite (5 ml/5 g) in banana packages made of LDPE and reported a seven-day extension of shelf life at 19–25 °C based on color change over the LDPE control packages without the sachets. Similarly, a shelf-life extension of banana packaged with bricks containing 50 ml of $KMnO_4$ adsorbed to clay was reported at 25 °C in another work [44]. Palladium-loaded zeolite was shown to scavenge ethylene more effectively than neat zeolite and inhibited ethylene production and ripening of bananas in glass jars [28]. Zhu et al. [45] stored bananas in a box lined with nanofibers of polyacrylonitrile with electrospun TiO_2 while exposed to UVA light at 23 °C. They observed a significant inhibition of yellowing and softening in the fruit stored with the nanofiber compared to the PP control. Packaging bananas in LDPE films incorporated with zeolites (2%) and halloysite nanotubes (5%) as ethylene adsorber resulted in better preservation, delayed ripening, and shelf-life extension [31, 46]. Packaging bananas in the presence of $KMnO_4$-based scavengers resulted in a slower increase in the pulp/peel ratio, a quality parameter that increases during ripening [47, 48]. Tirgar et al. [49] developed a membrane having $KMnO_4$ loaded onto alumina nanofibers and tested it in banana packaging. They found that the film had a great capacity to scavenge ethylene and inhibit softening and color change in bananas. Bananas stored in a polyethylene nanocomposite film with halloysite nanotubes showed slower ripening and better quality than the control films [31].

6.5.2 Mango

Mango is another climacteric fruit that can be preserved by controlling postharvest ripening. Active packaging with $KMnO_4$ sachets or films resulted in the inhibition of softening and decay in mango during storage [50]. Zeolite was incorporated into an LDPE-based film and used in mango packaging in another work [51]. The authors observed complete prevention of ethylene accumulation together with inhibitions of weight loss, growth of fungi, and softening, all resulting in a 20-day increase in shelf life compared to a control package. On the other hand, $KMnO_4$-based scavengers resulted in no significant effect on the loss of firmness of mangoes though they caused a lower weight loss in another study [52].

6.5.3 Tomato

Tomato is another climacteric fruit that is usually harvested at the mature green stage for longer storage life. Shelf-life extension of tomato fruit can be achieved by controlling the postharvest ripening process. This can be accomplished by controlling ethylene in the storage environment. The impact of ethylene-scavenging applications on tomato fruit shelf life and quality has been investigated in several studies, some of which are briefly explained here. Tomatoes packaged in films with TiO_2 (either coated or integrated) showed a reduction in ethylene production, delayed ripening, less color change, and delayed softening [53–55]. Paperboard boxes incorporated with activated carbon were shown to absorb ethylene effectively and result in delayed ripening of tomatoes in another study [56]. Sachets containing activated carbon alone or with palladium acetate resulted in the inhibition of color change, weight loss, and softening through reduced ethylene accumulation in tomatoes during storage [35, 57]. Tomatoes stored in PET containers with zeolites and $KMnO_4$ showed delayed ripening as assessed by reductions in softening, soluble solid content, and weight loss [58]. Alvarez-Hernandez et al. [59] developed antifungal and ethylene scavenger sachets based on thymol and $KMnO_4$-loaded sepiolite and tested them on cherry tomatoes. They found that the ethylene scavenger component contributed to the control of fungal growth in addition to its preserving effects on the fruit quality. A nanocomposite LDPE film with alkaline halloysite nanotubes was successfully developed, and the film performed well in preserving firmness and weight loss in cherry tomatoes [60]. Tas et al. [31] developed a halloysite nanotube-polyethylene nanocomposite film and found that tomatoes packaged in this film retained their firmness better than the ones in neat PE film. In addition, a novel ethylene scavenger that can be activated by visible light and is made of platinum-loaded WO3 nanorods deposited on ZSM-5 was developed, and the test done with tomatoes showed that the scavenger significantly inhibited ripening of mature green tomatoes during 12-day storage at 20–25 °C [20].

6.5.4 Apple

Apple is a popular climacteric fruit with several cultivars throughout the world. The fruit can be available throughout the year thanks to advanced postharvest treatments and storage technologies such as 1-MCP, ethylene scavenging, and controlled atmosphere storage. The postharvest technologies control the ripening process and postharvest diseases. Ethylene-scavenging systems are usually integrated into the technologies for the shelf-life extension of apples. One of the earliest applications of activated carbon as an ethylene absorber was performed in the CA storage of McIntosh apples, resulting in an extension of shelf life and inhibition of scald [61]. Physiological disorders such as superficial scald and bitter pit were reduced by removing ethylene using Purafil ($KMnO_4$ adsorbed to aluminum oxide) in PE bags of Granny Smith apples [62]. Sachets of $KMnO_4$ used in apple packages delayed the ripening of apples in other studies [63, 64]. Besides, various $KMnO_4$-based ethylene scavengers showed significant benefits in the maintenance of the quality of apples in different package formats [64, 65]. 1-MCP has been shown to have a great impact on preserving apple quality on a commercial scale as a pretreatment before storage [36]. It would have great potential and convenience if 1-MCP could be utilized as an active agent in the packaging of apples.

6.5.5 Others

Loss of chlorophyll, delayed yellowing, and shelf-life extension of broccoli were achieved by using sachets containing acidified activated carbon powder with palladium chloride and copper sulfate [23]. Esturk et al. [66] reported that broccoli packaged in LDPE film incorporated with zeolite (50%) had 20 days of shelf life with inhibition of ethylene accumulation (<1 ppm) in the headspace, while neat LDPE films resulted in only 15-day shelf life with 61.8 ppm ethylene accumulation during storage at 4 °C. DeEll et al. [67] found that the use of $KMnO_4$ sachets caused a delayed softening in "Marathon" broccoli. Strawberries packaged in LDPE film incorporated with halloysite nanotubes (5%) had lower weight loss during storage at 23 °C [31]. Packaging strawberries in LDPE films with anatase nano-TiO_2 and NDZ-201 resulted in lower ethylene production, weight loss, softening, and decay during dark storage at 4 °C for 14 days [68]. Active packages made of LDPE film incorporated with anatase and rutile TiO_2, kaolin, and nano-Ag brought about a decrease in fruit decay and oxidative enzymes and maintained total soluble solid ascorbic acid and titratable acidity in strawberries during storage at 4 °C [69]. The benefits of LDPE film with nano-TiO_2 in quality maintenance of strawberries during storage at 4 °C were also reported in another work [70]. Sha et al. [71] packaged strawberries in cardboard boxes coated with hydrated SiO_2, $KMnO_4$, and polylysine and found that the fruit's quality (weight loss, membrane integrity, firmness) was better maintained with the coated boxes.

Sachets of $KMnO_4$ in packages of guava resulted in inhibited ripening as assessed by reduced softening and loss of green color [72]. Similar benefits in the quality of guava were achieved with MA-packages containing KMnO4 sachets in another work [73]. An expanded polyethylene foam net was coated with gelatin-TiO_2 nanocomposite and used in the packaging of papaya in another work [74]. The authors found that this active packaging resulted in significantly lower ethylene accumulation and delayed ripening during 4-day storage under UVA light. Similarly, a $KMnO_4$-based scavenger resulted in delayed changes in color and soluble solid content in the "Sunrise Golden" papaya [75]. Emadpour et al. [76] found that nano-zeolites loaded with $KMnO_4$ resulted in lower softening and weight loss and delayed changes in titratable acidity and soluble solid content in cherries. Alvarez-Hernandez et al. [77] developed an ethylene scavenger with $KMnO_4$ supported on sepiolite and tested it on fresh apricots. They found that the scavenger performed very well in removing ethylene from the package headspace and maintaining the quality of the fruits. Bal and Celik [78] demonstrated that the ripening-related changes such as softening, loss of chlorophyll, and changes in titratable acidity and SSC were delayed by using $KMnO_4$-based scavengers in "Hayward" kiwi fruit. Red bell pepper was packaged under MAP with a $KMnO_4$-based scavenger, and it was found that the inclusion of the scavenger in the packages resulted in a lower respiration rate and positive effects on bioactive compounds [79].

Packaging mushrooms in paper coated with molecular sieves loaded with $KMnO_4$ and 1-MCP caused an absorption and reduction in ethylene, inhibition of softening, browning, and weight loss resulting in shelf-life extension [80]. Besides, Wu et al. [81] developed a multilayered biodegradable film having a mid-layer containing 1-MCP and a hydrophilic inner layer and showed that the film performed well in packaging mushrooms through release of 1-MCP and controlling humidity in the headspace. Zhang et al. [82] found positive effects of 1-MCP-containing paper included in the MAP of plums. The authors used a small piece of paper ($2.5\,cm^2$) containing 1-MCP in packages of plums and found that 1-MCP resulted in a reduction of respiration rate, ethylene production rate, weight loss, and maintenance of firmness in plums during storage. Besides, KMnO4-based scavengers in different forms resulted in the inhibition of ripening-related enzymes and shelf-life extension in plums [83].

6.6 Conclusion and Future Perspective

The use of ethylene scavengers in the active packaging of fresh produce can be a convenient tool to maintain quality and extend the shelf life of fresh fruits and vegetables. Fresh produce can synthesize ethylene as part of their metabolism, and the released ethylene induces various degradative changes in product quality. Thus, the removal of ethylene from the package atmosphere or prevention of ethylene signal transduction is required. Ethylene in the headspace of a package can be physically adsorbed onto some scavengers like zeolites, clays, MOFs, etc., or it can be chemically degraded by others such as $KMnO_4$ and TiO_2. The adsorption efficiency of the adsorbers under the storage and packaging conditions of the fresh produce needs to be improved. $KMnO_4$ as an oxidizer of ethylene has been widely used in the fresh produce industry and packaging. However, special care must be taken with $KMnO_4$ not to come into contact with the product, as it is toxic and not allowed. The activity of TiO_2 on ethylene requires UV exposure, which may not be convenient for packaging applications. These drawbacks stimulated the search for novel ethylene oxidizers or adsorbers. Another approach is to use ethylene receptor blockers like 1-MCP in an active packaging system. The challenge in 1-MCP active packages is to release the agent to the headspace of the package in a sufficient amount and in a controlled manner. Some progress has been made on this, but further work and developments are needed for successful commercial exploitation.

References

1 Awalgaonkar, G., Beaudry, R., and Almenar, E. (2020). Ethylene-removing packaging: basis for development and latest advances. *Compehensiver Reviews in Food Science and Food Safety* 19: 3980–4007. https://doi.org/10.1111/1541-4337.12636.

2 Gaikwad, K.K., Sing, S., and Negi, Y.S. (2020). Ethylene scavengers for active packaging of fresh food produce. *Environmental Chemistry Letters* 18: 169–284. https://doi.org/10.1007/s10311-019-00938-1.

3 Sadeghi, K., Lee, Y., and Seo, J. (2021). Ethylene scavenging systems in packaging of fresh produce: a review. *Food Reviews International* 37 (2): 155–176. https://doi.org/10.1080/87559129.2019.1695836.

4 Wei, H., Seidi, F., Zhang, T. et al. (2021). Ethylene scavengers for the preservation of fruits and vegetables: a review. *Food Chemistry* 337: 127750. https://doi.org/10.1016/j.foodchem.2020.127750.

5 Zhang, H., Han, M., Xie, Y. et al. (2022). Applications of ethylene-regulating packaging in post-harvest fruits and vegetables storage: a review. *Packaging Technology and Science* 35: 461–471. https://doi.org/10.1002/pts.2644.

6 Botton, A., Tonutti, P., and Ruperti, B. (2019). Biology and biochemistry of ethylene. In: *Postharvest Physiology and Biochemistry of Fruits and Vegetables* (ed. E.M. Yahia), 93–112. Elsevier Inc. https://doi.org/10.1016/B978-0-12-813278-4.00005-1.

7 Pareek, S. (2016). Ripening physiology: an overview. In: *Postharvest Ripening Physiology of Crops* (ed. S. Pareek), 1–48. CRC Press. https://doi.org/10.1201/b19043.

8 Yang, S.F. and Hoffman, N.E. (1984). Ethylene biosynthesis and its regulation in higher plants. *Annual Reviews in Plant Physiology* 35: 155–189.

9 Wills, R.B.H., Warton, M.A., and Ku, V.V.V. (2000). Ethylene levels associated with fruit and vegetables during marketing. *Australian Journal of Experimental Agriculture* 40: 465–470.

10 Martinez-Romero, D., Bailen, G., Serrano, M. et al. (2007). Tools to maintain postharvest fruit and vegetable quality through the inhibition of ethylene action: a review. *Critical Reviews in Food Science and Nutrition* 47: 543–560. https://doi.org/10.1080/10408390600846390.

11 Blankenship, S.M. and Dole, J.M. (2003). 1-Methylcyclopropene: a review. *Postharvest Biology and Technology* 28: 1–25. https://doi.org/10.1016/S0925-5214(02)00246-6.

12 Alvarez-Hernandez, M.H., Martimez-Hernandez, G.B., Avalos-Belmontes, F. et al. (2019). Potassium permanganate-based ethylene scavengers for fresh horticultural produce as an active packaging. *Food Engineering Reviews* 11: 159–183. https://doi.org/10.1007/s12393-019-09193-0.

13 Ebrahimi, A., Zabihzadeh, K.M., Ahmadi, S. et al. (2022). Novel strategies to sontrol ethylene in fruit and vegetables for extending their shelf life: a review. *International Journal of Environmental Science and Technology* 19 (5): 4599–4610. https://doi.org/10.1007/s13762-021-03485-x.

14 Magargee, R., Ahrens, W., Eyde, D., and Smith, B. (2006). Sodium permanganate ethylene absorption agent. US Patent 2006/0070523 A1.

15 de Matos Fonseca, J., dos Santos Alves, M.J., Soares, L.S. et al. (2021). A review on TiO_2-based photocatalytic systems applied in fruit postharvest: set-ups and perspectives. *Food Research International* 144: 110378.

16 Hussain, M., Bensaid, S., Geobaldo, F. et al. (2011). Photocatalytic degradation of ethylene emitted by fruits with TiO_2 nanoparticles. *Industrial & Engineering Chemistry Research* 50 (5): 2536–2543. https://doi.org/10.1021/ie1005756.

17 Pathak, N., Caleb, O.J., Geyer, M. et al. (2017). Photocatalytic and photochemical oxidation of ethylene: Potential for storage of fresh produce – a review. *Food and Bioprocess Technology* 10 (6): 982–1001. https://doi.org/10.1007/s11947-017-1889-0.

18 Zhang, W. and Rhim, J.W. (2022). Titanium dioxide (TiO_2) for the manufacture of multifunctional active food packaging films. *Food Packaging and Shelf Life* 31: 100806. https://doi.org/10.1016/j.fpsl.2021.100806.

19 Satter, S.S., Hirayama, J., Nakajima, K., and Fukuoka, A. (2018). Low temperature oxidation of trace ethylene over Pt nanoparticles supported on hydrophobic mesoporous silica. *Chemistry Letters* 47: 1000–1002. https://doi.org/10.1246/cl.180364.

20 Kim, S., Jeong, G.H., and Kim, S.W. (2019). Ethylene gas decomposition using ZSM-5/WO3-Pt-nanorod composites for fruit freshness. *ACS Sustainable Chemistry & Engineering* 7 (13): 11250–11257.

21 Na, J., Zheng, D., Kim, J. et al. (2019). Coating of Pt-loaded mesoporous silica layers on ceramics scaffolds for practical preservation system for greengrocery. *ACS Applied Materials & Interfaces* 11 (35): 32084–32089.

22 Gaikwad, K.K., Singh, S., and Lee, Y.S. (2018). High adsorption of ethylene by alkali-treated halloysite nanotubes for food packaging applications. *Environmental Chemistry Letters* 16: 1055–1062. https://doi.org/10.1007/s10311-018-0718-7.

23 Cao, J., Li, X., Wu, K. et al. (2015). Preparation of a novel $PDCL_2$-$CUSO^4$-based ethylene scavenger supported by acidified activated carbon powder and its effects on quality and ethylene metabolism of broccoli during shelf-life. *Postharvest Biology and Technology* 99: 50–57. https://doi.org/10.1016/j.postharvbio.2014.07.017.

24 Alver, B.E. (2013). A comparative adsorption study of C_2H_4 and SO_2 on clinoptilolite-rich tuff: effect of acid treatment. *Journal of Hazardous Materials* 262: 627–633. https://doi.org/10.1016/j.jhazmat.2013.09.014.

25 Patdhanagul, N., Rangsriwatananon, K., Siriwong, K., and Hengrasmee, S. (2012). Combined modification of zeolite NaY by phenyl trimethyl ammonium bromide and potassium for ethylene gas adsorption. *Microporous and Mesoporous Materials* 153: 30–34. https://doi.org/10.1016/j.micromeso.2011.12.007.

26 Sue-aok, N., Srithanratana, T., Rangsriwatananon, K., and Hengrasmee, S. (2010). Study of ethylene adsorption on zeolite NaY modified with group I metal ions. *Applied Surface Science* 56 (12): 3997–4002. https://doi.org/10.1016/j.apsusc.2010.01.065.

27 Smith, A.W.J., Poulston, S., Rowsell, L. et al. (2009). A new Palladium-based ethylene scavenger to control ethylene-induced ripening of climacteric fruit. *Platinum Metals Review* 53 (3): 112–122. https://doi.org/10.1595/147106709X462742.

28 Tzeng, J.H., Weng, C.H., Huang, J.W. et al. (2019). Application of palladium-modified zeolite for prolonging post-harvest shelf life of banana. *Journal of the Science of Food and Agriculture* 99 (7): 3467–3474. https://doi.org/10.1002/jsfa.9565.

29 Alver, B.E. and Sakizci, M. (2012). Ethylene adsorption on acid-treated clay minerals. *Adsorption Science & Technology* 30 (3): 265–273. https://doi.org/10.1260/0263-6174.30.3.265.

30 Sakizci, M. (2013). Effect of salt modification and acid activation on ethylene adsorption properties of sepiolite. *Adsorption* 19 (6): 1083–1091. https://doi.org/10.1007/s10450-013-9531-6.

31 Tas, C.E., Hendessi, S., Baysal, M. et al. (2017). Halloysite nanotubes/polyethylene nanocomposites for active food packaging materials with ethylene scavenging and gas barrier properties. *Food and Bioprocess Technology* 10 (4): 789–798. https://doi.org/10.1007/s11947-017-1860-0.

32 Kuppler, R.J., Timmons, D.J., Fang, Q.R. et al. (2009). Potential applications of metal-organic frameworks. *Coordination Chemistry Reviews* 253 (23–24): 3042–3066. https://doi.org/10.1016/j.ccr.2009.05.019.

33 Galan, E. (1996). Properties and applications of palygorskite-sepiolite clays. *Clay Minerals* 31 (4): 443–453. https://doi.org/10.1180/claymin.1996.031.4.01.

34 Liu, Z.X., Park, J.N., Abdi, S.H.R. et al. (2006). Nano-sized carbon hollow spheres for abatement of ethylene. *Topics in Catalysis* 39 (3–4): 221–226. https://doi.org/10.1007/s11244-006-0060-3.

35 Bailén, G., Guillén, F., Castillo, S. et al. (2007). Use of a palladium catalyst to improve the capacity of activated carbon to absorb ethylene, and its effect on tomato ripening. *Spanish Journal of Agricultural Research* 5 (4): 579–586. https://doi.org/10.5424/sjar/2007054- 5359.

36 Watkins, C.B. and Nock, J.F. (2012). Rapid 1-methylcyclopropene (1-MCP) treatment and delayed controlled atmosphere storage of apples. *Postharvest Biology and Technology* 69: 24–31.

37 Hu, Z., Tang, C., He, Z. et al. (2017). 1-Methylcyclopropene (MCP)-containing cellulose paper packaging for fresh fruit and vegetable preservation: a review. *BioResources* 12 (1): 2234–2248. https://doi.org/10.15376/biores.12.1.2234-2248.

38 Hotchkiss, J.H., Watkins, C.B., and Sanchez, D.G. (2007). Release of 1-methylcyclopropene from heat-pressed polymer films. *Journal of Food Science* 72 (5): E330–E334. https://doi.org/10.1111/jfpp.12197.

39 Lee, Y.S., Beaudry, R., Kim, J.N., and Harte, B.R. (2006). Development of a 1-methylcyclopropene (1-MCP) sachet release system. *Journal of Food Science* 71 (1): C1–C6.

40 Chen, M., Chen, X., Ray, S., and Yam, K. (2020). Stabilization and controlled release of gaseous/volatile active compounds to improve safety and quality of fresh produce. *Trends in Food Science and Technology* 95: 33–44. https://doi.org/10.1016/j.tifs.2019.11.005.

41 Pnevskaya, A.Y. and Bugaev, A.L. (2023). Theoretical screening of M3(btc)2 metal-organic frameworks for ethylene and 1-methylcyclopropene storage. *Results in Chemistry* 5: 100831.

42 Khosravi, R., Hashemi, S.A., Sabet, S.A., and Rezadoust, A.M. (2013). Thermal, dynamic mechanical, and barrier studies of potassium permanganate-LDPE nanocomposites. *Polymer-Plastics Technology and Engineering* 52 (2): 126–132. https://doi.org/10.1080/03602559.2012.719056.

43 Hassan, M.K., Shipton, W.A., Coventry, R.J., and Gardiner, C.P. (2005). Maintenance of fruit quality in organically-grown bananas under modified atmosphere conditions. *Asian Journal of Plant Sciences* 4: 409–412. https://doi.org/10.3923/ajps.2005.409.412.

44 Illeperuma, K., Chamara, D., and Galappatty, P.T. (2000). Effect of modified atmosphere and ethylene absorbers on extension of storage life of "Kolikuttu" banana at ambient temperature. *Fruits* 55 (6): 381–388.

45 Zhu, Z., Zhang, Y., Shang, Y., and Wen, Y. (2019). Electrospun nanofibers containing TiO_2 for the photocatalytic degradation of ethylene and delaying postharvest ripening of bananas. *Food and Bioprocess Technology* 12: 281–287. https://doi.org/10.1007/s11947-018-2207-1.

46 Li, D.L., Shi, Q.P., and Xu, W.C. (2012). Effects of zeolite modified LDPE film on banana fresh keeping. *Advanced Materials Research* 393–395: 724–728. https://doi.org/10.4028/www.scientific.net/AMR.393-395.724.

47 Chauhan, O., Raju, P., Dasgupta, D., and Bawa, A. (2006). Modified atmosphere packaging of banana (cv. Pachbale) with ethylene, carbon dioxide, and moisture scrubbers and effect on its ripening behavior. *American Journal of Food Technology* 1: 179–189. https://doi.org/10.3923/ajft. 2006.179.189.

48 Zewter, A., Woldetsadik, K., and Workneh, T. (2012). Effect of 1-methylcyclopropene, potassium permanganate and packaging on quality of banana. *African Journal of Agricultural Research* 7: 2425–2437. https://doi.org/10.5897/AJAR11.1203.

49 Tirgar, A., Han, D., and Steckl, A.J. (2018). Absorption of ethylene on membranes containing potassium permanganate loaded into alumina-nanoparticle-incorporated alumina/carbon nanofibers. *Journal of Agricultural and Food Chemistry* 66: 5635–5643. https://doi.org/10.1021/acs. jafc.7b05037.

50 Castro, J.V., Pfaffenbach, L.B., and Carvalho, C.R.L. (2005). Effects of film packaging and cold storage on postharvest quality of 'tommy atkins' mangoes. *Acta Horticulturae* 682: 1683–1688.

51 Boonruang, K., Chonhenchob, V., Singh, S.P. et al. (2012). Comparison of various packaging films for mango export. *Packaging Technology and Science* 25: 107–118.

52 Ezz, T. and Awad, R. (2011). Effect of some post harvest treatments under different low temperatures on two mango cultivars. *Australian Journal of Basic and Applied Sciences* 5: 1164–1174.

53 Kaewklin, P., Siripatrawan, U., Suwanagul, A., and Lee, Y.S. (2018). Active packaging from chitosan-titanium dioxide nanocomposite film for prolonging storage life of tomato fruit. *International Journal of Biological Macromolecules* 112: 523–529.

54 Maneerat, C. and Hayata, Y. (2008). Gas-phase photocatalytic oxidation of ethylene with TiO_2-coated packaging film for horticultural products. *Transactions of the ASABE* 51: 163–168.

55 Zhu, Z., Zhang, Y., Zhang, Y. et al. (2019). Preparation of PAN@TiO2 nanofibers for fruit packaging materials with efficient photocatalytic degradation of ethylene. *Materials* 12: 896–906. https://doi.org/10.3390/ma12060896.

56 Taechutrakul, S., Netpradit, S., and Tanprasert, K. (2009). Development of recycled paper-based ethylene scavenging packages for tomatoes. *Acta Horticulturae* 837: 365–370. https://doi.org/10.17660/ActaHortic.2009.837.49.

57 Bailén, G., Guillén, F., Castillo, S. et al. (2006). Use of activated carbon inside modified atmosphere packages to maintain tomato fruit quality during cold storage. *Journal of Agricultural and Food Chemistry* 54 (6): 2229–2235. https://doi.org/10.1021/jf0528761.

58 Salamanca, F., Balaguera-López, H., and Herrera, A. (2014). Effect of potassium permanganate on some postharvest characteristics of tomato 'Chonto' fruits (*Solanum lycopersicum* L.). *Acta Horticulturae* 1016: 171–176. https://doi.org/10.17660/ActaHortic.2014.1016.24.

59 Alvarez-Hernandez, M.H., Martínez-Hernandez, G.B., Castillejo, N. et al. (2021). Development of an antifungal active packaging containing thymol and an ethylene scavenger. Validation during storage of cherry tomatoes. *Food Packaging and Shelf Life* 29: 100734. https://doi.org/10.1016/j.fpsl.2021.100734.

60 Boonsiriwit, A., Xiao, Y., Joung, J. et al. (2020). Alkaline halloysite nanotubes/low-density polyethylene nanocomposite films with increased ethylene absorption capacity: applications in cherry tomato packaging. *Food Packaging and Shelf Life* 25: 100533. https://doi.org/10.1016/j.fpsl.2020.100533.

61 Southwick, F.W. and Smock, R. (1943). Lengthening the storage life of apples by removal of volatile materials from the storage atmosphere. *Plant Physiology* 18 (4): 716–717. https://doi.org/10.1104/pp. 18.4.716.

62 Shorter, A., Scott, K., Ward, G., and Best, D. (1992). Effect of ethylene absorption on the storage of Granny Smith apples held in polyethylene bags. *Postharvest Biology and Technology* 1 (3): 189–194. https://doi.org/10.1016/0925-5214(92)90001-6.

63 Chaves, M.A., Bonomo, R.C.F., Silva, A.A.L. et al. (2007). Use of potassium permanganate in the sugar apple post-harvest preservation. *Journal Ciencia and Tecnologia Alimentaria* 5 (5): 346–351. https://doi.org/10.1080/11358120709487711.

64 Sardabi, F., Mohtadinia, J., Shavakhi, F., and Jafari, A.A. (2014). The effects of 1-methylcyclopropen (1-MCP) and potassium permanganate coated zeolite nanoparticles on shelf life extension and quality loss of golden delicious apples. *Journal of Food Processing and Preservation* 38: 2176–2182.

65 Knee, M. and Hatfield, S. (1981). Benefits of ethylene removal during apple storage. *The Annals of Applied Biology* 98: 157–165. https://doi.org/10.1111/j.1744-7348.1981.tb00433.x.

66 Esturk, O., Ayhan, Z., and Gokkurt, T. (2014). Production and application of active packaging film with ethylene adsorber to increase the shelf life of broccoli (*Brassica oleracea* L. var. Italica). *Packaging Technology and Science* 27 (3): 179–191. https://doi.org/10.1002/ Pts.2023.

67 DeEll, J., Toivonen, P., Cornut, F. et al. (2006). Addition of sorbitol with $KMnO_4$ improves broccoli quality retention in modified atmosphere packages. *Journal of Food Quality* 29: 65–75. https://doi.org/10.1111/j.1745-4557.2006.00056.x.

68 Li, D., Ye, Q., Jiang, L., and Luo, Z. (2017). Effects of nano-TiO_2-LDPE packaging on postharvest quality and antioxidant capacity of strawberry (*Fragaria ananassa* Duch.) stored at refrigeration temperature. *Journal of the Science of Food and Agriculture* 97 (4): 1116–11123. https://doi.org/10.1002/jsfa.7837.

69 Yang, F.M., Li, H.M., Li, F. et al. (2010). Effect of nano-packing on preservation quality of fresh strawberry (*Fragaria ananassa* Duch. cv Fengxiang) during storage at 4 °C. *Journal of Food Science* 75 (3): 236–240. https://doi.org/10.1111/j.1750-3841.2010.01520.x.

70 Luo, Z.S., Ye, Q.Y., and Li, D.D. (2013). Influence of nano-TiO*2* modified LDPE film packaging on quality of strawberry. *Modern Food Science and Technology* 29 (10): 2340–2344.

71 Sha, L., Zhao, H., and Xiao, G. (2013). Effect of coating agents onto cardboard boxes on keeping freshness of strawberry. *Food Science and Technology Research* 19 (4): 553–560. https://doi.org/10.3136/fstr.19.553.

72 Murmu, S.B. and Mishra, H.N. (2018). Selection of the best active modified atmosphere packaging with ethylene and moisture scavengers to maintain quality of guava during low temperature storage. *Food Chemistry* 253: 55–62. https://doi.org/10.1016/j.foodchem.2018.01.134.

73 Singh, R. and Giri, S.K. (2014). Shelf-life study of Guava (*Psidium guajava* L.) under active packaging: an experiment with potassium permanganate salt as ethylene absorbent. *Journal of Food Safety and Food Quality-Archiv Fur Lebensmittelhygiene* 65 (2): 32–39. https://doi.org/10.2376/0003-925X-65-32.

74 de Matos Fonseca, J., Pabon, N.Y.L., Nandi, L.G. et al. (2021). Gelatin-TiO_2-coated expanded polyethylene foam nets as ethylene scavengers for fruit postharvest application. *Postharvest Biology and Technology* 180: 111602. https://doi.org/10.1016/j.postharvbio.2021.111602.

75 Silva, D., Salomão, L., Siqueira, D. et al. (2009). Potassium permanganate effects in postharvest conservation of the papaya cultivar Sunrise Golden. *Pesquisa Agropecuária Brasileira* 44: 669–675. https://doi.org/10.1590/S0100-204X2009000700003.

76 Emadpour, M., Ghareyazie, B., Rezai, K. et al. (2009). Effect of potassium permanganate-coated nano-zeolites on cherry quality and shelf life. *Journal of Agricultural Engineering Research* 10: 11–26.

77 Alvarez-Hernandez, M.H., Martimez-Hernandez, G.B., Avalos-Belmontes, F. et al. (2019). Postharvest quality retention of apricots by using a novel sepiolite-loaded potassium permanganate ethylene scavenger. *Postharvest Biology and Technology* 160: 111061. https://doi.org/10.1016/j.postharvbio.2019.111061.

78 Bal, E. and Celik, S. (2010). The effects of postharvest treatments of salicylic acid and potassium permanganate on the storage of kiwifruit. *Bulgarian Journal of Agricultural Science* 16: 576–584.

79 Cerit, I. and Demirkol, O. (2020). Effects of modified atmosphere packaging conditions and ethylene absorber on the quality of red bell pepper. *Journal of Food and Nutrition Research* 59 (1): 35–43.

80 Ni, X., Yu, J., Shao, P. et al. (2021). Preservation of *Agaricus bisporus* freshness with using innovative ethylene manipulating active packaging paper. *Food Chemistry* 345: 128757. https://doi.org/10.1016/j.foodchem.2020.128757.

81 Wu, W., Ni, X., Shao, P. And Gao, H. 2021. Novel packaging film for humidity-controlled manipulating of ethylene for shelf-life extension of *Agaricus bisporus*. *LWT- Food Science and Technology*, 145: 111331. https://doi.org/10.1016/j.lwt.2021.111331

82 Zhang, Q., Guan, W., Xiong, J. et al. (2023). Combined effects of paper-containing 1-methylcyclopropene with nitrogen-modified atmosphere packaging in the logistics process of e-commerce on fruit quality of plum (*Prunus salicina* cv. 'Qingcuili') during shelf life. *International Journal of Food Science and Technology* 58: 4144–4154. https://doi.org/10.1111/ijfs.16486.

83 Sharma, S., Sharma, R.R., Pal, R.K. et al. (2012). Ethylene absorbents influence fruit firmness and activity of enzymes involved in fruit softening of Japanese plum (*Prunus salicina* Lindell) cv. Santa Rosa. *Fruits* 67 (4): 257–266. https://doi.org/10.1051/fruits/2012021.

7

Moisture-Absorbent Food Packaging Systems and the Role of Chitosan

Srasti Yadav[1] *and Pradip Kumar Dutta*[2]

[1] *Division of Chemistry, School of Basic Sciences, Galgotias University, Greater Noida, Uttar Pradesh, India*
[2] *Department of Chemistry, Motilal Nehru National Institute of Technology, Allahabad, Prayagraj, India*

7.1 Introduction

Food loss is a significant economic problem for the food business, but foodborne outbreaks represent a serious hazard to public health. One of the main causes of food loss may be attributed to cross contamination, which led to the growth of spoilage bacteria on raw materials and processed foods. As a result, food undergoes oxidation, produces off flavors and odors, as well as unfavorable changes in texture and color, which reduce its nutritional and sensory qualities. In addition, illnesses in humans caused by consuming contaminated foods result in high medical costs and lost productivity, which cost over $77.1 billion annually in the United States alone [1]. Increasing consumer awareness of food safety and health has led to a demand for naturally high-quality, preservative-free foods with a sufficient expiration life [2, 3]. This has led to innovations in packaging technologies such as intelligent packaging, modified atmosphere packaging (MAP), and active packaging [3–5]. The primary function of packaging is to separate food from the environment, reducing exposure to degradation factors and preserving desirable compounds, thereby increasing expiration life [3]. Active and MAP technologies provide solutions for extending shelf life, preserving nutritional quality, and assuring food safety [6, 7].

Food packaging systems' main purpose is to create a barrier between the food and the outside world, shielding it from any physical, chemical, and biological dangers that might lead to contamination and deterioration. Packaging may also improve the look, retain the nutritional value, and increase the shelf life of food products [4, 8]. Controlling moisture levels within food packaging is essential for preventing microbial growth and preserving flavor characteristics. Dry foods are dependent on relative humidity (RH), hence, even low levels can result in considerable quality loss. Increased moisture makes food susceptible to microbial contamination and changes in textural and physical appearance, thereby decreasing shelf life [9, 10]. Increased water content activity-based products benefit from controlled high RH inside the package to prevent dehydrating, as extra liquid from muscular

Smart Food Packaging Systems: Innovations and Technology Applications, First Edition.
Edited by Avik Mukherjee, Santosh Kumar, Manjusri Misra, and Amar K. Mohanty.
© 2025 John Wiley & Sons Ltd. Published 2025 by John Wiley & Sons Ltd.

deterioration and variations in temperature can have an effect on product desirability [11]. Moisture absorbers are an efficient way of controlling excess moisture within containers and regulating water activity to inhibit microbial growth [10]. Common types of moisture absorbers include sachets, films, containers, and pads, which are used to manage fluid exudates and maintain RH within a product. Moisture-absorbing materials physically absorb moisture and are added to packages during stowage in the form of sachets and pads. Appropriate selection of packaging materials with high water vapor barrier properties helps in the regulation of RH levels within containers [12].

In an effort to produce safer foods, researchers have been examining novel technologies. Active packaging is a comparatively new innovation because the packaging material interacts constructively with the wrapped food, overcoming the passive nature of advertising and protection of food items from environment [6]. There are numerous types of active food packaging, including those that employ oxygen scavengers, ultraviolet (UV) barriers, moisture absorbers, and other mechanisms to confer antioxidant, flavoring, or antibacterial properties. In addition to desiccants, active packaging employs natural clays, calcium oxide, and silica gel as moisture-regulating agents for foods with high moisture content. Dry products are dehydrated with silica gel [13]. Moisture content and water activity have a significant impact on the safety and quality of various categories of food. Numerous dried items, for instance, are susceptible to humidity during storage, and even extremely low RH levels inside the containers can result in significant quality loss. Increased moisture reduces shelf life by increasing susceptibility to microbial deterioration and altering the texture and appearance of the product. Fresh fish, meat, fruits, and vegetables can be protected from dehydration by maintaining a controlled, high RH level within the packaging. In addition, it is common for perishable foods such as fish and meat to have excess moisture due to leakage loss [14]. In addition to their function in food packaging, moisture absorbers are used in a variety of other industries. They are used in electronic packaging to prevent moisture damage, in pharmaceutical packaging to preserve the efficacy of medications by preventing moisture interference, and in automotive packaging to prevent oxidation and corrosion caused by moisture.

Desiccants like silica gel, calcium oxide, activated clays, and minerals that are tear-resistant in permeable plastic sachets are commonly used as moisture absorbers for packaging dehydrated foods in the form of sachets, pads, linens, and comforters. Along with moisture absorber sachets to regulate humidity in packaged dehydrated foods, moisture drip absorbent cushions, linens, and comforters are used to control liquid water in foods with a high-water activity, such as meats, fish, poultry, and fruit. Between two layers of a microporous nonwoven polyethylene or polypropylene plastic sheet, a superabsorbent polymer that can absorb up to 500 times its own weight is inserted. Carboxymethylcellulose (CMC), polyacrylate salts, and starch copolymers are examples of typical superabsorbent polymers because of their high water affinity. Under packed fresh meats, fish, and poultry to catch unnoticed tissue exudates, moisture drop absorber cushions are frequently used. When transporting horticulture products over long distances or absorbing melting ice from refrigerated seafood during air freight, larger sheets and blankets are employed [15, 16].

Chitosan has emerged as a highly effective material for absorbent food packaging. It is an exceptional water molecule adsorbent due to its notable properties, which include a large surface area, hydrophilicity, and the capacity to form hydrogen bonds. Incorporating chitosan into packaging materials such as films or coatings reduces water vapor permeability and prevents moisture infiltration effectively. In addition, chitosan possesses antimicrobial properties that inhibit the growth of bacteria and fungi, thereby further enhancing the quality and safety of food products [17]. According to the literature, the most common active coatings generated by chitosan as the primary polymeric matrix are antioxidative and/or antimicrobial. In recent years, there has been a great deal of interest in the use of active constituents derived from natural sources, such as essential oils and phenolic compounds found in high concentrations in a variety of fruits, vegetables, legumes, and seeds, among other things. When combined with biodegradable polymeric matrices, these active constituents exhibit potent antibacterial and/or antioxidant properties, suggesting their use in the production of food coatings or films. These active compounds, which are directly added to food, extend its shelf life and assure its safety and quality. Chitosan films have demonstrated antibacterial and antioxidant properties when combined with essential oils such as bergamot, clove bud, cinnamon, or Eucalyptus globulus, among others [7, 17, 18].

According to the literature, oxygen absorbers, moisture absorbers, and ethylene absorbers are all examples of active scavenging systems that have been studied and found to be effective in the food packaging industry [4]. Oxygen scavengers in the films or pads form absorb oxygen from the packaging's atmosphere, thereby extending the shelf life and preserving the food's integrity [19]. Singh et al. synthesized chitosan films containing gallic acid with enhanced oxygen-scavenging properties [20]. Biopolymers like chitosan may be made to act as ethylene scavengers when TiO_2 nanoparticles are included in the filming process of the chitosan-based films [21]. In conclusion, moisture absorbers play a vital role in food packaging applications. They contribute to maintaining a stable environment within the packaging, shielding the product from the detrimental effects of moisture and other environmental factors. By preventing spoilage and degradation, they help to minimize food waste and enhance the sustainability of the food industry. Based on the above properties of chitosan, it can be considered as a good biopolymer for developing moisture-absorbent food packaging systems by extending product shelf life and enhancing product quality. Its unique properties make it an excellent choice for improving the moisture control of packaged foods and prolonging their freshness.

7.2 Classification of Moisture Absorbents

Based on their chemical composition and physical structure, moisture absorbents can be categorized. There are three fundamental categories of moisture absorbers: organic, inorganic, and polymeric, as shown pictorially (Figure 7.1).

7.2.1 Inorganic Absorbents

Inorganic absorbents play an essential role in food packaging systems, moderating moisture levels, absorbing odors, and maintaining the integrity of food products. Inorganic-based moisture absorbers, such as calcium oxide (CaO), silica gel, $CaCl_2$ (calcium chloride), potassium carbonate, potassium chloride, and bentonite, are commonly used in food packaging due to their non-toxic and anti-corrosive nature. These materials are incorporated

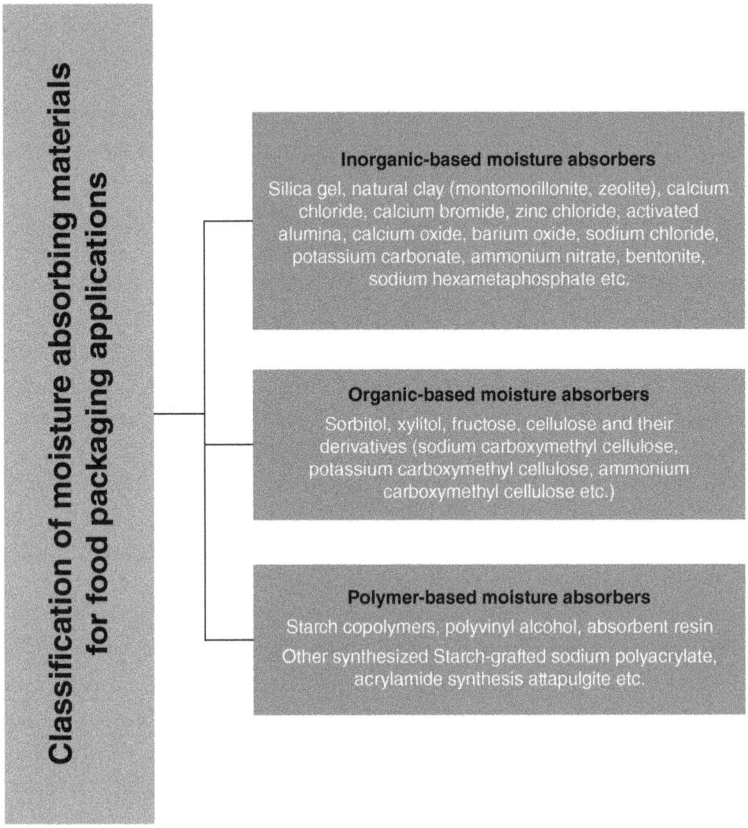

Figure 7.1 Categorization of moisture-absorbing materials for applications in food packaging.

into polymeric matrices and produce moisture-absorbing matrices. Commercial inorganic moisture-absorbing sachets (e.g., MOISTCATCH™, Dri-Loc®, AGELESS, CHUNWANG) and tray-based products (e.g., MoistPack, Luquasorb®, Fresh-R-Pax) are available and used for food packaging applications [4]. Bentonite clay is a natural clay that possesses a remarkable water-absorption capacity, making it valuable as a desiccant and absorbent for moisture, odors, and contaminants in food packaging. Zeolite-based films and low-moisture-absorbing sheets containing polyvinylidene chloride (PVDC) and calcium carbonate-coated acrylonitrile have also been developed. Zein films, with or without plasticizers like oleic acid, have shown reduced moisture uptake compared to zein powder. Xtend® films have been used to modify atmospheric composition and RH in packages containing fresh produce. Desiccants containing calcium oxide, sorbitol, and $CaCl_2$ have demonstrated moisture-holding capacities. Inorganic absorbents can be incorporated into various packaging materials, including films, sachets, or labels, to actively control moisture and prevent food spoilage. Nevertheless, it is crucial to ensure that the inorganic absorbents employed in food packaging are safe for consumption and do not release harmful substances into the food (Table 7.1).

Table 7.1 Inorganic absorbers used in food packaging applications.

Active substances	Application	Moisture-absorption property	Reference
Calcium oxide (CaO), silica gel	Sachets for very low relative humidity	Used as a moisture absorber in the food packaging	[4]
Zeolites	Moisture-absorbing films for food packaging	Attracts and releases moisture without changing crystalline structure	[22]
Bentonite/sorbitol/$CaCl_2$	Packaging of mushrooms	Decrease in moisture condensation; improved transparency of packaging film	[8]
Calcium oxide/calcium chloride/sorbitol	Desiccants	Moisture content of mixed desiccants evaluated over 5 days at 10 °C	[13]
Calcium carbonate/ acrylonitrile and polyvinylidene chloride (PVDC)	Low-moisture-absorbing sheet	Showed the highest moisture absorption	[23]
Zein powder and zein films	Packaging film and extruded film	Zein films with plasticizer exhibited reduced moisture uptake	[24]

7.2.2 Organic Absorbents

Organic absorbents, derived from natural sources, play a vital role in food packaging by effectively absorbing moisture, gases, and other contaminants. They are instrumental in maintaining the quality, freshness, and safety of food products. Here are some examples of organic absorbents commonly utilized in food packaging, as shown in Table 7.2. Researchers have recently focused on organic-based moisture absorbers for moisture-absorbing packaging, which offer potential for new packaging systems with desired qualities for consumer perception and sustainability. Sorbitol has been used in fresh products such as green tomatoes, mushrooms, and broccoli to extend shelf life and inhibit microbial growth. Incorporating sorbitol powder in MAP maintained the physical and nutritional quality of broccoli heads. Sorbitol has also been used as a moisture absorber for regulating humidity in mushroom packaging. Starch-based absorbents: Materials such as cornstarch, potato starch, and tapioca starch exhibit a high capacity for absorbing water and are employed as desiccants to control moisture levels in food packaging. Cellulose and its derivatives, like CMC, have been explored as moisture absorbers, but CMC has disadvantages like low resistance to decomposition, insects, and light. Coconut coir: Derived from natural coconut fiber, coconut coir has superior water retention properties and is used as a lining in the packaging of fresh produce. By absorbing moisture, it helps fruits and vegetables retain their quality and vitality. Activated carbon is efficient at absorbing pollutants and odors due to its highly porous structure. Commonly used as a filter to remove odors and contaminants from food packaging. Organic absorbents can be incorporated into a range of packaging materials, including coatings, cushions, and containers, to actively control moisture

Table 7.2 Organic absorbers used in food packaging applications.

Active substances	Application	Moisture-absorption property	References
Organic-based moisture absorbers (e.g., fructose, sorbitol)	Moisture-absorbing packaging	Fructose: initiates moisture absorption at about 55% RH; retains moisture for a long period even at low RH	[25]
Fructose	Packaging of fresh strawberries	Good moisture absorption containing 30% fructose (Fruitpad)	[12]
Sorbitol	Packaging of fresh produce (e.g., green tomatoes, mushrooms, broccoli), modified atmosphere packaging for broccoli	Improves maintenance of physical and nutritional quality; increases shelf life	[26, 27]
Cellulose and derivatives (e.g., carboxymethylcellulose – CMC)	Moisture absorber for food packaging	CMC absorbs nearly eight times more than standard cotton	[16]

and prevent food spoilage. Nevertheless, it is crucial that organic absorbents used in food packaging are safe for human consumption and do not introduce harmful substances into the food. Organic-based moisture absorbers are more expensive and have lower absorbing capacity compared to inorganic-based absorbers. Further research is needed to enhance technical and commercial possibilities, including stability during processing and moisture absorbance efficiency.

7.2.3 Polymeric Absorbents

Polymeric absorbents are versatile substances capable of absorbing and retaining water, hydrocarbons, or other contaminants within their structure. Their use in food packaging is essential for preserving the integrity and safety of food products. Polymer-based moisture absorbers are gaining attention in the packaging industry, offering a base polymer modified with moisture-absorbing polymers. According to the literature, various types of polymer-based moisture absorbers are used, like polyacrylic acid, PVA, cellulose, and cellulose derivatives. PVA is a biodegradable polymer used for the formation of food packaging films because of its good film-forming property and its good mechanical and oxygen-barrier properties. PVA films added to low-density polyethylene (LDPE) showed good moisture-absorption properties but poor mechanical properties. PVA films containing starch have been investigated for food packaging applications. Active packaging films with moisture-absorbing and antioxidant properties have been developed using PVA and green tea extract. Polyacrylic acid-based water-absorbent resin has also been developed with acceptable moisture-absorption properties. Cellulose and cellulose-based derivatives, such as CMC, have been used to prepare moisture-absorbing films for the formation of food packaging films. Nanocellulose films exhibited water vapor sorption performance, while cellulose-based NatureFlex™ films showed high water permeability for fresh produce packaging [28].

Here are some examples of polymeric absorbents commonly used in food packaging: Polyacrylamide (PAM): PAM is a water-soluble polymer that has an exceptional capacity to absorb and retain substantial quantities of water. It is commonly used as a desiccant in food packaging to regulate moisture levels and prevent spoilage. Polylactic acid (PLA) is a biodegradable polymer with absorption and moisture retention properties. It is commonly used as a breathable membrane in food packaging to control moisture and maintain the freshness of food products. Polyethylene glycol (PEG): PEG, a polymer that is water-soluble, has the ability to absorb and retain water. Its application in food packaging aids in preserving the texture and consistency of packaged foods. To actively control moisture and prevent food deterioration, polymeric absorbents can be incorporated into diverse packaging materials such as films, pads, and sachets. Nevertheless, it is essential that the polymeric absorbents used in food packaging are secure for consumption and do not introduce hazardous substances into the food (Table 7.3).

Table 7.3 Polymeric absorbers used in food packaging applications.

Active substance	Application	Moisture-absorption property	References
Starch, polyacrylic acid, polyvinyl alcohol, cellulose-based biopolymers	Packaging material modifiers	Provide moisture-absorption properties to base polymers	[29, 30]
Polyvinyl alcohol (PVA)	Moisture-absorbing film	Good moisture absorption but poor mechanical properties in LDPE matrix	[31, 32]
PVA and starch-based films	Active food packaging	Suitable for packaging applications with moisture sorption characteristics	[33]
PVA and green tea extract	Active packaging film with moisture absorption and antioxidant properties	PVA film with 2% green tea extract for packaging applications	[29]
Polyacrylic acid and water-absorbent resin	Moisture-absorbing polymer	Developed with acceptable moisture-absorption properties	[34]
Cellulose and derivatives (e.g., CMC)	Moisture-absorbing film for active food packaging	Cellulose films containing sulfuric-acid-treated cellulose nanocrystal	[35]
Cellulose-based NatureFlex™	Fresh product packaging	High water permeability, caution required to prevent excessive food moisture and weight loss	[36]

7.3 Mechanisms of Moisture Absorption

Moisture absorption in food packaging systems occurs through various mechanisms, depending on the type of absorbent material used and the design of the packaging. Here are some common mechanisms of moisture absorption in food packaging systems. Moisture absorbers are substances that physically absorb and retain water molecules from the surrounding environment, whereas moisture scavengers absorb water through a chemical reaction between two functional substances. Dry desiccants absorb moisture physically and chemically, reducing RH in sealed packs. Silica gel demonstrates physical moisture adsorption, while calcium chloride undergoes a chemical reaction. Moisture absorbers are commonly used in food packaging. As a moisture absorber, synthesized attapulgite was evaluated in a study by Choi and Lee. Attapulgite is a hydrated magnesium aluminum silicate mineral with exchangeable cations and reactive OH surface groups. When added to a polymeric matrix, particularly polyacrylic acid, attapulgite improves the moisture absorption and swelling rate relative to other clays [10, 37]. Cellulose is extensively studied as an organic-based moisture absorber due to its hygroscopic nature, with water adsorption sites at hydroxyl groups (—OH) and carboxyl groups (—COOH). Moisture is primarily absorbed in the amorphous regions of cellulose rather than the crystalline regions [35]. Derivatives of cellulose, like sodium CMC, are gaining attention for active packaging because of their good film-forming properties. The presence of polarity groups in CMC exhibits moisture-absorption properties. CMC, a polar macromolecule compound, exhibits moisture absorption based on RH and the presence of polarity groups. A higher number of polarity groups results in greater moisture absorption. Although moisture absorption can deform the macromolecular structure of cellulose and compromise dimensional stability, it offers unique potential as a moisture absorber in active packaging applications [16].

7.4 Incorporation of Moisture Absorbents in Food Packaging

In order to preserve the quality and safety of packaged foods, various techniques exist for incorporating moisture absorbents into food packaging. Small bags or containers containing moisture-absorbent substances, such as silica gel or activated carbon, are incorporated into food packaging to prevent degradation. Inside food packaging containers, such as platters or cartons, moisture-absorbent coverings made of cellulose fibers are intended to fit. These cushions absorb and retain moisture to prevent it from permeating the consumable product. Moisture-absorbent coatings, typically made of polymeric materials such as polyethylene, can be used as the interior layer of food packaging to create a moisture barrier. Coatings: Moisture-absorbent coatings, frequently composed of natural polymers like chitosan or starch-based materials, are applied to the surface of food packaging to prevent the entrance or egress of moisture. It is necessary to include moisture absorbents in food packaging for moisture control and to prevent food deterioration. It is crucial that moisture absorbents are safe for human consumption and do not introduce harmful substances into food. These techniques facilitate extending the shelf life of food products and reducing food waste.

Moisture-absorbing materials lower RH in packages, preventing condensation. Water absorption in closed packages depends on the type and volume of the absorbent material, the strength of water binding, the amount of water already absorbed, and temperature [12]. Common moisture-absorbing-based packaging systems, including sachets, pads, trays, and polymeric films, are used for the removal of excess moisture. Established technologies like sachets and pads can be used without affecting the structure of the exterior packaging materials. Sachets containing moisture absorbers are commonly used in dried food packaging, such as meat, nuts, spices, biscuits, crackers, sweets, instant coffee, and milk powder. These moisture absorbers help reduce excess moisture and inhibit bacterial growth [14]. Moisture absorbers in granule or powder form should not be exposed directly to high RH in the environment. When evaluating desiccant-based sachets, important factors to consider are the equilibrium adsorptive capacity, production rate, and storage temperatures. Sachets contain powder-absorbent materials sealed in a high-permeable material. The packaging material for the sachet must be resistant to handling and distribution in order to prevent failure and leakage. In an enclosed system, sachets are placed alongside food products and are separated from the outside environment by an outer packaging that meets certain permeability requirements [20]. Moisture-absorbent-based pads are placed at the bottom of trays or containers to absorb excess moisture from food during transportation and storage. This packaging system is commonly used for raw meat, poultry, and fish, aiming to extend shelf life and inhibit microbial growth (Figure 7.2).

However, moisture-absorbent pads are not frequently considered active packaging. According to EU guidance, pads made of natural constituents, such as 100% cellulose,

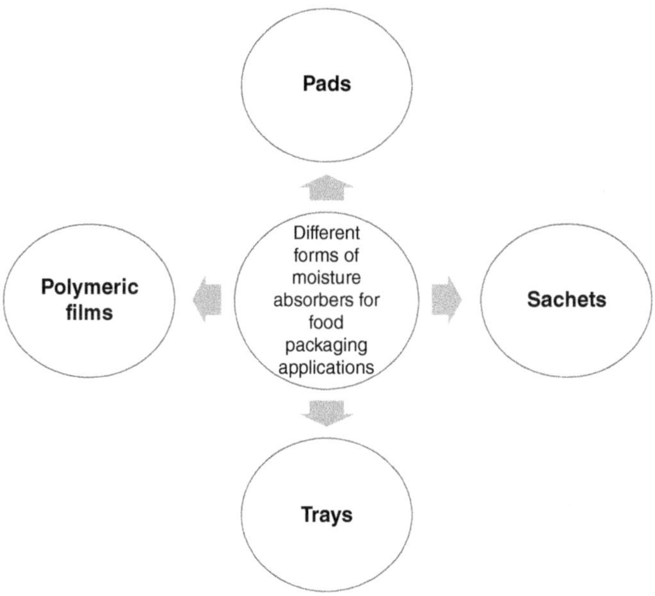

Figure 7.2 Various forms of moisture absorbers for effective food packaging applications.

are not considered active materials since they do not release or absorb substances intentionally. There are some drawbacks associated with moisture-absorbent pads in food preservation. The pad's absorbent capacity could be limited by the total molecular mass of the packaged food, preventing it from absorbing all the liquid exuded. Additionally, the outer layers of the pad can be perforated, and careful consideration is needed to prevent film tearing and the release of the absorbent material. Plastic materials offer a wide range of options for food packaging, allowing companies to tailor solutions to specific product needs. Polymeric-based films like polyethylene, polypropylene, polyvinyl chloride, and polyethylene terephthalate are commonly used in fresh produce packaging due to their low water vapor transmission rate (WVTR).

Various methods are employed to incorporate moisture absorbers, such as solution casting or melt extrusion. These absorbers can be integrated via hot melt extrusion, foil, and barrier laminates, or developed into pouches and blister packs. Examples include DesiMax®SLF, SuperDryFoil™, Activ-Blister®, and Nor® Absorbit. Researchers have also developed moisture-absorbing films using materials like PVA, sodium polyacrylate, and polyacrylic acid sodium salt, which have shown effective moisture-absorption capabilities and benefits for food preservation [14]. However, moisture-absorbing films have some drawbacks compared to desiccant-based sachets, including lower moisture-absorption capacity and slower absorption rates. Issues like potential migration of absorbing materials into food products and production losses are also being addressed by researchers.

7.5 Moisture-Absorbent-Based Active Food Packaging

Active food packaging containing moisture absorbents is a specialized packaging approach that incorporates active components designed to interact with the food or its environment. The goal is to extend the product's shelf life, preserve its quality, and guarantee its safety. Moisture absorbents play an important function among the active components typically employed in this type of packaging. Active food packaging comprising moisture absorbents functions by absorbing excess moisture, which has the potential to cause food to deteriorate or degrade. This packaging method is particularly advantageous for moisture-sensitive products, such as dried fruits, almonds, and cereals. The ability of active food packaging containing moisture absorbents to reduce food deterioration by extending the product's shelf life is a significant advantage of using such packaging. By regulating the moisture levels within the container, this packaging method prevents the growth of bacteria and other microorganisms responsible for deterioration.

Active food packaging containing moisture absorbents can be designed to release the absorbed moisture back into the environment when necessary. When the moisture levels in the package fall below a certain threshold, for example, the moisture absorbent can discharge the absorbed moisture back into the environment, thereby maintaining the ideal moisture levels for the food product.

Active food packaging containing moisture absorbents is a valuable tool for maintaining the quality and safety of food products, reducing food waste, and extending the shelf life of food products. By modulating the moisture levels within the container, this type of packaging can help preserve the freshness and essence of food products, thereby enhancing their consumer appeal. Innovative moisture-absorbing films like Nor® Absorbit have been developed for specific applications, such as flexible microwavable packaging. Researchers have explored different materials, such as PVA, superabsorbent sodium polyacrylate, and polyacrylic acid sodium salt, to develop moisture-absorbing films with varying properties. However, moisture-absorbing films have some drawbacks, including lower absorption capacity compared to desiccant-based sachets and potential migration of absorbing materials into food products. The production process may also result in significant mass and activity losses. Mistriotis et al. found that biodegradable PLA films had high water vapor permeability compared to polypropylene (OPP) films. Micro-perforated films made of different polymers such as oriented polypropylene, biaxially oriented polypropylene, LDPE, and polyvinyl chloride have been extensively studied for fresh produce packaging [38]. Almenar et al. investigated MAP in equilibrium for wild strawberries and found that films with one and three micro-perforations maintained quality through appropriate gas and water permeability. However, micro-perforated films have drawbacks such as higher cost, potential moisture and odor losses, and the possibility of microbial ingress during wet handling conditions [39].

7.6 Chitosan: A Better Substrate in Food Packaging

Chitosan is a natural polysaccharide with a high molecular weight obtained by deacetylating chitin [40–42]. Due to its unique chemical composition, positive charge, and the presence of reactive hydroxyl groups and amino groups, chitosan is a polycationic biopolymer with broad biological applications. It possesses desirable properties such as biodegradability, biocompatibility, non-toxicity, and excellent antimicrobial and film-forming abilities. Chitosan has numerous potential commercial and biomedical applications. It is commonly used in agriculture as a seed treatment and biopesticide to help plants combat fungal infections. In the medical field, it can expedite wound healing. As a result, chitosan has found application as a substrate in food packaging, particularly as absorbent materials (Figure 7.3). Various additives, such as essential oils and natural extracts, have been added to chitosan packaging to improve the film's properties. These additives aim to improve the film's antimicrobial effectiveness against foodborne microorganisms [43–46]. Due to its brittleness and weak mechanical properties, however, the use of chitosan in food packaging is restricted. Thus, chitosan can be combined with other polymers to expand its applications. Previous research has demonstrated that composite films composed of cellulose nanocrystals and chitosan have enhanced tensile strength as well as enhanced water and UV barrier properties. Recent research has focused on the development of pH-responsive smart films composed of chitosan/methylcellulose and chitosan/chitin nanocrystals incorporated with anthocyanin extract. These films have the potential to be used as quality sensors in food packaging [47, 48].

By using chitosan as a food packaging material, the expiration life of various food products can be extended, while synthetic plastics are used less frequently. Increasingly, edible or biodegradable materials, plant extracts, and nanomaterials are replacing synthetic additives, which pose hazards to the environment and human health. Chitosan-based

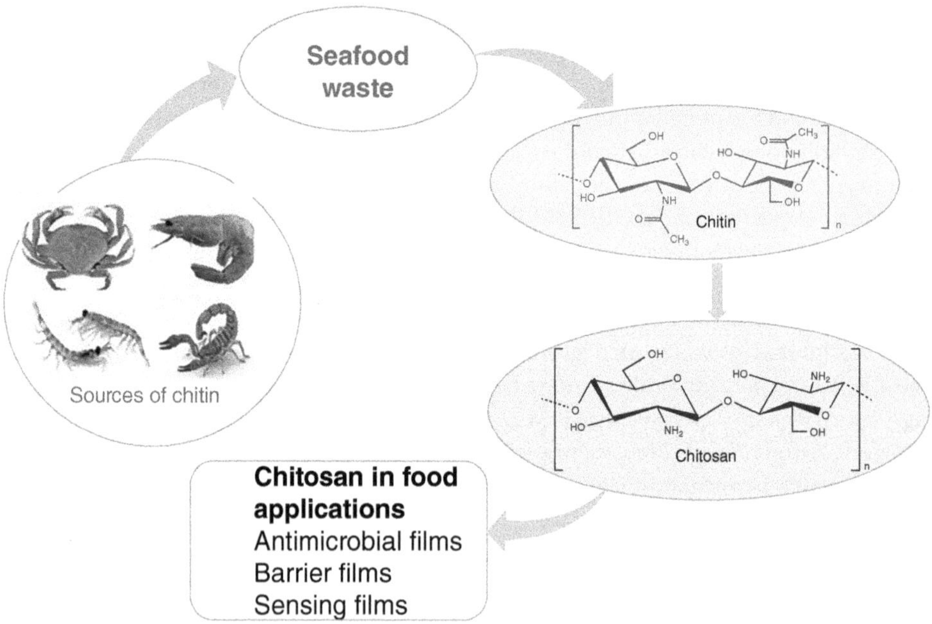

Figure 7.3 Overview of chitin/chitosan extraction methods and their applications.

films have made significant advancements in food packaging applications such as antibacterial films, barrier films, and sensing films [48, 49]. Chitosan materials, known for their barrier properties, play a crucial role in product storage and food preservation. To prevent the growth of bacteria and the decomposition of active ingredients in the desired products, it is necessary to minimize the permeation of moisture and oxygen through the packaging material. The barrier properties of chitosan-based materials are affected by factors including molecular weight and degree of deacetylation, as well as the modifications of the chitosan, additives, and plasticizers used in the production of the chitosan-based films. In evaluating the barrier properties of the chitosan-based films, the method of their fabrication is also important; the WVTR values of thermally dried films are less than those of air-dried films, according to research studies [50–53].

The incorporation of propolis extract into chitosan films reduces the permeability of water vapor and oxygen, with higher concentrations of propolis extract resulting in lower permeability. The WVTR is decreased when clay is added to chitosan film matrices. The dispersion of clay silicate layers within the polymer matrix hinders the diffusion of water molecules into composites. Depending on its concentration, the incorporation of the plasticizer glycerol influences the film's properties. A 20% weight concentration of glycerol decreases the water absorption of a substance. This is a result of the interaction of hydroxyl groups present in glycerol and the chitosan functional groups, which forms a network that inhibits further interaction between water and the hydroxyl groups of chitosan. Meanwhile, the addition of glycerol into the chitosan films enhances water absorption [54, 55].

The utilization of active films based on chitosan has been extensively investigated across various food products. Among these, antioxidant capacity has been the most extensively studied in the literature. Notably, chitosan-based films enriched with olive pomace and mango leaf extract have shown remarkable efficacy as extremely protective films against nut oxidation, with respective shelf lives of 31 and 28 days, respectively. Rambabu et al. and De Moraes Crizel et al. analyzed the peroxide value of the nut oils to determine the significant antioxidant property of the prepared chitosan active films [56, 57]. Evaluation of physical parameters, such as color, firmness, weight loss, and respiration rate, during storage of fresh fruits and vegetables demonstrates the beneficial effect of antioxidant chitosan coatings. Additionally, packaging films enriched with clove essential oil, banana peel extract, aloe vera, and procyanidins possess enhancements in various parameters during the transportation and preservation of apples, pomegranate arils, blueberries, mangoes, and litchi fruits. Similarly, films containing citric acid demonstrated the preservation of color and texture in green chili, whereas the mushroom films containing α-tocopherol enabled the preservation of firmness and the prevention of discoloration [20, 37, 58–61].

Additionally, the chitosan-based antioxidant films have been synthesized for the meat products. Moradi et al. found that sausages packaged with ordinary chitosan films showed a rapid increase in lipid oxidation, as measured by thiobarbituric acid, similar to the control samples without film. Although chitosan films prepared with grape seed extract and Zataria essential oil demonstrated a rapid decrease in oxidation during the first six days of refrigerated storage. For the purpose of extending the shelf life of refrigerated pork, chitosan coatings containing extracts of thyme, cloves, prickly ash, fennel, geranium, and cinnamon were evaluated. After seven days, the results demonstrated that the active coatings effectively reduced the total viable bacterial count and averted meat discoloration. In addition, chitosan coatings enriched with syringic acid had a preserving effect on quail eggs, resulting in a delayed decrease in weight loss rate and yolk index compared to uncoated eggs [62–64].

According to Tripathi et al., it has been proposed that incorporating chitosan as a biomaterial in nanocomposite films with silver oxide exhibits good antibacterial properties suitable for food packaging purposes [65]. It has been reported by Tripathi et al. that chitosan-based films with PVA have antimicrobial properties and showed promise as packaging materials [66]. In another study, chitosan/PVA/pectin films were successfully prepared via the solution casting method. The films exhibited positive antimicrobial activity against pathogenic bacteria, suggesting their potential as a universal antimicrobial food packaging material [67]. In our research studies, chitosan-based films containing gelatin and quercetin-starch (Q) and chitosan films with ZnO@gal (zinc oxide nanoparticles loaded with gallic acid) were developed using an eco-friendly method. They exhibited improved surface morphology, water vapor barrier, tensile strength, antioxidant activity, and antimicrobial properties, suggesting their potential for active food packaging [49, 68]. A summary of chitosan-based multicomponent nanocomposite films using various components for their physical and biological properties in food packaging applications is depicted in Table 7.4.

Table 7.4 Chitosan-based active films for food packaging applications.

Food product	Chitosan-based active films	Effects/results	References
Nuts	Chitosan films enriched with olive pomace and mango leaf extract	Preserved color and texture	[56, 57]
Fresh fruit and vegetables	Films containing banana peel extract, clove essential oil, ascorbic acid, aloe vera, procyanidins	Preserved color and texture	[37, 58, 61]
Green chili	Citric acid-based films	Preserved color and texture	[60]
Mushrooms	Films with α-tocopherol	Maintained firmness and alleviated browning	[37]
Meat products	Films containing essential oil and grape seed extract	Rapid decrease in oxidation during refrigerated storage	[62]
Pork	Films with extracts of cloves, and cinnamon	Decreased bacterial count and prevented meat discoloration after 7 days	[63]
Quail eggs	Films enriched with syringic acid	Slower decrease in weight loss rate compared to uncoated eggs	[64]
Fishery products	Films containing extract of red grape seed and essential oil	Delayed oxidation of lipids and proteins in minced trout fillet and grass carp fillets	[69]
Shrimp	Chitosan/gelatin film	Decreased total bacterial count, including L. monocytogenes	[70]
Chicken burger	Chitosan/anise essential oil films	Improved chemical properties, delayed lipid oxidation	[71]
Butter cake	Chitosan film with vanilla and potassium sorbate	Tripled shelf life of the product	[72]
Bread	Chitosan-based films containing apricot kernel essential oil	After 10 days, no fungal growth observed	[73]
Restructured chicken	Chitosan films	Lower counts of psychotropic microorganisms decreased microbial spoilage compared to control	[74]
Meat products ready-to-eat	Films with Thymus vulgaris essential oil	Enhanced meat color preservation	[75]
Cooked ham samples	Chitosan-based films containing essential oil	Enhancement of food pH for storage time analysis of 10 days	[76]
Fresh pasta	Chitosan films with chestnut extract	Inhibited growth of aerobic mesophilic bacteria, retrogradation process maintained hard texture after 9 days	[77]
Ground meat	Active chitosan films	Increased shelf life by 5–6 days due to O_2 scavenging capacity	[78]
Sliced sukuk	Active chitosan films	Delayed deterioration of meat color due to O_2 scavenger	[79]
Tomato fruit	Active chitosan films	Increased postharvest storage life, delayed ripening process	[80]
Mango	Active chitosan films	Increased postharvest storage life, delayed ripening process	[61]
Cooked pork	Thyme essential oil and chitosan film	Masking process of undesirable odors	[75]

7.7 Applications of Moisture Absorbents in Food Packaging

7.7.1 Fruits and Vegetables

Fruits and vegetables that are packaged frequently experience condensation, resulting in excess water. This is a result of the product's respiration and transpiration processes and the packaging material's low water vapor permeability. The trapped water vapor can result in condensation, promoting microbial growth and spoilage. To control humidity, moisture absorbers can be used in the packaging of fruits and vegetables. However, finding a balance between reducing condensation and minimizing weight loss in the products remains a challenge. Predictive mathematical modeling can help select the appropriate packaging system, but it also needs to account for the unique characteristics of each type of fresh product. Various studies have explored the use of moisture absorbers for different produce, such as silica gel for chilies, sodium carbonate for mushrooms, and sorbitol for broccoli. Feng et al. used tea polyphenols as active components combined with gelatin as a packaging material for fish fillets. Tornuk et al. incorporated thymol, carvacrol, and eugenol into linear LDPE to create nanoclays for packaging Turkish-type fermented sausage (sucuk). Ozogul et al. utilized nanoemulsions containing rosemary, laurel, thyme, and sage as active components in the packaging material for fish fillets [81–83]. Siragusa et al. incorporated nisin, an antimicrobial peptide, into polyethylene for packaging beef carcass tissue. Daeschel et al. applied a silicon coating to beef tissue as a packaging material. Camo et al. used rosemary and oregano extracts as active components in polystyrene packaging for lamb steaks. Balentine et al. utilized a cross-linked caseinate film containing thyme, rosemary, sage, and whey protein as a packaging material for ground beef. Cutter incorporated triclosan into a plastic matrix to target foodborne microorganisms associated with meat products. Ouattara et al. utilized chitosan, derived from chitin, as a packaging material for cooked ham [84–88].

7.7.2 Muscle Foods

Raw meat contains a high percentage of water, which can cause leakage and cellular damage during distribution and storage. The presence of exuding liquids in meat packaging is perceived as unhygienic and promotes microbial growth, reducing shelf life. To address these issues, absorbent pads are frequently used in the meat industry to absorb moisture and make aesthetically pleasing packaging. Recent studies have explored the use of N-halamine-coated absorbent pads, sorbent polymers, and nanosilver-based absorbing pads to reduce microbial growth [82]. The meat industry has shown a keen interest in technological advancements to extend the shelf life, enhance nutritional quality, and ensure consumer acceptance of meat-based products [89]. The active packaging system combines the preservative role of antimicrobials and other components with existing packaging concepts in the meat industry [90]. By incorporating active substances into the package through pads, tablets, or sachets and utilizing mechanisms such as absorption and evaporation, the interior environment of the packaging can be modified to inhibit microbial growth and other degradation processes. Food products actively packaged exhibit better quality attributes compared to conventionally preserved food [91].

Table 7.5 Packaging materials used in muscle foods.

Active components	Packaging materials	Substrates	Reference
Absorbent pads coated with N-halamine	N-Halamine antimicrobial compound	Raw beef	[94]
Sorbent polymer (polyacrylamide)	Superabsorbent pad	Chicken breast	[95]
Absorbent pads incorporated with silver nanoparticles	Silver nanoparticles with silver nitrate	Poultry meat	[92]
Polyvinyl chloride	Absorbent pad	Chicken breast	[93]

By incorporating high-moisture-absorbing materials into the pH indicator, it was determined that its potential to track the freshness of various types of meat products was outstanding. Fernández et al. developed absorbent pads containing silver nanoparticles, which led to a 40% reduction in total aerobic microbial growth in poultry meat. To maintain the quality of chicken, other packaging systems, such as polyvinyl chloride overwrap and absorbent pads, are used to maintain the freshness of the chicken. The selection of the appropriate moisture absorber, including type, size, and capacity, is critical in meat packaging to avoid undesirable effects like color changes and unpleasant odors caused by exudate saturation and microbial growth. Therefore, reducing microbial loads in absorbent pads and incorporating antimicrobial agents are necessary to ensure food safety and quality during packaging and storage [92, 93]. A brief description of packaging materials used in muscle food is enumerated in Table 7.5.

7.7.3 Dry Foods

Dry food products, despite having low moisture content, are susceptible to water vapor permeating through the packaging, which can negatively affect their quality and shelf life. Moisture can lead to issues such as crystallization, stickiness, and texture changes in these foods. The use of high-moisture barrier films or moisture absorbers in packaging is recommended to mitigate these effects. Researchers have developed LDPE films incorporated with moisture-absorbing materials like attapulgite and acrylamide for packaging dry-seasoned laver. Multifunctional packaging films with moisture-absorbing and antioxidant properties, such as those based on PVA and extracts of green tea, have been successfully employed for preserving dried eel. Superabsorbent polymers have been used to reduce aflatoxin contamination and aid in the drying process of maize. The results demonstrate the effectiveness of moisture control measures in maintaining the quality and safety of dry food products during storage. The selection of the specific type and size of moisture absorbent depends on the product and packaging requirements. Choosing the appropriate moisture absorbent is crucial to achieve optimal moisture control and preserve the quality of dry food products effectively [29, 95].

7.7.4 Other Foods

Moisture absorbents are extensively employed in food packaging across various food types to regulate moisture levels, prevent spoilage, and uphold product quality. Here are some applications of moisture absorbents in food packaging for different food categories: By applying moisture absorbents, food manufacturers can effectively control moisture levels and maintain the freshness, quality, and safety of a variety of food products throughout their shelf life. Including fish and prawns, seafood is extremely perishable and susceptible to moisture-related issues. To prevent these issues and prolong the shelf life of packaged seafood, it is common practice to include moisture absorbents. These absorbents assist in regulating moisture levels, inhibiting microbial growth, and preserving freshness. Moisture absorbents are also used in the packaging of sauces, dressings, and condiments to prevent moisture-related problems including separation, decomposition, and microbial growth. Typically, desiccant sachets or moisture-absorbing coatings are included in the packaging of these products in order to maintain their quality and consistency. In addition, moisture absorbents are used in the packaging of ready-to-eat foods, such as pre-packaged salads and microwaveable dishes, to regulate moisture levels, prevent sogginess, and preserve the texture and quality of the food throughout its shelf life. The choice of moisture absorbent depends on the specific requirements of the food product and packaging. Different types of absorbents, such as silica gel, molecular sieves, or absorbent pads, may be used to suit the moisture control needs of different food items.

The research on the utilization of moisture-based absorbers in the food packaging applications has investigated a variety of materials and packaging forms for controlling moisture and extending the expiration life of various food products. Chen et al. investigated the use of poly(vinyl alcohol) film as a moisture absorber for packaging dried eel. Murmu and Mishra studied the effectiveness of silica gel sachets in controlling moisture in guava packaging. [9, 29]. Other studies focused on the development of moisture-based absorbing materials and packaging forms for specific food applications. Bovi et al. examined the use of a polyethylene-based pad containing cellulose and fructose for moisture absorption in strawberry packaging. Chand and Kumar explored the coating of jaggery cubes with hydroxypropyl methyl cellulose and CMC as a moisture control measure. Wang et al. investigated the application of polyacrylate-based Tyvek film sachets for moisture absorption in packaging shiitake mushrooms. Kim et al. studied a PVA-based pH indicator as a moisture absorber in raw chicken breast packaging. Rux et al. developed the use of a linear LDPE-based tray containing sodium salt/hygroscopic ionomer to control moisture in strawberry and tomato packaging [11, 12, 40, 96, 97]. The research studies also explored the use of superabsorbent materials, desiccants, and other moisture-absorbing substances in various packaging forms. Lee and Lee investigated the application of superabsorbent sodium polyacrylate in polyurethane films for moisture absorption [98]. An studied the effectiveness of a superabsorbent polymer sachet in moisture control for shiitake mushrooms [99]. Sharma and Thakur explored the use of humectant salt in an aluminum laminated pouch for wild pomegranate packaging [100]. These studies demonstrate the wide range of moisture-absorbing materials, packaging forms, and food applications that have been investigated to enhance the quality and extend the shelf life of various products.

7.8 Conclusion and Future Perspective

Moisture-absorbing active packaging is a promising approach for developing food packaging suitable for high-water activity products. Using both organic and inorganic-based moisture absorbers in film form not only satisfies consumer demand for natural alternatives to the synthetic preservatives but also provides enhanced protection against excess moisture present in fresh meat and products. Incorporating a humidity indicator and sensor into packaging can provide an active and intelligent method for monitoring moisture levels [101]. Researchers are focused on studying the moisture absorbency and absorption of unwanted by-products of these systems, both in vitro and in vivo. However, while the physicochemical qualities of absorbent pads and sachets are usually evaluated, data on their sensory impacts on actual food quality is often lacking. Selecting the most appropriate packaging materials or formation of moisture control is crucial, and the physiological characteristics of the specific food product must be taken into account. Moisture-absorbing active packaging systems, particularly in the form of packaging films and pads, appear to have a bright future for food preservation and are consistent with an approach to food safety that enhances consumer transparency. Trays, sachets, films, and pads containing moisture absorbers are already extensively accepted on the commercial market. However, industry and academia continue to pay attention to the utilization of moisture absorbers in polymer-based packaging materials.

In order to address safety concerns and bridge existing knowledge gaps, chitosan polymer is anticipated to play a significant role in the development of moisture-absorbing-based packaging. The antibacterial and antioxidant properties have been significantly enhanced after the incorporation of phenolic compounds, essential oils, and food extracts into the chitosan films. Further investigations are required to evaluate the mechanical, barrier, and functional characteristics of alternative chitosan-based films, including those capable of scavenging ethylene, absorbing or releasing oxygen and CO_2, or absorbing or releasing flavor and odor compounds. However, more extensive research is necessary to develop chitosan-based packaging that can effectively compete with petroleum-based packaging while meeting practical requirements. Areas that require attention include: (i) continued research on chitosan and its combinations with other materials to meet practical demands. (ii) The assessment of potential health hazards associated with the use of chitosan films. (iii) The examination of which types of food products can come in contact with chitosan-based packaging.

References

1 Otoni, C.G., Espitia, P.J.P., Avena-Bustillos, R.J., and McHugh, T.H. (2016). Trends in antimicrobial food packaging systems: emitting sachets and absorbent pads. *Food Research International* 83: 60–73.

2 Singh, S., Nwabor, O.F., Syukri, D.M., and Voravuthikunchai, S.P. (2021). Chitosan-poly(vinyl alcohol) intelligent films fortified with anthocyanins isolated from *Clitoria ternatea* and *Carissa carandas* for monitoring beverage freshness. *International Journal of Biological Macromolecules* 182: 1015–1025.

3 Gaikwad, K.K., Singh, S., and Lee, Y.S. (2019). Antimicrobial and improved barrier properties of natural phenolic compound-coated polymeric films for active packaging applications. *Journal of Coating Technology and Research* 16 (1): 147–157.

4 Yildirim, S., Röcker, B., Pettersen, M.K. et al. (2018). Active packaging applications for food. *Comprehensive Reviews in Food Science and Food Safety* 17 (1): 165–199.

5 Biji, K.B., Ravishankar, C.N., Mohan, C.O., and Srinivasa Gopal, T.K. (2015). Smart packaging systems for food applications: a review. *Journal of Food Science and Technology* 52 (10): 6125–6135.

6 Ahmed, M.W., Haque, M.A., Mohibbullah, M. et al. (2022). A review on active packaging for quality and safety of foods: current trends, applications, prospects and challenges. *Food Packaging and Shelf Life* 33: 100913.

7 Eça, K.S., Sartori, T., and Menegalli, F.C. (2014). Films and edible coatings containing antioxidants – a review. *Brazilian Journal of Food Technology* 17 (2): 98–112.

8 Mahajan, P.V., Rodrigues, F.A.S., Motel, A., and Leonhard, A. (2008). Development of a moisture absorber for packaging of fresh mushrooms (*Agaricus bisporous*). *Postharvest Biology and Technology* 48 (3): 408–414.

9 Murmu, S.B. and Mishra, H.N. (2018). Selection of the best active modified atmosphere packaging with ethylene and moisture scavengers to maintain quality of guava during low-temperature storage. *Food Chemistry* 253: 55–62.

10 Choi, H.Y. and Lee, Y.S. (2013). Characteristics of moisture-absorbing film impregnated with synthesized attapulgite with acrylamide and its effect on the quality of seasoned laver during storage. *Journal of Food Engineering* 116 (4): 829–839.

11 Rux, G., Mahajan, P.V., Linke, M. et al. (2016). Humidity-regulating trays: moisture absorption kinetics and applications for fresh produce packaging. *Food and Bioprocess Technology* 9 (4): 709–716.

12 Bovi, G.G., Caleb, O.J., Klaus, E. et al. (2018). Moisture absorption kinetics of FruitPad for packaging of fresh strawberry. *Journal of Food Engineering* 223: 248–254.

13 Azevedo, S., Cunha, L.M., Mahajan, P.V., and Fonseca, S.C. (2011). Application of simplex lattice design for development of moisture absorber for oyster mushrooms. *Procedia Food Science* 1: 184–189.

14 Gaikwad, K.K., Singh, S., and Ajji, A. (2019). Moisture absorbers for food packaging applications. *Environmental Chemistry Letters* 17 (2): 609–628.

15 Vinay, K., Suneetha, W.J., and Bethapudi, A. (2018). Active packaging systems in food packaging for enhanced shelf life. *ResearchGate* 7: 2044–2046.

16 Pinming, C., Sukgorn, N., Suhatcho, T. et al. (2016). Humidity sensor using carboxymethyl cellulose hydrogel membrane. In: *2016 13th International Conference on Electrical Engineering/Electronics, Computer, Telecommunications and Information Technology (ECTI-CON) [Internet]. Chiang Mai*, 1–3. Thailand: IEEE. http://ieeexplore.ieee.org/document/7561435/ (accessed 5 June 2023).

17 Flórez, M., Guerra-Rodríguez, E., Cazón, P., and Vázquez, M. (2022). Chitosan for food packaging: recent advances in active and intelligent films. *Food Hydrocolloids* 124: 107328.

18 Hafsa, J., Smach, M.A., Ben Khedher, M.R. et al. (2016). Physical, antioxidant and antimicrobial properties of chitosan films containing *Eucalyptus globulus* essential oil. *LWT- Food Science and Technology* 68: 356–364.

19 Dutta, P.K., Yadav, S., and Mehrotra, G.K. (2021). Modified chitosan films/coatings for active food packaging. In: *Chitosan for Biomaterials III* (ed. R. Jayakumar and M. Prabaharan), 203–232. Cham: Springer International Publishing. (Advances in Polymer Science; vol. 287). https://link.springer.com/10.1007/12_2021_84 (accessed 14 June 2023).

20 Ozdemir, M. and Floros, J.D. (2004). Active food packaging technologies. *Critical Reviews in Food Science and Nutrition* 44 (3): 185–193.

21 Siripatrawan, U. and Kaewklin, P. (2018). Fabrication and characterization of chitosan-titanium dioxide nanocomposite film as ethylene scavenging and antimicrobial active food packaging. *Food Hydrocolloids* 84: 125–134.

22 Julkapli, N.M. and Bagheri, S. (2016). Magnesium oxide as a heterogeneous catalyst support. *Reviews in Inorganic Chemistry* 36 (1): 1–41.

23 Kia, H.G. (2010). Development of low moisture absorbing SMC. *Journal of Composite Materials* 44 (1): 55–74.

24 Wang, Y. and Padua, G.W. (2004). Water sorption properties of extruded zein films. *Journal of Agricultural and Food Chemistry* 52 (10): 3100–3105.

25 White, J.S. (2014). Sucrose, HFCS, and fructose: history, manufacture, composition, applications, and production. In: *Fructose, High Fructose Corn Syrup, Sucrose and Health* (ed. J.M. Rippe), 13–33. New York, NY: Springer New York. http://link.springer.com/10.1007/978-1-4899-8077-9_2 (accessed 5 June 2023).

26 Shirazi, A. and Cameron, A.C. (1993). Measuring transpiration rates of tomato and other detached fruit. *Hortscience* 28 (10): 1035–1038.

27 DeELL, J.R., Toivonen, P.M.A., Cornut, F. et al. (2006). Addition of sorbitol with KMnO4 improves broccoli quality retention in modified atmosphere packages. *Journal of Food Quality* 29 (1): 65–75.

28 Sousa-Gallagher, M.J., Mahajan, P.V., and Mezdad, T. (2013). Engineering packaging design accounting for transpiration rate: model development and validation with strawberries. *Journal of Food Engineering* 119 (2): 370–376.

29 Chen, C.W., Xie, J., Yang, F.X. et al. (2018). Development of moisture-absorbing and antioxidant active packaging film based on poly(vinyl alcohol) incorporated with green tea extract and its effect on the quality of dried eel. *Journal of Food Processing & Preservation* 42 (1): e13374.

30 Srinivasa, P.C., Ramesh, M.N., Kumar, K.R., and Tharanathan, R.N. (2003). Properties and sorption studies of chitosan–polyvinyl alcohol blend films. *Carbohydrate Polymers* 53 (4): 431–438.

31 Oun, A.A., Shin, G.H., Rhim, J.W., and Kim, J.T. (2022). Recent advances in polyvinyl alcohol-based composite films and their applications in food packaging. *Food Packaging and Shelf Life* 34: 100991.

32 Ding, W., Wei, S., Zhu, J. et al. (2010). Manipulated electrospun PVA nanofibers with inexpensive salts. *Macromolecular Materials and Engineering* 295 (10): 958–965.

33 Shi, R., Bi, J., Zhang, Z. et al. (2008). The effect of citric acid on the structural properties and cytotoxicity of the polyvinyl alcohol/starch films when molding at high temperature. *Carbohydrate Polymers* 74 (4): 763–770.

34 Matsuyama, H., Teramoto, M., Matsui, K., and Kitamura, Y. (2001). Preparation of poly(acrylic acid)/poly(vinyl alcohol) membrane for the facilitated transport of CO_2. *Journal of Applied Polymer Science* 81 (4): 936–942.

35 Guo, X., Liu, L., Hu, Y., and Wu, Y. (2018). Water vapor sorption properties of TEMPO oxidized and sulfuric acid treated cellulose nanocrystal films. *Carbohydrate Polymers* 197: 524–530.

36 Madonna, M., Caleb, O.J., Sivakumar, D., and Mahajan, P.V. (2018). Understanding the physiological response of fresh-cut cauliflower for developing a suitable packaging system. *Food Packaging and Shelf Life* 17: 179–186.

37 Zhang, G., Zhao, X., Li, X. et al. (2020). Challenges and possibilities for bio-manufacturing cultured meat. *Trends in Food Science & Technology* 97: 443–450.

38 Mistriotis, A., Briassoulis, D., Giannoulis, A., and D'Aquino, S. (2016). Design of biodegradable bio-based equilibrium modified atmosphere packaging (EMAP) for fresh fruits and vegetables by using micro-perforated poly-lactic acid (PLA) films. *Postharvest Biology and Technology* 111: 380–389.

39 Almenar, E., Del-Valle, V., Hernández-Muñoz, P. et al. (2007). Equilibrium modified atmosphere packaging of wild strawberries. *Journal of the Science of Food and Agriculture* 87 (10): 1931–1939.

40 Wang, H., Qian, J., and Ding, F. (2018). Emerging chitosan-based films for food packaging applications. *Journal of Agricultural and Food Chemistry* 66 (2): 395–413.

41 Yadav, M., Goswami, P., Paritosh, K. et al. (2019). Seafood waste: a source for preparation of commercially employable chitin/chitosan materials. *Bioresources and Bioprocessing* 6 (1): 8.

42 Leceta, I., Guerrero, P., and De La Caba, K. (2013). Functional properties of chitosan-based films. *Carbohydrate Polymers* 93 (1): 339–346.

43 Huq, T., Khan, A., Brown, D. et al. (2022). Sources, production and commercial applications of fungal chitosan: a review. *Journal of Bioresources and Bioproducts* 7 (2): 85–98.

44 Azmana, M., Mahmood, S., Hilles, A.R. et al. (2021). A review on chitosan and chitosan-based bionanocomposites: promising material for combatting global issues and its applications. *International Journal of Biological Macromolecules* 185: 832–848.

45 Michalik, R. and Wandzik, I. (2020). A mini-review on chitosan-based hydrogels with potential for sustainable agricultural applications. *Polymers* 12 (10): 2425.

46 Dutta, P.K., Ravikumar, M.N.V., and Dutta, J. (2002). Chitin and chitosan for versatile applications. *Journal of Macromolecular Science, Part C: Polymer Reviews* 42 (3): 307–354.

47 Qin, C., Li, H., Xiao, Q. et al. (2006). Water-solubility of chitosan and its antimicrobial activity. *Carbohydrate Polymers* 63 (3): 367–374.

48 Dutta, P.K., Tripathi, S., Mehrotra, G.K., and Dutta, J. (2009). Perspectives for chitosan based antimicrobial films in food applications. *Food Chemistry* 114 (4): 1173–1182.

49 Yadav, S., Mehrotra, G.K., Bhartiya, P. et al. (2020). Preparation, physicochemical and biological evaluation of quercetin based chitosan-gelatin film for food packaging. *Carbohydrate Polymers* 227: 115348.

50 Vargas, M., Albors, A., and Chiralt, A. (2011). Application of chitosan-sunflower oil edible films to pork meat hamburgers. *Procedia Food Science* 1: 39–43.

51 Kerch, G. (2015). Chitosan films and coatings prevent losses of fresh fruit nutritional quality: a review. *Trends in Food Science & Technology* 46 (2): 159–166.

52 Zehra, A., Wani, S.M., Bhat, T.A. et al. (2022). Preparation of a biodegradable chitosan packaging film based on zinc oxide, calcium chloride, nano clay and poly ethylene glycol incorporated with thyme oil for shelf-life prolongation of sweet cherry. *International Journal of Biological Macromolecules* 217: 572–582.

53 Ashrafi, A., Jokar, M., and Mohammadi, N.A. (2018). Preparation and characterization of biocomposite film based on chitosan and kombucha tea as active food packaging. *International Journal of Biological Macromolecules* 108: 444–454.

54 Kusmono, A.I. (2019). Water sorption, antimicrobial activity, and thermal and mechanical properties of chitosan/clay/glycerol nanocomposite films. *Heliyon* 5 (8): e02342.

55 Siripatrawan, U. and Vitchayakitti, W. (2016). Improving functional properties of chitosan films as active food packaging by incorporating with propolis. *Food Hydrocolloids* 61: 695–702.

56 De Moraes, C.T., De Oliveira, R.A.D., Alves, V. et al. (2018). Active food packaging prepared with chitosan and olive pomace. *Food Hydrocolloids* 74: 139–150.

57 Rambabu, K., Bharath, G., Banat, F. et al. Mango leaf extract incorporated chitosan antioxidant film for active food packaging. *International Journal of Biological Macromolecules* 126: 1234–1243.

58 Wang, W., Zhang, Y., Yang, Z., and He, Q. (2021). Effects of incorporation with clove (*Eugenia caryophyllata*) essential oil (CEO) on overall performance of chitosan as active coating. *International Journal of Biological Macromolecules* 166: 578–586.

59 Mannozzi, C., Tylewicz, U., Chinnici, F. et al. (2018). Effects of chitosan based coatings enriched with procyanidin by-product on quality of fresh blueberries during storage. *Food Chemistry* 251: 18–24.

60 Priyadarshi, R. and Rhim, J.W. (2020). Chitosan-based biodegradable functional films for food packaging applications. *Innovative Food Science & Emerging Technologies* 62: 102346.

61 Shah, S. and Hashmi, M.S. (2020). Chitosan–aloe vera gel coating delays postharvest decay of mango fruit. *Horticulture, Environment and Biotechnology* 61 (2): 279–289.

62 Moradi, M., Tajik, H., Razavi Rohani, S.M., and Oromiehie, A.R. (2011). Effectiveness of *Zataria multiflora* Boiss essential oil and grape seed extract impregnated chitosan film on ready-to-eat mortadella-type sausages during refrigerated storage: effects of chitosan film incorporated by antimicrobial plant agents on mortadella. *Journal of the Science of Food and Agriculture* 91 (15): 2850–2857.

63 Liu, T., Liu, L., Gong, X. et al. (2021). Fabrication and comparison of active films from chitosan incorporating different spice extracts for shelf life extension of refrigerated pork. *LWT* 135: 110181.

64 Yang, K., Dang, H., Liu, L. et al. (2019). Effect of syringic acid incorporation on the physical, mechanical, structural and antibacterial properties of chitosan film for quail eggs preservation. *International Journal of Biological Macromolecules* 141: 876–884.

65 Tripathi, S., Mehrotra, G.K., and Dutta, P.K. (2011). Chitosan–silver oxide nanocomposite film: preparation and antimicrobial activity. *Bulletin of Materials Science* 34 (1): 29–35.

66 Tripathi, S., Mehrotra, G.K., and Dutta, P.K. (2009). Physicochemical and bioactivity of cross-linked chitosan–PVA film for food packaging applications. *International Journal of Biological Macromolecules* 45 (4): 372–376.

67 Tripathi, S., Mehrotra, G.K., and Dutta, P.K. (2010). Preparation and physicochemical evaluation of chitosan/poly(vinyl alcohol)/pectin ternary film for food-packaging applications. *Carbohydrate Polymers* 79 (3): 711–716.

68 Yadav, S., Mehrotra, G.K., and Dutta, P.K. (2021). Chitosan based ZnO nanoparticles loaded gallic-acid films for active food packaging. *Food Chemistry* 334: 127605.

69 Shahbazi, Y. (2017). The properties of chitosan and gelatin films incorporated with ethanolic red grape seed extract and Ziziphora clinopodioides essential oil as biodegradable materials for active food packaging. *International Journal of Biological Macromolecules* 99: 746–753.

70 Mohebi, E. and Shahbazi, Y. (2017). Application of chitosan and gelatin based active packaging films for peeled shrimp preservation: A novel functional wrapping design. *LWT- Food Science and Technology* 76: 108–116.

71 Mahdavi, V., Hosseini, S.E., and Sharifan, A. (2018). Effect of edible chitosan film enriched with anise (*Pimpinella anisum* L.) essential oil on shelf life and quality of the chicken burger. *Food Science & Nutrition* 6 (2): 269–279.

72 Sangsuwan, J., Rattanapanone, N., and Pongsirikul, I. (2015). Development of active chitosan films incorporating potassium sorbate or vanillin to extend the shelf life of butter cake. *International Journal of Food Science and Technology* 50 (2): 323–330.

73 Priyadarshi, R., Sauraj, K.B., Deeba, F. et al. (2018). Chitosan films incorporated with Apricot (*Prunus armeniaca*) kernel essential oil as active food packaging material. *Food Hydrocolloids* 85: 158–166.

74 Serrano-León, J.S., Bergamaschi, K.B., Yoshida, C.M.P. et al. (2018). Chitosan active films containing agro-industrial residue extracts for shelf life extension of chicken restructured product. *Food Research International* 108: 93–100.

75 Quesada, J., Sendra, E., Navarro, C., and Sayas-Barberá, E. (2016). Antimicrobial active packaging including chitosan films with *Thymus vulgaris* L. *Essential Oil for Ready-to-Eat Meat Foods* 5 (4): 57.

76 Amor, G., Sabbah, M., Caputo, L. et al. (2021). Basil essential oil: composition, antimicrobial properties, and microencapsulation to produce active chitosan films for food packaging. *Foods* 10 (1): 121.

77 Kõrge, K., Bajić, M., Likozar, B., and Novak, U. (2020). Active chitosan–chestnut extract films used for packaging and storage of fresh pasta. *International Journal of Food Science and Technology* 55 (8): 3043–3052.

78 Chounou, N., Chouliara, E., Mexis, S.F. et al. (2013). Shelf life extension of ground meat stored at 4 °C using chitosan and an oxygen absorber. *International Journal of Food Science and Technology* 48 (1): 89–95.

79 Şahin, A., Çarkcıoğlu, E., Demirhan, B., and Candoğan, K. (2017). Chitosan edible coating and oxygen scavenger effects on modified atmosphere packaged sliced sucuk. *Journal of Food Processing & Preservation* 41 (6): e13213.

80 Kaewklin, P., Siripatrawan, U., Suwanagul, A., and Lee, Y.S. (2018). Active packaging from chitosan-titanium dioxide nanocomposite film for prolonging storage life of tomato fruit. *International Journal of Biological Macromolecules* 112: 523–529.

81 Feng, X., Ng, V.K., Mikš-Krajnik, M., and Yang, H. (2017). Effects of fish gelatin and tea polyphenol coating on the spoilage and degradation of myofibril in fish fillet during cold storage. *Food and Bioprocess Technology* 10 (1): 89–102.

82 Tornuk, F., Sagdic, O., Hancer, M., and Yetim, H. (2018). Development of LLDPE based active nanocomposite films with nanoclays impregnated with volatile compounds. *Food Research International* 107: 337–345.

83 Ozogul, Y., Yuvka, İ., Ucar, Y. et al. (2017). Evaluation of effects of nanoemulsion based on herb essential oils (rosemary, laurel, thyme and sage) on sensory, chemical and microbiological quality of rainbow trout (*Oncorhynchus mykiss*) fillets during ice storage. *LWT* 75: 677–684.

84 Camo, J., Beltrán, J.A., and Roncalés, P. (2008). Extension of the display life of lamb with an antioxidant active packaging. *Meat Science* 80 (4): 1086–1091.

85 Cutter, C.N. (2006). Opportunities for bio-based packaging technologies to improve the quality and safety of fresh and further processed muscle foods. *Meat Science* 74 (1): 131–142.

86 Siragusa, G.R., Cutter, C.N., and Willett, J.L. (1999). Incorporation of bacteriocin in plastic retains activity and inhibits surface growth of bacteria on meat. *Food Microbiology* 16 (3): 229–235.

87 Daeschel, M.A., Mcguire, J., and Al-Makhlafi, H. (1992). Antimicrobial activity of nisin adsorbed to hydrophilic and hydrophobic silicon surfaces. *Journal of Food Protection* 55 (9): 731–735.

88 Balentine, C.W., Crandall, P.G., O'Bryan, C.A. et al. (2006). The pre- and post-grinding application of rosemary and its effects on lipid oxidation and color during storage of ground beef. *Meat Science* 73 (3): 413–421.

89 Chiavaro, E., Vittadini, E., Musci, M. et al. (2008). Shelf-life stability of artisanally and industrially produced durum wheat sourdough bread ("Altamura bread"). *LWT- Food Science and Technology* 41 (1): 58–70.

90 Mauriello, G., Ercolini, D., La Storia, A. et al. (2004). Development of polythene films for food packaging activated with an antilisterial bacteriocin from *Lactobacillus curvatus* 32Y. *Journal of Applied Microbiology* 97 (2): 314–322.

91 Lee, D.S. and Han, J.H. (2011). Antimicrobial packaging. In: *Nonthermal Processing Technologies for Food [Internet]* (ed. H.Q. Zhang, G.V. Barbosa-Cánovas, V.M. Balasubramaniam, et al.), 462–471. Oxford: Wiley-Blackwell. https://onlinelibrary.wiley.com/doi/10.1002/9780470958360.ch30 (accessed 6 June 2023).

92 Fernández, A., Soriano, E., López-Carballo, G. et al. (2009). Preservation of aseptic conditions in absorbent pads by using silver nanotechnology. *Food Research International* 42 (8): 1105–1112.

93 Charles, N., Williams, S.K., and Rodrick, G.E. (2006). Effects of packaging systems on the natural microflora and acceptability of chicken breast meat. *Poultry Science* 85 (10): 1798–1801.

94 Praharasti, A.S., Kusumaningrum, A., Khasanah, Y. et al. (2019). Physicochemical properties and its relations of beef rendang inside retort pouch packaging in various temperature storage conditions. *IOP Conference Series: Earth and Environmental Science* 251: 012043.

95 Kim, D., Jung, J., Park, S., and il, Seo J. (2015). Preparation and characterization of LDPE/PVA blend films filled with glycerin-plasticized polyvinyl alcohol. *Journal of Applied Polymer Science* 132 (22): 1234–1243.

96 Kim, D., Lee, S., Lee, K. et al. (2017). Development of a pH indicator composed of high moisture-absorbing materials for real-time monitoring of chicken breast freshness. *Food Science and Biotechnology* 26 (1): 37–42.

97 Chand, K., Kumar, S., Shahi, N.C., and Post Harvest Process and Food Engineering, College of Technology, G.B.Pant University of Agriculture and Technology, Pantnagar Uttarakhand India Effect of active packaging and coating materials on quality parameters of jaggery cubes. *IJERT* V7 (1): IJERTV7IS010003.

98 Lee, J.H. and Lee, S.G. Preparation and swelling behavior of moisture-absorbing polyurethane films impregnated with superabsorbent sodium polyacrylate particles. *Journal of Applied Polymer Science* 133 (38): https://doi.org/10.1002/app.43973.

99 An, D.S. (2016). Effect of active master packaging system on preservation of fresh shiitake mushrooms in supply chain. *Journal of the Korean Society of Food Science and Nutrition* 45 (3): 402–408.

100 Sharma, A. and Thakur, N.S. (2016). Influence of active packaging on quality attributes of dried wild pomegranate (*Punica granatum* L.) arils during storage. *JANS* 8 (1): 398–404.

101 Ahvenainen, R. (2003). Active and intelligent packaging. In: *Novel Food Packaging Techniques*, 5–21. Elsevier. https://linkinghub.elsevier.com/retrieve/pii/B9781855736757500067 (accessed 6 June 2023).

8

Off-Flavor Absorbents in Active Food Packaging

Rishabh Goyal[1], Het Paraskumar Bhadania[2], Sushil Kumar Singh[1], and Poonam Singha[1]

[1] *Department of Food Process Engineering, National Institute of Technology Rourkela, Odisha, India*
[2] *Department of Chemical Engineering, Institute of Chemical Technology-Mumbai, IOC Campus, Bhubaneswar, Odisha, India*

8.1 Introduction

Food is a complex system of varied components, sometimes containing substances undesired by consumers because of their properties. One of the major undesirable food components is off-flavor. Off-flavor is not a single compound but a mixture of low molecular volatile organic compounds (VOCs) having hydrophobic properties. These off-flavors may be generated by degradation processes, which include microbial growth, oxidation, protein breakdown, enzymatic reactions, light-induced reactions, and contamination through packaging or related chemicals [1]. Unpleasant smell even with a trace amount of off-odor compound made food products suspected of deterioration and rejected by the consumer even if food is safe to consume. This reduces the desirability and demand for a product by consumers due to its flavor characteristics and causes loss to the food manufacturer. Rejection of safe food by consumers due to off-flavor causes waste generation and related issues. Formation of off-flavor is also a challenge while utilizing some processing techniques such as pressing and pasteurization of citrus juices which may cause the liberation of limonene [2]. To increase consumption, profitability, food security, and the overall flavor of the food, removal of off-flavor is essential. Active packaging interacts with food and the inside environment rather than merely providing a barrier to the external environment. Active packaging allows for the absorption and removal of the off-flavor compound present or generated during the distribution process [3–5]. Active packaging allows real-time control over off-flavor removal and has the potential to control food safety and increase food quality [6].

Off-flavor generation in fruits and vegetables is associated with respiration rate. Higher respiration rates cause faster deterioration in flavor due to changes in composition. Some fruits naturally contain bitter compounds in their peel, membrane, and seed, which become

Smart Food Packaging Systems: Innovations and Technology Applications, First Edition.
Edited by Avik Mukherjee, Santosh Kumar, Manjusri Misra, and Amar K. Mohanty.
© 2025 John Wiley & Sons Ltd. Published 2025 by John Wiley & Sons Ltd.

part of the juices during processing. For example, triterpenoid lactone, limonene and glycosidic flavanone naringin are present in bitter oranges and citrus juices, respectively [7]. Milk and milk products are nutritionally rich and have higher water activity, which supports the growth of a variety of microorganisms. This microbial growth causes the formation of off-flavor compounds such as aldehydes several weeks after heat processing of product such as ultrahigh temperature (UHT) milk [2]. Meat products produce an unpleasant odor due to microbial metabolites and protein breakdown, which cause the formation of amines and aldehydes and accumulate in the headspace of the package as putrid (hydrogen sulfide) [7]. The oxidation of amines in fish products is one example of protein degradation [8]. Foods rich in fat, such as peanuts, are susceptible to lipid oxidation, which leads to oxidative rancidity due to free radicals' generation, which is associated with an unpleasant flavor [7]. The formation of aldehydes occurs due to the oxidation of fatty acids in potato crisps, biscuits, fried foods, and cereals [9]. Table 8.1 provides a list of off-flavor compounds in different food products. Physical characteristics of packaging material, such as odor permeability, migration, and scalping, also play a huge role in odor presence. Off-odor may develop in the packaging material from processes such as extrusion, molding, film, sheet blowing, and casting. Some polyolefin components break down or oxidize into short-chain and odorous hydrocarbon compounds. Confinement odor development in packaged foods, especially in fresh poultry and cereal products, takes place due to gas-barrier packaging during the distribution processes. These odors are generally mild but are detected by consumers upon opening. They can arise from the breakdown of proteins and amino acids, resulting in sulfurous compounds, or from lipid oxidation and anaerobic glycolysis, leading to volatile aldehyde and ketone compounds. Although these odors are harmless and do not indicate any significant spoilage, they may lead to rejection of the product, even though they dissipate into the surrounding air within seconds and do not cause any harm to consumers on consumption [18, 19]. Removal of off-flavor has both beneficial and detrimental effects. Flavor removal may remove desirable components of the food including their desired characteristic flavor and reduce overall quality. Furthermore, off-flavor compounds are used as a detection mechanism by the consumer for food contamination or deterioration. Odor removal may cause consumers to consume food that is unfit for consumption and affects food safety; hence, it is not permitted by various nations, including the European Union [2].

8.2 Flavor Absorbents

Flavor absorbents are substances or systems that remove flavor compounds from food products, which affects consumer perception toward the food product. Different flavor absorbents can be selected based on the target flavor compound and the compatibility of the absorbent with the food matrix. Flavor absorbents are absorbents that are used to capture, neutralize, or eliminate unwanted flavors and odors. They are commonly employed in various industries, including food and beverage, pharmaceuticals, and household products, to maintain the quality and sensory attributes of the packaged products. The main purpose of flavor absorbents is to prevent off-flavors or off-odors from developing or spreading within

Table 8.1 Some food products and the off-flavor compounds.

Food product	Category	Off-flavor compound	Source	Reference
Citrus juices	Fruit and fruit product	Bitter compound naringenin	Naturally present in peel and seed	[10]
Apple juice	Fruit and fruit product	Phenolic compounds cause "musty," "earthy," or "like roots" odor	Microbial growth	[11]
Orange juices	Fruit and fruit product	Bitter compound limonene	Naturally present in peel and seed	[12]
Durian fruit	Fruit and fruit product	Ethyl disulfide – sulfide fruity flavor	High respiration	[13]
UHT milk	Milk product	Aldehyde compounds like hexanal and nonanal	Heat processing	[2]
Fish	Milk product	Amines compound "trimethylamine"	Protein breakdown	[14]
Meat	Milk product	Amines, aldehydes, and hydrogen sulfide	Microbial metabolites and protein breakdown	[7]
Peanuts	Fatty food	Aldehyde compounds like hexanal and nonanal	Lipid oxidation	[7]
Soybean products	Fatty food	Beany flavor from hexanal, 3-methylbutanal, 2-methylbutanal, hexanol, and various esters	Lipid oxidation	[15]
Snacks, crackers, biscuits, and cereal products	Fatty food	Hexanal and heptanal	Lipid oxidation	[7]
Alcohol-free beer	Beverages	Strecker aldehydes	Heat-induced reaction between a reducing sugar and an amino acid	[16]
Packaged food	Packaged product	Amines and aldehyde compounds	Microbial metabolites and protein breakdown products	[7]
Vacuum-packaged food product	Packaged product	Confinement odor and hydrogen sulfide	Reaction during distribution	[17]

the packaged items. These off-flavors can arise from a variety of sources, such as chemical reactions, microbial growth, oxidation, or interactions between different components of the product. By utilizing flavor absorbents, manufacturers can extend the shelf life, enhance the sensory experience, and ensure overall consumer satisfaction with their products.

8.2.1 Direct Flavor Absorbents

Direct flavor absorbents remove flavor compounds by binding, breakdown, or conversion into less volatile compounds. They act directly on flavor compounds. Ideally, they should not react with other flavor compounds and constituents of food and food packaging while actively removing the unwanted off-flavor compounds. Hence, it is important to select a suitable flavor absorber for food system. Some flavor absorbents are also used in combination to increase their overall effectiveness and reduce their individual undesired effect on food. Major direct flavor absorbents are activated carbon, zeolites, resins and polymers, enzymes, acidic compounds, and metals (Table 8.2).

8.2.1.1 Activated Carbon

Activated carbon has a large surface area, microporous structure, and variable surface chemical composition, which makes it the most versatile adsorbent (Figure 8.1) and hence used in many different fields. Activated carbon used in hard granules or hard dust-free pellets has a high adsorptive capacity and selectivity for volatile organic vapors due to its microporosity with surface areas of 1000–2000 $m^2\ g^{-1}$. In addition, activated carbon has high retention capacity and preferential adsorption of flavor compounds in the presence of moisture, which makes it suitable for food applications. Activated carbon can be prepared from hard materials like nutshells or fruit pits by crushing, carbonization, and activation, or by coal pulverization, briquettes with a binder, and crushed to size [24]. Activated carbon-based adsorbent pads reduce confinement odor in a vacuum-packaged food product along with their antimicrobial properties [17]. There are even some commercialized products such as MINIPAX1 and STRIPPAX1 (Multisorb Technologies, USA) available to absorb mercaptans and H_2S from food. 2-in-1™ of United Desiccants, a product of silica gel and activated carbon, which removes off-flavor from packaged food [25].

8.2.1.2 Zeolites

Zeolites are crystalline alumino-silicates having microporous structures with varied pore sizes and channels, which allow them to selectively absorb the flavor compounds from the food product, as shown in Figure 8.2. Their framework consists of corner-sharing TO_4 tetrahedral, where T is typically silicon (Si) or aluminum (Al) atoms, as shown in Figure 8.3. These tetrahedral coordinated Al atoms cause negative charges in the framework, which are compensated by extra-framework cationic organic or inorganic species located inside the pores and cavities of the zeolite framework. Based upon the Si/Al ratio, zeolites can be classified as highly polar low-silica zeolites (Si/Al ratios: <2), intermediate polarity medium-silica zeolites (Si/Al ratio 2 to 5), and high-silica zeolites (Si/Al >5). Besides these traditional zeolites, heterogeneous zeolite is also available with T molecules other than Si or Al, which cause varied physical and chemical properties. More than 250 different zeolitic structures are known to exist; some are natural, and some are synthetic. The preparation of zeolite is done by hydrothermal synthesis, which mimics the natural condition and causes crystallization of zeolites. Some novel methods are also available to produce zeolite, but in all post-synthesis modification is required for absorbent application. Zeolites are amazing shape- and size-selective adsorbents (Figure 8.2), and even slight changes in cation distribution or framework structure can be of high importance for their applicability as

Table 8.2 Flavor absorbents in food.

Flavor absorbent	Mechanism	Application	References
Absorbent pad with activated carbon	Adsorption	Confinement flavor removal from vacuum-packaged food product	[17]
Mixture of charcoal and nickel	Adsorption, diffusion	Fruity flavor removal from durian fruit	[13]
Silica gel and activated carbon in product "2-in-1™"	Adsorption, diffusion	Off-flavor control from packaged food	[9]
Polyolefinic resin with activated montmorillonite clay	Adsorption, chelation	Amine flavor removal from fish	[20]
Hydrophobic resins	Adsorption, hydrophobic interaction	Strecker aldehyde removal from alcohol-free beer	[16]
Ion exchange and adsorbent Resins	Adsorption, electrostatic interaction	Bitterness from orange juices	[12]
Functionalized ethylene-based copolymer in product "Bynel IXP101"	Adsorption	Aldehyde scavengers in cured poultry	[21]
Amine-functionalized polymers	Adsorption, hydrophobic interaction	Strecker aldehydes removal from alcohol-free beer	[16]
Zeolites	Adsorption, electrostatic interaction	Strecker aldehydes removal from alcohol-free beer	[16]
Cellulose acetate	Adsorption	Bitterness removal in grapefruit juice and orange juice	[9], [22]
Cellulose triacetate and acetylated layer	Adsorption	Bitterness removal in oranges	[7]
Nylon MXD6, D-sorbitol and cyclodextrin	Adsorption	Aldehyde scavengers	[23]
Ferrous salt and an organic acid in product "ANICO Bag"	Adsorption, oxidation, diffusion	Amine scavenger in fish	[21]
Sodium sulfate and other organic sulfates	Chemical reaction	Aldehyde scavengers	[2]
Tocopherols and vitamin E	Oxidation, chelation	Aldehyde scavengers	[2]
Naringinase	Enzymatic reaction	Bitterness removal from citrus juices	[10]
Aldehyde dehydrogenase	Enzymatic reaction	Beany flavor removal from soybean products	[15]
Laccase	Enzymatic reaction	Musty flavor removal of apple juice	[11]

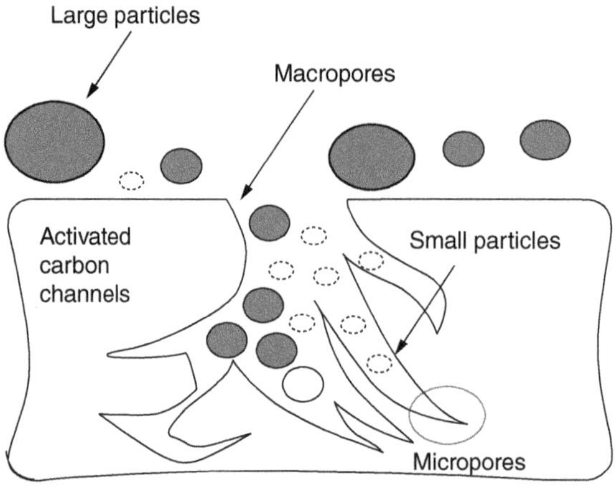

Figure 8.1 Adsorption and diffusion through activated carbon channels.

Figure 8.2 Cage structure of zeolite.

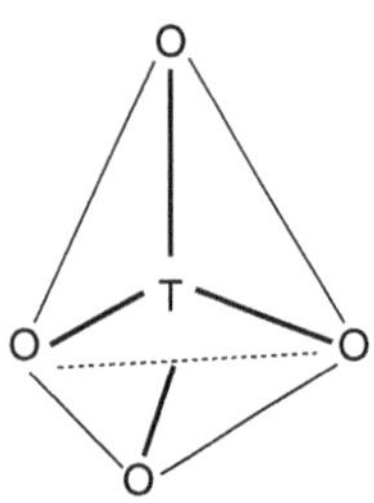

Figure 8.3 Tetrahedral structure of TO_4 unit.

adsorbents [25]. It is reported that zeolite has the highest potential for selectively absorbing wort flavor from alcohol-free beer among amine-functionalized polymers, hydrophobic resins, and zeolites. The suitability of the adsorbent is related to the matrix composition and careful selection of pore size and Si/Al ratio to achieve the desired selectivity. In this case, hydrophobic zeolite Socony mobil-5 (ZSM-5) type zeolites are found suitable as adsorbents [16]. Also, BMH™ powder, synthetic alumina-silicate zeolites developed by EKA Nobel, and Akzo are incorporated in packaging materials such as paper to absorb aldehyde through porous interstices [2].

8.2.1.3 Resins and Polymers

Resin and polymer-based adsorbents selectively adsorb or chemically trap VOCs that contribute to off-flavors or off-odors. Resins and polymer absorption depend on the volatility and affinity of flavor compounds. Resins are solid or semi-solid organic compounds that are insoluble in water, have high molecular weights, and are composed of repeating units called monomers. They have specific properties to efficiently adsorb or absorb VOCs and are derived from natural and synthetic sources. It is stated that selected volatiles generated in spent coffee grounds by thermal hydrolysis have been absorbed using non-polar resins (styrene or divinylbenzene or activated carbon) [26]. Biopolymers are polymers that are derived from renewable sources, such as plants, animals, or microorganisms. They have biodegradable and environmentally friendly characteristics. They work as a base material for the adsorbent matrix in active packaging, along with another absorbent such as citric acid. It is reported that the polymer absorbs the off-flavor while citric acid oxidizes to remove the off-flavor. Similarly, Dupont Polymers claimed that the Bynel IXP101 layer, a HDPE master batch, can remove aldehyde off-flavors such as hexanal and heptanal from package headspaces [9]. There are different biopolymers in the interest of flavor absorption, such as cellulose triacetate, acetylated paper, Nylon MXD6, D-sorbitol, and cyclodextrin. Cellulose triacetate and acetylated layer are used as limonene absorber-active packaging in bitter oranges [22]. Polyethylene terephthalate (PET) blended with Nylon MXD6, D-sorbitol, and cyclodextrin can act as aldehyde scavengers [2].

8.2.1.4 Enzymes

Enzymes are biological catalysts that selectively break down or modify specific compounds responsible for undesirable flavors or odors. Enzyme-based flavor adsorbents offer the advantage of specificity, as enzymes target only specific compounds or classes of compounds and remain unreacted to others. They can be incorporated into packaging materials or applied as coatings to the packaging surface to interact with off-flavor compounds during storage or transportation. Naringinase is an enzyme complex consisting of alpha-mannosidase and flavonoid-beta-glucosidase that converts naringin to naringenin and prunes in a two-step process. Naringenin is only one-third as bitter as naringin [10]. Another enzyme, crude diacetyl reductase from *Aerobacter aerogenes* or yeast was also utilized to remove diacetyl from beer. Complete removal of diacetyl from beer and 66% removal from orange juice was achieved in 96 h at 5°C by the activity of the enzyme [27]. Similarly, limonoate dehydrogenase oxidized limonene, a bitter compound of grapefruit, with sulfhydryl oxidase used in the treatment of UHT milk and enzymes of cell-free extract of baker's yeast for decomposition of hexanal. Laccases and copper-containing polyphenol oxidases eliminated off-flavors such as polyphenols, methoxy-substituted phenols, and diamines in apple juice [11]. Aldehyde dehydrogenase (ALDH) purified from bovine liver mitochondria removes the beany flavor from defatted soybean extracts, and soybean milk converts aldehydes into acids [15].

8.2.1.5 Acid Compounds

Acid compounds are used as flavor absorbents due to their ability to interact and neutralize unwanted flavors by either masking or counteracting the off-flavors, thereby improving the overall sensory perception of the product. In food, generally, organic acid compounds are used, such as citric acid, but some inorganic acids are also used, such as sulfuric acid compounds. When using acid compounds as flavor absorbents, it is crucial to consider the desired flavor profile, the concentration of the acid, and any regulatory restrictions or guidelines that may apply. Additionally, the sensory impact of acid compounds should be carefully evaluated to ensure they do not overpower or negatively affect the overall taste and balance of the product. Acids can interact with off-flavor compounds through chemical reactions, forming complexes or esters that reduce their volatility or sensory impact. Acidic compounds such as citric acid and ascorbic acid have the application of oxidizing off-flavor compounds, especially amine, when incorporated with polymers and ferrous salts (Japanese company, Anico Co.) [9]. Other salts such as sodium sulfate and other organic sulfates were reported to remove off-flavor aldehyde compounds from food. Even tocopherols, such as vitamin E, are capable of absorbing aldehydes [2]. These compounds are naturally occurring and can be used in conjugation of other flavor absorbents to enhance the absorption capacity while keeping the food safe. Overall, acid compounds can be effective flavor absorbents, providing a way to modify or neutralize unwanted flavors and improve the sensory experience of various food and beverage products.

8.2.1.6 Metal

Metals, in their elemental form or as metal ions, can be employed as flavor absorbents due to their adsorption properties. They can interact with flavor compounds through various mechanisms, including physical adsorption, chemical reactions, or complexation. It is important to ensure metal safety, compatibility with the product and packaging, and the targeted off-flavors when selecting metals as flavor absorbents. Regulatory requirements and maximum allowable levels for metals in food contact materials should also be considered. Metals are commonly used in conjugation with other flavor absorbents, as flavor compounds diffuse more rapidly in metallic salts, which allow higher flavor removal (Table 8.2). Metal also acts as catalyst for flavor removal reaction. Mixture of charcoal and nickel is used to remove ethyl disulfide, a fruity flavor from durian fruit [13]. Generally, metals are used in conjugation with organic acids or acid salts because of their high affinity toward amines such as ANICO bags, while they also absorb the oxygen, which acts as precursor in off-flavor development in some foods [21].

8.2.2 Indirect Off-Flavor Absorbent

Indirect flavor absorbents are materials or components that modify the factors responsible for off-flavor development in food products. Instead of directly interacting with the flavor compounds, indirect absorbents target the factors that contribute to off-flavors. These factors include changes in oxygen levels, removal of ethylene, and moisture control by oxygen absorbers, ethylene scavengers, and moisture absorbers. Indirect off-flavor absorbents focus on addressing the underlying factors that contribute to off-flavors rather than directly interacting with the flavor compounds themselves. By controlling factors such as oxygen, moisture, light, microbial activity, or pH, these absorbents help maintain the integrity and quality of the product, indirectly preventing the development or intensification of off-flavors.

8.2.2.1 Oxygen Absorbers

Oxygen absorbers help to reduce the oxygen within the package, which reduces oxidation and the development of off-flavors. These oxygen absorbers generally contain reducing agents along with a catalyst and moisture-absorbing materials, which together cause oxygen absorption. Off-flavors from oxidation arise from lipid oxidation, protein degradation, or other oxidative reactions. Generally, oxygen absorbers are based on iron powder, which oxidizes into ferrous oxide to remove oxygen. Some oxygen absorbers are reducing agents, such as ascorbic acid, and some are based on enzymatic reactions [28].

8.2.2.2 Ethylene Scavengers

Ethylene is a plant hormone that accelerates the ripening and senescence process, leading to the formation of off-flavors. This off-flavor development occurs in horticulture products due to physiological disorders. Ethylene scavengers remove or neutralize ethylene gas and, hence, reduce the development of off-flavors in food products. Biofresh is a Grofit Plastics which absorbs gases such as ethylene, ammonia, and hydrogen sulfide, known to contribute toward the ripening of fruits [2]. There are various ethylene scavengers used such as potassium permanganate immobilized on an inert mineral substrate, such as alumina or silica gel, powdered pumice stone films in fruits, zeolites in packages, activated carbon, or specific catalysts [28]. EverFresh bags by EverFresh USA utilize oya, a natural mineral, to absorb ethylene in fresh produce [2].

8.2.2.3 Moisture Absorbers

Moisture absorbers primarily prevent as microbial growth and product degradation, which indirectly contribute to reducing off-flavors in packaged products. Excess moisture favors microbial growth, leading to spoilage and the development of off-flavors. Moisture absorbers mostly come in the form of sachets, pads, sheets, or blankets. One product named Tyvek™ (Dupont Chemicals, Wilmington, Delaware, USA), a tear-resistant permeable plastic sachets uses desiccants such as silica gel, calcium oxide, and activated clays and minerals for packaged dried food applications. This allows dual action: activated carbon serves as odor adsorption, and iron powder serves as oxygen scavenger [21]. Moisture absorbers utilize compounds with high affinity for water, such as silica gel, activated clay, molecular sieves, or desiccants, that adsorb or absorb water molecules [28].

8.3 Mechanisms of Off-Flavor Absorbents

The mechanism of off-flavor absorbents in active packaging is found to be different based on the nature of the targeted off-flavor compound and the type of absorbent used. These different off-flavor absorbents pose different absorption capacities, selectivity, and interaction mechanisms with off-flavor compounds. The mechanisms of off-flavor absorbents are described in the following sections.

8.3.1 Adsorption

Adsorption is the process in which liquid or gaseous molecules from the mixture bind to the solid surface. In adsorption, the adsorbate (the substance being absorbed) accumulates on the surface of the adsorbent (the material onto which adsorption occurs) rather than being absorbed into it. Intermolecular attraction (physical adsorption), electrostatic attraction (ion exchange), and chemical affinity (chemical adsorption or chemisorption) are responsible for the binding of molecules [29]. In physical adsorption, off-flavor adsorbents cause the physical attachment or trapping of off-flavor molecules at the surface, as shown in Figure 8.1. When the off-flavor molecules encounter active sites of the adsorbent surface, interaction with off-flavor molecules takes place due to functional groups or microstructural features that cause attachment. These adsorption/active sites can be surface defects, pores, crevices, or functional groups that cause the formation of London-van der Waals forces of attraction. Physical adsorption is reversible under certain conditions, such as changes in temperature, pressure, or humidity. It has low selectivity and relatively low levels of adsorption energy. Activated carbon, zeolite, and silica gel are based on physical adsorption. In ion-exchange adsorption, a special type of adsorbent is used termed as ion exchangers, which are charged insoluble minerals or synthetic polymers (polyions). These charged adsorbents attract the oppositely charged off-flavor molecules, as shown in Figure 8.4. The adsorbent matrix can be anionic (negatively charged) or cationic (positively charged) based on the targeted off-flavor. Ion exchange resins are found effective in removing acid and bitterness from citrus juices [12, 30]. In chemical adsorption, adsorbents form strong chemical bonds through dipole–dipole interactions, hydrogen bonding, or π–π interactions responsible for attachment depending upon the nature of the off-flavor compounds and the adsorbent material. This type of adsorption is not reversible and is highly specific. Limonene adsorption on the silica surface may occur due to hydrogen bonding and hydrophobic interactions [31].

Different off-flavor absorbents, such as activated carbon, zeolites, or molecular sieves, possess a porous structure of different pore sizes. These pores provide a large surface area and a network of channels that facilitate the adsorption process. Off-flavor molecules diffuse into these pores and can be adsorbed onto the internal surfaces of the absorbent. The adsorption capacity of these adsorbents depends on factors like the surface area, pore

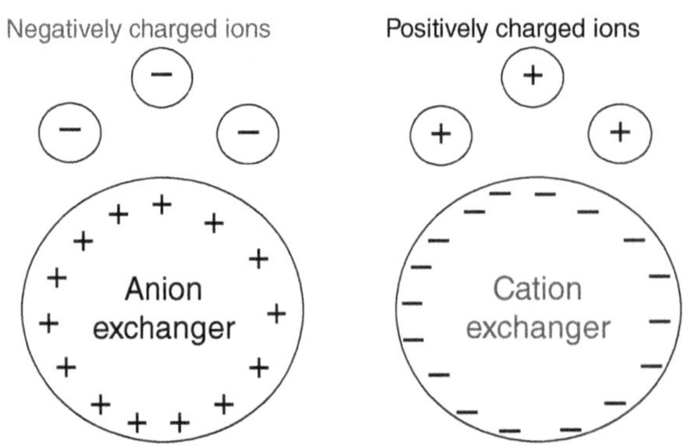

Figure 8.4 Mechanism of ion exchange.

volume, and the presence of functional groups on the absorbent material. A higher adsorption capacity allows more off-flavor molecules to be removed from the food. It is important to understand that the adsorption capacity of the absorbent can diminish over time as the adsorption sites become occupied or saturated. The adsorption capacity of the off-flavor adsorption depends upon the following factors:

1) Surface area: Off-flavor adsorbents having a porous structure with a multitude of interconnected pores provide a significantly large surface area per unit volume, which enhances the adsorption capacity of the absorbent. The large surface area provides more interaction sites between the absorbent and off-flavor molecules.
2) Adsorption sites: Higher adsorption sites within the porous structure become more available for interaction with off-flavor molecules. These sites include the internal surface of the pores, including inner walls, edges, or defects that can physically attach, adhere, or be trapped on these adsorption sites.
3) Distribution of pore size: Adsorbents consists of porous structure with a wide range of pore sizes, which allows for the accommodation of off-flavor molecules with different sizes and shapes. This distribution of pore size affects the adsorption rate and selectivity of the adsorbent. Smaller pores absorb smaller off-flavor molecules, while larger pores can capture larger molecules.
4) Pore volume: The total volume of the pores within the absorbent material, known as pore volume, influences the overall adsorption capacity as higher pore volume corresponds to a greater amount of available space to trap off-flavor molecules, leading to increased adsorption capacity [29].

Some off-flavor adsorbents exhibit selective adsorption, where they preferentially adsorb certain molecules or classes of compounds while leaving other desirable components of the packaged product unaffected. This selectivity is based on factors such as molecular size, shape, polarity, or chemical properties of the adsorbent. The adsorbent material is designed to have active sites or functional groups that can recognize and interact with the target off-flavor molecules. This selective adsorption mechanism relies on the affinity and specificity of the adsorbent toward off-flavor compounds compared to other components of the food product. The pore size and structure of the adsorbent play a crucial role in selective adsorption. The adsorbent material is tailored to have pore sizes that allow the passage and adsorption of specific off-flavor compounds while excluding other larger molecules present in the food product as such. The surface chemistry of the adsorbent material can be modified to enhance selectivity. Functional groups or chemical modifications on the adsorbent surface can facilitate specific interactions with the target off-flavor compounds, leading to their preferential adsorption. By leveraging the selective adsorption mechanism, off-flavor adsorbents can effectively remove specific off-flavor compounds while preserving the desired flavor, aroma, and quality of the food product. The development of selective adsorbents requires a deep understanding of the off-flavor compounds present in different food matrices and their interactions with the adsorbent materials. Different adsorption has a different affinity toward different off-flavors. In comparison to 21 adsorbents (amine-functionalized polymers, hydrophobic resins, and zeolites) for adsorption of off-flavor aldehydes, hydrophobic ZSM-5 type zeolites show the best affinity in alcohol-free beer [16].

8.3.2 Chemical and Enzymatic Reaction

Absorbent reacts with off-flavor compounds, which causes binding, breakdown, or conversion of off-flavor molecules. This results in the removal or neutralization of off-flavors and improves product quality.

8.3.2.1 Oxidation and Reduction

Redox reactions are defined as reactions involved in electron transfer. Loss of electrons is considered as oxidation, while gain of electrons is considered as reduction. Off-flavor compounds undergo oxidation or reduction on the surface of the adsorbent, which causes the transformation of off-flavor compounds into less volatile or odorous compounds.

$$X - 2e^- \rightarrow X^{2+}$$

In this reaction, the substance X loses two electrons ($2e^-$) and is oxidized to form a positively charged species X^{2+}.

$$Y^+ + e^- \rightarrow Y$$

In this reaction, the positively charged species Y^+ gains one electron (e^-) and is reduced to form the neutral species Y.

Sulfide scavengers in poultry, aldehyde scavengers in biscuits, and amine scavengers in fish products are reported oxidation-based off-flavor absorbents. Acidic compounds such as citric acid and ascorbic acid are known well to oxidize the off-flavor compounds when incorporated with an absorbent such as polymers or ferrous salt [9].

8.3.2.2 Hydrolysis and Enzymatic Reactions

In general, hydrolysis refers to breakdown of a compound into smaller components using water molecule [32, 33]. In this case, the reactions involve the breaking of the off-flavor molecule through the addition of water, which results in the degradation of the off-flavor compound into less odorous forms. Generally, off-flavor enzymatic adsorbents facilitate the hydrolysis of off-flavor compounds, such as the enzyme naringinase used as an off-flavor adsorbent to hydrolyze bitter component naringenin in citrus fruit juices [9]. There are varieties of enzyme-based absorbents with the application of off-flavor absorbents, as addressed in Section 8.2.1.4.

$$AB + H_2O \rightarrow AOH + BH$$

In this reaction, the compound AB reacts with water (H_2O) to produce two products in the presence of enzyme: AOH and BH.

8.3.2.3 Hydrophobic and Hydrophilic Interactions

The hydrophobic-hydrophilic interaction mechanism relies on the preference of hydrophobic off-flavor compounds to be with no polar substances but not with water molecules. Hydrophobic absorbents are non-polar or low-polarity materials, which include activated carbon, silica gels, certain clays, or polymers with hydrophobic properties. Hydrophilic interactions are intermolecular attractive forces that are much weaker than covalent bonds.

But hydrogen bonds increase the adsorption capacity of the off-flavor absorbent towards the selective off-flavor compounds such as those containing hydroxyl (—OH), carbonyl (C=O), or amino (—NH_2) groups. Activated carbon, activated clay, and silica gel were found to be effective off-flavor adsorbents for hydrogen bonding.

8.3.2.4 Electrostatic Interaction

Electrostatic interaction between the absorbent and off-flavor compounds involves the interaction between opposite electric charges on the adsorbent and off-flavor compounds. This electrostatic interaction causes dipole–dipole interaction or ionic bonds. Dipole–dipole interaction utilizes chemical adsorption, while ionic bonding utilizes ion exchange adsorption, as shown in Figure 8.4. Some modified and activated carbon or ion exchange resins possess charged surfaces that carry positive charges (cationic) or negative charges (anionic), depending on the material and the specific functional groups present in the adsorbent. Off-flavor compounds carry charges or have polar functional groups, which cause electrostatic interaction based on their charge or polarity [29].

8.3.2.5 Chelation

Chelators or chelating agents form stable coordination complexes with off-flavor compounds (Figure 8.5). Chelating agents are used in combination with activated carbon or silica gels, which result in sequestration of off-flavor compounds on the adsorbent surface. These complex compounds are quite stable and remain bound to the absorbent [34]. Di-ethylene-tri-amine (DETA), one of the chelating agents have been used to remove off-flavor aldehyde from alcohol-free beer [16].

8.3.3 Diffusion

In diffusion, off-flavor compounds absorb or migrate to the absorbent due to the concentration gradient between the package and the absorbent, which facilitates the adsorption, as shown in Figure 8.6. The off-flavor compounds move from an area of higher concentration (food product) to a lower concentration (adsorbent material) to establish equilibrium. Off-flavor compounds possess kinetic energy that causes random motion

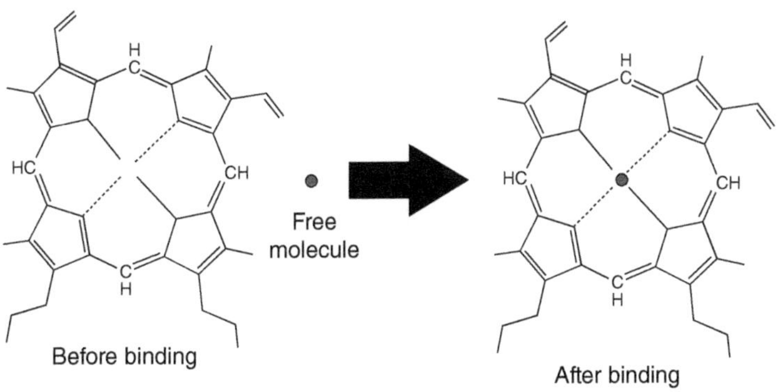

Figure 8.5 Binding by chelating agent.

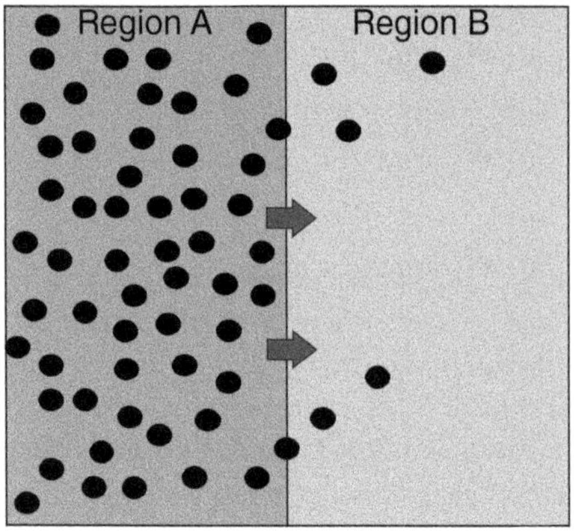

Figure 8.6 Diffusion phenomena (diffusion of molecules from Region A to Region B).

and allows diffusion through the adsorbent material. These diffused off-flavor compounds then undergo the binding mechanism as defined in Section 8.3.1. Mass transfer of off-flavor compounds due to diffusion into the adsorbent material depends on the concentration gradient, the molecular size of the off-flavor compounds, temperature, pressure, and the nature of the adsorbent material used. But after a certain period, the off-flavor adsorbent is saturated, and its capacity to adsorb additional off-flavor compounds is reduced due to the established equilibrium between the concentration of off-flavor compounds within the food product and the adsorbent material as it reaches a steady state [35]. Flavor absorbents, which allow more diffusion, are combined with other absorbents that allow little or no diffusion of flavor compounds in them. The combination of diffusion mechanism with other mechanisms increases the overall effectiveness of the flavor absorption system. Such as mixture of charcoal and nickel in durian fruit [13], silica gel and activated carbon product "2-in-1TM" in packaged food [9], and ferrous salt with organic acid "ANICO Bag" in fish products [21].

8.4 Off-Flavor Absorbent-Based Active Food Packaging

Active packaging has the potential to improve food product preservation in several ways. It is new and fascinating in the food industry that sustains popularity and trust, and the reason behind that is its latest advancement in the packaging industry. Before using any kind of active packaging option, it is necessary to comprehend how food deteriorates. By assisting in an increase in shelf life, conserving product quality, and assuring food safety, active food packaging can offer several advantages to customers, retailers, and food manufacturers [2]. Off-flavors are often produced by oxidation, suggesting to the customer that the food's quality has deteriorated. Consumers may detect certain unpleasant scents when the container is opened, even though the product looks to be safe [36]. Unpleasant compounds such as aldehydes, amines, and sulfides build up within the packaging as a consequence of food deterioration and can be specifically scavenged [8]. Off-flavor absorbent-based active packaging refers to incorporation of flavor absorbent in the packaging matrix, which allows packages to interact with off-flavor throughout the storage [19]. Different off-flavor absorbents were developed by the researchers in the past few years that can be incorporated as off-flavor absorbents in active packaging. However, only a few are considered for practical implementation, which includes removal of sulfide, amine, and aldehyde from the food products. Hence, these sulfide scavengers, amine scavengers, and aldehyde scavengers can only be considered as off-flavor absorbent-based active food packaging until new off-flavor absorbents are developed and used [8].

8.4.1 Sulfide Scavengers

Sulfide scavengers, as off-flavor absorbents, are substances or compounds that are specifically designed to capture or remove sulfurous compounds that contribute to off-flavors in various products. Sulfurous off-flavors can arise from the presence of sulfur-containing compounds, such as hydrogen sulfide (H_2S), mercaptans, or disulfides, which can be produced during processing, storage, or microbial activity. Sulfide scavengers work by chemically reacting with or adsorbing these sulfurous compounds, thereby mitigating their impact on the sensory attributes of the product. Sulfide scavengers are designed to specifically target sulfurous compounds, ensuring their selectivity for sulfur-based off-flavors. This allows for the effective removal or reduction of the off-flavors without significantly affecting the other desirable flavors or aromas in the product. Sulfurous odor compounds such as hydrogen sulfide are produced during protein degradation and cause putrid smell [19]. In some cases, sulfurous odor is produced due to microbial metabolites that accumulate in the headspace [7]. Sulfurous odor is one of the confinement odors in fresh poultry and cereal packaged products due to high barrier properties of packaging [19].

Odor-absorbing polymers have been successfully commercialized to remove the mercaptans and hydrogen sulfide produced during the transmission of protein-containing meals. Profresh is a master batch for commercial odor control and freshness preservation products [35]. Minipax and Strippax are two different products from Multisorb Technologies that are capable of absorbing mercaptans and hydrogen sulfide. Another product, Bynel IXP 101, a film developed by Dupont Polymers, was used to eliminate hydrogen sulfide from packages of cured chicken [2]. The active component known as ADI50 (composition unknown) has been reported to remove H_2S [9] and can be used as sulfide scavenger active packaging. Biofresh is a Grofit Plastics which absorbs gases including hydrogen sulfide known to contribute toward the ripening of fruits [2]. Citric acid and ascorbic acid are used to oxidize the sulfurous compounds when incorporated with an absorbent such as polymers or ferrous salt [9]. Charcoal and nickel mixture is used to remove ethyl disulfide, a fruity flavor from durian fruit [13].

8.4.2 Amine Scavengers

Amine scavengers capture or remove amines that contribute to off-flavors such as fishy, ammonia-like, or putrid smells. Amine scavengers work by chemically reacting with or adsorbing these amines, reducing their concentration, and mitigating their impact on the sensory attributes of the product. Fish products are one of the most common food products that produce huge number of undesirable odors and flavors, and elimination of these undesirable odors and flavors is one of the major requirements in active packaging.

The main reason associated with unpleasant smell is oxidation of amines [8]. Hence, applications of amine scavengers are required for elimination of amines generated by the oxidation of foods high in protein, such as fish [2]. The ANICO bags (Japan) oxidized the amines since they are constructed of film incorporating ferrous salt and citric acid or ascorbic acid, which is highly reactive toward amino compounds. Generally, metal conjugation with acid or acid salts has high affinity toward amines such as ANICO bags [21]. Adsorption-based compounds such as activated carbon, zeolites, and certain resins display a high attraction for amines. These polymers can absorb and hold amines, inhibiting their diffusion into food that has been packaged. Laccases and copper-containing polyphenol oxidases eliminate diamine off-flavors in apple juice [11].

8.4.3 Aldehyde Scavengers

Food rich in fat goes through lipid oxidation, causing formation of free radicals, which ultimately generate off-flavor compounds including aldehydes [7]. This aldehyde formation occurs in potato crisps, biscuits, fried foods, and cereals due to the oxidation of fatty acids [9]. While anaerobic glycolysis generates aldehydes and ketones in food products [6]. Even aldehyde is one of the major confinement odors in packaged products [18, 19] and accumulates in head space, especially in meat products [7]. Aldehyde compounds are also produced due to microbial growth in several weeks after heat processing of milk products such as UHT milk [2]. Polyethylene imine has been demonstrated to interact with and efficiently eliminate aldehydes produced during peroxide cracking of unsaturated fatty acids in lipids [37]. Antioxidants such as Vitamin E are added to the resin before extrusion to avoid lipid oxidation in dry meals such as snacks, cereal, and bread items. Dupont Polymers claims that the layer Bynel IXP101, which is an HDPE master batch, may remove aldehydes such as hexanal and heptanal from package [9]. Also, BMH™ powder, a synthetic alumina-silicate zeolite by EKA Nobel and Akzo, is used as aldehyde scavenger by incorporation in packaging materials such as paper by porous interstices [2]. ALDH, an enzyme purified from bovine liver mitochondria, has tendency to remove the bean flavor from defatted soybean extracts and soybean milk by converting aldehydes into acids [15]. Twenty-one adsorbents (amine-functionalized polymers, hydrophobic resins, and zeolites) are used to adsorb of off-flavor aldehyde in which hydrophobic ZSM-5-type zeolites show the best affinity in alcohol-free beer. DETA, chelating agents, is also utilized as an aldehyde scavenger in alcohol-free beer [16]. Sodium sulfate and other organic sulfates, PET-blended Nylon MXD6, D-sorbitol, and cyclodextrin were also reported as aldehyde scavengers in active packaging [2].

8.5 Applications of Off-Flavor Absorbing System in Food Packaging

8.5.1 Fruits and Vegetable Products

Freshly or barely treated fruits and vegetables are very susceptible to bacterial deterioration and experience unexpected biochemical changes during transit and preservation. Even higher metabolism of freshly picked vegetables and fruits is associated with higher respiration rate, which causes faster deterioration in flavor due to changes in composition [7]. The quantity of carbon dioxide inside the group can rise significantly and more than carbon dioxide, leading to harm to tissues, discoloration, and pack collapse, as well as off-flavors and smells and the proliferation of anaerobic microorganisms [2]. Fruits and veggies absorb O_2 from the outside environment for oxidative reduction of respiratory substrates to carbon dioxide, water, and heat. This energy raises the internal temperature of the produce [38]. In simple terms, the constant respiration of fresh products causes the carbon dioxide concentration of the container to rise. Peel, membrane, and seed of some fruits contain bitter compounds [39, 40] released in the product during processing techniques such as pressing and pasteurization of citrus juices release limonene [2], triterpenoid lactone limonene in bitter oranges, and glycosidic flavanone naringin in citrus juices [7].

There are different types of off-flavor absorbents used for different fruits and vegetable products due to wide variety of flavor compounds and food product characteristics, such as charcoal and nickel mixture used to remove fruity flavor from durian fruit [13]. Polymers are commonly used for off-flavor removal, such as cellulose triacetate and acetylated layer used in bitter oranges as limonene absorber-active packaging. Other polymers such as cellulose triacetate, acetylated paper, Nylon MXD6, D-sorbitol, and cyclodextrin also have applications in fruits and vegetables off-flavor removal [22]. Even resins are found effective toward off-flavor removal from citrus juices [30]. There are also few enzymatic off-flavor absorbents available, which include naringinase to remove naringin [10], limonoate dehydrogenase to oxidize limonene, laccases to remove polyphenols, methoxy-substituted phenols, and diamines [11]. Ethylene, a plant hormone, accelerates the formation of off-flavors in horticulture produce due to physiological disorders [2]. Even at low concentrations, ethylene is an inherent plant hormone for growth that increases the respiration of veggies and fruits and stimulates the ripening of fruits, softening, and aging [8]. Ethylene scrubbers are the most popular devices used to manage ethylene concentrations throughout post-harvest storage. To reduce the deterioration of crops and fruits, active packaging solutions that include oxygen scavengers, CO_2 emitters, desiccants, and antibacterial agents can be used [7]. Some market sachets based on activated carbon that is that can scavenge ethylene are SedoMate® (Japan), NeupalonTM (Japan), and Hatofresh® (Japan) [41]. Chemical ethylene scrubbers are one of the most often used equipment for controlling the level of ethylene during storage after harvest [7].

8.5.2 Meat and Meat Products

Meat, poultry, and fish are highly susceptible to fast microbial deterioration due to their high protein and fat content. Fresh meat is packaged to minimize infection, postpone spoilage, allow some enzyme activity in order to promote softness, limit weight loss, and, if appropriate, to assure an oxygenated hemoglobin or cherry red color in red meats at the retail or consumer level [7]. Color is one of the most important factors influencing the customer's choice. Myoglobin, a meat protein, normally appears in one of three chemical states: (i) deoxymyoglobin, a purple pigment, was oxygenated to produce cherry red oxymyoglobin. (ii) oxymyoglobin is then oxidized to (iii) metmyoglobin. This metmyoglobin causes discoloration and a loss of freshness. As a result, scavengers of oxygen have been utilized in the packaging of meat to reduce the negative effects of oxygen [42]. The main reason for the unpleasant odor in meat products is due to the formation of amines and aldehydes and accumulated in the headspace of the package as putrid (hydrogen sulfide) [7]. The breakdown of proteins and amino acids causes formation of sulfurous compounds, from lipid oxidation or/and anaerobic glycolysis leading to volatile aldehydes and ketone compounds [18, 19].

Fish and fishery products have received a lot of praise for being an excellent supply of proteins, vital vitamins and minerals, and polyunsaturated fats [43]. Bad smell and spoilage in seafood are caused by the decomposition of lipids and proteins, which is followed by the production of organic and volatile amines due to the oxidation of amines in fish products [8]. The consequences of an iron-based O_2 absorber on the nutritional value and security of freshwater catfish and seer fish have been examined during cold storage [43] using an O_2 absorber increased the life span of catfish to 20 days, instead of just 10 days for air-filled samples. In the instance of seer fish, O_2 scavenger packing materials were shown to be suitable for as long as 20 days, whereas standard air-bagged samples were discovered to be okay for 12 days. In conjunction with IRTA, Nanopack Technology and Packaging SL is creating a line of interleaves (Sanic films) capable of extending the life expectancy of vacuum-packaged meat products [44]. Application of off-flavor absorbers in meat products include sulfide scavengers in poultry and amine scavengers in fish products [9]. Activated carbon-based adsorbent pads reduce confinement odor [17], MINIPAX1 and STRIPPAX1 (Multisorb Technologies, USA) absorb mercaptans, and 2-in-1TM of United Desiccants removes off-flavor from packaged meat [25]. Metals are commonly used with organic acid or acid salts due to their high affinity toward amines such as ANICO bags [21].

8.5.3 Milk and Milk Products

Fresh milk is spoiled by the growth of microorganisms, and the taste of milk becomes impossible when the number of bacteria reaches 10^7 CFU/ml. Pasteurized milk has a very short shelf life in tropical countries, even at refrigeration temperatures. Pasteurized milk could benefit from antimicrobial packaging [21]. Dairy products, particularly cheeses, are sensitive to microbial development due to the greater water activity. Milk and milk products have higher water activity and thus support the growth of a variety of microorganisms, which cause the formation of off-flavor aldehydes. These off-flavor compounds can form several weeks after heat processing of products such as UHT milk [2]. The physiochemical parameters of milk pomade sweet-sherbet packaged in various materials for packaging including iron-based oxygen scavengers were assessed prior to and following packing and preservation. Bi-axially oriented polyethylene terephthalate (BOPET)/PE and aluminum film packaging, when combined with an oxygen scavenger, might be an effective packaging material for sherbet packing and for long-term storage. Another study used multiple-layer co-extruded films composed of extremely dense polyethylene (added with titanium dioxide), ethylene vinyl alcohol, and a layer of lower-density polyethylene involving antioxidants such as butylated hydroxyanisole (1.5%), butylated hydroxytoluene (1.5%), and tocopherol (4%) [45]. According to the findings of this research, multilayer co-extruded films offer an acceptable optical barrier for whole milk powder, and the film containing tocopherol contributed to the prevention of vitamin A degradation. Enzymes are included in one type of active packaging to aid in chemical processing. These make use of bioactive chemicals like enzymes and peptides to immobilize the component by trapping or physical adsorption. It shows how this technique works in eliminating lactose from milk. Lactase might be integrated into the product container's wall. The introduction of lactase into pasteurized or sterilized milk prior to packing to break lactose after packaging is described in a British patent awarded to Tetra Pak International AB. There are different off-flavor removal methods used in milk and milk products including sulfhydryl oxidase used in the treatment of UHT milk [11] and ALDH, which remove the beany flavor from soybean milk [15].

8.5.4 Fatty Foods

Fatty foods like oils, meat products, dairy products, and fried snacks are prone to oxidation and lipid breakdown, which can result in off-flavors and rancidity. This is due to oxidative rancidity generated by free radicals' generation, which is associated with an unpleasant flavor [7]. Moisture can hasten lipid oxidation and, as a result, the production of off-flavors in fatty meals [46]. The formation of aldehyde is common due to the oxidation of fatty acids in food containing fat such as potato crisps, biscuits, fried foods, and cereals [9]. Desiccant-containing packaging supplies or sachets, such as silica gel or molecular sieves, can remove excessive moisture, producing an environment that is more secure and reducing off-flavor production. Toppan sheetTM (Toppan Printing Co. Ltd., Japan), ThermariteTM (Thermarite Pty Ltd., Australia), LuquasorbTM (BASF, Germany), and other moisture absorbers are currently utilized extensively as sheets, blankets, or trays to increase the shelf life of food goods [42]. Packaging films with high barrier qualities may be required for fatty foods with strong flavors, such as some cheeses or flavored oils, to avoid the movement of aromatic off-flavor chemicals. Barrier films having poor permeability to oxygen, moisture, and other flavor-active chemicals can aid in the preservation of the product's natural flavor character [36]. ALDH removes the beany flavor from defatted soybean extracts and soybean milk [15]. Other off-flavor absorbents used in fatty foods are sodium sulfate and other organic sulfates including tocopherols, such as vitamin E [2]. Activated charcoal is another off-flavor absorbent that incorporates iron powder in packaging to serve as oxygen scavenging and odor scavenging, respectively [21]. Moisture absorbers such as silica gel, activated clay, molecular sieves, or desiccants adsorb or absorb moisture molecules and utilize compounds with high affinity for water [28].

8.6 Conclusion and Perspective

Off-flavors can significantly impact the sensory quality and overall acceptability of food products, reducing food security and causing economic losses for manufacturers. Active packaging, with its ability to interact with the food environment and modify it, offers a promising solution to address off-flavor issues. The utilization of off-flavor absorbents in active packaging holds promise, but there are several areas that require attention. Firstly, more studies are needed to understand the interaction between off-flavor compounds and absorbent materials under different packaging conditions. Secondly, the long-term stability and reuse of off-flavor absorbents need to be investigated. Continuous exposure to off-flavor compounds may lead to saturation and reduced efficacy of the absorbents over time. Developing methods to regenerate or replace absorbents within active packaging systems would ensure their sustained performance throughout the product's shelf life. Moreover, active packaging solutions should not compromise the nutritional, microbiological, or physical properties of food products. It is essential to conduct comprehensive studies to evaluate the overall safety and compatibility of off-flavor absorbents with different food matrices. Furthermore, the scalability and cost-effectiveness of incorporating off-flavor absorbents in commercial packaging systems need to be considered. The production and integration of absorbent materials into existing packaging processes should be feasible and economically viable for widespread industry adoption. Most importantly, legislative control over unethical use of off-flavor absorbent systems is required. Unethical approaches such as removal of off-flavor compounds to sell unsafe food are one of the major challenges for governments. Hence, there are restrictions on the use of off-flavor absorbents by the governments. We explored different flavor absorbents, which include activated carbon, zeolites, resins and polymers, enzymes, acidic compounds, and metals, and their mechanisms to remove off-flavor. There are different patents on these off-favor absorbents and mostly utilize the combination of absorbents to increase efficacy. Although very few off-flavor absorbent-based active packaging systems are commercially available even they have wide applications for different food products. Sulfide scavengers, aldehyde scavengers, and amine scavengers have potential applications.

References

1 Reineccius, G. (1991). Off-flavors in foods. *Critical Reviews in Food Science and Nutrition* 29 (6): 381–402. https://doi.org/10.1080/10408399109527534.

2 Day, B.P.F. and Potter, L. (2011). Active packaging. In: *Food and Beverage Packaging Technology*, 251–262. https://doi.org/10.1002/9781444392180.ch9.

3 Dash, D.R., Singh, S.K., and Singha, P. (2024). Bio-based composite active film/coating from deccan hemp seed protein, taro starch and leaf extract: characterizations and application in grapes. *Sustainable Chemistry and Pharmacy* 39. https://doi.org/10.1016/j.scp.2024.101609.

4 Gorde, P.M., Dash, D.R., Singh, S.K., and Singha, P. (2024). Advancements in sustainable food packaging: a comprehensive review on utilization of nanomaterials, machine learning and deep learning. *Sustainable Chemistry and Pharmacy* 39. https://doi.org/10.1016/j.scp.2024.101619.

5 Pavani, M., Singh, S.K., and Singha, P. (2024). Chitosan from agro-waste for food packaging applications. In: *Agro-Waste Derived Biopolymers and Biocomposites: Innovations and Sustainability in Food Packaging* (ed. S. Kumar, A. Mukherjee, and V. Katiyar), 267–294. Scrivener Publishing LLC & Wiley. https://doi.org/10.1002/9781394175161.ch10.

6 Brody, A.L., Bugusu, B., Han, J.H. et al. (2008). Scientific status summary. *Journal of Food Science* 73 (8): R107–R116. https://doi.org/10.1111/j.1750-3841.2008.00933.x.

7 Aastha Bhardwaj, T.A., Alum, T., and Talwar, N. (2019). Recent advances in active packaging of agri-food products: a review. *Journal of Postharvest Technology* 7 (1): 29. http://jpht.in/MenuscriptFile/5b19803b-c444-4b72-b3b6-af89c03cc802.pdf.

8 Biji, K.B., Ravishankar, C.N., Mohan, C.O., and Srinivasa Gopal, T.K. (2015). Smart packaging systems for food applications: a review. *Journal of Food Science and Technology* 52 (10): 6125–6135. https://doi.org/10.1007/s13197-015-1766-7.

9 Vermeiren, L., Devlieghere, F., van Beest, M. et al. (1999). Developments in the active packaging of foods. *Trends in Food Science & Technology* 10 (3): 77–86. https://doi.org/10.1016/s0924-2244(99)00032-1.

10 Puri, M. and Banerjee, U.C. (2000). Production, purification, and characterization of the debittering enzyme naringinase. *Biotechnology Advances* 18 (3): 207–217. https://doi.org/10.1016/s0734-9750(00)00034-3.

11 Schroeder, M., Pöllinger-Zierler, B., Aichernig, N. et al. (2008). Enzymatic removal of off-flavors from apple juice. *Journal of Agricultural and Food Chemistry* 56 (7): 2485–2489. https://doi.org/10.1021/jf073303m.

12 Kola, O., Kaya, C., Duran, H., and Altan, A. (2010). Removal of limonin bitterness by treatment of ion exchange and adsorbent resins. *Food Science and Biotechnology* 19 (2): 411–416. https://doi.org/10.1007/s10068-010-0058-2.

13 Chairat, R., Lertsiri, S., and Tananuwong, K. (2022). Volatile compound production and quality characteristics of durian fruit Cv. Monthong as affected by 1-methylcyclopropene and modified atmosphere storage. *International Journal of Fruit Science* 22 (1): 481–494. https://doi.org/10.1080/15538362.2022.2060418.

14 Day, B.P.F. (2008). Active packaging of food. In: *Smart Packaging Technologies for Fast Moving Consumer Goods*, 1–18. https://doi.org/10.1002/9780470753699.ch1.

15 Chiba, H., Takahashi, N., and Sasaki, R. (2014). Enzymatic improvement of food flavor II. Removal of beany flavor from soybean products by aldehyde dehydrogenase. *Agricultural and Biological Chemistry* 43 (9): 1883–1889. https://doi.org/10.1080/00021369.1979.10863730.

16 Gernat, D.C., Penning, M.M., Swinkels, F.M. et al. (2020). Selective off-flavor reduction by adsorption: a case study in alcohol-free beer. *Food and Bioproducts Processing* 121: 91–104. https://doi.org/10.1016/j.fbp.2019.12.007.

17 Versteylen, S., Stoll, B.N., and Michaels, M.H. (2014). Absorbent Pad to Reduce Confinement Odor in Food Packages. Patent Application no. 14/497,973.

18 Awulachew, M.T. (2022). A review of food packaging materials and active packaging system. *International Journal of Health Policy Planning* 1 (1). https://doi.org/10.33140/ijhpp.01.01.03.

19 Vinay Pramod Kumar, K., Suneetha, J., and Bethapudi, A.K. (2018). Active packaging systems in food packaging for enhanced shelf life. *Journal of Pharmacognosy and Phytochemistry* 7 (6): 3. https://www.phytojournal.com/archives/2018/vol7issue6/PartAJ/7-5-604-866.pdf.

20 Takase, I., Kakie, T., and Harasawa, A. (1991). Deodorizing Packaging Material. Japan Patent No JPH04246444A.

21 Rooney, M.L. (1995). Overview of active food packaging. In: *Active Food Packaging* (ed. M. L. Rooney), 1–37, Springer; 1995th edition.

22 Chandler, B.V. and Johnson, R.L. (1977). Cellulose acetate as a selective sorbent for limonin in orange juice. *Journal of the Science of Food and Agriculture* 28 (10): 875–884. https://doi.org/10.1002/jsfa.2740281002.

23 Suloff, E.C., Marcy, J.E., Blakistone, B.A. et al. (2003). Sorption behavior of selected aldehyde-scavenging agents in poly(ethylene terephthalate) blends. *Journal of Food Science* 68 (6): 2028–2033. https://doi.org/10.1111/j.1365-2621.2003.tb07013.x.

24 Rodríguez-Reinoso, F. (2001). Activated carbon and adsorption. In: *Encyclopedia of Materials: Science and Technology*, 22–34. https://doi.org/10.1016/B0-08-043152-6/00005-X.

25 Pérez-Botella, E., Valencia, S., and Rey, F. (2022). Zeolites in adsorption processes: state of the art and future prospects. *Chemical Reviews* 122 (24): 17647–17695. https://doi.org/10.1021/acs.chemrev.2c00140.

26 Reineccius, G.A. (2007). Flavour-isolation techniques: chemistry, bioprocessing and sustainability. In: *Flavours and Fragrances*, 409–426. Springer Berlin Heidelberg. https://doi.org/10.1007/978-3-540-49339-6_18.

27 Chase, T. (1974). Flavor enzymes. In: *Food Related Enzymes*, 241–266. https://doi.org/10.1021/ba-1974-0136.ch010.

28 Ozdemir, M. and Floros, J.D. (2004). Active food packaging technologies. *Critical Reviews in Food Science and Nutrition* 44 (3): 185–193. https://doi.org/10.1080/10408690490441578.

29 Kammerer, J., Carle, R., and Kammerer, D.R. (2010). Adsorption and ion exchange: basic principles and their application in food processing. *Journal of Agricultural and Food Chemistry* 59 (1): 22–42. https://doi.org/10.1021/jf1032203.

30 Johnson, R.L. and Chandler, B.V. (1985). Ion exchange and adsorbent resins for removal of acids and bitter principles from citrus juices. *Journal of the Science of Food and Agriculture* 36 (6): 480–484. https://doi.org/10.1002/jsfa.2740360609.

31 Dąbrowski, A. (2001). Adsorption – from theory to practice. *Advances in Colloid and Interface Science* 93 (1–3): 135–224. https://doi.org/10.1016/s0001-8686(00)00082-8.

32 Asaithambi, N., Singha, P., and Singh, S.K. (2022). Comparison of the effect of different desugarization techniques on the functionality of egg white protein hydrolysates. *Applied. Food Research* 2 (2): https://doi.org/10.1016/j.afres.2022.100152.

33 Asaithambi, N., Singha, P., and Singh, S.K. (2022). Recent application of protein hydrolysates in food texture modification. *Critical Reviews in Food Science and Nutrition* 1–32. https://doi.org/10.1080/10408398.2022.2081665.

34 Wang, L., Green, L., Li, Z., McCabe Dunn, J., Bu, X., Welch, C. J., ... Cui, J. (2011). Screening binary systems of chelating agents combined with carbon or silica gel adsorbents: the development of a cost-effective method to remove palladium from pharmaceutical intermediates and APIs. *Organic Process Research & Development*, 15(6), 1371–1376. https://doi.org/10.1021/op2001657

35 López-Rubio, A., Almenar, E., Hernandez-Muñoz, P. et al. (2004). Overview of active polymer-based packaging technologies for food applications. *Food Reviews International* 20 (4): 357–387. https://doi.org/10.1081/fri-200033462.

36 Adhiaman, A.S., Ramalakshmi, K., and Madhushalini, D. (2021). Industry-ready active packaging technologies in food sector. *International Journal of Current Research and Review* 13 (22): 6. https://doi.org/10.31782/IJCRR.2021.132202.

37 DelNobile, M.A., Buonocore, G.G., Fava, P., and Piergiovanni, L. (2002). Modeling of hexanal sorption kinetic in an aldehydes scavenger film intended for food packaging applications. *Journal of Food Science* 67 (7): 2687–2691. https://doi.org/10.1111/j.1365-2621.2002.tb08799.x.

38 Nayik, G.A. and Muzaffar, K. (2014). Developments in packaging of fresh fruits- shelf life perspective: a review. *American Journal of Food Science and Nutrition Research* 1 (5): 6.

39 Barbhuiya, R.I., Singh, S.K., and Singha, P. (2022). Mangosteen wastes: chemistry, processing, and utilization. In: *Handbook of Fruit Wastes and By-Products* (ed. K. Muzaffar, S.A. Sofi, and S.A. Mir), 113–124. Boca Raton: CRC Press.

40 Barbhuiya, R.I., Singha, P., and Singh, S.K. (2022). Pomelo wastes: chemistry, processing, and utilization. In: *Handbook of Fruit Wastes and By-Products* (ed. K. Muzaffar, S.A. Sofi, and S.A. Mir), 19–38. Boca Raton: CRC Press.

41 Rooney, M.L. (2005). Introduction to active food packaging technologies. In: *Innovations in Food Packaging* (ed. M. L. Rooney), 63–79, Springer; 1995th edition.

42 Kerry, J.B. and Butler, P. (2008). *Smart Packaging Technologies for Fast Moving Consumer Goods*. Chichester: Wiley-Blackwell. https://www.cabdirect.org/cabdirect/abstract/20123125620.

43 Mohan, C.O., Ravishankar, C.N., and Srinivasa Gopal, T.K. (2010). Active packaging of fishery products: a review. *Fishery Technology* 47 (1): 1–18. http://krishi.icar.gov.in/jspui/handle/123456789/49112.

44 Maryam Sardarodiyan, E.M. (2016). Active packaging systems for a modern society. *International Journal of PharmTech Research* 9 (7): 7. https://sphinxsai.com/2016/ph_vol9_no7/2/(357-363)V9N7PT.pdf.

45 Granda-Restrepo, D., Peralta, E., Troncoso-Rojas, R., and Soto-Valdez, H. (2009). Release of antioxidants from co-extruded active packaging developed for whole milk powder. *International Dairy Journal* 19 (8): 481–488. https://doi.org/10.1016/j.idairyj.2009.01.002.

46 Ravindran, N., Singh, S.K., and Singha, P. (2024). Nanovehicles for essential polyunsaturated lipids. In: *Advances in Biopolymers for Food Science and Technology* (ed. K. Pal, P. Sarkar, and M.Â. Cerqueira), 291–307. https://doi.org/10.1016/B978-0-443-19005-6.00012-8.

9

Biopolymer-Based Nanocomposites for Active Food Packaging

Soubani Baidya and R. Anandalakshmi

Department of Chemical Engineering, Indian Institute of Technology Guwahati, Guwahati, Assam, India

9.1 Introduction

The management of food safety is an essential aspect in all the stages of the food supply chain. Ensuring safety encompasses safeguarding against biological, chemical, and physical hazards during the entire "farm to fork" journey of the food product. Post manufacturing of the food products and their effective packaging have thus become essential for storage and transportation in food supply chain, playing a pivotal role in ensuring food safety, minimizing food waste, and extending the overall shelf life of the food products. Furthermore, in the case of active food packaging systems, the retention of food quality is closely intertwined with the assurance of food safety through the transformation of active components from the packaging materials by several active packaging systems. In this regard, polymers have received greater attention as the main matrix for the incorporation of active components because of the reduction in chances of accidental breakage, higher flexibility, good gas barrier properties, excellent sealing capabilities, etc. [1]. Within the spectrum of polymers, petrochemical-based polymers in general are widely used in food packaging systems because of their cost-effectiveness and other remarkable properties like optimal stiffness, good barrier against oxygen, and good heat-sealing property [2, 3]. However, in recent years, the increase in awareness in the society toward the creation of sustainable environment has imposed the necessity to reduce the utilization of petrochemical-based polymers as the main substrate to prepare polymeric packaging material. This is because most of the petrochemical-based polymers are not biodegradable, and these polymers significantly contribute to the increasing levels of greenhouse gas emissions. The utilization of biopolymers is the ultimate choice to replace petrochemical-based polymers for creating a sustainable solution for the environment. A significant proportion of biobased polymers exhibit biodegradability, contributing to their eco-friendliness. Besides environmental friendliness, the additional advantage of biopolymers is that these polymers mostly come from agricultural sources and are edible, therefore they can be used as a primary packaging layer in edible food packaging [4, 5].

Smart Food Packaging Systems: Innovations and Technology Applications, First Edition.
Edited by Avik Mukherjee, Santosh Kumar, Manjusri Misra, and Amar K. Mohanty.
© 2025 John Wiley & Sons Ltd. Published 2025 by John Wiley & Sons Ltd.

Despite all the advantages associated with biopolymers, certain drawbacks, including low mechanical, thermal, and barrier properties, throw up challenges in bringing advanced technologies to improve the properties of the biopolymers. Compatibility of these polymers with the functional active ingredients is another important factor to consider while preparing bio-based active food packaging materials. Ongoing research initiatives are actively exploring ways to address these inherent challenges. Utilization of nanotechnology to prepare biopolymer-based nanocomposite is one such novel approach to overcome the existing challenges of biopolymer-based packaging material. In this approach, nanomaterials are incorporated into the polymeric matrix, exerting a pronounced positive influence on the macroscopic properties of biopolymers. The impact of nanotechnology across diverse sectors within food domain, including food additives and ingredients, nutritional supplements, post-harvest management, and food packaging, has been phenomenal. This chapter thus highlights the developments in different biopolymer nanocomposite active food packaging systems and their applications across various food categories, including fruits, vegetables, dairy products, muscle foods, bakery items, and other food products. Furthermore, the regulation and safety aspects associated with the biopolymer nanocomposite-based packaging are also discussed.

9.2 Biopolymers

In the simplest terms, biopolymers can be defined as polymers obtained from renewable sources [6]. While a majority of the biopolymers exhibit biodegradability, it is noteworthy that some nonbiodegradable polymers are also prepared from renewable sources. Polymers like biopolyethylene, biopolypropylene, and biopolyethylene terephthalate are the examples of nonbiodegradable biopolymers, according to the European Bioplastic Association [7]. Biopolymers that are biodegradable in nature can be classified into three different categories based on their origin, as shown in Figure 9.1 [8, 9].

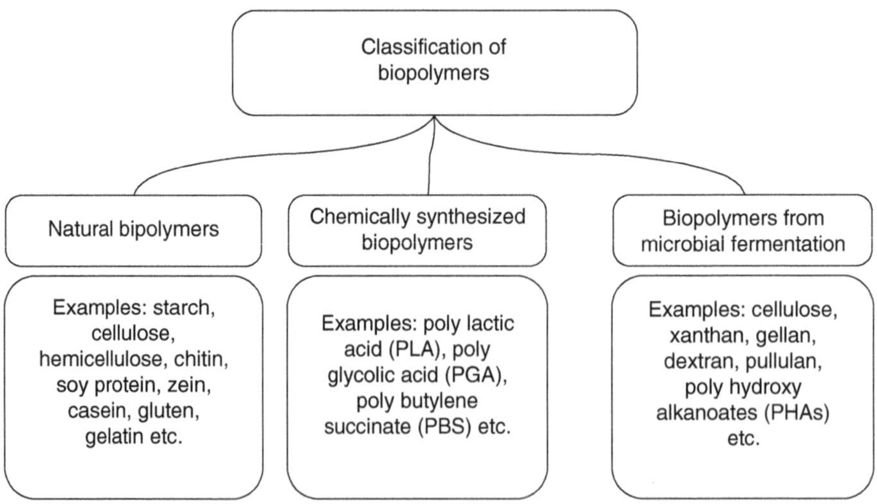

Figure 9.1 Classification of biopolymers.

9.2.1 Natural Biopolymers

Polymers belonging to the first category are easily accessible from a broad spectrum of origins. For example, starch is one of the most common biopolymers used in packaging and comes from various botanical entities: potatoes, peas, bananas, or other cereals or grains like rice, corn, wheat, etc. Cellulose, recognized as the most abundant polymer on earth, is prominently present in plant materials including cotton, ramie, kenaf, jute, several fruits and vegetables, grains, nuts, legumes, and seeds. Cellulose also originates from bacterial sources [10, 11]. Chitin is the second most abundant biopolymer just after cellulose [12, 13]. It is generally obtained from fungal origins, notably within the cellular structure of fungi. Additionally, it is also obtained from marine organisms, which encompass lobster, shrimp, prawns, etc.

Other than polysaccharide-based polymers, protein-based polymers, such as soy protein, zein, gluten, and whey protein, have gained attention for their utilization as packaging materials because of their high nutritional value and good film-forming capability. Some proteins owing to their distinctive amino acid composition and complex molecular structures exhibit better gas barrier properties than many polysaccharide-based polymers and many other synthetic polymers [14]. This property makes protein-based polymers particularly appealing in food packaging applications where enhanced gas barrier properties are crucial.

9.2.2 Synthetic Biopolymers

Synthetic biopolymers are those polymers that are obtained through a multistage chemical synthesis of the raw materials obtained from natural sources [15]. A notable example is polylactic acid (PLA), an aliphatic polyester that comes from agricultural sources like corn, wheat, and sugar beet by fermentation to form lactic acid. Lactic acid is then converted to lactide and further polymerized to form PLA [16–18]. Some other examples of synthetic biopolymers include polybutylene succinate (PBS), polyglycolic acid (PGA), and polyhydroxy alkanoates (PHAs). In contrast to the polymers directly obtained from natural sources, these polymers are highly stable in terms of water vapor barrier properties, and thermal and mechanical properties [19]. However, the cost of production of such polymers is high because of the requirement for advanced equipment and technologies [18]. As a result, there are still challenges in transitioning the production of these polymers from laboratory scale to the industrial scale.

9.2.3 Biopolymers from Microbial Fermentation

Many distinct extracellular polymeric substances (EPSs) are produced by microorganisms, predominantly composed of the carbohydrate moieties including cellulose, gellan, pullulan, and xanthan. Another noteworthy polymer derived from microbial fermentation is PHAs, a class of polymers produced by over 300 microbial species, mostly bacteria [20]. Several research studies are persistently directed toward the development of effective microbial strain through the application of advanced molecular tools to increase the efficiency of PHA production [21]. However, one important factor hindering wider industrial application of these biopolymers lies in the substantial financial investments required in the research of strain optimization, increasing fermentation efficiency, and downstream processing.

9.3 Nanomaterials and Their Functional Properties

Nanomaterials are formed when the materials in their bulk are reduced to a size within the nanoscale range. The fundamental features of nanomaterials are enhanced as a result of size reduction from materials in bulk form, making them suitable for various novel applications. Some of these special features include enhancement in the surface-to-volume ratio, higher chemical, mechanical, thermal, and kinetic stability, as well as high adsorption capacity [22, 23]. The segmentation of nanomaterials is done based on different criteria, such as phase composition, shapes, dimension, synthesis process, and functional property. The schematic diagram of the different criteria of classification is given in Figure 9.2. Based on the dimensions, there are zero-dimensional (quantum dots), one-dimensional (nanofibers), two-dimensional (nano-coatings), and three-dimensional (nanocrystalline) structures. Synthesis is typically conducted either by a top-down approach (such as mechanical milling, laser ablation, sputtering, and nanolithography) or by a bottom-up approach (such as sol–gel and chemical-vapor deposition, etc.). Nanomaterials can further be classified based on their organic or inorganic sources [24, 25]. Various organic nanomaterials exist in the form of dendrimers, micelles, cyclodextrin, and several lipid-based nanoparticles (nanoemulsions, nanoliposomes, solid-lipid nanoparticles, and nanostructured lipid carriers). Inorganic nanomaterials typically include metals such as aluminum, copper, iron, zinc, gold, and silver, as well as metal oxides like zinc oxide (ZnO), copper oxide (CuO), titanium oxide (TiO_2), etc. Features and preparation methods of several types of nanomaterials are given in Table 9.1. The specific functional properties of nanomaterials arise from differences in their morphology, aspect ratio (length-to-diameter ratio), and chemical nature. Some of the functional properties of nanomaterials are discussed here.

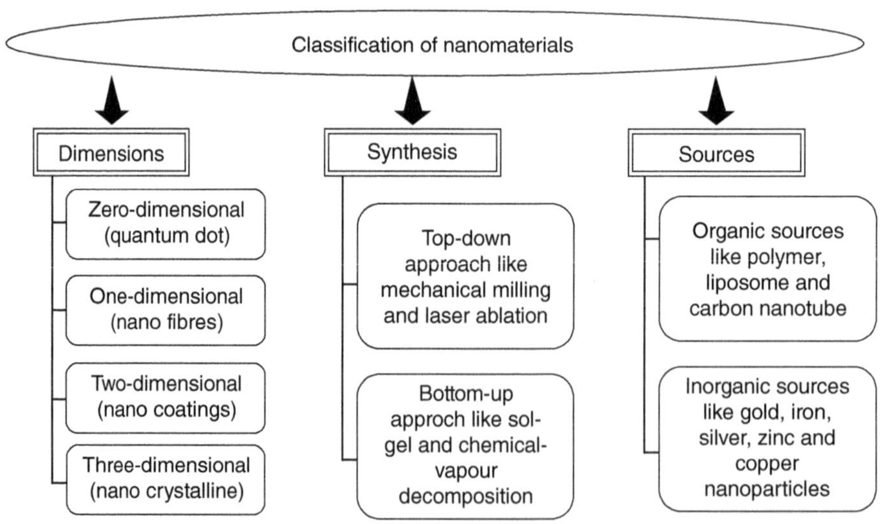

Figure 9.2 Different classifications of nanomaterials.

Table 9.1 Features and preparation methods of several types of nanoparticles.

Nanocarriers and nanoparticles	Types	Methods of preparation	Features	References
ZnO, TiO_2, CuO, Fe_2O_3, etc. nanoparticles	Inorganic metal oxides	Physical methods include physical vapor deposition, arc plasma method, thermal evaporation, or chemical methods like sol–gel and chemical vapor decomposition	Higher stability, mechanical stiffness, and biocompatibility when compared to organic nanomaterials	[26, 27]
Cyclodextrin	Oligosaccharides (biopolymer)	Nanoprecipitation, emulsion, solvent evaporation	Truncated cone-shaped cyclic oligosaccharides having amphiphilic characteristics	[28]
Nanoemulsion	Lipid	High-energy methods like high-pressure homogenization, ultrasonication or low-energy methods such as micro-emulsification, phase inversion technique	Spherical-sized droplets consist of two immiscible liquids in the form of oil-in-water; water-in-oil; oil–water–oil; water–oil–water emulsions	[29, 30]
Nanoliposomes	Lipid	Physical methods such as sonication, extrusion, or micro fluidization or chemical methods like film dehydration, ethanol injection, reverse phase evaporation	An aqueous interior surrounded by a lipid bilayer membrane	[29, 30]
Solid-lipid nanoparticles	Lipid	Hot and cold homogenization techniques	Oil-in-water emulsion, in which oil phase is solid, has greater control over the release of functional moieties	[29, 30]
Halloysite nanotubes	Clay	Purification of raw halloysite followed by size reduction and modification of its properties through physical methods (such as solution processing or melt processing) or chemical methods (non-covalent or covalent modification) for its better dispersion in the polymer matrix	Aluminosilicate clay material having ultra-tiny hollow tube structure with negatively charged exterior	[31]
Biopolymeric nanohydrogel	Biopolymer	Physical cross-linking, chemical cross-linking, radiation cross-linking, grafting	Nanosized biopolymeric linkages hold liquids inside the structure	[32]
Biopolymeric nanofiber	Biopolymer	Electrospinning, flash spinning, template synthesis, melt blowing, bicomponent leaching, salt spinning	Fibrous material in nanosize has high surface area for better interfacial interaction	[32]

9.3.1 Antimicrobial Property of Nanomaterials

The precise mechanism of antimicrobial action of the nanomaterials remains a subject of ongoing investigation. Two possible mechanisms are assumed. First one is due to free metal ion toxicity arising from the dissolution of metals from the surface of the nanoparticles. Another mechanism might be due to oxidative stress via the generation of the reactive oxygen species (ROS) on crystal surfaces of some nanoparticles that can damage components present in the cells of microorganisms [33, 34]. Commonly used inorganic nanomaterials having inherent antibacterial properties are metal ions like silver, zinc, copper, gold, etc. Among all the nanoparticles, silver has received significant interest due to its comparatively low nontoxicity toward the eukaryotic cells and its efficacy against a wide range of microorganisms, including bacteria, fungi, and viruses [35–38]. In addition to inorganic materials, various organic compounds, such as chitosan derived from deacetylation of chitin, exhibit antimicrobial activity. Several types of nanocarriers including nanoemulsions, nanoliposomes, solid-lipid nanoparticles, lipid nanocapsules, and other biopolymeric and cyclodextrin-based nanomaterials are used to enhance antimicrobial efficacy of different natural antimicrobial agents like essential oils, lysozyme, carvacrol, eugenol, thymol, etc. [29, 32, 39]. In situations where targeting specific types of microorganisms is imperative, active packaging systems are equipped with compounds that exhibit a narrow spectrum of activity, such as nanoencapsulated bacteriocin. For example, nisin-loaded nanoparticles are a popular example of bacteriocins used to selectively inhibit the growth of pathogenic gram-positive bacteria in various food items.

9.3.2 Ethylene and Oxygen-Scavenging Property of Nanomaterials

In a nanoscale range, several metal oxide nanoparticles, like TiO_2, show enhanced photocatalytic mechanisms. This enhanced mechanism arises from the photoinduced processes when the materials in the nanoscale range are exposed to ultraviolet (UV) light with photo energy exceeding 3.2 electron volt (eV) or 388 nm wavelength. It causes charge separation due to the promotion of the electron from the valence to the conduction band. As a result, electrons accumulate on the surface of the nanomaterials. The crucial aspect of the photocatalytic process lies in the subsequent transfer of these electrons to oxygen in the surrounding air. The pronounced propensity of the nanoparticles to undergo such photocatalytic reaction mechanism makes them better ethylene and oxygen-scavenging agents than conventional scavengers [40]. In a study conducted by Mu et al. [41], iron nanoparticles were utilized as the oxygen-scavenging agent in roasted sunflower seeds and walnuts. Iron nanoparticles showed 1.4 times higher oxygen-scavenging capacity than the conventional iron powder, thus inhibiting lipid oxidation at an accelerated rate. In another study conducted by Upadhyay et al. [42], corn starch and gum acacia nanocomposite film were prepared by incorporation of sepiolite nanoclay. A notable improvement in the ethylene-scavenging property of the film was observed with an increasing concentration of sepiolite nanoclay. The application of the enzyme-based oxygen-scavenging system is impeded by the inherent instability of enzymes under normal conditions due to their high sensitivity to water activity, temperature, pH, and substrate present within the sachet. Thus, efforts have been made to enhance the activity of oxygen-scavenging enzymes such as glucose oxidase, laccase, and ethanol oxidase through the incorporation of nanoparticles [43, 44]. Incorporation of nanoparticles serves to immobilize these enzymes, thereby augmenting functional stability. This immobilization gives some additional advantages, like thermal stability, high selectivity, and high resistance to pH.

9.3.3 Antioxidant Property of Nanomaterials

Due to their large surface area-to-volume ratio, nanomaterials like metals and metal oxide-based substances such as titanium, copper, iron, selenium, silica, cerium oxide, and copper oxide have higher ratios of active electrons on their outer surface, leading to increased antioxidant capacity. As an example, metal oxide nanoparticles may participate in efficient ROS scavenging and/or deoxygenating processes [45]. Besides metal oxide nanoparticles, several bioactive components, including vitamins and polyphenols, possess inherent antioxidant properties. However, some of these antioxidant components are highly volatile in nature because of their molecular structure. As a result, these compounds are often encapsulated with the help of several nanocarriers to mitigate this volatility, consequently increasing bioavailability. Controlled release of these components from the nanocarriers imparts gradual antioxidant benefits to the food products.

9.3.4 Flavor-Releasing Property of Nanomaterials

The correlation between enhanced stability in the flavor of the food products and increased consumer acceptability is a pivotal consideration. Despite the incorporation of flavor-releasing substances within the active food packaging systems, preserving the stability of these flavors throughout the entire food supply chain proves to be a formidable challenge. This challenge arises from the intrinsic high volatility of these flavor components and the potential for degradation during storage, influenced by factors such as exposure to air, fluctuations in temperature, and exposure to light. Additionally, a considerable proportion of the flavor components are insoluble within the food matrix due to the differences in their polarities. Nanoencapsulation of flavors with the help of several nanocarriers emerges as a promising strategy to mitigate this challenge by reducing their volatility and chances of degradation under external factors such as air, temperature, and light. As a result, the integrity of the flavors in food is well maintained within the food supply chain. Furthermore, nanoencapsulation offers an additional advantage of improving the solubility of the flavoring compounds within the food matrix. This increased solubility contributes to a more uniform distribution of flavors in the food products. It can also lead to reduced reliance on surfactants, which are commonly employed to enhance the solubility of the hydrophobic compounds in the food products. In addition, innovative approaches like emulsion and coaxial approaches are occasionally employed to nanoencapsulate these flavor components with diverse solubility characteristics [46]. Several biopolymers like starch, sodium alginate, maltodextrin, carrageenan, etc. are used as the nanoencapsulating agents of the flavors as they are generally regarded as safe (GRAS) for human health [47]. Additionally, several lipid-based nanocarriers exhibit superior efficacy as nanoencapsulating agents for flavoring compounds. This is because many flavoring agents, being essential oils, are non-polar in nature, making them easier to entrap in lipid-based carriers such as liposomes [48].

9.3.5 UV-Blocking Property of Nanomaterials

Materials like TiO_2 exhibit increased UV-blocking properties when reduced in size to the nanoscale range. This is because their small particle size and high surface area per unit mass-to-volume ratio enable photoexcited electrons and holes to diffuse toward the surface before recombining [49]. One additional advantage of utilizing nanomaterials as the UV-blocking property is the capacity to engineer them for transparency, enabling them to develop colorless UV-blocking formulations. It helps to maintain the optical clarity and overall integrity of the packaging materials. For example, De Moraes et al. [48] used graphene oxide nanoparticles in the film as a UV-shielding material, concurrently enhancing optical properties. Additionally, graphene oxide being non-catalytic in nature, did not induce photodegradation of polymers as an adverse impact, unlike other metal oxide-based UV-shielding materials.

9.4 Biopolymer-Nanocomposites Active Packaging Systems

A composite is a multiphase material where the phases are chemically dissimilar and separated by a distinct interphase. However, the interphase should be good enough for effective stress transfer. In nanocomposite active packaging systems, the polymer matrix serves as the continuous phase, while the nanomaterial acts as the discontinuous phase dispersed in the polymeric matrix. There are several methods of synthesizing biopolymer-based nanocomposites including solvent evaporation, melt processing, and *in situ* polymerization. In case of *in situ* polymerization, a suitable solvent is mixed with nanomaterials and monomers, or several monomers, and then polymerization is carried out using the reagent. In melt processing, the incorporation of clay elements or nanofillers into the polymer matrix is achieved through a specific time of severe mixing followed by high-temperature annealing. After that, a homogeneous distribution of nanoparticles is obtained to promote the intercalation and exfoliation of silicates, clay, or nanofillers into the polymeric matrix. The traditional process for creating polymer thin films with different thicknesses is solvent evaporation. Using this method, the polymer is continuously stirred to dissolve it in a particular solvent, and then the nanofillers are mixed into the solution to create a homogenous mixture. After that, the mixture is poured into a mold to remove the solvent and eventually produce thin films with intercalated clay layers that are polymer-oriented [50]. A schematic diagram of development of active nanocomposite system is shown in Figure 9.3.

Strong interfacial interaction between nanomaterials and the polymer matrix is crucial for the effective retention of active agents within the polymer matrix and controlled release of them to the food matrix. There are several factors that influence the interfacial interaction of the nanomaterials with the polymer matrix, for example, size, shape, morphology, surface charge, concentration of the nanomaterials, etc. Several other processing conditions, such as degree of dispersion of nanomaterials, mixing time, and temperature, also play a pivotal role in influencing the interfacial interaction between the nanomaterials and

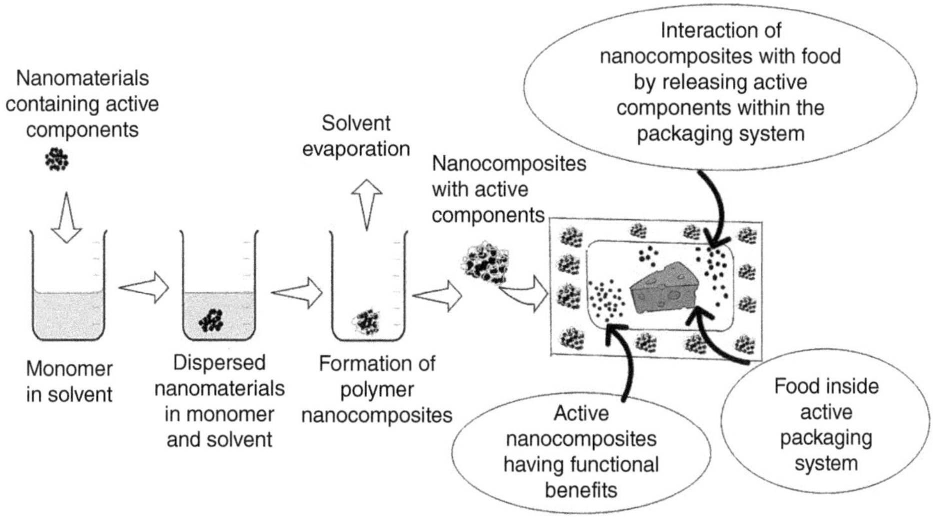

Figure 9.3 Preparation of biopolymer nanocomposite active packaging system.

the polymer matrix. For example, if the concentration of nanomaterials is not optimized or if their dispersion in the polymeric matrix is not uniform, there can be generation of weak Van der Waals interactions between the nanoparticles, leading to the formation of clusters. The formation of clusters prevents the polymer chain entanglement and their penetration into the aggregated nanomaterials, consequently diminishing the adhesion between the polymers and the nanomaterials [51–53]. In addition, many polymers are not chemically compatible with the nanomaterials because of the differences in surface functionalities, chemical structures, and polarities of both nanomaterials and the polymers. It further leads to the weakening of the interfacial interactions between the nanomaterials and the polymer matrix. In this regard, several pretreatments are given to the nanomaterials to improve their adhesion to the polymer matrix. Some of these pretreatments are adsorption via cationic interaction, chemical grafting by the addition of coupling agents, alkali treatment, acetylation, and peroxide treatment. Table 9.2 shows several pretreatments of nanoparticles before preparing bio-nanocomposite systems.

9.4.1 Starch-Based Active Nanocomposite System

Starch comprises two primary glucans: amylose and amylopectin. Amylose is characterized by its linear structure resulting from the polymerization of glucose through α-1–4 glycosidic linkage. It is present in the concentration of 20–30% of the starch composition. Conversely, amylopectin, another component of starch, is also derived from glucose. But it differs from amylose due to its highly branched structure because of the presence of additional α-1-6 glycosidic linkage. It accounts for approximately 70–80% of the starch composition. The coexistence of amylose and amylopectin in varying proportions influences thermal, mechanical, and several functional properties of the starch molecules. Starch in its native form is brittle with poor mechanical strength. The presence of a higher content of amylose

Table 9.2 Several pretreatment methods of nanomaterial before the preparation of nanocomposite.

Bio-nanocomposite systems	Pretreatment methods	Reference
Cassava starch/cellulose nanocrystals	Carboxymethylation of cellulose nanocrystals	[54]
Soy protein isolate/halloysite nanotubes	Silane grafting of halloysite nanotubes	[55]
Soy protein/nanoclay montmorillonite (MMT)	Coating MMT with tannic acid to improve the interfacial interaction	[56]
Soy protein isolate/cellulose nanofiber/nanosilica	Silane grafting to improve adhesion	[57]
PLA/nanocrystalline cellulose	Graft modification of nanocrystalline cellulose by 3-amino propyl tri-ethoxy silane	[58]
PLA/cellulose nanocrystals/ ZnO nanoparticles	Acetylation of cellulose nanofibers to improve dispersion in the polymer matrix	[59]
Chitosan/cellulose nanofiber	Curcumin grafted to 2,2,6,6-tetramethylpiperidine-1-oxyl (TEMPO)-oxidized cellulose nanofiber	[18]
Chitosan/cellulose nanofiber/ benzyl isothiocyanate	Cross-linking of cellulose nanofiber with chitosan via genipin	[60]

in starch from certain botanical sources exacerbates this brittleness due to the restricted chain mobility and high crystallinity of the amylose region [61]. This inherent property of starch poses significant hurdles for subsequent processing and use as a packaging material without pretreatments. Hence, starch is primarily gelatinized or chemically plasticized to disrupt the crystalline amylose region [61, 62].

Various methods, such as chemical cross-linking of starch, mixing with other polymers, and addition of functional nanomaterials, are implemented for further processing of starch-based food packaging films with improved functional properties and thermal stability. For example, Mondragón et al. [63] compared the nanocomposites prepared from three types of starch: high amylose starch, native starch, and high amylopectin starch. Plasticized high-amylopectin starch molecules exhibited greater susceptibility to intercalation or exfoliation by nanoclay, thereby showing the highest increment in the Young's modulus of the film. The blending of starch with other polymers was explored for the preparation of active nanocomposite-based food packaging systems. Mathew et al. [64] conducted research to prepare nanocomposite film by blending starch and polyvinyl alcohol (PVA) and further incorporating silver nanoparticles. Besides the improvement in overall mechanical properties, silver nanoparticles showed antimicrobial action against gram-positive *S. aureus* and gram-negative *S. typhimurium*.

9.4.2 Cellulose-Based Active Nanocomposite System

Cellulose is formed by the polymerization of glucose molecules joined together by β, 1–4 glycosidic linkage. The main limitation associated with cellulose is its limited solubility in both water and several organic solvents [65]. Due to this, cellulose undergoes several modifications, such as esterification, etherification, and carboxymethylation. These modifications broaden the application of cellulose in the food packaging domain. Carboxymethylcellulose (CMC), one of the most promising modified forms of cellulose, has received great recognition in the fabrication of food packaging film [66–68]. In research conducted by Ebrahimi et al. [69], the addition of ZnO and Ag-based nanomaterials into CMC film matrix significantly showed inhibitory effects against *S. sureus* and *E. coli*. Additionally, these nanoparticles also showed UV-absorbing property without compromising the transparency of CMC-based film. In another study, Wang et al. [70] reported the effect of the addition of halloysite nanotubes in CMC-based polymer matrix. This study involved the individual surface functionalization of halloysite nanotubes, followed by dispersion into various saturated metallic salt solutions. The surface functionalization of halloysite nanotubes demonstrated variations in the creation of inhibitory effects against microorganisms, contingent upon the type of metal ions present on the surface of distinct nanotubes. In addition to serving as polymeric matrix in the nanocomposite film, cellulose finds its application as a nanomaterial for the fabrication of completely organic nanocomposite film [71]. This is accomplished through various physical methods, including high-pressure homogenization, microfluidization, and chemical methods like acid hydrolysis. Notably, enzymatic hydrolysis utilizing cellulase enzymes represents another viable approach for selective degradation of cellulose, leading to the production of cellulose-based nanomaterials [72–74].

9.4.3 Chitosan-Based Active Nanocomposite System

Chitosan, which is essentially made up of (1–4)-linked 2-amino-2-deoxy-β-D-glucose monomers, is a deacetylated derivative of chitin. The specific attributes of chitosan such as molecule size, nitrogen content, degree of acetylation (DA), and polydispersity of chitosan are contingent upon the source of chitin. One important characteristic of chitosan is that it possesses inherent antimicrobial property. Several assumptions are already predicted to understand the mechanism of antimicrobial action of chitosan. Among these, the most plausible theory involves an electrostatic interaction between protonated amino groups (NH^{3+}) of glucosamine in the backbone of chitosan and the negatively charged components of microbial cell membranes. This interaction impacts the integrity of the cell membrane, leading to the leakage of several intracellular components of the bacterial cell [75, 76]. In order to further augment the antimicrobial efficacy of chitosan, several metal oxide nanoparticles are embedded into the matrix of chitosan. For example, Rodrigues et al. [77] reported that in chitosan-based film, reinforced with sodium montmorillonite (MMT) and zinc nanoparticles, synergistic inhibitory effect against *E. coli* and *S. aureus* was contributed by both chitosan and ZnO nanoparticles. In another study conducted by Yang et al. [78], a synergistic action of lignin nanoparticles and chitosan not only inhibited the growth of microorganisms but also showed antioxidative responses. This response of chitosan and lignin nanoparticles was confirmed by DPPH radical scavenging activity. The antioxidative activity of chitosan was also corroborated by several other researchers [79–81]. Therefore, it has been established that the residual free amino (NH_2) groups and two hydroxyl groups of chitosan can react with free radicals to produce stable radicals, and then NH_2 groups take hydrogen ions from the solution to form ammonium (NH^{3+}) groups [82].

9.4.4 Gelatin-Based Active Nanocomposite System

Gelatin comes from partial hydrolysis of fibrous protein, collagen, present in the bones and skin of animal tissue. A distinctive characteristic of this polymer is that it can form thermo-reversible gel with water, exhibiting a reversible transition between gel and sol states in response to temperature variations [83]. This thermo-reversible gelling property makes gelatin suitable for the fabrication of food packaging film at relatively low temperatures, which can then solidify upon cooling. However, it is crucial to note that the properties of gelatin-based films differ based on the source, as gelatin exhibits variation in molecular weight distribution and amino acid composition depending on its sources [84]. For example, Ninan et al. [85] reported in their study that water vapor and oxygen barrier properties were superior in case of carp skin gelatin film when compared with mammalian skin gelatin film. Moreover, the study highlighted the differences in gel strengths between gelatin-based films derived from carp species and mammalian species, attributing these variations to distinct mechanical properties in each case.

The gelatin film was reinforced by copper sulfide nanoparticles by Roy and Rim [86], and enhanced UV-blocking phenomena were observed. Surprisingly, the control film, lacking additional reinforcement, also exhibited some degree of UV protection. It was attributed to the presence of aromatic amino acids such as tyrosine and phenylalanine in gelatin, which possess UV absorption capability. In another study conducted by Liu et al. [87] tea polyphenol nanoparticles were used as the antioxidative agent in the gelatin-based film. The process involved initially adding aqueous tea polyphenol solution to sulfobutyl ether-β-cyclodextrin sodium solution. Then the nanoparticles were formed by cross-linking sulfobutyl ether-β-cyclodextrin sodium with chitosan hydrochloride via ionic gelation mechanism. The resulting nanoparticle suspension was added to the gelatin film-forming solution, and then the film was formed by solution casting. Encapsulation of tea polyphenol and then preparation of the nanocomposite film showed better antioxidative potential than free, homogenously distributed tea polyphenol. This enhancement of antioxidative efficacy was attributed to the protective effect of encapsulation, preventing the oxidation of tea polyphenol through direct environmental exposure over time.

9.4.5 Casein and Whey Protein-Based Active Nanocomposite System

Casein and whey protein are the two types of milk proteins, in which casein accounts for about 80% of the total protein concentration and whey accounts for 20% of the total concentration in milk. Casein is a calcium-binding phosphoprotein, assuming a structural arrangement in the form of micelles within the milk matrix. The formation of these micelles, coupled with the proclivity of calcium binding to the negatively charged region within the casein molecule, prevents casein from thermal denaturation. Additionally, casein distinguishes itself by the absence of disulfide bridges, and the α-helix and β-strands are interrupted by the proline peptide present in casein. Consequently, casein exhibits a comparatively low secondary or tertiary structure, which further plays a substantial role in augmenting its thermal stability [88]. This higher thermal stability of casein along with its noteworthy emulsifying property and solubility in diverse range of polar and non-polar solvents, makes it a promising polymer in the context of food packaging.

Ma et al. [89] stated the mechanism of casein-based ZnO latexes. The findings of the study revealed a significant alternation in the performance of casein-based nanocomposites when ZnO nanoparticles are added using the twofold in situ approach. This was because the latex particles started to consolidate among themselves when the moisture level of the film reduced during the drying process. The study further reported excellent inhibitory effect against *E. coli*, reinforcing the potential antimicrobial properties of the developed nanocomposite film. In another interesting study by Wang et al. [90], a distinct methodology was employed for creating self-assembling nanoparticles loaded with curcumin. The subsequent preparation of nanocomposite films involved the combination of curcumin-loaded zein nanoparticles and sodium caseinate (NaCas) through an organic-free solvent method, employing pH-driven method. The study delineated the process by which dissociation of NaCas protein occurs at high pH, facilitating the entry of curcumin-loaded zein nanoparticles into the open network structure of NaCas matrix. Neutralization of pH again leads to the reassociation of NaCas with folded protein. Furthermore, the drying process during film formation was observed to intensify the NaCas/zein nanoparticle interaction. It was attributed to the increased concentration of sodium casein protein and its migration into the film. However, when the concentration of the zein nanoparticle increased, it resulted in the nanomaterial aggregation rather than interaction with the polymer matrix. The study also reported the effectiveness of zein nanoparticles to enhance the stability of curcumin against photodegradation. This method of preparation of film conferred superior antioxidative benefits to the film. In contrast to casein, whey protein is prone to denaturation under heat treatment. The flexible nature of whey protein-based film is contingent upon this denaturation process. This is because denaturation leads to increase in protein solubility with water. It also instigates the formation of three-dimensional structure due to several intermolecular interactions (disulfide bridges, Van der Waals interactions, hydrophobic and electrostatic interactions) formed by the exposed groups as a result of unfolding of proteins [14, 91].

Several researchers have emphasized the escalating prominence of whey protein-based film because of its cost-effectiveness and excellent gas barrier properties. However, the film made of whey protein generally lacks good thermal and mechanical stability. In order to combat these challenges, several research studies have already been conducted to prepare whey protein-based nanocomposite film. Mohammadi et al. [92] prepared whey protein-isolate-based nanocomposite film using chitosan nanofiber, which contributed to the increase in the mechanical strength of the film. Unexpectedly, chitosan nanofiber did not show antimicrobial activity because the primary amine group of chitosan was engaged in the interaction with the biopolymer to enhance its miscibility with the biopolymer. However, when chitosan nanofiber was added in addition to nano-formulated cinnamon oil, the film showed superior antimicrobial activity when compared with the film containing only cinnamon oil. This could be the result of cinnamon oil being released gradually and continuously by nanostructured lipid carrier throughout the incubation period of the film. This subtle strategy helped whey protein nanocomposite film to mitigate the antimicrobial challenges.

9.4.6 Polylactic Acid-Based Active Nanocomposite System

As discussed earlier, PLA formation happens through multistage chemical processes after the extraction of raw materials from agricultural sources. PLA has grabbed attention of the food packaging industry due to its similarity in mechanical features with several petrochemical-based polymers [93]. As a result, it can also undergo several processing techniques, like extrusion molding, blow molding, etc., for the fabrication of food packaging materials. However, despite these advantageous features, this polymer contends with certain drawbacks that impede its competitive standing against petrochemical-based polymers. Some of these drawbacks are the partial crystallization, poor gas, and water vapor barrier properties [94]. Several research studies on PLA-based nanocomposite films have been reported to improve several properties of PLA-based films. For example, Li et al. [95] reported overall increase in the water vapor permeability barrier after the addition of titanium dioxide and silver nanomaterials. The film also exhibited inhibitory effect against *E. coli* and *L. monocytogenes*. However, the increase in the concentration of nanoparticles enhanced the crystallization tendency of the nanocomposite film. This was attributed to the phenomenon of heterogenous nucleation. Similar enhancement of the nucleating effect was also found by Fortunati et al. [72] after adding surfactant-modified cellulose nanocrystal into PLA matrix. This phenomenon not only augmented crystallization propensity but also endowed the film with antibacterial efficacy against *S. aureus* and *E. coli* due to combined incorporation of cellulose nanocrystals and silver nanoparticles into the film matrix.

9.4.7 Other Biopolymer-Based Active Nanocomposite System

Several other biopolymer-based nanocomposite systems, like those from algae sources, are enriched with diverse bioactive compounds, responsible for conferring numerous health benefits when utilized in the form of active nanocomposite systems [96, 97]. Other than that, the biopolymers that are extracted from microbes, offer high purity levels with excellent mechanical strength and large surface area. Hence, these are one of the ideal classes of biopolymers for the fabrication of nanocomposite systems [98, 99]. In recent years, edible nanocomposite systems have been widely used for the preservation of food. Edible nanocomposite packaging systems come in the form of edible coating or edible film [100]. As discussed earlier, the distinctive feature of edible packaging systems, particularly edible coatings, lies in their direct interaction with food substances by maintaining continuous contact with the food materials. This characteristic enables more pronounced enhancement in the sensory attributes and overall quality of the food product by releasing nanoencapsulated flavoring components and other bioactive components directly into the food materials they are in contact with Trajkovska Petkoska et al. [101]. Several types of biopolymer-based nanocomposite active packaging systems along with their functional benefits are being addressed in Table 9.3.

Table 9.3 Biopolymer-based active nanocomposite systems.

Polymer matrix	Nanocarrier/ nanomaterials	Bioactive components inside the nanocarrier	Functional benefits	Reference
Starch and sodium caseinate	Soy and rapeseed nanoliposomes	Orange essential oil and limonene	Antimicrobial activity against *L. monocytogenes*	[102]
Methyl cellulose	Nanoemulsion	Clove bud (*S. aromaticum*) essential oil, Oregano (*O. vulgare*) essential oil	Antibacterial activity against yeasts and molds	[103]
Gelatin	Nanoemulsion, montmorillonite	Ginger essential oil	Antioxidant activity	[104]
Starch	Halloysite nanotube	Antimicrobial peptides (nisin or pediocin)	Antimicrobial activity against *L. monocytogenes* and *C. perfringens*	[105]
Agar	Silver–copper alloy nanoparticle	—	Antimicrobial activity against *L. monocytogenes* and *S. enterica sv typhimurium*	[106]
Gelatin/ K-carrageenan	Nanosilicon dioxide (SiO_2)	—	UV-absorbing capacity, thereby delaying UV light-driven reactions	[107]
Hydroxypropyl methyl cellulose/ whey protein concentrate	Chitosan nanoparticles	—	Antibacterial effect against *L. monocytogenes* and *E. coli*	[108]
PLA	Carbon nanotubes	Cinnamaldehyde	Antibacterial activity against *S. aureus* and *E. coli*	[109]
Wheat gluten	ZnO nanoparticles	—	Antioxidant activity, inhibitory effect against *E. coli* and *A. niger*	[110]
Hemicellulose/ pectin	Nanocellulose	Mango peel polyphenol	Enhancement in antioxidant activity	[111]

9.5 Application of Nanocomposites in Food Packaging Systems

Nanotechnology has already made substantial impact on various processed and unprocessed foods. Table 9.4 shows the remarkable results of bio-nanocomposite-based packaging on various food products.

9.5.1 Fruits and Vegetables

Fruits and vegetables generally have 70–90% moisture content. The higher moisture content not only enhances the respiration rate but also welcomes yeast and mold to form visible colored rot bringing about postharvest losses. Dehydration of fruits and vegetables is a common preservation technique, but nutritional quality and sensory quality are sacrificed. Several alternative approaches, such as edible packaging, modified atmosphere packaging, controlled atmosphere packaging, and active packaging are encountered to counteract quality degradation by reducing moisture loss, controlling respiration rate, and protecting from mechanical injury and oxygen penetration inside the food system. Conventional essential oil emulsions in edible packaging systems can alter the color and flavor of fruit depending on the concentration and can also cause deterioration of essential oils when subjected to harsh environmental conditions (temperature, pH, oxygen, light, and moisture). Another drawback is the rapid release of the volatile compounds, which can undermine their biological influence on fruit preservation. Thus, the application of essential oils in the form of nanoemulsions when used in edible packaging film has shown greater effectiveness in providing functional benefits to the food products when compared with conventional emulsions. This is because the essential oils in the form of nanoemulsion make them much more stable and enhance their functionality by increasing the surface area of the oil droplets. The controlled release of essential oils when applied as nanoemulsions to the fruits brings antibacterial, antifungal, and antioxidant effects without bringing any strong undesirable odor, flavor, or color. A schematic diagram of fabrication of nanoemulsion-based edible coatings in fruits is shown in Figure 9.4.

Robledo et al. [136] used nanoemulsion that contained thymol as the bioactive component on quinoa protein/chitosan-based edible films for preservation of cherry tomatoes. The film showed higher increase in the retardation in the growth of mold *Botrytis cineria* in cherry tomatoes. In another study by Prakash et al. [137] sodium alginate-based edible coatings were prepared containing citral nanoemulsion for increasing the shelf life of fresh-cut pineapple. It led to the reduction of two artificially inoculated bacteria, *L. monocytogenes* and *S. enterica*. These findings thus contribute valuable insights into the effectiveness of nanoemulsions on edible food packaging systems for fruits and vegetables.

The inherent antimicrobial activity of some biopolymers, like chitosan, is not always sufficient to ensure the prolonged shelf life of fruits and vegetables when used as packaging films. Addition of nanomaterials is thus essential to offer comprehensive protection against the loss of moisture and softening during storage besides ensuring inhibitory effect against the microorganisms. Several research studies have been conducted in this regard to not only augment the antimicrobial efficacy of the packaging film but also enhance its barrier properties to prevent mass loss, gas permeation, and mechanical injury [134, 138–140]. Researchers have also focused on the utilization of carbon nanotubes to prolong the shelf

Table 9.4 Remarkable outcomes of active bio-nanocomposite food packaging materials.

Active bio-nanocomposite food packaging materials	Food products	Remarkable outcomes	Reference
PLA/copper nanoparticle-based coating	Fiordilatte cheese	Inhibitory effects against microorganisms such as *P. Spp*. But no retardation of the proliferation of lactic acid bacteria in cheese and a delayed kinetic of the organoleptic degradation of the nanocomposite film compared to the control film.	[112]
Starch/cashew tree gum/MMT nanoclay-based coating	Cashew nut	Reduction of moisture absorption and oxidation of nanocomposite-coated samples.	[113]
Potato starch/thyme essential oil/MMT-based packaging film	Spinach leaves	Higher inhibitory effect against bacteria *E. coli* and *S. typhimurium* because of increasing the concentration of thyme essential oil rather than changing the concentration of MMT. No negative change in the sensory property of the spinach leaves in storage.	[114]
CMC/ZnO nanoparticle-based coating	Pomegranates	Higher inhibitory action against microorganisms, especially against yeast and mold, reduction in the loss of moisture, retardation of oxidation reaction, and higher retention of bioactive components like vitamin C and anthocyanin.	[115]
Cellulose-based paper (p)/TiO_2/Ag/zeolite-based nanocomposite film	Bread	Positive effect on the hydrophobicity, photocatalytic activity, and opacity of the bio-nanocomposite-based packaging films when compared to the plain paper-based film. The poor air and grease barrier properties are due to the free channel inside the structure of zeolite in p-Ag-TiO_2-zeolite-based nanocomposite film, higher antimicrobial activity of p-Ag-TiO_2-zeolite-based film, and higher nutritional retention of p-Ag-TiO_2-based film than other films.	[116]
PLA/TiO_2/Ag nanoparticle-based film	Refrigerated cottage cheese	Higher concentration of lactic acid bacteria counts in control films than films containing titanium dioxide and silver nanoparticles, higher score of overall sensory attributes of nanocomposite film. Higher inhibitory effects against pathogenic bacteria, yeast, and mold than the control film.	[117]

(Continued)

Table 9.4 (Continued)

Active bio-nanocomposite food packaging materials	Food products	Remarkable outcomes	Reference
Chitosan/MMT/ *Z. clinopodioides* essential oil-based film Or CMC/MMT/ *Z. clinopodioides* essential oil-based film	Minced camel meat	Higher reduction in pathogenic microorganisms such as *P. Spp.*, Enterobacteriaceae family, lactic acid bacteria, and *B. thermosphacta* than the control film, higher retardation of oxidative rancidity due to low peroxide value, lower formation of total volatile base nitrogen (TVB-N) value, and overall higher sensory attributes of bio-nanocomposite film than the control film.	[118]
Cellulose nanocomposite film	Bread	Decrease in total plate count after wrapping the bread with the nanocomposite film in comparison to the control film	[119]
Chitosan/cellulose acetate phthalate/ZnO nanoparticle-based packaging film	Black grapes	Improvement in thermal, mechanical, and barrier properties, prevention of the penetration of the UV light by blocking UV light by the ZnO nanoparticles, thereby preventing discoloration of fruit, and higher antibacterial property against two tested microorganisms, *E. coli* and *S. aureus*.	[120]
Soybean protein isolate/ cinnamaldehyde/ZnO nanomaterial-based edible coating	Bananas	Increase in the firmness of fruits, which delays the ripening, reduces the loss of weight, and inhibits the growth of fungi.	[121]
Semolina flour/ZnO nanoparticles/nano kaolin-based film	Mozzarella cheese	Higher inhibitory effect against pathogenic bacteria *E. coli* and *S. aureus,* yeast *C. albicans,* and mold *A. niger* because of the presence of ZnO nanoparticles rather than kaolinite nanoparticles, reduction in the loss of moisture, and higher retention in overall sensory attributes with time than the control film.	[122]
Chitosan/MMT/ rosemary essential oil-based film	Fresh poultry meat	Prevention of meat discoloration, retardation of lipid oxidation, inhibition against the growth of microorganisms, and prevention of the increase in the pH value are due to the combined effect of inherent antioxidant and antimicrobial properties of rosemary essential oil and good barrier properties against oxygen of the nanocomposite film.	[123]

Table 9.4 (Continued)

Active bio-nanocomposite food packaging materials	Food products	Remarkable outcomes	Reference
Sodium alginate/ *M. spicata* essential oil/cellulose nanoparticle-based edible coating	Raw silver carp fillets	Significant reduction of peroxide value and total volatile base nitrogen value after increase in the concentration of both essential oil and cellulose nanoparticles, reduction in total viable count, psychrotrophic count, and Enterobacteriaceae family after increase in the concentration of both essential oil as well as cellulose nanoparticles.	[124]
Gelatin/chitosan nanofiber/ZnO nanoparticle-based film	Chicken fillet and cheese	Significant reduction in the growth of total bacterial count and growth of inoculated microorganisms (*S. aureus* and *E. coli* in chicken fillet sample, *P. aeruginosa* and *E. coli* in the cheese sample), prevention in the increase of pH, weight loss, and discoloration in both the food samples.	[125]
CMC/guanidinylated chitosan/TiO_2 nanoparticle-based edible coating	Green bell peppers	Improvement in thermal and mechanical properties, reduction in water vapor permeability, and high UV barrier properties are due to the high UV-shielding effect of TiO_2 nanoparticles.	[126]
PLA/iron nanoparticle-based film	Goat cream cheese	Higher reduction of bacteria *E. coli, B. subtilis, S. epidermidis,* and yeast *R. rubra*, and mold *G. candidum* by the addition of zero-valent iron but no significant result of inhibition by the addition of iron oxide.	[127]
Chitosan/potato protein/linseed oil/ZnO nanoparticle-based film	Raw meat	Good water vapor barrier capacity, lower pH values in bio-nanocomposite film when compared to control film, higher reduction in bacterial count due to combined effects of inherent antimicrobial property of chitosan film, presence of phenolic compound, unsaturated fatty acids in linseed oil, and release of ROS after the incorporation of ZnO nanoparticles.	[128]
PVA/chitosan/SiO_2 nanoparicle-doped with ZnO nanoparticle-based film	Bread	The least fungal growth and bacterial growth in bread samples wrapped with nanocomposite film carrying 5% ZnO-SiO_2 nanoparticles. The sample with the highest fungal and bacterial growth was wrapped in a normal, nonbiodegradable polyethylene film.	[129]

(Continued)

Table 9.4 (Continued)

Active bio-nanocomposite food packaging materials	Food products	Remarkable outcomes	Reference
Chitosan/clay nanocomposite film carrying immobilized papain	Wine	Reduction of the haze potential and the amount of protein content of the two wine samples by covalently immobilized papain on nanocomposite film.	[130]
Chitosan/MMT/ginger essential oil-based film	Chilled beef	Reduction in the changes in pH value, percentage of moisture content, and hue angle value of the meat packed in bio-nanocomposite film. But reduction in the antioxidant and antimicrobial capacities after the incorporation of MMT resulted in the slow migration of the active components.	[131]
CMC/ZnO nanoparticles/grape seed extract	High-fat beef	Good inhibitory effect against pathogenic bacteria *L. monocytogenes* and *E. coli* during refrigerated storage after the incorporation of ZnO nanoparticles. Good antioxidant capability and UV-protecting capacity due to the incorporation of grape seed extract.	[132]
Whey protein/cocoa nanoemulsion-based coating	Muffin	Decrease in water solubility and water vapor permeability with the incorporation of cocoa liquor. Better prevention of moisture loss with better texture and retention of antioxidant capacity of the coated film than the control film.	[133]
Chitosan/litchi peel extract/TiO_2 nanoparticle-based film for coating	Apples	Better retention of the firmness of the fruit, slowdown of respiration rates and weight loss, and inhibition of the activity of the enzyme polyphenol oxidase, which inhibited browning. Improvement in other properties such as low water vapor permeability and high thermal and mechanical properties.	[134]
Pectin/pracaxi oil nanoemulsion-based film	Butter	Greater improvement in oxidation stability and better barrier properties against water vapor after the incorporation of nanoemulsion.	[135]

life of fruits and vegetables. Multiwalled carbon nanotubes were used by Liu et al. [141] to reduce brittleness and improve heat stability, all while preserving the inhibitory effects of chitosan/PLA film against microorganisms. It is essential to retard the ethylene deterioration effect on fruits and vegetables that causes premature ripening. Several metal oxide nanoparticles have been reported to effectively control the production of ethylene by

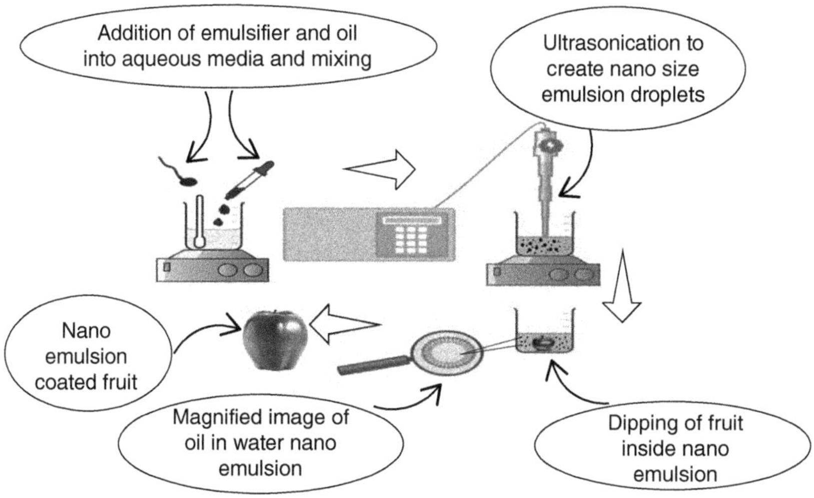

Figure 9.4 Nanoemulsion-based coated fruit.

photocatalytic oxidation of ethylene gas [142, 143]. For example, Riahi et al. [144] prepared titanium nanotubes by doping titanium oxides with copper oxides to reduce photon energy bandgap of nanoparticles. It was used in banana packaging to increase the shelf life of bananas. Several benefits were observed such as ethylene-scavenging activity, reduction in weight loss, and retardation of enzymatic browning. All these benefits together helped to maintain the freshness of bananas for a longer period of time.

9.5.2 Dairy Products

Milk by nature is one of the most perishable food items, creating a suitable medium for the growth of microbes. Although a variety of milk products with altered organoleptic characteristics are accessible, they only partially guarantee extended shelf life for dairy products. Hence, dairy industries encounter persistent challenges in meeting the packaging needs tailored to the surface chemistry and microbiological history of diverse range of products spanning from yogurt, condensed milk, and butter to soft and hard cheese. Applications of several bio-nanocomposites are reported in cheese packaging. During storage of cheese, the breakdown of components such as protein, fat, and sugar occurs releasing flavor components that modify the taste and texture of cheese. Therefore, active packaging must be designed to exert inhibitory effects against microorganisms while preserving the development of desired flavor and texture of cheese during storage. Youssef et al. [145] reported the inhibitory effect of chitosan/PVA/TiO_2 bio-nanocomposite film on the microorganisms in soft cheese. However, there are not many significant changes in certain parameters, like increase in moisture, pH value, and acidity, as compared to control film. Almost similar results were obtained by Youssef et al. [146], while using CuO as the nanomaterial in CMC/PVA matrix. Only a slight rate of change was observed in the enhancement of certain parameters, including acidity and fat content, during storage as compared to control film. These observations suggest that the introduction of nanomaterials in packaging did not interfere significantly with the desirable physical and chemical changes that contribute to the distinctive flavor and texture of cheese during ripening. This indicates a positive outcome for the utilization of nanocomposites in cheese packaging as they effectively show their functional benefits without compromising the essential characteristics of the cheese. Cheese while stored can facilitate the growth of various psychrotrophic pathogenic microorganisms, such as *L. monocytogenes*, which pose a persistent risk, especially during cold storage. In order to prevent the growth of these microorganisms, several studies have investigated the nanoencapsulation of bacteriocins, such as nisin – an antimicrobial peptide – and their application in edible packaging or coating of cheese [147, 148].

Nanotechnology has grabbed the attention of several fat-based milk products. The fat-based products like ghee and butter are generally microbiologically stable due to their low water activity and high fat content. But some yeasts and molds are still able to penetrate through the fat phase of butter, especially in cases of faulty pasteurization or the use of unclean equipment, leading to bacterial contamination of cream used in production. However, the primary concern in the spoilage of fat-based products lies in chemical spoilage, particularly rancidity from the oxidation of unsaturated fat. Oxidative and microbial stability were measured by Asdagh and Pirsa [149] after the preparation of active packaging film based on heteropolysaccharide pectin, MMT nanoclay, carum capticum essential oils, and β-Carotene. Carum capticum essential oils, β-Carotene, possess inherent antibacterial and antioxidant properties. Synergistic action of these two components along with the incorporation of nanoclay successfully delayed oxidation reaction by penetrating the structure of butter and absorbing free radicals and hydroperoxide from the surface. Another concern associated with the packaging for high-oily food is that the packaging based on paper in causes the formation of greasiness on the surface of paper because of the porous nature of paper, allowing absorption of oil via capillary force [150]. Apjok et al. [151] reported that incorporation of Ag/TiO_2 nanoparticles in paper-coated chitosan-based packaging film can remove the formation of greasy spots on the surface of the film. It is done by creating surface barrier with the nanoparticles to reduce the porosity of paper for grease penetration. The nanoparticles also reduced rancid flavor of the clarified butter by providing suitable barrier against oxygen, UV light, and moisture.

9.5.3 Muscle Foods

Muscle foods are the important source of proteinaceous diet to meet global food hunger. But the neutral pH, high water activity, and presence of large amounts of non-protein nitrogen such as free amino acids and nucleotides cause rapid invasion of microorganisms inside the meat matrix. Another common important chemical spoilage is the formation of metmyoglobin by the oxidation of deoxy or oxymyoglobin, thereby changing fresh red color of meat to the brown color. Alizadeh-Sani et al. [152] found that reinforcement of cellulose nanofiber/whey protein-based film with TiO_2 nanoparticles and rosemary essential oil as the functional ingredient successfully inhibited the rapid metmyoglobin formation when compared with the control film. The inhibition was evident through higher "a" value in the color parameter, signifying increased redness in the meat. The presence of various antioxidative compounds in the rosemary oil has contributed to the inhibition of rapid metmyoglobin formation. In addition to this, the presence of nanoparticles may act as a barrier to entry of oxygen inside the meat matrix. This enhanced barrier against oxygen not only contributed to inhibiting metmyoglobin formation but also resulted in a substantial retardation in rancidity. Similar result was obtained by Pires et al. [153], where chitosan-based film was reinforced by MMT nanoclay carrying rosemary essential oil or ginger oil to protect fresh poultry meat. The study also demonstrated that the incorporation of nanoclay alongside essential oils enhanced the antimicrobial activity of chitosan, further contributing to the protection of the poultry meat.

Other than plain meat, active packaging has been developed for several meat-based products. Products like sausage are the preserved forms of meat products, utilizing all the edible parts of the carcass having low water activity loaded with salts and spices. Some of them have tangy flavor due to desirable fermentation by lactic acid bacteria. Some highly resistant bacteria, yeast, and mold can nevertheless flourish in meat-based products with low water activity. For example, several gram-positive microorganisms (*L. monocytogenes*) have the ability to grow in refrigerated conditions when foods are stored. Microorganisms such as *Staphyllococcus* and *Streptococcus* species have high salt tolerance. That is why sausage needs active packaging with higher functional benefits, including antifungal and antibacterial properties, to avoid the growth of these microorganisms. Mathew et al. [38] prepared biodegradable active nanocomposite pouches containing silver nanoparticles to increase the shelf life of chicken sausages. Antibacterial properties were enhanced against *S. typhimurium* and *S. aureus*. In another study, nanoliposomes has been used as the nanocarrier of essential oil and added to chitosan-based film. Significant reductions in *S. aureus* and psychotropic microorganisms were observed in chicken breast fillets by Kamkar et al. [154].

Other than metmyoglobin formation, rancidity, and general microbial spoilage, there are some other signs of the spoilage of muscle foods like formation of nitrogenous compounds, which are trimethyl amine (TMA), ammonia, or increase in the free amino acids by the break down of high-molecular-weight proteins at neutral pH. This process initiates the action of microorganisms to further break down the larger molecular weight nitrogenous compound into smaller molecular weight volatile nitrogenous compound, thereby reducing the freshness of fish and meat. Shokri et al. [155] reported lower formation of total volatile basic nitrogen (TVB-N) in chitosan-essential oil nanoemulsion-coated rainbow trout fillet samples due to enhanced antimicrobial action of chitosan after the incorporation of essential oil-coated nanoemulsion, highlighting its potential to mitigate the formation of spoilage-related nitrogenous compounds. Similar result was obtained by Gharibzahedi and Mohammadnabi [156], where the value of TVB-N along with free fatty acid and peroxide value, significantly decreased after the incorporation of jujube gum and nettle oil-coated nanoemulsion to enhance the shelf life of Beluga Sturgeon Fillets.

9.5.4 Bakery and Other Products

Products such as bread, biscuits, cakes, cookies, candies, and chocolates fall under Intermediate Moisture Food (IMF) products. Therefore, these kinds of products have higher microbial stability when compared with fish, meat, poultry, and dairy products. The main cause of spoilage is due to yeast and mold causing ropiness and moldiness in bread.

Another important spoilage, especially in bakery products unrelated to microbes, is staling, which is caused by starch retrogradation on storage reducing softness, elasticity, flavor, and overall consumer acceptability. Noshirvani et al. [157] reported that no visible fungal development was observed even after 15 days of storage after the incorporation of ZnO nanoparticles in CMC-chitosan-based film. The packaging samples containing ZnO nanoparticles showed lower value of firmness in bread when compared with the control samples which indicated higher rate of retrogradation of starch in control samples. This suggests that ZnO nanoparticles contributed to reduction in staling. Similar result was obtained by Sahraee et al. [158] while analyzing the texture of cake. Gelatin-based packaging that was not incorporated with the nanoparticles showed higher increase in the hardness of the cake when compared with cake packed in polyethylene film. The higher barrier of polyethylene film against moisture evaporation, contributed to the softness and better texture of the cake, thereby inhibiting staling. However, hardness in cake decreased in gelatin-based film after the incorporation of chitin and ZnO-based nanoparticles. The film incorporated with nanoparticles also exhibited antifungal properties reducing the growth of mold [158].

9.6 Regulation and Safety Aspects

Incorporation of nanomaterials has become a new frontier in the field of active food packaging systems because of their unique novel properties. Despite the rapid development and use of nanotechnology, there are few relevant rules and regulations on the utilization of nanomaterials [159]. The reason might be due to the lack of better understanding of the properties of these materials in terms of safety based on their size, shape, surface charge, and concentration added to the packaging matrix. Nanomaterials are equipped with higher reactivity than normal macroscopic particles because size reduction enables them to move freely. A common challenge being faced is the migration of these nanomaterials due to higher reactivity from the packaging matrix to the food matrix. Furthermore, if the solubility of the nanoparticles in food simulants is high, their stability embedded into the polymer matrix gets affected, and their reactivity to the food matrix is enhanced, thereby causing the migration. Other than this problem of migration, inappropriate disposal of the packaging materials might lead to the unwanted accommodation of nanomaterials in the food chain [160]. After the consumption of these foods, the nanoparticles might get accumulated in different biological systems through blood circulation. Numerous research studies have shown the hazardous side effects of nanoparticles on the environment and on living things. An interesting review by Ema et al. [161] presented the alternations in reproductive and developmental processes of laboratory-tested animals caused by zinc nanoparticles. Some of these alternations are genotoxicity in testis and sperm in male rats, ovarian and embryonic toxicity in female rats, growth inhibition in offspring, behavioral changes in offspring, biochemical changes in the brain of offspring, and transplacental genotoxicity.

In order to get rid of the limitations and other safety-related issues, regulations for proper analysis of the nanomaterials and their application in the packaging become essential. General regulatory rules of Codex Alimantarius, the European Union (EU), the Food and Drug Administration (FDA), etc. show high concern over general food safety, overlooking the diversities in the properties of the different nanomaterials depending on the way the surfaces of these nanomaterials are functionalized. There are few EU regulations regarding the substances coming in contact with the food where the nanomaterials are not directly mentioned, for example, Regulation (EC) No. 450/2009 states, "substances deliberately engineered to particle size which exhibit functional physical and chemical properties that significantly differ from those at a larger scale"; therefore, assessment of active and intelligent materials must be done on a case-by-case basis [162]. However, the countries still follow the existing conventional law to specify the level of migration of several components in the food products which is measured in terms of mg components per kg of food. According to Regulation (EU) No. 10/2011, substances that are in nanoform can only be used if listed in Annex I of the regulation [162]. The Annex I included several permitted additives or polymers such as starch, cellulose, gums, pectin, gelatin, etc. This regulation documented about the limit of migration of the substances that might come into contact with food where for a cubic packaging of 1 kg of food, migration limits of 10 mg/dm^2 were set, which is equivalent to 60 mg/kg of food. However, these migration limits are not applicable to engineered nanomaterials [163].

Numerous research activities are still going on for proper regulations of nanomaterials based on their properties. Side by side, several analytical techniques have been established to properly detect, characterize, and quantify nanomaterials in different food products and packaging materials. Several conventional analytical techniques like light scattering, chromatography, scanning electron microscopy, X-ray microscopy, etc. are already used for the characterization of nanomaterials, but lack of analytical standards and correct analytical sequence do not ensure higher accuracy in detection, characterization, and quantification of nanomaterials in food packages and products [164, 165]. The improvement of the knowledge regarding several physicochemical properties of nanomaterials and understanding the mechanisms of toxicological effects of different types of nanomaterials in a better way on various species or systems can help different regulatory bodies to fulfill the deficiencies of regulations in nanomaterials.

9.7 Conclusion and Perspective

In this chapter, the importance of biopolymers in active food packaging is discussed. It is clear that utilization of biopolymers in food packaging is a reliable strategy to lessen the negative impact on environment caused by the synthetic polymers. However, the lack of the performance characteristics of the biopolymers when compared with traditional plastics limits their utilization on large scale. The use of nanomaterials or nanocarriers in biopolymer-based active packaging is thus a novel promise to solve the inherent weaknesses of the biopolymers. When utilized in the active packaging, these nanomaterials provide a shielding effect on the bioactive components against the outside environment by keeping them immobilized and stable when embedded in the biopolymeric matrix. Some of these nanomaterials are already equipped with inherent functional properties, and their incorporation provides functional characteristics in the active food packaging. Although numerous reports published to date proclaim the suitability of the nanomaterials as the potential candidate in terms of maintaining quality and prolonged shelf life of active food packaging, several other important aspects, such as bioavailability, of bioactive components, digestibility, and toxicity must be explored thoroughly. Proper concentration, degree of dispersion of nanomaterials, and good biocompatibility with the polymer matrix are necessary to obtain the optimum functional performance of the active nanocomposite-based packaging. The knowledge gaps related to variations in nanoparticle characteristics and toxicity must be filled. The dose limit of different nanoparticles in packaging material should differ depending on the amount of cytotoxicity or genotoxicity created by different nanoparticles. Thus, worldwide regulatory bodies still have long journey to cover to come up with the proper regulatory rules to ensure safety.

References

1 Huang, T., Qian, Y., Wei, J. et al. (2019). Polymeric antimicrobial food packaging and its applications. *Polymers*. https://doi.org/10.3390/polym11030560.

2 Siracusa, V., Rocculi, P., Romani, S. et al. (2008). Biodegradable polymers for food packaging: a review. *Trends in Food Science and Technology* 634–643. https://doi.org/10.1016/j.tifs.2008.07.003.

3 Tang, X.Z., Kumar, P., Alavi, S. et al. (2012). Recent advances in biopolymers and biopolymer-based nanocomposites for food packaging materials. *Critical Reviews in Food Science and Nutrition* 426–442. https://doi.org/10.1080/10408398.2010.500508.

4 Alsaggaf, M.S., Moussa, S.H., and Tayel, A.A. (2017). Application of fungal chitosan incorporated with pomegranate peel extract as edible coating for microbiological, chemical and sensorial quality enhancement of Nile tilapia fillets. *International Journal of Biological Macromolecules* 99: 499–505. https://doi.org/10.1016/j.ijbiomac.2017.03.017.

5 Oyom, W., Xu, H., Liu, Z. et al. (2022). Effects of modified sweet potato starch edible coating incorporated with cumin essential oil on storage quality of early crisp. *LWT* 153. https://doi.org/10.1016/j.lwt.2021.112475.

6 Flaris, V. and Singh, G. (2009). Recent developments in biopolymers. *Journal of Vinyl and Additive Technology* 15 (1): 1–11. https://doi.org/10.1002/vnl.20171.

7 Rahman, M.H. and Bhoi, P.R. (2021). An overview of non-biodegradable bioplastics. *Journal of Cleaner Production* https://doi.org/10.1016/j.jclepro.2021.126218.

8 Bordes, P., Pollet, E., and Avérous, L. (2009). Nano-biocomposites: Biodegradable polyester/nanoclay systems. *Progress in Polymer Science (Oxford)* 125–155. 10.1016/j.progpolymsci.2008.10.002.

9 Rhim, J.W., Park, H.M., and Ha, C.S. (2013). Bio-nanocomposites for food packaging applications. *Progress in Polymer Science* 1629–1652. https://doi.org/10.1016/j.progpolymsci.2013.05.008.

10 Esa, F., Tasirin, S.M., and Rahman, N.A. (2014). Overview of bacterial cellulose production and application. *Agriculture and Agricultural Science Procedia* 2: 113–119. https://doi.org/10.1016/j.aaspro.2014.11.017.

11 Mohammadkazemi, F., Azin, M., and Ashori, A. (2015). Production of bacterial cellulose using different carbon sources and culture media. *Carbohydrate Polymers* 117: 518–523. https://doi.org/10.1016/j.carbpol.2014.10.008.

12 Cazón, P., Velazquez, G., Ramírez, J.A. et al. (2017). Polysaccharide-based films and coatings for food packaging: a review. *Food Hydrocolloids* 68: 136–148. https://doi.org/10.1016/j.foodhyd.2016.09.009.

13 Nešić, A., Cabrera-Barjas, G., Dimitrijević-Branković, S. et al. (2020). Prospect of polysaccharide-based materials as advanced food packaging. *Molecules*. https://doi.org/10.3390/molecules25010135.

14 Chen, H., Wang, J., Cheng, Y. et al. (2019). Application of protein-based films and coatings for food packaging: a review. *Polymers* 11 (12). https://doi.org/10.3390/polym11122039.

15 Balaji, A.B., Pakalapati, H., Khalid, M. et al. (2017). Natural and synthetic biocompatible and biodegradable polymers. In: *Biodegradable and Biocompatible Polymer Composites: Processing, Properties and Applications* (ed. N.G. Shimpi), 3–32. Elsevier. http://doi.org/10.1016/B978-0-08-100970-3.00001-8.

16 Jamshidian, M., Tehrany, E.A., Imran, M. et al. (2010). Poly-lactic acid: production, applications, nanocomposites, and release studies. *Comprehensive Reviews in Food Science and Food Safety* 9 (5): 552–571. https://doi.org/10.1111/j.1541-4337.2010.00126.x.

17 Tawakkal, I.S.M.A., Cran, M.J., Miltz, J. et al. (2014). A review of poly(lactic acid)-based materials for antimicrobial packaging. *Journal of Food Science* 79 (8). https://doi.org/10.1111/1750-3841.12534.

18 Zhang, Q., Song, M., Xu, Y. et al. (2021). Bio-based polyesters: Recent progress and future prospects. *Progress in Polymer Science*. https://doi.org/10.1016/j.progpolymsci.2021.101430.

19 Martinez Villadiego, K., Arias Tapia, M.J., Useche, J. et al. (2022). Thermoplastic starch (TPS)/polylactic acid (PLA) blending methodologies: a review. *Journal of Polymers and the Environment* 75–91. https://doi.org/10.1007/s10924-021-02207-1.

20 Anjum, A., Zuber, M., Zia, K.M. et al. (2016). Microbial production of polyhydroxy-alkanoates (PHAs) and its copolymers: a review of recent advancements. *International Journal of Biological Macromolecules* 161–174. https://doi.org/10.1016/j.ijbiomac.2016.04.069.

21 Saratale, R.G., Cho, S.-K., Saratale, G.D. et al. (2021). An overview of recent advancements in microbial polyhydroxyalkanoates (Pha) production from dark fermentation acidogenic effluents: a path to an integrated bio-refinery. *Polymers*. https://doi.org/10.3390/polym13244297.

22 Chausali, N., Saxena, J., and Prasad, R. (2022). Recent trends in nanotechnology applications of bio-based packaging. *Journal of Agriculture and Food Research* 7. https://doi.org/10.1016/j.jafr.2021.100257.

23 Naghdi, M., Taheran, M., Pulicharla, R. et al. (2019). Pine-wood derived nanobiochar for removal of carbamazepine from aqueous media: Adsorption behavior and influential parameters. *Arabian Journal of Chemistry* 12 (8): 5292–5301. https://doi.org/10.1016/j.arabjc.2016.12.025.

24 Majhi, K.C. and Yadav, M., Inamuddin, Boddula, R., Ahamed, M. I., & Asiri, A. M. (Eds.). (2021) Synthesis of inorganic nanomaterials using carbohydrates. *Green Sustainable Process for Chemical and Environmental Engineering and Science: Green Inorganic Synthesis*, pp. 109–135. https://doi.org/10.1016/B978-0-12-821887-7.00003-3.

25 Sekhon, B.S. (2010). Food nanotechnology-an overview. *Nanotechnology, Science and Applications* 1–15. https://doi.org/10.2147/nsa.s12187498.

26 Jamkhande, P.G., Ghule, N.W., Bamer, A.H. et al. (2019). Metal nanoparticles synthesis: An overview on methods of preparation, advantages and disadvantages, and applications. *Journal of Drug Delivery Science and Technology*. https://doi.org/10.1016/j.jddst.2019.101174.

27 Król, A., Pomastowski, P., Rafińska, K. et al. (2017). Zinc oxide nanoparticles: synthesis, antiseptic activity and toxicity mechanism. *Advances in Colloid and Interface Science* 37–52. https://doi.org/10.1016/j.cis.2017.07.033.

28 Varan, G., Varan, C., Erdoğar, N. et al. (2017). Amphiphilic cyclodextrin nanoparticles. *International Journal of Pharmaceutics* 531 (2): 457–469. https://doi.org/10.1016/j.ijpharm.2017.06.010.

29 Bahrami, A., Delshadi, R., Assadpour, E. et al. (2020). Antimicrobial-loaded nanocarriers for food packaging applications. *Advances in Colloid and Interface Science*. https://doi.org/10.1016/j.cis.2020.102140.

30 Faridi Esfanjani, A., Assadpour, E., and Jafari, S.M. (2018). Improving the bioavailability of phenolic compounds by loading them within lipid-based nanocarriers. *Trends in Food Science and Technology* 56–66. https://doi.org/10.1016/j.tifs.2018.04.002.

31 Cheng, C., Song, W., Zhao, Q. et al. (2020). Halloysite nanotubes in polymer science: purification, characterization, modification and applications. *Nanotechnology Reviews* 9 (1): 323–344. https://doi.org/10.1515/ntrev-2020-0024.

32 Rehman, A., Jafari, S.M., Aadil, R.M. et al. (2020). Development of active food packaging via incorporation of biopolymeric nanocarriers containing essential oils. *Trends in Food Science and Technology* 106–121. https://doi.org/10.1016/j.tifs.2020.05.001.

33 Besinis, A., De Peralta, T., and Handy, R.D. (2014). The antibacterial effects of silver, titanium dioxide and silica dioxide nanoparticles compared to the dental disinfectant chlorhexidine on *Streptococcus mutans* using a suite of bioassays. *Nanotoxicology* 8 (1): 1–16. https://doi.org/10.3109/17435390.2012.742935.

34 Priyadarshi, R., Roy, S., Ghosh, T., et al. (2022) 'Antimicrobial nanofillers reinforced biopolymer composite films for active food packaging applications – a review', Sustainable Materials and Technologies. https://doi.org/10.1016/j.susmat.2021.e00353.

35 Fortunati, E., Rinaldi, S., Peltzer, M. et al. (2014). Nano-biocomposite films with modified cellulose nanocrystals and synthesized silver nanoparticles. *Carbohydrate Polymers* 101 (1): 1122–1133. https://doi.org/10.1016/j.carbpol.2013.10.055.

36 Istiqola, A. and Syafiuddin, A. (2020). A review of silver nanoparticles in food packaging technologies: Regulation, methods, properties, migration, and future challenges. *Journal of the Chinese Chemical Society* 1942–1956. https://doi.org/10.1002/jccs.202000179.

37 Lazić, V., Vivod, V., Peršin, Z. et al. (2020). Dextran-coated silver nanoparticles for improved barrier and controlled antimicrobial properties of nanocellulose films used in food packaging. *Food Packaging and Shelf Life* 26. https://doi.org/10.1016/j.fpsl.2020.100575.

38 Mathew, S., Snigdha, S., Mathew, J. et al. (2019). Biodegradable and active nanocomposite pouches reinforced with silver nanoparticles for improved packaging of chicken sausages. *Food Packaging and Shelf Life* 19: 155–166. https://doi.org/10.1016/j.fpsl.2018.12.009.

39 Chen, G. and Liu, B. (2016). Cellulose sulfate based film with slow-release antimicrobial properties prepared by incorporation of mustard essential oil and β-cyclodextrin. *Food Hydrocolloids* 55: 100–107. https://doi.org/10.1016/j.foodhyd.2015.11.009.

40 Mihindukulasuriya, S.D.F. and Lim, L.T. (2014). Nanotechnology development in food packaging: A review. *Trends in Food Science and Technology* 40 (2): 149–167. https://doi.org/10.1016/j.tifs.2014.09.009.

41 Mu, H., Gao, H., Chen, H. et al. (2013). A nanosised oxygen scavenger: preparation and antioxidant application to roasted sunflower seeds and walnuts. *Food Chemistry* 136 (1): 245–250. https://doi.org/10.1016/j.foodchem.2012.07.121.

42 Upadhyay, A., Kumar, P., Kardam, S.K. et al. (2022). Ethylene scavenging film based on corn starch-gum acacia impregnated with sepiolite clay and its effect on quality of fresh broccoli florets. *Food Bioscience* 46. https://doi.org/10.1016/j.fbio.2022.101556.

43 Abbasi, M., Amiri, R., Bardbar, A.-K. et al. (2016). Improvement of the stability and activity of immobilized glucose oxidase on modified iron oxide magnetic nanoparticles. *Applied Surface Science* 364: 752–757. https://doi.org/10.1016/j.apsusc.2015.12.120.

44 Li, D., He, Q., Cui, Y. et al. (2007). Immobilization of glucose oxidase onto gold nanoparticles with enhanced thermostability. *Biochemical and Biophysical Research Communications* 355 (2): 488–493. https://doi.org/10.1016/j.bbrc.2007.01.183.

45 Amani, H., Habibey, R., Hajmiresmail, S.J. et al. (2017). Antioxidant nanomaterials in advanced diagnoses and treatments of ischemia reperfusion injuries. *Journal of Materials Chemistry B* 5 (48): 9452–9476. https://doi.org/10.1039/c7tb01689a.

46 Saifullah, M., Shishir, M.R.I., Ferdowsi, R. et al. (2019). Micro and nano encapsulation, retention and controlled release of flavor and aroma compounds: a critical review. *Trends in Food Science and Technology* 230–251. https://doi.org/10.1016/j.tifs.2019.02.030.

47 Estevinho, B.N. and Rocha, F. (2017). A key for the future of the flavors in food industry: nanoencapsulation and microencapsulation. In: *Nanotechnology Applications in Food: Flavor, Stability, Nutrition and Safety* (ed. A.E. Oprea and A.M. Grumezescu), 1–19. Elsevier Inc. https://doi.org/10.1016/B978-0-12-811942-6.00001-7.

48 De Moraes, A.C.M., Andrade, P.F., Faria, A.F. et al. (2015). Fabrication of transparent and ultraviolet shielding composite films based on graphene oxide and cellulose acetate. *Carbohydrate Polymers* 123: 217–227. https://doi.org/10.1016/j.carbpol.2015.01.034.

49 Zhang, Y., Yu, L., Ke, S. et al. (2011). TiO2/SiO2 hybrid nanomaterials: synthesis and variable UV-blocking properties. *Journal of Sol-Gel Science and Technology* 58 (1): 326–329. https://doi.org/10.1007/s10971-010-2395-2.

50 Ravichandran, K., Praseetha, P.K., Arun, T. et al. (2018). Synthesis of nanocomposites. *Synthesis of Inorganic Nanomaterials: Advances and Key Technologies* 141–168. https://doi.org/10.1016/B978-0-08-101975-7.00006-3.

51 Jagadeesh, P., Puttegowda, M., Rangappa, S.M. et al. (2021). Influence of nanofillers on biodegradable composites: a comprehensive review. *Polymer Composites* 5691–5711. https://doi.org/10.1002/pc.26291.

52 Österberg, M., Henn, K.A., Farooq, M. et al. (2023). Biobased nanomaterials—the role of interfacial interactions for advanced materials. *Chemical Reviews* 2200–2241. https://doi.org/10.1021/acs.chemrev.2c00492.

53 Schnepp, Z. (2013). Biopolymers as a flexible resource for nanochemistry. *Angewandte Chemie – International Edition* 1096–1108. https://doi.org/10.1002/anie.201206943.

54 Ma, X., Cheng, Y., Qin, X. et al. (2017). Hydrophilic modification of cellulose nanocrystals improves the physicochemical properties of cassava starch-based nanocomposite films. *LWT* 86: 318–326. https://doi.org/10.1016/j.lwt.2017.08.012.

55 Kang, H., Liu, X., Zhang, S. et al. (2017). Functionalization of halloysite nanotubes (HNTs) via mussel-inspired surface modification and silane grafting for HNTs/soy protein isolate nanocomposite film preparation. *RSC Advances* 7 (39): 24140–24148. https://doi.org/10.1039/c7ra02987j.

56 Wang, Z., Kang, H., Zhang, W. et al. (2017). Improvement of interfacial interactions using natural polyphenol-inspired tannic acid-coated nanoclay enhancement of soy protein isolate biofilms. *Applied Surface Science* 401: 271–282. https://doi.org/10.1016/j.apsusc.2017.01.015.

57 Qin, Z., Mo, L., Liao, M. et al. (2019). Preparation and characterization of soy protein isolate-based nanocomposite films with cellulose nanofibers and nano-silica via silane grafting. *Polymers* 11 (11). https://doi.org/10.3390/polym11111835.

58 Jin, K., Tang, Y., Zhu, X. et al. (2020). Polylactic acid based biocomposite films reinforced with silanized nanocrystalline cellulose. *International Journal of Biological Macromolecules* 162: 1109–1117. https://doi.org/10.1016/j.ijbiomac.2020.06.201.

59 Yu, F., Fei, X., He, Y. et al. (2021). Poly(lactic acid)-based composite film reinforced with acetylated cellulose nanocrystals and ZnO nanoparticles for active food packaging. *International Journal of Biological Macromolecules* 186: 770–779. https://doi.org/10.1016/j.ijbiomac.2021.07.097.

60 Jiang, J., Chen, X., Zhang, G.-L. et al. (2022). Preparation of chitosan-cellulose-benzyl isothiocyanate nanocomposite film for food packaging applications. *Carbohydrate Polymers* 285. https://doi.org/10.1016/j.carbpol.2022.119234.

61 Zhong, Y., Amiri, R., Bordbar, A.-K. et al. (2023). High-amylose starch: Structure, functionality and applications. *Critical Reviews in Food Science and Nutrition* 8568–8590. https://doi.org/10.1080/10408398.2022.2056871.

62 Liu, C., Xiong, H., Chen, X. et al. (2015). Effects of nano-tio2 on the performance of high-amylose starch based antibacterial films. *Journal of Applied Polymer Science* 132 (32). https://doi.org/10.1002/app.42339.

63 Mondragón, M., Mancilla, J.E., and Rodríguez-González, F.J. (2008). Nanocomposites from plasticized high-amylopectin, normal and high-amylose maize starches. *Polymer Engineering and Science* 48 (7): 1261–1267. https://doi.org/10.1002/pen.21084.

64 Mathew, S., Jayakumar, A., Kumar, V.P. et al. (2019). One-step synthesis of eco-friendly boiled rice starch blended polyvinyl alcohol bionanocomposite films decorated with in situ generated silver nanoparticles for food packaging purpose. *International Journal of Biological Macromolecules* 139: 475–485. https://doi.org/10.1016/j.ijbiomac.2019.07.187.

65 Wang, J., Wang, L., Gardner, D.J. et al. (2021). Towards a cellulose-based society: opportunities and challenges. *Cellulose* 4511–4543. https://doi.org/10.1007/s10570-021-03771-4.

66 Hashmi, M., Ullah, S., Ullah, A. et al. (2021). Carboxymethyl cellulose (CMC) based electrospun composite nanofiber mats for food packaging. *Polymers* 13 (2): 1–12. https://doi.org/10.3390/polym13020302.uy.

67 Riahi, Z., Rhim, J.-W., Bagheri, R. et al. (2022). Carboxymethyl cellulose-based functional film integrated with chitosan-based carbon quantum dots for active food packaging applications. *Progress in Organic Coatings* 166. https://doi.org/10.1016/j.porgcoat.2022.106794.

68 Roy, S. and Rhim, J.W. (2020). Carboxymethyl cellulose-based antioxidant and antimicrobial active packaging film incorporated with curcumin and zinc oxide. *International Journal of Biological Macromolecules* 148: 666–676. https://doi.org/10.1016/j.ijbiomac.2020.01.204.

69 Ebrahimi, Y., Peighambardoust, S.J., Peighambardoust, S.H. et al. (2019). Development of antibacterial carboxymethyl cellulose-based nanobiocomposite films containing various metallic nanoparticles for food packaging applications. *Journal of Food Science* 84 (9): 2537–2548. https://doi.org/10.1111/1750-3841.14744.

70 Wang, L.F. and Rhim, J.W. (2017). Functionalization of halloysite nanotubes for the preparation of carboxymethyl cellulose-based nanocomposite films. *Applied Clay Science* 150: 138–146. https://doi.org/10.1016/j.clay.2017.09.023.

71 Zhang, X., Li, Y., Guo, M. et al. (2021). Antimicrobial and UV blocking properties of composite chitosan films with curcumin grafted cellulose nanofiber. *Food Hydrocolloids* 112. https://doi.org/10.1016/j.foodhyd.2020.106337.

72 Fortunati, E., Armentano, I., Zhou, Q. et al. (2012). Multifunctional bionanocomposite films of poly(lactic acid), cellulose nanocrystals and silver nanoparticles. *Carbohydrate Polymers* 87 (2): 1596–1605. https://doi.org/10.1016/j.carbpol.2011.09.066.

73 Roy, S., Kim, H.C., Panicker, P.S. et al. (2021). Cellulose nanofiber-based nanocomposite films reinforced with zinc oxide nanorods and grapefruit seed extract. *Nanomaterials* 11 (4). https://doi.org/10.3390/nano11040877.

74 Tarrés, Q., Saguer, E., Pelach, M.A. et al. (2016). The feasibility of incorporating cellulose micro/nanofibers in papermaking processes: the relevance of enzymatic hydrolysis. *Cellulose* 23 (2): 1433–1445. https://doi.org/10.1007/s10570-016-0889-y.

75 Dutta, P.K., Tripathi, S., Mehrotra, G.K. et al. (2009). Perspectives for chitosan based antimicrobial films in food applications. *Food Chemistry* 1173–1182. https://doi.org/10.1016/j.foodchem.2008.11.047.

76 Haghighi, H., Licciardello, F., Favaa, P. et al. (2020). Recent advances on chitosan-based films for sustainable food packaging applications. *Food Packaging and Shelf Life*. https://doi.org/10.1016/j.fpsl.2020.100551.

77 Rodrigues, C., Muneron de Mello, J.M., Dalcanton, F. et al. (2020). Mechanical, thermal and antimicrobial properties of chitosan-based-nanocomposite with potential applications for food packaging. *Journal of Polymers and the Environment* 28 (4): 1216–1236. https://doi.org/10.1007/s10924-020-01678-y.

78 Yang, W., Fortunati, E., Bertoglio, F. et al. (2018). Polyvinyl alcohol/chitosan hydrogels with enhanced antioxidant and antibacterial properties induced by lignin nanoparticles. *Carbohydrate Polymers* 181: 275–284. https://doi.org/10.1016/j.carbpol.2017.10.084.

79 Avelelas, F., Horta, A., Pinto, L.F.V. et al. (2019). Antifungal and antioxidant properties of chitosan polymers obtained from nontraditional Polybius henslowii sources. *Marine Drugs* 17 (4). https://doi.org/10.3390/md17040239.

80 Chang, S.H., Wu, C.H., and Tsai, G.J. (2018). Effects of chitosan molecular weight on its antioxidant and antimutagenic properties. *Carbohydrate Polymers* 181: 1026–1032. https://doi.org/10.1016/j.carbpol.2017.11.047.

81 Yen, M.T., Yang, J.H., and Mau, J.L. (2008). Antioxidant properties of chitosan from crab shells. *Carbohydrate Polymers* 74 (4): 840–844. https://doi.org/10.1016/j.carbpol.2008.05.003.

82 Abd El-Hack, M.E., El-Saadonyb, M.T., Shafic, M.E. et al. (2020). Antimicrobial and antioxidant properties of chitosan and its derivatives and their applications: a review. *International Journal of Biological Macromolecules* 2726–2744. https://doi.org/10.1016/j.ijbiomac.2020.08.153.

83 Karim, A.A. and Bhat, R. (2008). Gelatin alternatives for the food industry: recent developments, challenges and prospects. *Trends in Food Science and Technology* 19 (12): 644–656. https://doi.org/10.1016/j.tifs.2008.08.001.

84 Ramos, M., Valdés, A., Beltrán, A. et al. (2016). Gelatin-Based Films and Coatings for Food Packaging Applications. *Coatings* 6 (4): 41. https://doi.org/10.3390/coatings6040041.

85 Ninan, G., Joseph, J., and Abubacker, Z. (2010). Physical, mechanical, and barrier properties of carp and mammalian skin gelatin films. *Journal of Food Science* 75 (9). https://doi.org/10.1111/j.1750-3841.2010.01851.x.

86 Roy, S. and Rhim, J.W. (2021). Gelatin-based film integrated with copper sulfide nanoparticles for active packaging applications. *Applied Sciences (Switzerland)* 11 (14): 6307. https://doi.org/10.3390/app11146307.

87 Liu, F., Antoniou, J., Li, Y. et al. (2015). Preparation of gelatin films incorporated with tea polyphenol nanoparticles for enhancing controlled-release antioxidant properties. *Journal of Agricultural and Food Chemistry* 63 (15): 3987–3995. https://doi.org/10.1021/acs.jafc.5b00003.

88 Khan, M.R., Volpe, S., Valentino, M. et al. (2021). Active casein coatings and films for perishable foods: structural properties and shelf-life extension. *Coatings*. https://doi.org/10.3390/coatings11080899.

89 Ma, J., An, W., Xu, Q. et al. (2019). Antibacterial casein-based ZnO nanocomposite coatings with improved water resistance crafted via double in situ route. *Progress in Organic Coatings* 134: 40–47. https://doi.org/10.1016/j.porgcoat.2019.05.007.

90 Wang, L., Xue, J., and Zhang, Y. (2019). Preparation and characterization of curcumin loaded caseinate/zein nanocomposite film using pH-driven method. *Industrial Crops and Products* 130: 71–80. https://doi.org/10.1016/j.indcrop.2018.12.072.

91 Schmid, M. and Müller, K., Deeth, H. C., & Bansal, N. (Eds.). (2018) Whey protein-based packaging films and coatings. *Whey Proteins: From Milk to Medicine*, pp. 407–437. https://doi.org/10.1016/B978-0-12-812124-5.00012-6.

92 Mohammadi, M., Mirabzadeh, S., Shahvalizadeh, R. et al. (2020). Development of novel active packaging films based on whey protein isolate incorporated with chitosan nanofiber and nano-formulated cinnamon oil. *International Journal of Biological Macromolecules* 149: 11–20. https://doi.org/10.1016/j.ijbiomac.2020.01.083.

93 Jacob, J., Lawal, U., Thomas, S. et al. (2020). Biobased polymer composite from poly(lactic acid): processing, fabrication, and characterization for food packaging. In: *Processing and Development of Polysaccharide-Based Biopolymers for Packaging Applications* (ed. Y. Zhang), 97–115. Elsevier. https://doi.org/10.1016/B978-0-12-818795-1.00004-6.

94 Ojijo, V. and Sinha Ray, S. (2013). Processing strategies in bionanocomposites. *Progress in Polymer Science* 1543–1589. https://doi.org/10.1016/j.progpolymsci.2013.05.011.

95 Li, W., Zhang, C., Chi, H. et al. (2017). 'Development of antimicrobial packaging film made from poly(lactic acid) incorporating titanium dioxide and silver nanoparticles. *Molecules* 22 (7). https://doi.org/10.3390/molecules22071170.

96 Hentati, F., Tounsi, L., Djomdi, D. et al. (2020). Bioactive polysaccharides from seaweeds. *Molecules*. https://doi.org/10.3390/molecules25143152.

97 Wijesekara, I., Pangestuti, R., and Kim, S.K. (2011). Biological activities and potential health benefits of sulfated polysaccharides derived from marine algae. *Carbohydrate Polymers* 14–21. https://doi.org/10.1016/j.carbpol.2010.10.062.

98 Poddar, M.K. and Dikshit, P.K. (2021). Recent development in bacterial cellulose production and synthesis of cellulose based conductive polymer nanocomposites. *Nano Select* 2 (9): 1605–1628. https://doi.org/10.1002/nano.202100044.

99 Torres, F.G., Arroyo, J.J., and Troncoso, O.P. (2019). Bacterial cellulose nanocomposites: An all-nano type of material. *Materials Science and Engineering C* 1277–1293. https://doi.org/10.1016/j.msec.2019.01.064.

100 Jayakumar, A., Radoor, S., Kim, J. T., et al. (2022) Recent innovations in bionanocomposites-based food packaging films – a comprehensive review. *Food Packaging and Shelf Life*. https://doi.org/10.1016/j.fpsl.2022.100877.

101 Trajkovska Petkoska, A., Daniloski, D., D'Cunha, N.M. et al. (2021). Edible packaging: sustainable solutions and novel trends in food packaging. *Food Research International*. https://doi.org/10.1016/j.foodres.2020.109981.

102 Jiménez, A., Sánchez-González, L., Desobry, S. et al. (2014). Influence of nanoliposomes incorporation on properties of film forming dispersions and films based on corn starch and sodium caseinate. *Food Hydrocolloids* 35: 159–169. https://doi.org/10.1016/j.foodhyd.2013.05.006.

103 Otoni, C.G., Pontes, S.F.O., Medeiros, E.A.A. et al. (2014). Edible films from methylcellulose and nanoemulsions of clove bud (*Syzygium aromaticum*) and oregano (*Origanum vulgare*) essential oils as shelf life extenders for sliced bread. *Journal of Agricultural and Food Chemistry* 62 (22): 5214–5219. https://doi.org/10.1021/jf501055f.

104 Alexandre, E.M.C., Lourenço, R.V., Bittante, A.M.Q.B. et al. (2016). Gelatin-based films reinforced with montmorillonite and activated with nanoemulsion of ginger essential oil for food packaging applications. *Food Packaging and Shelf Life* 10: 87–96. https://doi.org/10.1016/j.fpsl.2016.10.004.

105 Meira, S.M.M., Zehetmeyer, G., Werner, J.O. et al. (2017). A novel active packaging material based on starch-halloysite nanocomposites incorporating antimicrobial peptides. *Food Hydrocolloids* 63: 561–570. https://doi.org/10.1016/j.foodhyd.2016.10.013.

106 Arfat, Y.A., Ahmed, J., and Jacob, H. (2017). Preparation and characterization of agar-based nanocomposite films reinforced with bimetallic (Ag-Cu) alloy nanoparticles. *Carbohydrate Polymers* 155: 382–390. https://doi.org/10.1016/j.carbpol.2016.08.097.

107 Hashemi Tabatabaei, R., Jafari, S.M., Mirzaei, H. et al. (2018). Preparation and characterization of nano-SiO2 reinforced gelatin-k-carrageenan biocomposites. *International Journal of Biological Macromolecules* 111: 1091–1099. https://doi.org/10.1016/j.ijbiomac.2018.01.116.

108 Shojaei, M., Eshaghi, M., and Nateghi, L. (2019). Characterization of hydroxypropyl methyl cellulose–whey protein concentrate bionanocomposite films reinforced by chitosan nanoparticles. *Journal of Food Processing and Preservation* 43 (10). https://doi.org/10.1111/jfpp.14158.

109 Cui, R., Jiang, K., Yuan, M. et al. (2020). Antimicrobial film based on polylactic acid and carbon nanotube for controlled cinnamaldehyde release. *Journal of Materials Research and Technology* 9 (5): 10130–10138. https://doi.org/10.1016/j.jmrt.2020.07.016.

110 Rezaei, M., Pirsa, S., and Chavoshizadeh, S. (2020). Photocatalytic/antimicrobial active film based on wheat gluten/ZnO nanoparticles. *Journal of Inorganic and Organometallic Polymers and Materials* 30 (7): 2654–2665. https://doi.org/10.1007/s10904-019-01407-6.

111 Mugwagwa, L.R. and Chimphango, A.F.A. (2022). Physicochemical properties and potential application of hemicellulose/pectin/nanocellulose biocomposites as active packaging for fatty foods. *Food Packaging and Shelf Life* 31. https://doi.org/10.1016/j.fpsl.2021.100795.

112 Conte, A., Longano, D., Costa, C. et al. (2013). A novel preservation technique applied to fiordilatte cheese. *Innovative Food Science and Emerging Technologies* 19: 158–165. https://doi.org/10.1016/j.ifset.2013.04.010.

113 Pinto, A.M.B., Santos, T.M., Caceres, C.A. et al. (2015). Starch-cashew tree gum nanocomposite films and their application for coating cashew nuts. *LWT* 62 (1): 549–554. https://doi.org/10.1016/j.lwt.2014.07.028.

114 Issa, A., Ibrahim, S.A., and Tahergorabi, R. (2017). Impact of sweet potato starch-based nanocomposite films activated with thyme essential oil on the shelf-life of baby spinach leaves. *Foods* 6 (6): 1–13. https://doi.org/10.3390/foods6060043.

115 Koushesh Saba, M. and Amini, R. (2017). Nano-ZnO/carboxymethyl cellulose-based active coating impact on ready-to-use pomegranate during cold storage. *Food Chemistry* 232: 721–726. https://doi.org/10.1016/j.foodchem.2017.04.076.

116 Mihaly-Cozmuta, A., Peter, A., Craciun, G. et al. (2017). Preparation and characterization of active cellulose-based papers modified with TiO_2, Ag and zeolite nanocomposites for bread packaging application. *Cellulose* 24 (9): 3911–3928. https://doi.org/10.1007/s10570-017-1383-x.

117 Li, W., Li, L., Zhang, H. et al. (2018). Evaluation of PLA nanocomposite films on physicochemical and microbiological properties of refrigerated cottage cheese. *Journal of Food Processing and Preservation* 42 (1). https://doi.org/10.1111/jfpp.13362.

118 Khezrian, A. and Shahbazi, Y. (2018). Application of nanocompostie chitosan and carboxymethyl cellulose films containing natural preservative compounds in minced camel's meat. *International Journal of Biological Macromolecules* 106: 1146–1158. https://doi.org/10.1016/j.ijbiomac.2017.08.117.

119 Sinaga, M.Z.E., Gea, S., Panindia, N. et al. (2018). The preparation of cellulose nanocomposite film from isolated cellulose of corncobs as food packaging. *Oriental Journal of Chemistry* 34 (1): 562–567. https://doi.org/10.13005/ojc/340166.

120 Indumathi, M.P., Saral Sarojini, K., and Rajarajeswari, G.R. (2019). Antimicrobial and biodegradable chitosan/cellulose acetate phthalate/ZnO nano composite films with optimal oxygen permeability and hydrophobicity for extending the shelf life of black grape fruits. *International Journal of Biological Macromolecules* 132: 1112–1120. https://doi.org/10.1016/j.ijbiomac.2019.03.171.

121 Li, J., Sun, Q., Sun, Y. et al. (2019). Improvement of banana postharvest quality using a novel soybean protein isolate/cinnamaldehyde/zinc oxide bionanocomposite coating strategy. *Scientia Horticulturae* 258. https://doi.org/10.1016/j.scienta.2019.108786.

122 Jafarzadeh, S., Rhim, J.-W., Alias, A.K. et al. (2019). Application of antimicrobial active packaging film made of semolina flour, nano ZnO and nano-kaolin to maintain the quality of low-moisture mozzarella cheese during low-temperature storage. *Journal of the Science of Food and Agriculture* 99 (6): 2716–2725. https://doi.org/10.1002/jsfa.9439.

123 Souza, V.G.L., Pires, J.R.A., Vieira, É.T. et al. (2019). Activity of chitosan-montmorillonite bionanocomposites incorporated with rosemary essential oil: From in vitro assays to application in fresh poultry meat. *Food Hydrocolloids* 89: 241–252. https://doi.org/10.1016/j.foodhyd.2018.10.049.

124 Shahbazi, Y. and Shavisi, N. (2019). Effects of sodium alginate coating containing *Mentha spicata* essential oil and cellulose nanoparticles on extending the shelf life of raw silver carp (*Hypophthalmichthys molitrix*) fillets. *Food Science and Biotechnology* 28 (2): 433–440. https://doi.org/10.1007/s10068-018-0486-y.

125 Amjadi, S., Emaminia, S., Nazari, M. et al. (2019). Application of reinforced ZnO nanoparticle-incorporated gelatin bionanocomposite film with chitosan nanofiber for packaging of chicken fillet and cheese as food models. *Food and Bioprocess Technology* 12 (7): 1205–1219. https://doi.org/10.1007/s11947-019-02286-y.

126 Salama, H.E. and Abdel Aziz, M.S. (2020). Optimized carboxymethyl cellulose and guanidinylated chitosan enriched with titanium oxide nanoparticles of improved UV-barrier properties for the active packaging of green bell pepper. *International Journal of Biological Macromolecules* 165: 1187–1197. https://doi.org/10.1016/j.ijbiomac.2020.09.254.

127 Ligaj, M., Tichoniuk, M., Cierpiszewski, R. et al. (2020). Efficiency of novel antimicrobial coating based on iron nanoparticles for dairy products' packaging. *Coatings* 10 (2). https://doi.org/10.3390/coatings10020156.

128 Wang, C., Chang, T., Dong, S. et al. (2020). Biopolymer films based on chitosan/potato protein/linseed oil/ZnO NPs to maintain the storage quality of raw meat. *Food Chemistry* 332. https://doi.org/10.1016/j.foodchem.2020.127375.

129 Al-Tayyar, N.A., Youssef, A.M., and Al-Hindi, R.R. (2020). Antimicrobial packaging efficiency of ZnO-SiO_2 nanocomposites infused into PVA/CS film for enhancing the shelf life of food products. *Food Packaging and Shelf Life* 25. https://doi.org/10.1016/j.fpsl.2020.100523.

130 Benucci, I., Lombardelli, C., Cacciotti, I. et al. (2020). Papain covalently immobilized on Chitosan–Clay nanocomposite films: application in synthetic and real white wine. *Nanomaterials* 10 (9): 1–11. https://doi.org/10.3390/nano10091622.

131 Zhang, Y.P., Wang, X., Shen, Y. et al. (2021). Preparation and characterization of bio-nanocomposites film of chitosan and montmorillonite incorporated with ginger essential oil and its application in chilled beef preservation. *Antibiotics* 10 (7). https://doi.org/10.3390/antibiotics10070796.

132 Priyadarshi, R., Kim, S.M., and Rhim, J.W. (2021). Carboxymethyl cellulose-based multifunctional film combined with zinc oxide nanoparticles and grape seed extract for the preservation of high-fat meat products. *Sustainable Materials and Technologies* 29. https://doi.org/10.1016/j.susmat.2021.e00325.

133 de Calva-Estrada, S.J., Jimenez-Fernandez, M., Vallejo-Cardona, A.A. et al. (2021). Cocoa nanoparticles to improve the physicochemical and functional properties of whey protein-based films to extend the shelf life of muffins. *Foods* 10 (11). https://doi.org/10.3390/foods10112672.

134 Liu, Z., Du, M., Liu, H. et al. (2021). Chitosan films incorporating litchi peel extract and titanium dioxide nanoparticles and their application as coatings on watercored apples. *Progress in Organic Coatings* 151. https://doi.org/10.1016/j.porgcoat.2020.106103.

135 Candido, G.S., Natarelli, C.V.L., Carvalho, E.E.N. et al. (2022). Bionanocomposites of pectin and pracaxi oil nanoemulsion as active packaging for butter. *Food Packaging and Shelf Life* 32. https://doi.org/10.1016/j.fpsl.2022.100862.

136 Robledo, N., Vera, P., López, L. et al. (2018). Thymol nanoemulsions incorporated in quinoa protein/chitosan edible films; antifungal effect in cherry tomatoes. *Food Chemistry* 246: 211–219. https://doi.org/10.1016/j.foodchem.2017.11.032.

137 Prakash, A., Baskaran, R., and Vadivel, V. (2020). Citral nanoemulsion incorporated edible coating to extend the shelf life of fresh cut pineapples. *LWT* 118. https://doi.org/10.1016/j.lwt.2019.108851.

138 Basumatary, I.B., Mukherjee, A., Katiyar, V. et al. (2021). Chitosan-based antimicrobial coating for improving postharvest shelf life of pineapple. *Coatings* 11 (11). https://doi.org/10.3390/coatings11111366.

139 Deng, Z., Jung, J., Simonsen, J. et al. (2017). Cellulose nanocrystal reinforced chitosan coatings for improving the storability of postharvest pears under both ambient and cold storages. *Journal of Food Science* 82 (2): 453–462. https://doi.org/10.1111/1750-3841.13601.

140 Ghosh, T., Nakano, K., and Katiyar, V. (2021). Curcumin doped functionalized cellulose nanofibers based edible chitosan coating on kiwifruits. *International Journal of Biological Macromolecules* 184: 936–945. https://doi.org/10.1016/j.ijbiomac.2021.06.098.

141 Liu, Y., Wang, S., Lan, W. et al. (2019). Fabrication of polylactic acid/carbon nanotubes/chitosan composite fibers by electrospinning for strawberry preservation. *International Journal of Biological Macromolecules* 121: 1329–1336. https://doi.org/10.1016/j.ijbiomac.2018.09.042.

142 Hussain, M., Bensaid, S., Geobaldo, F. et al. (2011). Photocatalytic degradation of ethylene emitted by fruits with TiO_2 nanoparticles. *Industrial and Engineering Chemistry Research* 50 (5): 2536–2543. https://doi.org/10.1021/ie1005756.

143 Kaewklin, P., Siripatrawan, U., Suwanagul, A. et al. (2018). Active packaging from chitosan-titanium dioxide nanocomposite film for prolonging storage life of tomato fruit. *International Journal of Biological Macromolecules* 112: 523–529. https://doi.org/10.1016/j.ijbiomac.2018.01.124.

144 Riahi, Z., Priyadarshi, R., Rhim, J.-W. et al. (2021). Titania nanotubes decorated with Cu(I) and Cu(II) oxides: antibacterial and ethylene scavenging functions to extend the shelf life of bananas. *ACS Sustainable Chemistry and Engineering* 9 (19): 6832–6840. https://doi.org/10.1021/acssuschemeng.1c01348.

145 Youssef, A.M., El-Sayed, S.M., Salama, H.H. et al. (2015). Evaluation of bionanocomposites as packaging material on properties of soft white cheese during storage period. *Carbohydrate Polymers* 132: 274–285. https://doi.org/10.1016/j.carbpol.2015.06.075.

146 Youssef, A.M., Assem, F.M., El-Sayed, H.S. et al. (2020). Synthesis and evaluation of eco-friendly carboxymethyl cellulose/polyvinyl alcohol/CuO bionanocomposites and their use in coating processed cheese. *RSC Advances* 10 (62): 37857–37870. https://doi.org/10.1039/d0ra07898k.

147 Cui, H., Wu, J., Li, C. et al. (2017). Improving anti-listeria activity of cheese packaging via nanofiber containing nisin-loaded nanoparticles. *LWT* 81: 233–242. https://doi.org/10.1016/j.lwt.2017.04.003.

148 Divsalar, E., Tajik, H., Moradi, M. et al. (2018). Characterization of cellulosic paper coated with chitosan-zinc oxide nanocomposite containing nisin and its application in packaging of UF cheese. *International Journal of Biological Macromolecules* 109: 1311–1318. https://doi.org/10.1016/j.ijbiomac.2017.11.145.

149 Asdagh, A. and Pirsa, S. (2020). Bacterial and oxidative control of local butter with smart/active film based on pectin/nanoclay/Carum copticum essential oils/β-carotene. *International Journal of Biological Macromolecules* 165: 156–168. https://doi.org/10.1016/j.ijbiomac.2020.09.192.

150 Sheng, J., Li, J., and Zhao, L. (2019). Fabrication of grease resistant paper with non-fluorinated chemicals for food packaging. *Cellulose* 26 (10): 6291–6302. https://doi.org/10.1007/s10570-019-02504-y.

151 Apjok, R., Cozmuta, A.M., Peter, A. et al. (2019). Active packaging based on cellulose-chitosan-Ag/TiO_2 nanocomposite for storage of clarified butter. *Cellulose* 26 (3): 1923–1946. https://doi.org/10.1007/s10570-018-02226-7.

152 Alizadeh-Sani, M., Mohammadian, E., and McClements, D.J. (2020). Eco-friendly active packaging consisting of nanostructured biopolymer matrix reinforced with TiO2 and essential oil: application for preservation of refrigerated meat. *Food Chemistry* 322. https://doi.org/10.1016/j.foodchem.2020.126782.

153 Pires, J.R.A., de Souza, V.G.L., and Fernando, A.L. (2018). Chitosan/montmorillonite bionanocomposites incorporated with rosemary and ginger essential oil as packaging for fresh poultry meat. *Food Packaging and Shelf Life* 17: 142–149. https://doi.org/10.1016/j.fpsl.2018.06.011.

154 Kamkar, A., Molaee-aghaee, E., Khanjari, A. et al. (2021). Nanocomposite active packaging based on chitosan biopolymer loaded with nano-liposomal essential oil: its characterizations and effects on microbial, and chemical properties of refrigerated chicken breast fillet. *International Journal of Food Microbiology* 342. https://doi.org/10.1016/j.ijfoodmicro.2021.109071.

155 Shokri, S., Parastouei, K., Taghdir, M. et al. (2020). Application an edible active coating based on chitosan-*Ferulago angulata* essential oil nanoemulsion to shelf life extension of Rainbow trout fillets stored at 4 °C. *International Journal of Biological Macromolecules* 153: 846–854. https://doi.org/10.1016/j.ijbiomac.2020.03.080.

156 Gharibzahedi, S.M.T. and Mohammadnabi, S. (2017). Effect of novel bioactive edible coatings based on jujube gum and nettle oil-loaded nanoemulsions on the shelf-life of Beluga sturgeon fillets. *International Journal of Biological Macromolecules* 95: 769–777. https://doi.org/10.1016/j.ijbiomac.2016.11.119.

157 Noshirvani, N., Ghanbarzadeh, B., Rezaei Mokarram, R. et al. (2017). Novel active packaging based on carboxymethyl cellulose-chitosan-ZnO NPs nanocomposite for increasing the shelf life of bread. *Food Packaging and ShelfLife* 11: 106–114. https://doi.org/10.1016/j.fpsl.2017.01.010.

158 Sahraee, S., Milani, J.M., Ghanbarzadeh, B. et al. (2020). Development of emulsion films based on bovine gelatin-nano chitin-nano ZnO for cake packaging. *Food Science and Nutrition* 8 (2): 1303–1312. https://doi.org/10.1002/fsn3.1424.

159 Lai, R.W.S., Yeung, K.W.Y., Yung, M.M.N. et al. (2018). Regulation of engineered nanomaterials: current challenges, insights and future directions. *Environmental Science and Pollution Research* 25 (4): 3060–3077. https://doi.org/10.1007/s11356-017-9489-0.

160 Kabir, E., Kumar, V., Kim, K.-H. et al. (2018). Environmental impacts of nanomaterials. *Journal of Environmental Management* 261–271. https://doi.org/10.1016/j.jenvman.2018.07.087.

161 Ema, M., Okuda, H., Gamo, M. et al. (2017). A review of reproductive and developmental toxicity of silver nanoparticles in laboratory animals. *Reproductive Toxicology* 149–164. https://doi.org/10.1016/j.reprotox.2017.01.005.

162 Jain, A., Ranjan, S., Dasgupta, N. et al. (2018). Nanomaterials in food and agriculture: an overview on their safety concerns and regulatory issues. *Critical Reviews in Food Science and Nutrition* 297–317. https://doi.org/10.1080/10408398.2016.1160363.

163 Garcia, C.v., Shin, G.H., and Kim, J.T. (2018). Metal oxide-based nanocomposites in food packaging: Applications, migration, and regulations. *Trends in Food Science and Technology* 21–31. https://doi.org/10.1016/j.tifs.2018.09.021.

164 Henkler, F., Tralau, T., Tentschert, J. et al. (2012). Risk assessment of nanomaterials in cosmetics: a European union perspective. *Archives of Toxicology* 1641–1646. https://doi.org/10.1007/s00204-012-0944-x.

165 Sadik, O.A., Du, N., Kariuki, V. et al. (2014). Current and emerging technologies for the characterization of nanomaterials. *ACS Sustainable Chemistry and Engineering* 1707–1716. https://doi.org/10.1021/sc500175v.

10

Introduction to Intelligent Food Packaging

Krishna Kamal Hazarika [1,2], *Pallabi Hazarika* [1,2], *and Swapnali Hazarika* [1,2]

[1] *Chemical Engineering Group and Centre for Petroleum Research Division, CSIR-North East Institute of Science and Technology, Jorhat, Assam, India*
[2] *Academy of Scientific and Innovative Research (AcSIR), Ghaziabad, Uttar Pradesh, India*

10.1 Introduction

In the year 2023, contaminated food supplies were estimated to have caused 600 million illnesses, or nearly one in 10 cases, according to the World Health Organization (WHO). Additionally, each year, 4 20 000 people die as a result of food contamination. Every year, 1 25 000 people die from foodborne diseases, including children under the age of 5. Furthermore, diarrheal diseases, which are mostly caused by consuming contaminated food, cause roughly 550 million illnesses annually [1]. The overall quality of food is influenced by a number of factors, such as food safety, consumer expectations, nutritional data, uniformity, and other factors that differ by nation. In the consumer-packaged goods sector, food packaging is one of the most important kinds of packaging because it protects the product with less damage, facilitates transportation, allows for safe storage, and functions as an additional form of product messaging. A new technique called intelligent packaging (IP) takes advantage of the communication function of the package to accelerate decision-making and enhance food safety and quality [2]. Following transit, distribution, storage, and retail, it is a planned system for making things usable. The desire among consumers for excellent and healthful foods has grown since the start of the globalization and industrialization eras. Food can remain fresh for several weeks or more when packaged appropriately, depending on its composition and a number of external circumstances. During the handling, preparation, or transportation of food products, biological, chemical, or physical substances may come into contact with them. Packaging can help food products retain their taste and freshness [3]. The primary goal of packaging is to shield the food from oxidative damage, biological injury, moisture gain or loss, and physical harm. Enabling the consumer to receive the items with accurate information is its secondary goal. It also eliminates the possibility of contamination at any point in the food supply chain that could arise from coming into contact with other foods or material

Smart Food Packaging Systems: Innovations and Technology Applications, First Edition.
Edited by Avik Mukherjee, Santosh Kumar, Manjusri Misra, and Amar K. Mohanty.
© 2025 John Wiley & Sons Ltd. Published 2025 by John Wiley & Sons Ltd.

components [4]. IP also acts as a precautionary measure to make sure the product is sealed throughout the entire distribution process until the end user opens it, preventing spills and identifying potential tampering problems. Importantly, food packaging acts as a means of identification and interaction between the manufacturer and the customer by providing information about the product, including nutritional facts, health benefits, usage and storage instructions, the location of manufacture, the date of manufacture, the expiration date, and best-before dates, as well as other pertinent details. To meet these standards, IP is employed in the food industry [5]. These systems can communicate customer information while continuously monitoring a product's quality state. Food waste can be decreased and customer satisfaction can be increased in this way. It is patently obvious that food packaging places a significant focus on the product's safety. Systems for packaging products maintain their quality for longer periods of time while somehow reducing the amount of waste. The attempts to create IP for various items will be discussed here.

10.2 Types of Intelligent Packaging

In order to extend the shelf life of packaged food and improve its quality, safety, and nutritional integrity, novel intelligent and active packaging (AP) techniques are continuously being developed [6]. AP modifies the food in order to extend the shelf life, improve food safety, or improve sensory elements at the price of the product's worth. IP systems monitor the condition of packed objects during storage and transportation, giving valuable insights into their quality. The word "IP" typically refers to a packaged system that can identify changes in its internal environment and notify users of those changes. There are three categories into which these devices fall. The first group of indicators comes from outside sources, indications for temperature and time, exterior vivid indications, and other indicators attached to the outside of the package. The second group consists of internal indicators that are placed inside the packing material. Indicators of CO_2, O_2, and microbiology, as well as pathogens, can typically be attached to the lid or placed in the headspace of the container. The third type of gadget is an indicator, which improves data transfer efficiency and interactivity between the outcome and the client [7]. The following order will be followed by the application systems for IP.

- Products with a high level of convenience.
- Those that enhance the product's quality and worth.
- Products that alter the features of gas permeability.
- Those that offer protection against theft, forgeries, and tampering.

These packaging systems must be simple to use, affordable, and able to perform a variety of functions to be useful. For instance, color-indicating labels ought to include color shifts that are permanent, clear, and simple for customers to comprehend. Depending on the packaged product – medicine, food, drink, home goods – "smartness" in packaging can mean many different things and encompass a wide range of functionalities [8]. A brief overview of different IP systems is shown in Figure 10.1.

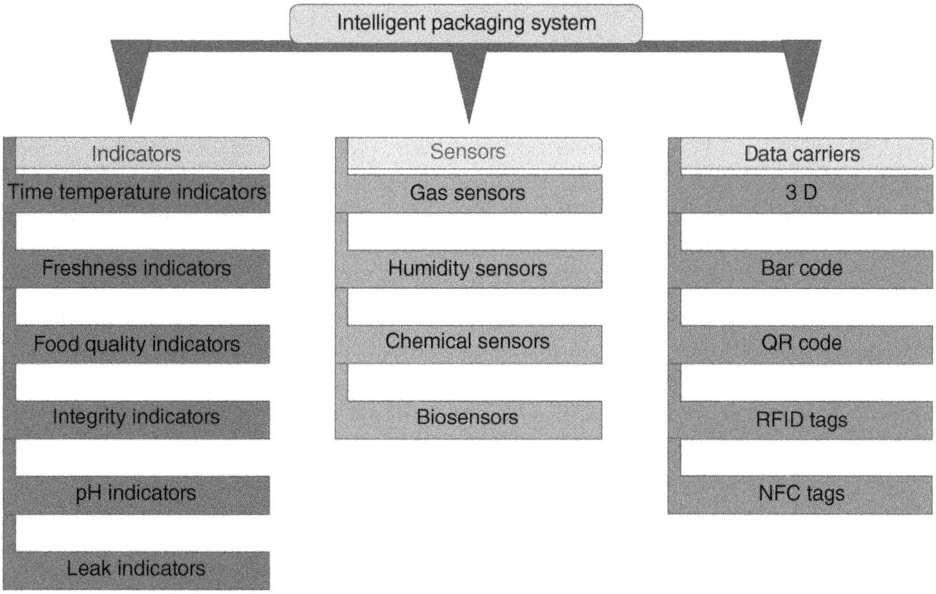

Figure 10.1 Intelligent packaging systems.

10.3 Indicators

Indicators' primary purposes are to communicate to consumers information about the availability of a specific drug, to determine whether a reaction involving more than one component is taking place, or to monitor the level of a single material. This type of data is reconstituted as signals, which are typically instantaneous visual changes. These signals then contain qualitative or quantitative data. Most of the time, an indicator's fundamental requirement is that these color or intensity changes be permanent. The indicators are a class of IP that is remarkably constant, and the most often-used classification, which also applies to this research, is based on the kind of variable regulated.

10.3.1 Time–Temperature Indicators

The increasing interest in IP depending on chemical, physical, or enzymatic functioning in the food product will follow the time and temperature indicator, which will give precise information about the product's quality, shelf life, and safety condition. Throughout their storage, distribution, and consumption phases, perishable goods are monitored for residual shelf life, and their temperature history is recorded using time–temperature indicators, or TTIs [9]. This is accomplished by the irreversible color shift brought about by the combined influences of temperature and time. TTI can be used in IP to give easily accessible visual information [10]. This allows consumers to determine whether food has degraded based on the information, ensuring food safety and quality. Based on various theories of color change, TTI can be divided into a number of categories, such as chemical, physical, biological, and other novel types. These types of TTI can, in theory, offer useful information regarding food quality, but there still exist a number of issues with them in the supply chain of food, including the migration of hazardous materials, the imprecision of temperature monitoring, and the high cost of commercial application. The aforementioned TTI issues can nevertheless be resolved by using certain scientific techniques [11]. The working mechanisms of the TTIs that are currently on the market are based on various principles. The fundamental mechanism underlying TTI operations is a chemical, mechanical, enzymatic, irreversible, or microbiological change that typically manifests as a visible reaction in the form of color changes and mechanical deformation. Physical and chemical indication is based on chemical reactions as well as physical changes in relation to temperature and time, for example, acid–base reactions, polymerization, melting, and so on. Whereas biological reactions alter based on biological activity, for example, microorganisms and enzymes, in relation to time and temperature. The changing rate depends on temperature, and it rises at very high temperatures exactly in a similar manner to the deteriorative reactions that cause products to lose quality. The visual response of the TTI indicator thus progressively represents the product's time–temperature history. TTIs must be simple to initiate and show repeatable time- and temperature-dependent changes that are simple to monitor [12]. TTIs are crucial to the perishable food product industry's supply chain, from product manufacture and packing to delivery to retail locations and client purchases. The movement of hazardous materials, the imprecision of temperature monitoring, and the high cost of commercial applications are only a few of the ongoing issues. The aforementioned issues restrict the use of TTI to some extent. Time and temperature indicators are typically firmly affixed to food packaging and may contain chemical or inorganic elements that are harmful. In the course of food distribution, the chemicals may seep into food and contaminate it. The color change of a TTI, regardless of how well it performs, is entirely due to environmental factors rather than precisely capturing the quality of food because it can only measure the outermost temperature by adhering to the particular surface of food packages; it cannot measure the actual temperatures of food. When it comes to practical applications, the TTI is a component of food-related packaging that establishes a maximum limit on cost. One important factor that influences people's decision to use or not utilize TTI is its cost [13]. Figure 10.2 represents a flow diagram showing different types of TTI based on different parameters like function, operating principle, and temperature.

10.3.1.1 Chemical Time and Temperature Indicators

TTI will chemically react with the progress of time and the rise in temperature, and this will change, resulting in a specific color shift. As a result, it will be used to indicate packaged product quality. Chemical-based TTI currently comprises photochromic, oxidation–reduction reactions, and polymerization-based time and temperature indicators [12]. The photochromic-based chemical TTI works by activating a photochromic compound with specific light wavelengths to produce a specific color. To prepare TTI, the reverse reaction technique is employed, based on the idea of thermally induced fading. As time and temperature increase, the reaction rate and degree of fading alter, making the shelf life clearly visible. In order to prevent unwanted photo-recharging or photo-bleaching throughout the reverse reaction, which may inhibit the thermochromic reverse reaction's ability to change color, the photochromic compound for TTI needs to be resistant to ambient light [10]. With the use of either the light-induced or redox reaction, redox reaction-based TTI displays

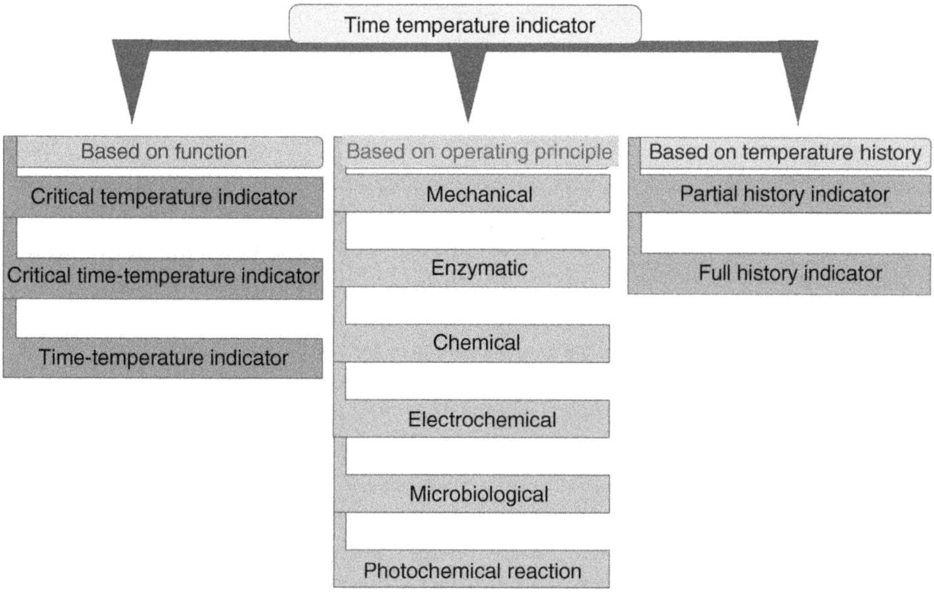

Figure 10.2 Types of time and temperature indicators.

color change. The substance interacts with atmospheric oxygen. Temperature and time accumulation have a direct impact on the reaction rate. The TTI's color change can be used to estimate a product's shelf life [14]. The solid-state polymerization process of a monomer containing an acetylene group (R1C≡C—C≡CR2) is the basis for the first kind of chemical TTI. The monomers undergo a 1,4-addition polymerization process when exposed to strong radiation and temperature to form a polymer with structure R1 $(C{=}C{-}C{=}C)_n$ R2. At the visible absorption peaks, the absorption spectrum changes from a high band to a low band due to the irreversible nature of the polymerization reaction and the increasing reaction rate. As a result, the indicator's color shifts [13].

10.3.1.2 Physical Time–Temperature Indicator

Physical TTI is classified as diffusion-based TTI, nanocarrier-based TTI, electronics TTI, and other types of TTI, depending on various operating principles. The number of new TTI systems based on the discoloration of physical characteristics is growing. The very simple idea of diffusion-based TTIs is that colored material diffuses through porous mediums when ambient temperature rises. The broad range of applications in commerce can be attributed to the simplicity of production and broad working temperature range of diffusion-based TTI, like Monitor Mark and Tempix. Nevertheless, the accuracy and safety of diffusion-based TTI will be negatively impacted by the exudation of colored material and aging porous substances [15]. The thermochromic feature of nanomaterials has been utilized in TTIs in recent investigations. In most cases, Ag, Au, or Ag/Au nanoparticles – which are considered noble metals – are utilized. As a result, the price of TTI could rise dramatically. In real-world applications, nanorods can contaminate food if they are not properly packaged or secured, endangering the consumers' health. It has been demonstrated in the realm of fundamental research that metal nanorods have a far higher extinction coefficient than other traditional dyes. The thermal sensor in an electronic TTI, which translates temperature readings into electrical signals and finally into a final visual output, is primarily responsible for the warning function of the device in general. The majority of electronic TTI types are generally heavy-duty yet have comparatively high precision. A smaller size is therefore required in order to fit them into a label. It is easy to determine that electronic TTI is expensive, which is most likely the reason why these devices, which have other benefits as well, have not yet been brought to market [14].

10.3.1.3 Time–Temperature Indication for Biological Systems

In recent years, biological TTI systems have made extensive use of yeast and lactic acid bacteria. The TTI systems may potentially contain other microorganisms, like Streptococcus [16]. The biological TTI system known as the yeast-based TTI is based on the anaerobic respiration process of yeast, which produces acid under specific conditions, particularly at specific temperatures, changing the pH indicator's color. The TTI system consists of two sections: the first part is filled with reactants, such as microorganisms, and a color indicator made of aqueous ink with the proper carrier separating them; both sections are adhered to a transparent plastic film; and the second section is filled with an activator and an aqueous adhesive layer. When the reactant makes surface contact with the activator, the TTI can be triggered. The TTI will activate by making contact between the activator surface (the second component) and the reactive agent surface (the first part). It is important to remember that each ingredient's quantity needs to be carefully regulated. Since the TTI relies on the yeast's aerobic respiration, which produces acid, the system's oxygen and glucose levels must be carefully regulated [17]. In order to alter pH under specific circumstances and cause a color shift that indicates the cumulative effect of temperature and time, lactobacillus-based TTIs use the lactic acid produced by lactic acid bacteria. In addition to being able to swiftly and precisely assess if a product has spoiled under specific circumstances, the TTI system's acid–base color indicator also has no effect on the growth of bacteria. To fix the microorganisms' material nutrients, a variety of materials, including polyacrylamide, carrageenan, agar, and alginic acid, are combined inside the TTI system [13, 18].

10.3.2 Freshness Indicators

The chemical reactions between the target bacteria and their metabolites that lead to food spoilage are analyzed by freshness indicators, which are direct indicators of food quality. The core idea is that as target microorganisms proliferate, they bring about irreversible changes like pH variation (which indirectly causes a color change), providing clear information about the quality of the food contained in the container. These indicators are used in a variety of products, such as fresh meals, fruits, and seafood. The total volatile basic nitrogen (TVB-N) content is considered a crucial indicator of deterioration in fish and meat [19]. The prevalent method of determining the TVB-N value is a time-spending method of identifying chemical interactions between perchloric acid and volatile amine groups, followed by an acid–acid titration study. Most of the time, they operate using dyes, which are basically pH-sensitive, colorimetric analysis, or naturally available biofilm. Based on colorimetric methods, freshness indicators can identify total volatile amines by detecting an increase in pH brought on by rotting. It has been demonstrated that food freshness indicators that depend on pH may be made to work efficiently with naturally available bio-based extracts as dyes because they can be readily operated since total color difference values are tools with maximum sensitivity as well as being biodegradable [20]. Locally available fruits emit a distinct aroma when they begin to ripen. This has been employed in a few instances; the sensor of the element reacts with the scent molecules and changes color as a result. The freshness indicator monitors the quality of food goods throughout preservation and transportation. As an outcome, the freshness indicator displays data on product chemical changes, microbial growth, and the existence of microbial alteration. For example, the metabolites that imply quality include organic acids, glucose, biogenic amines (BAs), ethanolic compounds, carbon dioxide compounds, volatile nitrogen compounds, ATP breakdown products, as well as sulfuric components. Freshness indicators might be positioned within the container in the form of a heat-sealable package to be in contact with the components [13]. Figure 10.3 shows cellulose nanofiber and natural dye-based biodegradable bag as a freshness indicator for meat and fish packaging.

10.3.3 Food Quality Indicators

Certain indicators that accurately depict the condition of food products, either qualitatively or quantitatively, are required to build sensors that are integrated into food packaging. Food rotting indicators are frequently connected to either physical or chemical alterations in the particular food's qualities. The following section discusses often-used food quality and safety indicators [21].

10.3.3.1 Oxygen and Carbon Dioxide

Combustion requires oxygen availability and some biological functions. Oxidation of fat, pigment, and processes of browning are some examples of the oxidation reactions that play a role in degrading food quality. In contrast, carbon dioxide is an odorless, colorless gas that can lower the pH of a food environment while inhibiting bacterial and fungal growth [22]. Gram-negative microorganisms that quickly cause microbial deterioration, including pseudomonas species, are inhibited by carbon dioxide. CO_2 has antibacterial properties because of its capacity to produce a stifling environment that hinders enzymatic decarboxylation. Reducing CO_2 emissions is consequently critical for extending the storage life of packaged foods. A modified-atmosphere packaged for non-respiring processed foods normally has a low concentration of O_2 and a high concentration of CO_2, with some variations depending on the product type. As a result, any fluctuations in air concentration levels could signal poor food quality and warn consumers about the probability of unfavorable microbial growth [23].

10.3.3.2 pH Change

The impact of microbial metabolites on the pH of the food environment may require monitoring pH changes, which is an useful means of detecting food spoilage. Some aerobic and anaerobic microbes can flourish within food packaging during the storage term of

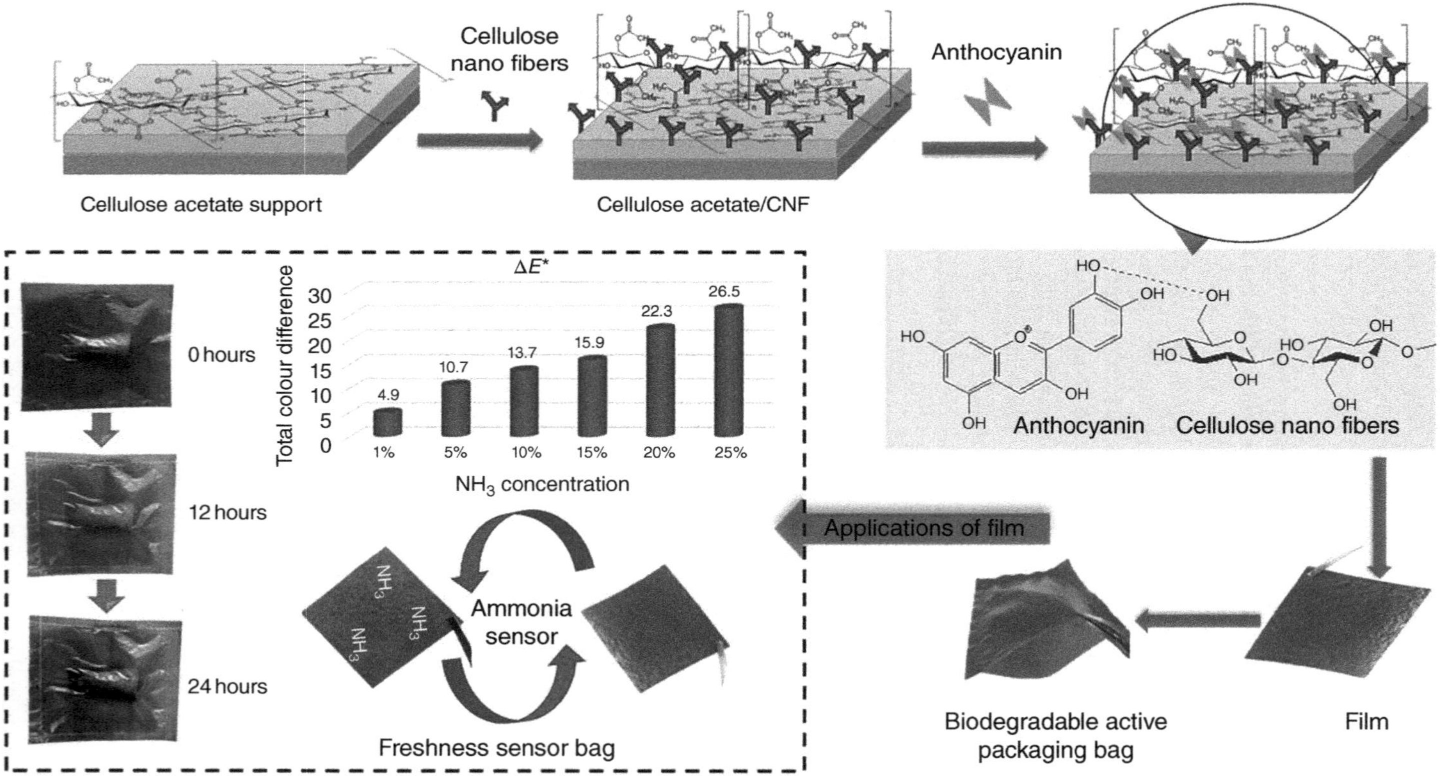

Figure 10.3 Cellulose nanofiber-mediated natural dye-based biodegradable bag with freshness indicator for packaging of meat and fish. *Source:* Reproduced from Hazarika et al. [13]/with permission of Elsevier.

the food goods. The main products of glucose fermentation are organic acids (such as acetic acid and lactic acid). These metabolites have the potential to lower a dietary sample's pH. Furthermore, ethanol itself is a by-product of lactic acid metabolism. The pH change caused by ethanol, on the other hand, is minor, while it is just possibly slightly higher than water in terms of basicity. Microbial development generates CO_2 gas, which may dissolve in the packaged food while producing carbonic acid. The carbonic acid's hydrogen ions may then dissociate to generate hydrogen as well as bicarbonate ions. As a proton, a hydrogen ion joins with a molecule of water to generate a hydronium ion, thereby lowering the pH of such a sample [24].

10.3.3.3 Humidity

To preserve the specific condition and texture of product foods and extend their shelf life, it is crucial to maintain stable relative humidity inside the food packaging. The impact of relative humidity is extremely essential in many aspects of the food product industry, including meat, fish, dairy, and dry food. The humidity within the package could be changed in a variety of ways. One prevalent issue is the shattering of the food's sealed package because of poor manipulation. Prolonged exposure to high-temperature changes is another factor that may cause humidity levels in the inside compartments of the package to change. This can lead to condensation, jeopardizing the packaging system's integrity. Humidity is the critical indication that is used in evaluating food quality since increasing moisture creates a suitable growing habitat for bacteria as well as fungi, posing a safety concern for product use. Along with microbial activity, humidity accelerates product deterioration by causing softening and perhaps dampening of the product [25].

10.3.4 Gas and Volatile Indicators

It is challenging to maintain the safety and quality of food items inside packaging material because the composition of gases in a package headspace is easily altered by fresh food respiration, leaks, the infiltration of air through the packaging film, or gas generated through microbial spoilage. Gas indicators, which can be labels or printed on packing films, can track changes in the gas composition inside a container, aiding in the monitoring of product safety and quality. These indicators undergo a chemical or enzymatic process that causes them to change color, usually revealing the existence of both CO_2 and O_2. These gas and volatile indicators, which keep an eye on the gas-filled environment inside the package, must be in close proximity to the food. These types of indicators are used for the packaging of food products in modified or controlled environments [13]. Although oxygen and carbon dioxide are the most commonly used gases in MAP products, other gases such as carbon monoxide, sulfur dioxide, nitrous oxide, ozone, and chlorine are also used in some sectors. The presence of oxygen on food products affects the developing process of aerobic microorganisms; fruits and vegetables go through browning processes catalyzed by enzymes and the process of oxidation of lipids, acids, and a variety of flavors [26]. Oxygen indication is the most prevalent form of gas indicator used in applications involving food packaging. The most popular oxygen indicators are based on color changes caused by redox reactions, light-activated redox reactions, or oxygen-binding reactions. These indicators can be purchased as labels, printed layers, tablets, or laminated polymer films. These indicators not only provide information about the gas concentration inside the package during transit and storage, but they also identify the early stages of food spoilage. This class of indicators may comprise indicators of food deterioration and quality loss or aging, often known as freshness indicators, as well as leakage indicators based on carbon dioxide and/or oxygen detection. Despite patents and academic literature describing freshness indicators and detectors, they are not implemented commercially as extensions of TTIs. They detect volatile metabolites created during food decomposition, including amines, ammonia, carbon dioxide, ethanol, and diacetyl, as well as hydrogen sulfide [27]. The phrase "ripening" refers to the physiological changes that occur together with the development of the fruit's textural and sensory qualities, making it suitable for consumption. The breakdown of starch by organic acids and subsequent conversion to sugars causes an increase in sweetness; the breakdown of pectins and breakdown of the components of cell walls results in a decrease in firmness; the generation of aromatic volatile compounds gives rise to an increase in flavor; and the production and breakdown of pigments cause color changes [28, 29].

10.3.5 pH Indicators

A pH indicator on packaging signifies among the more recent innovations in this rapidly growing industry that indicate the freshness of food goods. A pH change occurs when a specific meal begins to spoil. One of the most significant markers of food quality is this alteration. The hue of the indication in the package changes with a change in pH. In this instance, the packaging of the product, which illustrates the pH of the food before consumption, is a great approach to ensuring its safety and quality as a consumer's product [30]. The objective of smart packaging is to maintain constant monitoring of the food's quality inside the package and communicate that information via sensors attached to the package or colorimetric changes. pH sensors are typically made up of two different parts: a pH-sensitive dye and a solid base. A pH change occurs when certain foods begin to deteriorate. This variation is one of the product quality indicators. The pH changes at the commencement of the breakdown process, causing a color shift in the sensor or container. So, a package that specifies the pH of food before purchase or consumption is a measure that promises the user about the product's safety and quality [31]. When natural polymers are used, they become environmentally beneficial, biodegradable, and edible. A method of guaranteeing the safety of packaged food, pH monitoring also informs the user as to whether the product is fit for eating. Researchers are studying and testing the development of colorimetric sensors that can be used to indicate food pH in packaging [32]. Color is a key factor in identifying and perceiving the quality and appearance of well-known foods. It is arguably one of the most remarkable and alluring features of food products, and it has a direct impact on customer choice, acceptance, and inclination to consume. Utilizing natural food coloring, IP tracks the pH variations in food and looks for visual changes to keep an eye on freshness, microbiological development, and other quality aspects of the product in real time. It offers a quick and accurate, on-the-spot evaluation of the safety and quality of food. However, future research should consider important factors that are economically viable, like cost, consumer acceptance and confidence, regulatory issues (like labeling), and the multifunctional performance of smart films [33].

10.3.5.1 Meat Products

Meat and its derivatives are nourishing, which makes them highly sensitive to microbial infection. One of the main issues with meat and meat products is microbial contamination. This is particularly true with today's highly popular ready-to-eat and ready-to-cook meats, where the food's surface is mostly contaminated by bacteria during processing and handling. In addition to causing spoilage, the contamination also results in product recalls and outbreaks of foodborne illnesses, all of which have a negative impact on public health and create significant financial losses. Many methods have been used in both conventional and innovative processing to lessen contamination. Although directly adding food preservatives can minimize pollutants while preserving food freshness and quality, there is a chance that they will react with the meal's constituent ingredients [34]. Fresh meat spoils quickly and is susceptible to physiological changes. Even when refrigerated, most meat starts deteriorate within three to four days, just after the animal is killed. Meat contains vital nutrients that are beneficial for microbial growth, such as amino acids, sugars, and nucleotides. These substances undergo microbial metabolism during decay, which results in the production of ammonia, biological amines, indole, hydrogen sulfide, and organic acids, as well as unpleasant aromas, off flavors, and discoloration. The creation of all these volatile microbial metabolites causes a shift in dietary pH, which can be detected by natural colorant-based pH sensors, which show an irreversible color change [35]. The purpose of packaging fresh meat is to prevent cross-contamination, postpone rotting, allow some enzymatic activity to enhance tenderness, minimize weight loss, and, if relevant, guarantee that the red meats being sold at retail or customer level have an oxymyoglobin or simply cherry-red color. Dehydration, lipid oxidation, discoloration, and loss of aroma are all things to consider when evaluating processed meat products. There are numerous meat packaging systems on the market today, each with unique features and uses. These systems include vacuum packing, bulk-gas flushing, or MAP systems using 100% carbon dioxide for long-term chilled storage. Overwrap packaging is also an option for short-term chilled storage and/or retail display, as are a variety of specified MAP techniques for longer-term chilled storage and/or display. Because there are many different product attributes as well as fundamental needs and applications for meat packaging, any packaging technology that may provide additional product along with quality control in a variety of affordable ways will be much appreciated [36, 37].

10.3.5.2 Seafood and Fish Products

Fish is one of those foods that spoils easily. Due to its extremely low acidity, this item is particularly prone to the development of bacteria that can cause food poisoning. Fish can decompose as a result of bacterial, oxidative, or enzymatic spoilage. Omega-3 fatty acids, sometimes referred to as polyunsaturated fatty acids, are found in fish and are thought to offer protection against heart-related illnesses. They can, however, also break down into odorous substances like peroxides, ketones, and aldehydes when they come into contact with oxygen due to their extreme instability [38]. Fish and seafood decay through three distinct processes: breakdown through enzymatic reactions, microbiological spoilage processes, and lipid oxidation mechanisms. The most significant factor contributing to the decline in seafood quality is microbial deterioration. When seafood is subjected to microbial breakdown, products of microbial metabolism are created, much like meat-based products. These products come in a variety of forms, such as trimethylamine, volatile basic nitrogen, sulfide compounds, ketones, esters, and aldehydes. Trimethylamine and volatile basic nitrogen are the most often employed target molecules for shelf life and decay indicators among these metabolic chemical by-products. Because of the alkaline nature of these chemicals, they build up inside the headspace of the packaging product and change the pH of both the indicators attached to it [39]. With fewer additives and softer processing methods (such as pasteurization rather than sterilization), packaging's preservation role as the last line of defense before seafood is distributed and stored gains importance. The use of MAP and vacuum has enhanced the shelf life of seafood [40].

10.3.5.3 Milk and Dairy Products

For vegans in particular, milk products like butter, cheese, and yogurt are necessary foods. Many different cattle species produce milk, and each of the milk has a different makeup. For instance, cow milk has 89% water, 5% carbs, 4% protein, 3% fat, and the remaining vitamins and minerals. On the market, each of these things comes in a variety of forms with various features. The most difficult problem for packaging researchers is, therefore, packing dairy products. Most dairy products are transported, handled, and kept chilled by using heat-sealable packaging material. As a result, IP is critical in determining the shelf life and quality of the milk during allocation and consumption. The pH of milk dropped from 6.7 to 4.6 because the temperature exceeded its freezing point [41]. A number of critical nutrients can be found in milk and other dairy-based meals. A dairy product's quality, safety, cost, and even marketing are all greatly impacted by its packaging. Dairy products need to be protected from physical harm, environmental contaminants, changes in scent, and moisture loss by packaging techniques. Due to the great perishability of dairy products, especially pasteurized milk, packaging must maintain its protective qualities throughout the whole supply chain until the product is consumed by the consumer. Aside from providing consumers with information about the product's full history, IP is one of the many types of packaging that may be used to preserve dairy goods [42].

10.3.6 Leak Indicators

Food quality indicators, such as carbon dioxide and oxygen levels, can be used as leak indicators. They can be used to detect leaks or to test the efficiency of an oxygen scavenger, for example. Some sort of chemical or enzymatic process causes the color of many of these markers to shift. Methylene blue, which changes color as a result of an oxidation–reduction reaction, is the stain that is most frequently used to detect leaks [43]. The exclusion of oxygen (O_2) is frequently preferred in modified-atmosphere packaging (MAP) because O_2 is necessary for aerobic organisms that spoil like fungi and can induce rancidity via oxidation and color change. The most significant antibacterial effects of the altered atmosphere are brought on by inert nitrogen and carbon dioxide, which are routinely employed in packaging, and the inert nitrogen atmosphere. This happens by slowing the growth rate of the bacteria and lengthening the lag period of their development. Low O_2 levels (0–2%), together with high CO_2 levels (20–80%), are typical of modified-environment packaging for non-respiring food. For instance, a high CO_2 concentration degrades the fresh vegetable's quality. As a result, the modified atmosphere needs to be tailored to each product [44].

Inside a package with a high CO_2 concentration, leakage often results in an increase in O_2 concentration and a decrease in CO_2 concentration. However, it is important to keep in mind that microbial growth, which is likely to happen after a leak, will lower the O_2 concentration and raise the CO_2 concentration, aggravating the situation. Leakage frequently leads to microbiological contamination in addition to the loss of the protective altered environment caused by the package's lack of integrity [45].

10.3.6.1 O_2 Indicators

Most indicators require an O_2 concentration of –0.1% to cause a hue change. Due to the difficulty in handling an indicator caused by this exceptional sensitivity, it is necessary to provide an anaerobic condition both at the time of the manufacturing of the indicator and during the packaging process. During the packaging process, residual O_2 is frequently trapped in modified-atmosphere packages, with a typical concentration of 0.5–2.0%. The indication is protected from oxidation during shipping by being enclosed in a film material. The majority of the abovementioned O_2 indicators have reversible color changes that occur in response to ambient O_2 concentrations. This reversal is useful for ensuring the indicator works properly when it is used to package O_2-sensitive products using an O_2 absorber. The presence of reducing substances in the packaged product, such as reducing sugars, ascorbic acid, and various inorganic salts, may also result in the dye's re-reduction and reversible color shift, producing an unreliable outcome [46].

10.3.6.2 CO_2 Indicators

Food items are frequently packaged under MAP conditions to reduce the rate of microbial development and ensuing spoilage. In order to create a protective atmosphere surrounding food within a pack and prevent or minimize oxygen content, carbon dioxide (CO_2) gas can be used 100% of the time, or it can be used in different combinations with other gases like nitrogen (N_2) and oxygen (O_2). This helps extend the shelf life of food by inhibiting aerobic spoilage microbes. Additionally, it has been proposed that oxygen plays a significant role in the lipid oxidation process, which degrades the quality of meat. Because it lowers microbial metabolic rates in the presence of oxygen, carbon dioxide is regarded as an active packing gas [47]. The composition of the protective environment varies according to the type of food and the stage of delivery. Leakage detection is a crucial component of MAP technology since the integrity of the package ultimately determines the quality of food packaged in MAP. CO_2 concentrations can be used to determine the freshness and safety of food in modified-atmosphere food packaging. Several by-products are generated and accumulate inside a sealed container when food begins to degrade under standard environmental packing circumstances. Consequently, it is potentially feasible to identify spoilage by looking for one or more of these by-products. When food deteriorates, it frequently produces pressure, heat, acidity, and carbon dioxide as by-products [48]. CO_2 sensors can be broadly classified into two types: optical and electrochemical, depending on the kind of transducer used. There are three other subcategories of electrochemical CO_2 sensors: conductometric, amperometric, and potentiometric. Conventionally, commercial gaseous CO_2 detection uses non-distributive infrared (NDIR) sensors and Severinghaus-type sensors. Despite having a high degree of precision, the NDIR instrument is costly, bulky, susceptible to contamination, and affected by water vapor. In summary, food packaging indication applications can benefit from the use of optical carbon dioxide gas sensors. Dry optical sensors with pH-sensitive dye indicators in particular are more appropriate for use as spoiling indicators in food packaging applications and/or for CO_2 monitoring [49].

10.3.6.3 Indicators of Microbial Growth

Even if it is challenging to detect microbial development in modified-atmosphere packages that already include a high concentration of CO_2, the microbial growth indicator method can be used to identify microbial contamination in different types of items. The increase of CO_2 in a package headspace is regarded as a marker of microbial growth because CO_2 is also created during microbial metabolism. A package leak (reduction in CO_2) is frequently followed by microbial growth (resulting in an increase in CO_2); consequently, in the worst-case state, the concentration of CO_2 will stay consistent despite leakage and microbial degradation [50]. Due to the distinct physiologies of pathogenic bacteria, antimicrobial drugs exhibit varying activity against them. To preserve the product's quality and safety, which results in an extended shelf life, antimicrobial agents are either integrated directly into food particles or added to packaging materials where they are released gradually. Microorganism characterization is a valuable tool in the selection of antimicrobial agents. This can include the makeup of the cell walls (both Gram-positive and Gram-negative), the oxygen requirements of aerobes and anaerobes, the growth stage (spores and vegetative cells), resistance to acid and osmosis, and the best-growing temperatures (mesophilic, thermophilic, and psychoactive). Certain antimicrobial substances block bacteria's vital metabolic pathways as well as parts of their cell walls and membranes. Thus, antimicrobial compounds serve two crucial purposes: they have antimicrobial cidal and antimicrobial static actions. In order to preserve the antimicrobial substance's ability to block the development of target microorganisms for the duration of the storage period or shelf life, it must have the active function of keeping the concentration above the limit of inhibition during microbial static effects [51].

10.3.7 Thermochromic Ink

There are inks on the market that react to temperature by changing color. The consumer can be given a message based on the color of the ink they are viewing when these inks are printed on packages or labels. Consumers can use thermochromic inks to determine whether a package is too hot to handle or contains a cool enough beverage. Drink packaging is increasingly using thermochromic inks. Consumers should not completely rely on the message of the inks when determining the right time to consume a meal because the inks employed can be negatively impacted by UV light and temperatures exceeding 121 °C [52]. The most recent advancement in IP that can increase user convenience is temperature-sensitive thermochromic ink. Labels or containers that have been heated or cooled before use can be printed with thermochromic ink, which indicates the appropriate temperature range for using the product. Thermochromic inks can change color in a way that is either irreversible or reversible. Once irreversible thermochromic ink has changed its color as a sign of a temperature change, the change is irreversible and cannot be reversed. This ink is invisible until it is subjected to a specific temperature. When heated or cooled, reversible thermochromic ink changes color; upon returning to the original temperature, it takes on its original color [53].

10.4 Sensors

An analytical tool that locates or quantifies issue signals or visual cues to gauge chemical or physical qualities is referred to as a sensor. Gas sensors, oxygen sensors based on fluorescence, and biosensors are some common sensor types that advance technologies for judging food quality. When a gas is present, gas sensors can react by changing their physical restrictions, which can then be verified by an outside instrument. Prior research revealed that the application of gas sensor technology could be a quick and/or conventional substitute for determining the rancidity of meat products and for the use of biosensors to identify carbamate insecticides in fruits and vegetables [54].

10.4.1 Gas Sensors

During the microbiological deterioration of food goods, gases are generated that give off rancid odors. In order to indicate food quality by sensing gas composition, a variety of food sensors were created for use in packing as well as processing. Carbon dioxide and oxygen are the most common markers for gas-targeting food sensors, followed by BAs and volatile organic compounds (VOCs). In high concentrations, carbon dioxide, a by-product of bacterial metabolism, is used as a preservative in food packaging in changing environments. Using pH indicator tactics, carbon dioxide can easily be detected as it produces carbonic acid in solution. Polypeptide (poly-L-lysine, PLL), anthocyanins, and an amino acid (L-lysine) were used to create a CO_2 sensor. Once exposed to CO_2 (2.5%), the sensor's color changes from azure to purple over the course of five minutes. A paper-based device that has been enhanced by using a hydrophilic silicon spray to arrange the array and direct sample flow for dissolved CO_2 and NH_3 detection utilizing pH-responsive dyes [55]. In a novel method, create an opaque (pH 7.0) and transparent (pH 6.5) CO_2 sensor by taking advantage of chitosan's pH-dependent solubility in water. Like how carbon dioxide concentration may be used to determine food quality and package integrity, oxygen content in food packaging can do the same. High oxygen levels in modified-environment packaging point to a leak, rapid food degradation, and a reduction in the anticipated shelf life of a product. As a viable food packaging detection approach, irreversible gas absorption indicators alter color in reaction to oxygen [56].

10.4.2 Humidity Sensors

The quality and shelf life of food have been connected to the humidity at which it is stored, making it a target for numerous food sensors. Furthermore, temperature and humidity influence the ability of foodborne bacteria to survive. By using packing, humidity may be more easily regulated than temperature, and its impact on food is slower and less harmful. As a result, humidity sensors are rarely used to monitor the storage conditions of individual servings of food and are more frequently employed to do so. Humidity sensors can be made using a variety of techniques, and they typically monitor the optical or electrical properties of various materials, such as organic polymers, ceramics, and hybrid composites, that are affected by moisture [57]. Recent developments have led to the creation of humidity sensors using biopolymer-based transducers that are safe for food, which may make it easier to integrate them into food packaging. Incorporating humidity sensors into clothing is a promising application for systems with multiplexes. The equipment makes use of field-effect transistors that are chemically sensitive and built on silicon that are embellished with sensing components (silver, ruthenium, and silicon oxide) that generate a sensitive and focused electronic response to the desired gases. The sensors are assembled into a specially manufactured silicon-based printed circuit board to form a portable device, and the authors demonstrated the sensor's efficacy by keeping track of the deterioration of the meat [58, 59].

10.5 Radio Frequency Identification

RFID is a well-established technology that enables the identification of things via wireless transmission in a specific frequency band. The tag, the reader, and the software are the three key elements of any RFID system. The tag is an integrated circuit that has an electronic product code (EPC), a special tracking identifier [60]. The reader receives the broadcast signal and establishes a network connection between the system software and the tag data. The entire system is managed by the software. RFID tags can be grouped into three categories based on how they are used for tracking and tracing. Passive tags, which are located in the reader's interrogation range, are battery-free and run entirely on the reader's power. A battery is incorporated into active tags to extend their range for data collection, tag-to-tag communication, etc. These, however, cost a lot more money. Batteries are employed in semi-active tags to store data and memory backups [61, 62]. The use of RFID is to keep information like temperature and relative humidity measurements, dietary statistics, and recipe directions. Some RFID tags are used to support operations and are helpful for inventory control, traceability, labor cost savings, and promoting quality and safety as the tagged objects move through the supply chain. The application of RFID tags to a variety of perishable items that must be shipped under tightly regulated circumstances.

10.6 3D Printing and Intelligent Packaging

In order to create customized food classes that meet the various consumer needs and have adequately complex structures, 3D printing technology has been widely applied in the field of food processing, with positive effects on both the economy and the environment. In essence, 3D technology offers an engineering solution for the creation of customized, personalized provisions and makes it easier to create food products. It may also be used to reconfigure the machinery in the supply chain for customized foods [63]. Additionally, the ability to analyze meat freshness degradation in real time is made possible using 3D printing technology; the pH of meat may be tested to identify its freshness. As a result, we may create composite materials employing chemicals like anthocyanins that are pH-sensitive and may indicate a pH change. The creation of food packaging film utilizing 3D printing technology can not only yield speedy results when determining the current state of meat but can also be used in the manufacture of packaging film for protecting food from air [64]. 3D printing is an appealing new technology with several benefits for the food industry. The idea of 4D printing is an expansion of 3D printing, and merging 3D printing with indications is an innovative idea. By adding components that change color in response to external stimuli, the process of 3D printing could modify the hue of food by transforming its physical and chemical properties, making it simpler for consumers to assess the freshness and perishability of food [65]. Notably, fish products could be packaged using composite material-based food packaging films that are 3D-printed. Given the issues with fish products, including parasites, spoilage, and dehydration, it would be viable to add antimicrobial compounds or ingredients that hold water as composites to improve the product's marketability. Furthermore, compared to control films, the food packaging films made with 3D printing and chitosan not only had improved antibacterial capabilities but also had better elasticity and elongation, indicating that there are still more materials awaiting research into their suitability for use in creating biocomposites. In general, 3D printing is still expensive for huge applications, particularly in low-profit industries [66]. However, new opportunities for intelligent food packaging are being supported by the present development of unique 3D printing technology. Such 3D printing technologies are widely employed in the food industry, and the fusion of 3D printing technology and indications with the food industry could become more prevalent in the future, providing a roadmap for further food research and development [67].

10.7 Legal Aspects of Intelligent Packaging

Emerging food packaging techniques are developing in response to consumer demands or industrial production trends toward moderately preserved, fresh, flavorful, and accessible food products with increased shelf lives while maintaining high standards. The food packaging industry faces numerous challenges because of market globalization, which has led to longer food distribution and spent very little time cooking and purchasing fresh food. These factors have also acted as catalysts for the introduction of novel and enhanced packaging ideas for increasing shelf life by maintaining and observing the quality and safety of food. The aim of these conventional food packaging items is that the food should be as inert as possible, which means there should be as little contact as possible between the food and the packaging. However, in recent years, "active and intelligent" (A&I) packaging, which is focused on intentional interactions with the food or the food environment, has become one of the most creative developments in the field of food packaging. "AP" is meant to maintain or even enhance the food's quality while also extending the food's shelf life. While the goal of "IP" is to monitor and indicate the food's level of freshness, recent studies on the use of A&I packaging methods for food have been extensively covered in scientific journals and reviews. The current article's objectives focused on providing a general historical perspective on major categories and concepts, worldwide and emerging ideas, legal considerations, and compliance examinations of cutting-edge methods for storing food.

10.8 Conclusion and Perspective

Food packaging has undergone innovation with the introduction of IP, which enables customers to check the food being sold for quality. With the presence of a wide variety of food and agricultural goods, many applications of the development of IP systems are not yet finished. The IP systems created have a gap between the research stage and the commercialization stage. To make IP systems commercially feasible, future research must consider a few obstacles. The complexity of food systems and wide variations in physical and chemical components limit the capacity to commercialize intellectual property (ILP) systems established on an in vitro or laboratory scale in constrained developmental settings. It is anticipated that single IP systems will be combined to provide extra information regarding the food product standard parameters in a cheap, small-scale machine. Collaboration in these disciplines of mechanical, food, and electronics engineering is particularly beneficial considering future successful IP developments. The functionality of IP systems can be enhanced by the combination of several sensors, making it possible to track intricate product deterioration. In this situation, it is important to carefully consider how the sensors' functionality interfered. The current IP devices should be replaced with more sophisticated systems. Because of complex manufacturing processes or significant expenses for raw materials, IP methods are expensive today, making the development of less expensive IP systems vital. Therefore, it is necessary to increase the technological and economic viability of such novel systems. Thus, future research can concentrate on the fusion of AP and IP to provide a complete smart packaging system for the application of actual food materials during distribution, storage, and consumption. IP devices could keep an eye on the food's quality or circumstances and make decisions regarding the AP device's corrective activities. Many foods can use the combined IP–AP technique, but more research is required to develop models and identify the unique IP–AP design for each food product. It is possible to determine whether a good is suitable for the consumer by looking into how AP and IP technologies affect the food's sensory attributes as it is packaged. Food packaging has a lot of potential for synergistic AP–IP techniques; however, the final price of systems that are economically feasible must be considered.

References

1 World Health Organization (2023). Food safety.

2 Yam, K.L. (2012). Intelligent packaging to enhance food safety and quality. In: *Emerging Food Packaging Technologies* 137–152. Woodhead Publishing. https://doi.org/10.1533/9780857095664.2.137.

3 Kalpana, S., Priyadarshini, S.R., Leena, M.M. et al. (2019). Intelligent packaging: trends and applications in food systems. *Trends in Food Science & Technology* 93: 145–157. https://doi.org/10.1016/j.tifs.2019.09.008.

4 Han, J.H. (ed.) (2005). *Innovations in Food Packaging*. Elsevier. https://doi.org/10.1016/B978-012311632-1/50041-3.

5 Vanderroost, M., Ragaert, P., Devlieghere, F., and De Meulenaer, B. (2014). Intelligent food packaging: The next generation. *Trends in Food Science & Technology* 39 (1): 47–62. https://doi.org/10.1016/j.tifs.2014.06.009.

6 Mirza Alizadeh, A., Masoomian, M., Shakooie, M. et al. (2021). Trends and applications of intelligent packaging in dairy products: a review. *Critical Reviews in Food Science and Nutrition* 62 (2): 383–397. https://doi.org/10.1080/10408398.2020.1817847.

7 De Jong, A.R., Boumans, H., Slaghek, T. et al. (2005). Active and intelligent packaging for food: is it the future? *Food Additives and Contaminants* 22 (10): 975–979. https://doi.org/10.1080/02652030500336254.

8 Müller, P. and Schmid, M. (2019). Intelligent packaging in the food sector: a brief overview. *Foods* 8 (1): 16. https://doi.org/10.3390/foods8010016.

9 Pavelková, A. (2013). Time temperature indicators as devices intelligent packaging. *Acta Universitatis Agriculturae et Silviculturae Mendelianae Brunensis* 61 (1): 245–251. https://doi.org/10.11118/actaun201361010245.

10 Zhang, C., Yin, A.X., Jiang, R. et al. (2013). Time–Temperature indicator for perishable products based on kinetically programmable Ag overgrowth on Au nanorods. *ACS Nano* 7 (5): 4561–4568. https://doi.org/10.1021/nn401266u.

11 Mohammadian, E., Alizadeh-Sani, M., and Jafari, S.M. (2020). Smart monitoring of gas/temperature changes within food packaging based on natural colorants. *Comprehensive Reviews in Food Science and Food Safety* 19 (6): 2885–2931. https://doi.org/10.1111/1541-4337.12635.

12 Wang, S., Liu, X., Yang, M. et al. (2015). Review of time temperature indicators as quality monitors in food packaging. *Packaging Technology and Science* 28 (10): 839–867. https://doi.org/10.1002/pts.2148.

13 Hazarika, K.K., Konwar, A., Borah, A. et al. (2023). Cellulose nanofiber mediated natural dye based biodegradable bag with freshness indicator for packaging of meat and fish. *Carbohydrate Polymers* 300: 120241. https://doi.org/10.1016/j.carbpol.2022.120241.

14 Liu, Y., Li, L., Yu, Z. et al. (2023). Principle, development and application of time–temperature indicators for packaging. *Packaging Technology and Science* 36 (10): 833–853. https://doi.org/10.1002/pts.2761.

15 Cheng, H., Xu, H., McClements, D.J. et al. (2022). Recent advances in intelligent food packaging materials: principles, preparation and applications. *Food Chemistry* 375: 131738. https://doi.org/10.1016/j.foodchem.2021.131738.

16 Ellouze, M. and Augustin, J.C. (2010). Applicability of biological time temperature integrators as quality and safety indicators for meat products. *International Journal of Food Microbiology* 138 (1-2): 119–129. https://doi.org/10.1016/j.ijfoodmicro.2009.12.012.

17 Ye, B., Chen, J., Ye, H. et al. (2022). Development of a time–temperature indicator based on Maillard reaction for visually monitoring the freshness of mackerel. *Food Chemistry* 373: 131448. https://doi.org/10.1016/j.foodchem.2021.131448.

18 Firouz, M.S., Mohi-Alden, K., and Omid, M. (2021). A critical review on intelligent and active packaging in the food industry: research and development. *Food Research International* 141: 110113. https://doi.org/10.1016/j.foodres.2021.110113.

19 Lee, E.J. and Shin, H.S. (2019). Development of a freshness indicator for monitoring the quality of beef during storage. *Food Science and Biotechnology* 28: 1899–1906. https://doi.org/10.1007/s10068-019-00633-5.

20 Becerril, R., Nerín, C., and Silva, F. (2021). Bring some colour to your package: freshness indicators based on anthocyanin extracts. *Trends in Food Science & Technology* 111: 495–505. https://doi.org/10.1016/j.tifs.2021.02.042.

21 Kuswandi, B., Wicaksono, Y., Abdullah, A. et al. (2011). Smart packaging: sensors for monitoring of food quality and safety. *Sensing and Instrumentation for Food Quality and Safety* 5: 137–146. https://doi.org/10.1007/s11694-011-9120-x.

22 Vasile, C. and Baican, M. (2021). Progresses in food packaging, food quality, and safety—controlled-release antioxidant and/or antimicrobial packaging. *Molecules* 26 (5): 1263. https://doi.org/10.3390/molecules26051263.

23 Dodero, A., Escher, A., Bertucci, S. et al. (2021). Intelligent packaging for real-time monitoring of food-quality: current and future developments. *Applied Sciences* 11 (8): 3532. https://doi.org/10.3390/app11083532.

24 Luo, Q., Hossen, A., Sameen, D.E. et al. (2023). Recent advances in the fabrication of pH-sensitive indicators films and their application for food quality evaluation. *Critical Reviews in Food Science and Nutrition* 63 (8): 1102–1118. https://doi.org/10.1080/10408398.2021.1959296.

25 Bumbudsanpharoke, N., Kwon, S., Lee, W., and Ko, S. (2019). Optical response of photonic cellulose nanocrystal film for a novel humidity indicator. *International Journal of Biological Macromolecules* 140: 91–97. https://doi.org/10.1016/j.ijbiomac.2019.08.055.

26 Eilamo, M., Kinnunen, A., Latva-Kala, K., and Ahvenainen, R. (1998). Effects of packaging and storage conditions on volatile compounds in gas-packed poultry meat. *Food Additives & Contaminants* 15 (2): 217–228. https://doi.org/10.1080/02652039809374633.

27 Tholl, D., Hossain, O., Weinhold, A. et al. (2021). Trends and applications in plant volatile sampling and analysis. *The Plant Journal* 106 (2): 314–325. https://doi.org/10.1111/tpj.15176.

28 Jiang, S. and Liu, Y. (2020). Gas sensors for volatile compounds analysis in muscle foods: a review. *TrAC Trends in Analytical Chemistry* 126: 115877. https://doi.org/10.1016/j.trac.2020.115877.

29 Ma, Q., Lu, X., Wang, W. et al. (2021). Recent developments in colorimetric and optical indicators stimulated by volatile base nitrogen to monitor seafood freshness. *Food Packaging and Shelf Life* 28: 100634. https://doi.org/10.1016/j.fpsl.2021.100634.

30 Romero, A., Sharp, J.L., Dawson, P.L. et al. (2021). Evaluation of two intelligent packaging prototypes with a pH indicator to determine spoilage of cow milk. *Food Packaging and Shelf Life* 30: 100720. https://doi.org/10.1016/j.fpsl.2021.100720.

31 Balbinot-Alfaro, E., Craveiro, D.V., Lima, K.O. et al. (2019). Intelligent packaging with pH indicator potential. *Food Engineering Reviews* 11: 235–244. https://doi.org/10.1007/s12393-019-09198-9.

32 Janjarasskul, T. and Suppakul, P. (2018). Active and intelligent packaging: the indication of quality and safety. *Critical Reviews in Food Science and Nutrition* 58 (5): 808–831. https://doi.org/10.1080/10408398.2016.1225278.

33 Alizadeh-Sani, M., Mohammadian, E., Rhim, J.W., and Jafari, S.M. (2020). pH-sensitive (halochromic) smart packaging films based on natural food colorants for the monitoring of food quality and safety. *Trends in Food Science & Technology* 105: 93–144. https://doi.org/10.1016/j.tifs.2020.08.014.

34 Cenci-Goga, B.T., Iulietto, M.F., Sechi, P. et al. (2020). New trends in meat packaging. *Microbiology Research* 11 (2): 56–67. https://doi.org/10.3390/microbiolres11020010.

35 Kim, D., Lee, S., Lee, K. et al. (2017). Development of a pH indicator composed of high moisture-absorbing materials for real-time monitoring of chicken breast freshness. *Food Science and Biotechnology* 26: 37–42. https://doi.org/10.1007/s10068-017-0005-6.

36 Gil, M. and Rudy, M. (2023). Innovations in the packaging of meat and meat products—A review. *Coatings* 13 (2): 333. https://doi.org/10.3390/coatings13020333.

37 Khodaei, S.M., Gholami-Ahangaran, M., Karimi Sani, I. et al. (2023). Application of intelligent packaging for meat products: a systematic review. *Veterinary Medicine and Science* 9 (1): 481–493. https://doi.org/10.1002/vms3.1017.

38 Abedi-Firoozjah, R., Salim, S. A., Hasanvand, S., Assadpour, E., Azizi-Lalabadi, M., Prieto, M. A., & Jafari, S. M. (2023). Application of smart packaging for seafood: a comprehensive review. *Comprehensive Reviews in Food Science and Food Safety* 22 (2), 1438–1461. https://doi.org/10.1111/1541-4337.13117

39 Pacquit, A., Frisby, J., Diamond, D. et al. (2007). Development of a smart packaging for the monitoring of fish spoilage. *Food Chemistry* 102 (2): 466–470. https://doi.org/10.1016/j.foodchem.2006.05.052.

40 Almeida, C., Loubet, P., da Costa, T.P. et al. (2022). Packaging environmental impact on seafood supply chains: a review of life cycle assessment studies. *Journal of Industrial Ecology* 26 (6): 1961–1978. https://doi.org/10.1111/jiec.13189.

41 Goodarzi, M.M., Moradi, M., Tajik, H. et al. (2020). Development of an easy-to-use colorimetric pH label with starch and carrot anthocyanins for milk shelf life assessment. *International Journal of Biological Macromolecules* 153: 240–247. https://doi.org/10.1016/j.ijbiomac.2020.03.014.

42 Rejeesh, C.R. and Anto, T. (2023). Packaging of milk and dairy products: approaches to sustainable packaging. *Materials Today Proceedings* 72: 2946–2951. https://doi.org/10.1016/j.matpr.2022.07.467.

43 Bai, Y. and Bai, Q. (2014). Leak detection systems. *Subsea Pipeline Integrity and Risk Management* 125–143. https://doi.org/10.1016/B978-0-12-394432-0.00006-8.

44 Smolander, M., Hurme, E., and Ahvenainen, R. (1997). Leak indicators for modified-atmosphere packages. *Trends in Food Science & Technology* 8 (4): 101–106. https://doi.org/10.1016/S0924-2244(97)01017-0.

45 Sari, A.N.A., Warsiki, E., and Kartika, I.A. (2021). Innovation of oxygen indicator for packaging leak detector: a review. In: *IOP Conference Series: Earth and Environmental Science*, vol. 749, No. 1, 012009. IOP Publishing. http://doi.org/10.1088/1755-1315/749/1/012009.

46 Lambrecht, A., Maier, E., Pernau, H.F. et al. (2017). Gas leak detection by dilution of atmospheric oxygen. *Sensors* 17 (12): 2804. https://doi.org/10.3390/s17122804.

47 Al Obaidi, A., Karaca, I.M., Ayhan, Z. et al. (2022). Fabrication and validation of CO_2-sensitive indicator to monitor the freshness of poultry meat. *Food Packaging and Shelf Life* 34: 100930. https://doi.org/10.1016/j.fpsl.2022.100930.

48 Hu, H., Bai, Y., Li, H. et al. (2023). An intelligent indicator for warning of short-term high CO_2 damage to garlic scapes. *Packaging Technology and Science* 36 (7): 557–567. https://doi.org/10.1002/pts.2727.

49 Mills, A., McDonnell, L., and Yusufu, D. (2023). Colorimetric CO_2 Indicators. *Accounts of Materials Research* https://doi.org/10.1021/accountsmr.2c00226.

50 Ghazi, C.J. and Marshall, J.S. (2014). A CO_2 tracer-gas method for local air leakage detection and characterization. *Flow Measurement and Instrumentation* 38: 72–81. https://doi.org/10.1016/j.flowmeasinst.2014.05.015.

51 Karanth, S., Feng, S., Patra, D., and Pradhan, A.K. (2023). Linking microbial contamination to food spoilage and food waste: the role of smart packaging, spoilage risk assessments, and date labeling. *Frontiers in Microbiology* 14: 1198124. https://doi.org/10.3389/fmicb.2023.1198124.

52 Arman Kandirmaz, E., Ozcan, A., and Er Ulusoy, D. (2020). Production of thermochromic microcapsulated inks for smart packaging and examination of printability properties. *Pigment & Resin Technology* 49 (4): 273–281. https://doi.org/10.1108/PRT-12-2019-0116/full/html.

53 Jamnicki Hanzer, S., Kulčar, R., Vukoje, M., and Marošević Dolovski, A. (2023). Assessment of thermochromic packaging prints' resistance to UV radiation and various chemical agents. *Polymers* 15 (5): 1208. https://doi.org/10.3390/polym15051208.

54 Mustafa, F. and Andreescu, S. (2020). Nanotechnology-based approaches for food sensing and packaging applications. *RSC Advances* 10 (33): 19309–19336. https://doi.org/10.1039/D0RA01084G.

55 Unander, T. and Nilsson, H.E. (2009). Characterization of printed moisture sensors in packaging surveillance applications. *IEEE Sensors Journal* 9 (8): 922–928. https://doi.org/10.1109/JSEN.2009.2024866.

56 Bowman, L. and Meindl, J.D. (1986). The packaging of implantable integrated sensors. *IEEE Transactions on Biomedical Engineering* 2: 248–255. https://doi.org/10.1109/TBME.1986.325807.

57 Lee, C.Y. and Lee, G.B. (2005). Humidity sensors: a review. *Sensor Letters* 3 (1–2): 1–15. https://doi.org/10.1166/sl.2005.001.

58 Chen, Z. and Lu, C. (2005). Humidity sensors: a review of materials and mechanisms. *Sensor Letters* 3 (4): 274–295. https://doi.org/10.1166/sl.2005.045.

59 Kulwicki, B.M. (1991). Humidity sensors. *Journal of the American Ceramic Society* 74 (4): 697–708. https://doi.org/10.1111/j.1151-2916.1991.tb06911.x.

60 Han, J.W., Ruiz-Garcia, L., Qian, J.P., and Yang, X.T. (2018). Food packaging: a comprehensive review and future trends. *Comprehensive Reviews in Food Science and Food Safety* 17 (4): 860–877. https://doi.org/10.1111/1541-4337.12343.

61 Juels, A. (2006). RFID security and privacy: a research survey. *IEEE Journal on Selected Areas in Communications* 24 (2): 381–394. https://doi.org/10.1109/JSAC.2005.861395.

62 Pala, Z. and Inanc, N. (2007). Smart parking applications using RFID technology. In: *2007 1st Annual RFID Eurasia*, 1–3. IEEE. https://ieeexplore.ieee.org/abstract/document/4368108.

63 Tracey, C.T., Predeina, A.L., Krivoshapkina, E.F., and Kumacheva, E. (2022). A 3D printing approach to intelligent food packaging. *Trends in Food Science & Technology* https://doi.org/10.1016/j.tifs.2022.05.003.

64 Ma, Y., Yang, W., Xia, Y. et al. (2022). Properties and applications of intelligent packaging indicators for food spoilage. *Membranes* 12 (5): 477. https://doi.org/10.3390/membranes12050477.

65 Zhou, W., Wu, Z., Xie, F. et al. (2021). 3D printed nanocellulose-based label for fruit freshness keeping and visual monitoring. *Carbohydrate Polymers* 273: 118545. https://doi.org/10.3390/molecules26123543.

66 Vasileiadis, A., Exarchos, D.A., Tsapara, G. et al. (2023). 3D-printed flexural sensor devices for smart packaging of agrifood products. In: *Nano-, Bio-, Info-Tech Sensors, and Wearable Systems 2023*, 12485: 88–93. SPIE. https://doi.org/10.1117/12.2663134.

67 Chen, S., Brahma, S., Mackay, J. et al. (2020). The role of smart packaging system in food supply chain. *Journal of Food Science* 85 (3): 517–525. https://doi.org/10.1111/1750-3841.15046.

11

Time–Temperature Indicators (TTIs) for Monitoring Food Quality

Kandeepan Gurunathan

Packaging Lab, ICAR-National Meat Research Institute, Hyderabad, Telangana, India

11.1 Introduction

The most important factor in maintaining food quality is temperature control in the supply chain. Changes in temperature at any time during transport and subsequent storage in the supply chain will cause food safety problems and reduce product acceptance or total rejection due to severe quality changes resulting from temperature disturbances. As hygiene and safety issues become more stringent in fresh and processed foods, it is important to establish a system to monitor temperature violations in the supply chain. Closely monitored temperature effects during distribution and optimal circulation at the retail level depending on the temperature of each product can lead to better quality control and a significant reduction in food waste than expiration dates that are often meaningless. Use-by date does not provide information about the individual condition of each package, because it does not report mishandling during distribution or differences in end-user performance. The time–temperature indicator (TTI) can be a potential tool to achieve this goal. Therefore, to check food spoilage in the supply chain, an efficient system that can monitor all changes in predetermined temperatures in the supply chain or a system that can detect the exposure of food products above a certain temperature is required.

Temperature disruptions in the supply chain can be tracked using time–temperature integrators (TTIs). A cheap and easy-to-install gadget, a time–temperature integrator (TTI) integrates all or a portion of the food product's cumulative temperature-time history. It can be mounted to the package's surface [1]. According to Giannakourou et al. [2], TTIs are straightforward, inexpensive instruments that can be readily measured by compiling the product's time–temperature history from the manufacturer to the customer. They can also display fluctuations in temperature and time that are dependent on both factors. Time–temperature integrators are useful as informational, control, and decision-making tools at every stage of the distribution chain. As soon as they are applied, they begin to respond. Over the course of their working lives, they progressively change color, achieving their end point – full-color change – in a predetermined window of time and temperature. TTIs,

Smart Food Packaging Systems: Innovations and Technology Applications, First Edition.
Edited by Avik Mukherjee, Santosh Kumar, Manjusri Misra, and Amar K. Mohanty.
© 2025 John Wiley & Sons Ltd. Published 2025 by John Wiley & Sons Ltd.

which are commonly defined as the visible reaction of color development or color movement associated with the shelf life of food goods at a specific temperature, make it simple for consumers to assess the quality of food [3]. The United States Department of Agriculture [4] states that certain microorganisms grow most quickly between 40 and 140 °F (4.5 and 60 °C), where they can double in 20 minutes. It is common to refer to this temperature range as the "Danger Zone." Foods kept longer than an hour at room temperature (above 90 °F/32 °C) ought not to be eaten [5]. High temperatures (above 21 °C) promote physical changes and chemical reactions that accelerate the deterioration of food quality, in addition to the growth of microbes and foodborne diseases [6]. According to Taoukis and Labuza [7], it is crucial to keep an eye on and document temperature conditions during distribution and storage.

Food pathogen kinetic models, data on internal product attributes, and the capacity to continuously monitor temperature history and time–temperature integrators that help reduce product losses, with high risk to customers, have all been combined recently in the development of safety management and assurance systems. TTIs are a great way to tell customers about the quality of foods that are packaged. Through color changes in TTI, it also aids in monitoring the food storage life of the product. These days, people focus less on checking and examining the finished product and more on tracking and managing critical variables throughout the life cycle of food products [1]. According to Nuin et al. [8], the time–temperature indicator (TTI) is a straightforward quality recording tool that can provide irreversible visual feedback for the purpose of monitoring and documenting crucial parameters at every stage of the food chain, such as production, storage, and internal storage.

One useful intelligent packaging (IP) tool that works well for a wide range of products is the time–temperature indicator [9]. Although smart packaging has a place in supply chain management, commercial applications have not made much progress in implementing it. Commercial development in the United States and Europe did not start until the mid-1990s, even though the initial concept originated in Japan in the 1970s. In developed nations like Europe and North America, industries like milk, food, and meat have begun to use this label. However, in developing nations like Asia Pacific, the Middle East, and Africa, there is very little knowledge of this technology, which is a major problem and limits market growth. Over the course of the forecast period, the Asia Pacific market is anticipated to grow at the fastest rate. According to Coherent Market Insights [10], the Asia Pacific market is anticipated to reach $44.5 million by 2028.

The rise in retail stores, e-commerce companies, and supermarkets in developing nations like China and India is the cause of this growth. For instance, the India Brand Equity Fund (IBEF) projects that the e-commerce market will reach $64 billion by 2020, up from $38.5 billion in 2017. The e-commerce sector in India is expanding. By 2034, the United States is predicted to rank second globally in terms of e-commerce sales. The region's need for temperature control technology, IP, and cold storage is therefore anticipated to rise as a result. The packaging industry is adopting time–temperature indicators more and more to preserve the quality of prepared foods, meat, poultry, and seafood. Coherent Market Insights' analysis indicates that the region sold 2.091 billion food packaging units in 2014, and by 2019, that number is predicted to rise to 2.578 billion units. The market for time–temperature indicators is therefore anticipated to grow faster as a result of the rise in packaged food products. The 3M Company, Temptime Corporation, Thin Film Electronics ASA, DeltaTrak, Inc., NiGK Corporation, Freshpoint Quality Assurance Ltd., LCR

Hallcrest, LLC, CCL Industries Inc., Biosynergy, Inc., and American Thermal Instruments are major companies in the global temperature indicator label market. TTIs are used to display the temporal and temperature history of frozen and chilled foods, including meat and poultry products [11], seafood [12], and seafood products. TTIs have been used to assess food products [13], dairy products [14], fruits and vegetables [15], pasteurization and sterilization processes, and the amount of time that fungi have left on their shelf [16].

11.2 Types of TTIs

TTI can be categorized into three types based on its mode of operation: biological, chemical, and physical systems [17]. Practically speaking, two or three systems – physical and biological – have served as the foundation for the development of numerous TTIs. Depending on the response mechanism, TTIs can be categorized as either complete history or partial history indicators. Three primary groups comprise the time–temperature indicator system (TTI). Three indicators have been identified: critical temperature, critical temperature time, and full history [7]. CTIs show the impact of temperature variations from a reference value. An additional significant event for which CTI will be helpful is the evolution of a crucial protein above the critical temperature or the development of a pathogenic microorganism. When temperature differences from reference values occur over a specific period, CTIs can reveal information about food impairment and changes in its organoleptic properties. This sensor makes use of an indicator's color change that is both visible and irreversible. A significant temperature-sensing system, for instance, is depicted in Figure 11.1a. It is based on an enzyme reaction that, when exposed to heat, changes color (orange color indicates product degradation) [18].

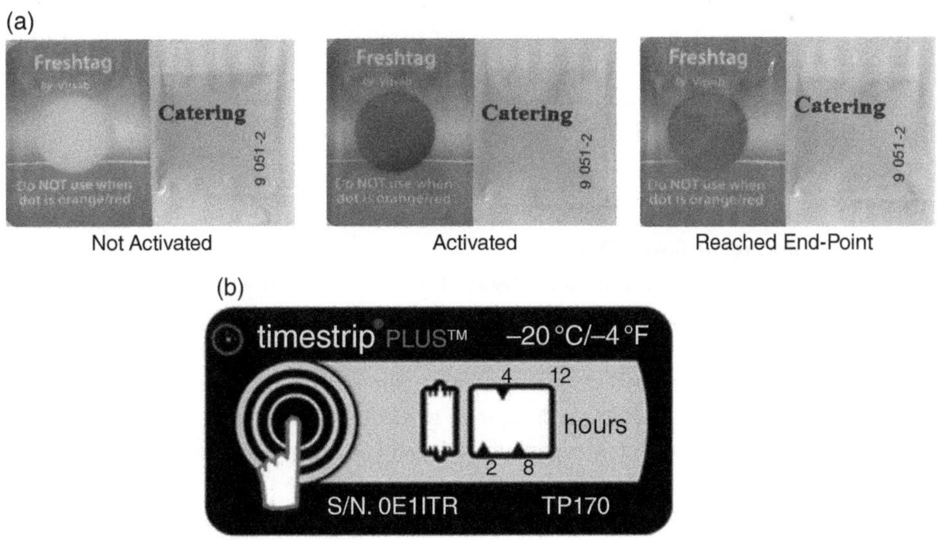

Figure 11.1 (a) An enzymatic reaction-based critical temperature-sensing system that can detect product degradation. *Source:* Vitsab International AB/https://vitsab.com//last accessed 12 April 2024 [18]. (b) The temperature-sensing system's critical time function indicates the duration of the disturbance to the threshold temperature. *Source:* With permission of Timestrip [19].

When a product is exposed to a temperature above the reference level for a predetermined amount of time, the critical time–temperature indicator provides an indication of the cumulative effect of both time and temperature on the product's quality or safety. An effective tool for identifying bottlenecks in products and supply chains where critical reactions to safety or quality are measured or happen above critical temperatures is the critical temperature/time integrator (CTTI). Microbial development and enzyme activity sustained below a critical temperature are two examples of these reactions. For instance, a commercially significant TTI system for tracking frozen food is shown in Figure 11.1b. The window turns blue if there is a change in the package's temperature. Together with the timestamps, the length of one or more interruptions shows how much blue has changed [19]. Consequently, all past data is documented regarding permanent modifications that elucidate food safety and quality for customers while also persistently tracking variations in sample temperature. Throughout the life of the product, they have exhibited a constant, temperature-dependent response. Because of their small size, inexpensive cost, and ease of use in comparison to other food monitoring devices, they are typically the only sensors used in the food industry. The alteration must be permanent, ideally comparable to or easily related to the rate of food deterioration and shelf life, according to Taoukis and Labuza [1]. Additionally, they reported that all TTIs had calculated activation energies between 10 and 40 kcal/mol and an Arrhenius temperature dependence.

11.2.1 Chemical TTIs

11.2.1.1 Polymerization-Based TTIs

The foundation of polymer-based systems, like Fresh-Check TTI by Temp Time, is the solid polymerization of thinly coated, colorless acetylene monomers, which at a temperature-dependent rate transform into clear-colored polymers [8]. Polymerization-based TTIs are used to observe the quality transition and foresee the shelf life of food products. These TTIs function by altering color based on a polymerization reaction. The faster the quality deteriorates or temperature escalates, the faster the TTI color changes. The polymer-based TTIs need more activation energy to transform color than other commercially existing TTIs.

11.2.1.2 Photochromic-Based TTIs

Photochromic-based TTIs are utilized to observe the food under preservation. These TTIs are generally used on food products to specify exposure to undue temperature and time during storage. OnVu™ TTI B1 is an irreversible TTI based on a photochromic reaction. OnVu™ TTI, a solid-state reaction system based on photosensitive compounds like benzylpyridines, is produced by Ciba. They turn colored when exposed to a certain wavelength of light, and as the temperature rises, this colored state eventually returns to its original colorless state [12].

11.2.1.3 Redox Reaction-Based TTIs

A redox reaction-based TTI modifies color through a redox or light-induced reaction. This type of TTI reaction system contains a colorimetric redox material (leucomethylene blue) and a redox reaction repressor (L-ascorbic acid/L-cysteine). It has been reported that in alkaline, solvent, or moisture barrier environments, redox-based TTIs contain oxidized redox dyes and significant amounts of reducing agents. In a sealed chamber, oxidized redox dyes undergo reduction to their reduced forms in response to oxygen flux, which is influenced by storage temperature.

11.2.1.4 Diffusion-Based TTIs

The basis of diffusion-based TTI, like the 3M-marketed Monitor Mark TTI, is the temperature-dependent diffusion reaction of colored fatty acid esters on premium, porous paper. The distance from the forward diffusion front's origin is the measurable response [3]. Additional patents focus on the use of bacteria or enzymes that exhibit color-changing properties through cracking after freezing and thawing and releasing liquid; spreading or mechanical deformation of oil because of temperature rise or freezing and thawing; two-color cards that blend their colors when melted, etc. In order to track the microbiological quality of non-pasteurized angelica (NPA) juice in response to temperature fluctuations, a time–temperature indicator (TTI) based on the diffusion of isopropyl palmitate (IPP) was created [20]. After 12.5 hours of storage at 25 °C, the aerobic number in NPA juice rose from the initial load of 3.7–6 log CFU/ml. The TTI system's IPP 7.0 mm prevalence is thought to represent the highest point for the quality of NPA juice bacteria.

11.2.1.5 Nanoparticle-Based TTIs

Metal nanoparticles vary in size, shape, and surface morphology depending on the exposure time and temperature conditions. When exposed to a specific temperature for a predetermined amount of time, nanoparticles exhibit an irreversible color change, which makes them a valuable tool for temperature-time sensors. A temperature history indicator (THI) based on gelatin and gold nanoparticles (AuNP) has been created; it shows a light-colored signal for six hours following exposure to 30 °C [21]. The length of exposure to the target temperature determines how strong the color signal is. Furthermore, the peak color intensity of AuNPs was observed at a concentration of 2% gelatin. Nevertheless, compared to alginate/AuNP-based THI, gelatin/AuNP-based time–temperature indicators are specifically made for low-temperature storage and offer several benefits, such as reduced sensitivity to color change, more restricted temperature control, and less preparation required than THI matrix. This characteristic makes plasmonic THI more useful and practical for end use. It does this by taking advantage of the surface plasmon resonance of AuNPs synthesized in alginate, which can be converted into a solid hydrogel by adding dissociated calcium ions.

11.2.2 Enzymatic TTIs

The food industry uses a few TTI enzymes. The first enzyme indicator was created by Ramstad and Volz [22] using the reaction of an oxidative enzyme with a colorless phenolic compound. An earlier commercial product is called I-Point [23]. This is predicated on the color shift brought on by the pH dropping as a result of the lipid substrate's regulated enzymatic hydrolysis. The lipid substrates and lipase are in two distinct compartments prior to activation, one of which has an open window. The barrier separating them dissolves when they are activated, allowing the enzyme and substrate to mix and causing pH fluctuations as well as the onset of color change when an additional pH indicator is present. The term "TTI" refers to a class of enzymes that have been developed recently. These include amylase TTI [24, 25], polyphenol oxidase TTI [26], glucoamylase TTI [27], and laccase TTI [28]. According to Kerry et al. [3] and Tsironi et al. [12], enzyme systems like VITSAB Check Point TTI are based on TTI color changes that result from a pH drop brought on by regulated enzymatic hydrolysis of lipid substrates. This causes the chromatic indicator to turn yellow–red.

Fungi, higher plants, and certain bacteria contain laccase [29, 30]. Interestingly enough, laccase can stain a variety of dyes with substrate specificity, including diamines, aromatic amines, phenolic compounds, benzenediol, and even some inorganic compounds like iodine [31]. Laccases can sometimes enhance color formation when they contain diazo derivatives of guaiacol, hydroquinone, vanillic acid, tannic acid, and syringaldazine [32]. Laccase is used in the food industry to enhance or alter the color of food and beverages to indicate safety [33]. The enzyme will benefit from laccase application in several ways, including the potential to produce TTI with various activation energies. With a simple catalytic mechanism, laccase is safe and environmentally friendly. Laccase uses sodium acetate, molecular oxygen, and enzyme stabilizer and develops color by the oxidation of guaiacol. It can oxidize guaiacol in a very simple reaction mixture that turns brown. The color response seems to be only affected by the amount of enzyme added to TTI. In this case, the color of the reaction mixture changes to a colorless mixture due to the oxidation of the phenolic hydroxyl group in the aromatic monophenolic structure of guaiacol, which causes absorption of light at a wavelength of 470 nm [34]. At 4, 10, 15, 25, and 37 °C, TTI needed to reach a hue value of 50 in the following times: 28.9, 14.4, 9.6, 5.8, and 3.9 [28].

The amylase enzyme finds extensive application in various industries such as baking, detergents, drain cleaners, natural sweeteners, pharmaceuticals, and animal feed [35]. The hydrolysis of α-1,4-glucosidic bonds in α-1,4-glucans, including starch, glycogen, and dextrin, is catalyzed by α-amylase. According to Pandey et al. [36], *Bacillus licheniformis, Bacillus amyloliquefaciens, Bacillus subtilis,* and *Aspergillus oryzae* are the primary sources of microbial α-amylase used in industry. Depending on the level of starch polymerization, clathrate compounds can be formed by iodine dyes and starch. This compound can be broken down by amylase into a colorless product. The temperature of the reaction and the enzyme concentration are the two main factors affecting the rate of this reaction. As a result, the amount of amylase and substrate can be adjusted to alter the indicator's color change time, which may work with various nutrients that have varying retention times. Additionally, he stated that this reaction's activation energy falls within the range of activation energies of a few significant food degradation reactions. Temperature-time integrator development is thus the best application for it. A glucoamylase-type TTI was created by Qian et al. [37], in which glucoamylase catalyzes the hydrolysis of dextrin and an iodine solution serves as an indicator.

Few studies have been done on the creation of urease-based temperature-time indicators [38]. The naturally occurring group of enzymes known as urease is found in bacteria, fungi, plants, and animals. Its ability to hydrolyze urea into carbamic acid and ammonia raises the pH of the urea solution. The urea solution will turn yellow to red if the pH is right, for instance, if phenol red is present. Anbukarasu and Sauvageau [39] developed a method of directly altering color by using the depolymerase enzyme to break down polyhydroxybutyrate (PHB)-dyed film. Enzymes break down the PHB film, releasing the dye into the solution, which then goes through an optically transparent transition. According to kinetic analysis, the dye was released in six hours at 37 °C, and the activation energy of the dye extraction process was 74 kJ mol^{-1}. Nonetheless, TTI needed 168 hours (7 days) of incubation at 4 °C in order to release all of the dye. The potential of TTI to monitor fresh food products with an ideal shelf life of approximately 4 °C is indicated by these kinetic values.

Some dyes can produce colored products in an enzyme or substrate reaction, which can then be used as a temperature indicator over time. The most widely used chromogen for peroxidase assays is ABTS (2,20-azinobis-(3-ethyl benzthiazoline-6 sulfonic acid), a well-liked complex molecule due to its high water solubility and safety. In the presence of H_2O_2, plant peroxidases (1 : 11 : 1 : 7), an oxidoreductase enzyme, can break down a variety of aromatic compounds due to their diverse substrate specificities. *Arbuscula* spp. roots are utilized in the synthesis of peroxidases. When peroxidase and ABTS are combined with H_2O_2 and left at room temperature for longer than 15 minutes, the substrate oxidizes, giving the mixture an irreversible blue color and turning it green [40]. A prototype TTI enzyme was created using peroxidase as a basis. There are three layers to TTI. The middle zone, made up of solid fat that functions as a partition, is substrated by H_2O_2, while the lowest layer is formed by the enzyme 2,2′-azinobis-(3-ethyl benzthiazoline-6-sulfanic acid). This enzyme's activity has been found to be increased by metal ions like Hg^{2+}, K^+, and Ca^{2+}. The prototype can be made into the TTI phase by introducing metal ions.

An enzymatic TTI was developed based on the interaction between porcine esterase and tripropionin, which lowers pH and changes the pH indicator's color from blue to bright green [41]. Casanova et al. [42] assessed a novel enzyme-colorimetric temperature-time integrator composed of lipase carb, butyrin, and a pH indicator for the purpose of lowering the temperature of refrigerated goods. Degradation occurred in 15, 10, 0.5, and 0.3 days with storage at 10, 15, 20, and 25 °C, respectively, according to the kinetics of the enzymes. Among them, commercialized and theoretically interpreted biological TTIs based on enzymatic reactions have been established. Finding the activation energy and the reaction rate constant under isothermal conditions are the two main steps in the development of the TTI reaction. The accuracy with which the activation energy, duration of action, and lifetime are predicted determines the rate of action. The primary shortcoming of the TTI enzyme for refrigerated food shelf-life monitoring is not only its restricted range of activation energy but also the fact that its value is not commensurate with food degradation and spoilage [28]. The accuracy drops if the activation energy values of food and TTI differ by more than 25 kJ/mol [13]. It is advised that food spoilage can be predicted using any TTI device whose activation energy is within 25 kJ/mol of food quality factors. According to Park et al. [43], respiration rates (29–79 kJ/mol), lipid oxidation (41–104 kJ/mol), nutrient loss (84–126 kJ/mol), and microbial growth (84–251 kJ/mol) would all be in harmony with TTI with 1.0 mM NaN when it comes to food quality changes [44, 45].

11.2.3 Microbial TTIs

The French company CRYOLOG provides microbiological TTI. Microbiological TTIs, TRACEO (Figure 11.2a), and eO are derived from specific strains of lactic acid bacteria. In order to stop bacteria from growing in the TTI medium, the TTIs were stored frozen (−18 °C)

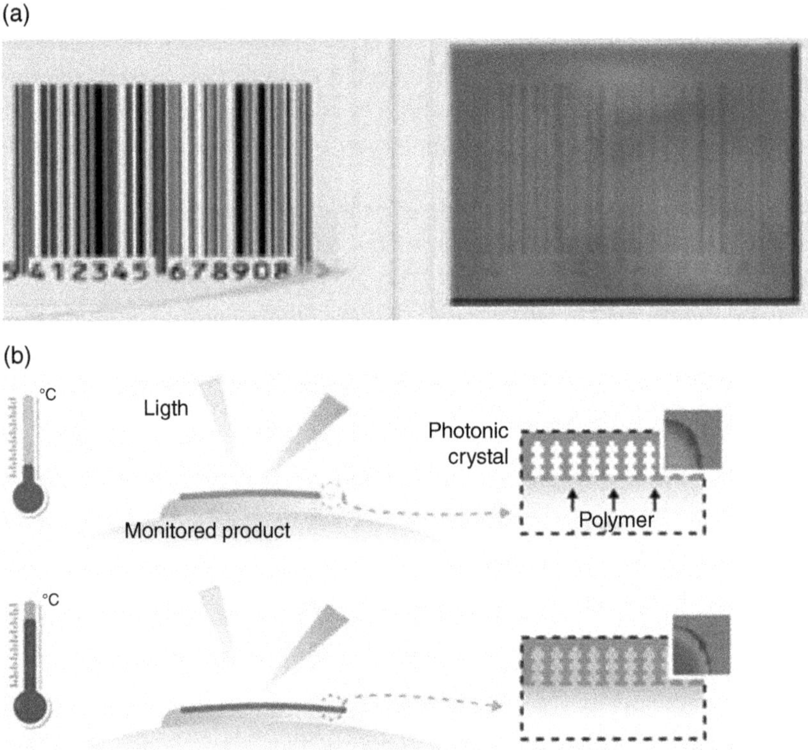

Figure 11.2 (a) CRYOLOG'S TRACEO TTI. *Source:* Ellouze et al. [46]/with permission of Elsevier. (b) An indicator of time and temperature based on the optical response of photonic crystals impregnated with polymers. *Source:* Chencha et al. [47]/with permission of Elsevier.

prior to use. It must be thawed at room temperature for a few minutes to activate due to its thinness. The temperature-dependent growth of TTI microorganisms causes the pH of the label to drop after storage in food, and when the temperature drops or the product reaches its use-by date. This results in an irreversible color change, with the medium chromatic index turning red [46]. In order to monitor food quality in the refrigerated food chain, Vaikousi [48] developed a time/temperature indicator (TTI) system based on the growth and metabolic activity of *Lactobacillus sakei* strains. Because of microbial growth and metabolism in the chosen medium, the chemical chromatic indicator (red-yellow) undergoes an irreversible color change that happens gradually as the pH drops. The microbial population was consistently between 10^7 and 10^8 CFU/ml, according to a correlation between the *L. sakei* TTI response (color change) and growth and metabolic activity (glucose consumption, lactic acid production, pH decrease). An activation energy of 97.7 kJ/mol was obtained by reducing the endpoint from 27 days at 0 °C to 2.5 days at 16 °C. This activation energy is extremely similar to the activation energy of the growth rate of *L. sakei* on the TTI substrate (103.2 kJ/mol). The microbial TTI prototype, primarily by lactic acid bacteria or other bacteria that exhibit comparable kinetic effects and spoiling potential, can be used as an efficient tool to monitor the shelf life of spoiled food products in distribution and storage. Grancher [49] created a TTI system that is activated through contact with two layers of plastic film: an adhesive layer made of particular microorganism activators and a reactive ink layer made of yeast and microorganisms that is water printed on the film. TTI is activated and attached to the product to track color changes brought on by microorganisms fermenting sugar.

11.2.4 Other Innovative TTIs

11.2.4.1 TTIs Based on Thermochromic Polymer/Dye Blends

According to Lu et al. [50], a novel temperature-time indicator was created using lipase, glycerol triacetate, bromothymol blue, mannitol, polyethylene glycol 400, and polyvinyl alcohol as its primary raw materials. The enzyme enhances the functionality of the original TTI by combining proven enzyme technology with the principles of chemical reaction and diffusion. The length of the colored band dispersion shows the combined effect of temperature and time. The acidity of fresh milk varies at different storage temperatures, according to ongoing temperature tests. Milk kept at 4, 15, and 25 °C reached the end of its life in 226, 63, and 26 hours, respectively, using acid as an indicator. An indicator pack, according to Nopwinyuwong et al. [51], contains a combination of pH-sensitive dyes, methyl red and bromothymol blue, which react by changing color to carbon dioxide, a harmful metabolite. Regarding the total color difference of the mixed pH-dye indicator, the color change is well correlated with the medium-moisture dessert's CO_2 level. In order to enable real-time monitoring of spoiling at various constant temperatures or with temperature fluctuations, the gold drop experiment looks into the relationship between indicator responses and microbial growth patterns in dessert samples.

11.2.4.2 TTIs Based on Photonic Lattice Change

Cencha et al. [47] created a time–temperature sensor in a mesoporous photonic crystal that uses thermoplastic polymer capillary absorption to detect visible light wavelengths (Figure 11.2b). Through the electrochemical deposition of silicon, photonic crystals are created, with microscale layers of alternating nanoscale pores forming a periodic structure. The effective refractive index of the crystal is irreversibly altered by polymer penetration, which results in constant variations in the reflected light that is visible to the unaided eye. Crucially, the temperature-dependent viscosity of the thermoplastic polymer (polyethylene vinyl acetate) is also explained by the Arrhenius law, which is frequently applied to ascertain the temperature dependence of quality parameters. As a result, the temperature of the system, which indicates the level of degradation of the product under observation, is correlated with each reflected color.

11.2.4.3 Time–Temperature Sensors Based on Electronics

An electronic time–temperature sensor is characterized as an apparatus that uses a heat sensor to convert an electrical signal from a temperature signal into a final visual output. This device can also be used to indicate the quality of a product. This type of time–temperature sensor is typically expensive and may require consumer training (by the manufacturer, consumer, and retailer), which may or may not be practical in the market. Commercialization may also limit the scope of their programs, as the reading device for the time–temperature sensor is complex and based on special electronics. But when it comes to tracking and documenting the thermal history of products, electronic-based time–temperature sensors are generally more advanced and more accurate than other kinds. Furthermore, a wide variety of reusable and eco-friendly electronic temperature sensors are available. Numerous new electronic time–temperature sensors that don't require specialized reading equipment or qualified personnel to take test readings have been developed as a result of the advancement of electronic-based sensors. Customers will find this technology more convenient, and the market will demand more IP. However, time–temperature indicators should be made of smaller, less expensive, and recyclable electronics in order to enable the use of electronics-based time–temperature sensors in the global market without sacrificing accuracy and safety.

11.3 Applications of TTIs for Monitoring Food Quality

11.3.1 Fruits and Vegetables

TTI has been developed to monitor fruit maturity levels. A TTI tag developed for this purpose is called RipeSense™ (Figure 11.3a).

11.3.2 Dairy Products

A TTI enzyme prototype was created by Gogou et al. [53], and it can be used to regulate sterilization processes like milk pasteurization within the 100–130 °C temperature range. Xylanase, which can boost thermostability at low water activity levels, is used in the TTI

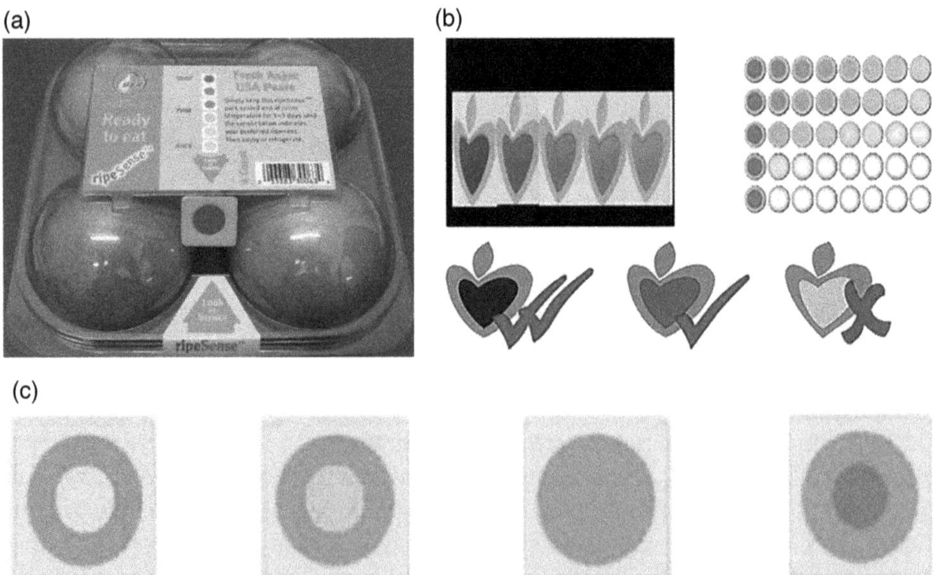

Figure 11.3 (a) RipeSense™ TTI for fruit. *Source:* Lee et al. (2014)/with permission of Elsevier. (b) OnVu™ TTI [52]. (c) Fresh-Check™ TTI. *Source:* Nuin et al. [8]/with permission of Elsevier.

prototype. The thermophilic fungus *Thermomyces lanuginosus* is the source of this xylanase solution. Thermophilic fungi can release a variety of thermozymes, including xylanase, and grow best at 48–52 °C. Khairunnisa et al. [54] assessed the usefulness of TTI as an indicator of microbiological deterioration of pasteurized milk (PM) using mixed vegetable oil (VOB) as a TTI indicator. In addition to 0.25% (m/v) red dye (CI 16255), VOB is made from a blend of palm, canola, and olive oils in the following ratios: 50 : 40 : 10 (C1), 50 : 25 : 25 (C2), and 50 : 10 : 40% (v/v). At isothermal temperatures, the diffusion length, diffusion kinetics, and model prediction accuracy of VOB were evaluated for each parameter. To keep an eye on the general microbial growth in PM, utilize the VOB indicator. In PM, the activation energy value (E_a) of microbial degradation is 58.105 kJ/mol. The Ea values of each indicator are 34.289 (C1), 35.294 (C2), and 40.646 kJ/mol (C3). Less than 25 kJ/mol separated each parameter's value from the E_a PM value. It demonstrates the high accuracy of the prediction model. Following a 24-hour storage period at 29 °C, the largest PM degradation was observed, with diffusion length indices of 6.50 (C1), 6.25 (C2), and 5.5 cm (C3).

11.3.3 Muscle Foods

According to Nuin et al. [8], time–temperature indicators like Monitor Mark®, Fresh-Check® (Figure 11.3c), and Check Point® have been used as quality and suitability indicators for fresh turbot (*Psetta maxima*). All fish from cobia samples kept on ice had their time and temperature histories tracked using the photochromic TTI [55]. Temperature-time indicator with OnVu® B1 label monitoring and the explanation of TTI's response were evident. The findings indicate that the smart indicator, which was activated for six seconds by the UV light, displays a comparable degree of deterioration in the visual response of the product under analysis. This smart indicator is a dynamic suitability indicator that can quickly and affordably assure customers of the highest quality point of all the cobia. To track the microbiological quality of foods that are refrigerated, a microbial prototype time/temperature indicator was created based on the development and metabolic activity of *Lactobacillus sakei* [48]. Lean beef packaged in a modified atmosphere has been studied to monitor the deterioration process using the microbial TTI, which is based on the growth and metabolic activity of *Lactobacillus sakei* [56, 57]. During isothermal cold storage, the sensory rejection point of beef was in line with the TTI color change endpoint. The total color change of TTI (ΔE) was estimated to have an activation energy of 112.77 kJ/mol, indicating that TTI microbes are a useful tool for quality control when chilled meat products are distributed and stored. Bacteria that display comparable kinetic reactions and degradation potential, such as lactic acid bacteria, are responsible for their degradation.

Cooked chicken slices packaged in modified atmosphere packaging (MAP) were evaluated for quality and safety using the biological temperature-time integrator (eO®) [11]. Fresh cod loins were stored in separate packages with varying conditions, and photochromic TTI was utilized to track the quality and shelf life of the fish [58]. Sodium azide was used to create a laccase-based temperature integrator (TTI) [43]. The urea-urease time–temperature indicator (TTI) system was the subject of an initial investigation [38]. Refrigerated chicken-based products are dynamically monitored for shelf life using an

OnVu™ TTI B1 UV label (Figure 11.3b), an intelligent photochromic indicator [52]. Sakai et al. [59] developed TTI based on the Maillard reaction to visualize temperature history during long-term storage and distribution of stored beef. Disodium hydrogen phosphate solution, 500 μl D-xylose, and 1000 μl glycine were added to the Maillard reaction-based TTI that was developed. The Maillard reaction starts when the two solutions are combined. Given that Australian chilled beef has a shelf life of 77 days in Japan, one TTI (TTI-chocolate) is light chocolate after that time. After 150 days, TTI-yellow turns yellow when the chilled beef emits a nice aroma, as determined by sensory evaluation.

In order to track the thermal use history of frozen chicken meat (−18 °C) throughout the supply chain, an amylase-based enzyme time–temperature indicator (TTI) was created using an amylase enzyme and starch substrate [60, 61]. At 15, 25, and 35 °C, TTI displayed an irreversible color shift from the starting blue–black to medium-light yellow, and finally, an irreversible yellow–white, except for 5 °C. The application of TTI as a fitness indicator was tested in a national livestock chain based on the well-validated OnVu™ TTI kinetics [62]. TTI faithfully captures variations in temperature that happen on actual trails. The product's microbiological shelf life is followed by the shelf life determined by TTI color.

11.3.4 Other Foods

Using mushrooms (*Agaricus bisporus Pilát*), a feasibility study was carried out to assess the potential of time–temperature integrators (TTI) as a quality indicator of horticultural products [16]. The preferred product is commercially available TTI enzyme with kinetic activity (Vitsab A.B., Malmö, Sweden). Using a spectrophotometer, the responses of fungi and particular TTIs were assessed under both constant and fluctuating temperature profiles.

11.4 Assessment of Applicability and Commercialization

11.4.1 Assessment of Applicability of TTI

The operating mechanisms of TTIs that are currently on the market are founded on various principles. According to Taoukis [63], TTI operates on the basis of irreversible changes in mechanical, chemical, enzymatic, or microbiological processes that manifest as observable reactions such as color development, movement, or deformation of the material. Chemical reactions are the foundation for processes like dissolution, polymerization, acid-base reactions, and others that result in chemical responses. Variations in the biological activity of microorganisms like spores or temperature-sensitive enzymes determine a biological response [64, 65]. The negative effect that causes the product quality to deteriorate is similar to the rate of transformation increasing at a higher temperature. The product's time–temperature history is summed up by the TTI's apparent response [63]. TTIs should be simple to activate and exhibit easily measured temperature-dependent rise time variations. These modifications ought to be permanent, ideally equal to or directly connected to the rate at which food spoils and the residue's shelf life. By varying the charging time during activation, the TTI can be tailored to various products, producing a modified intensity.

Using the CIE Lab color system, the squared value (SV) of the TTI response was calculated in order to assess the TTI's color change and ascertain the label retention time.

$$\mathrm{SV} = \left(L^2 + a^2 + b^2\right)^{1/2}$$

where "*a*" denotes the label's red and green component, "*b*" is its yellow and blue component, "*L*' is its lightness, and "SV" is its square value.

A logistic model can explain the label's color change or the SV's evolution over time [66]. The TTI staining time (expiration time) can be computed using the equation that follows.

$$\mathrm{SV} = A/\left(1 + e^{-k(t-\mathrm{tr})}\right)$$

where A is the amplitude, k is the reaction rate (h^{-1}), tr is the reversal point (h), t is the time (h), and SV is the square value of the TTI.

The temperature dependence of the label color is explained by the Arrhenius equation [66]. Plotting the reaction rate as a function of temperature is therefore possible.

$$\mathrm{In}(k) = \mathrm{In}(k_0) - E_\mathrm{a} / R \times 1/T$$

where T is the absolute temperature (K), R is the ideal gas constant ($8.314\,\mathrm{J\,mol^{-1}\,K^{-1}}$), E_a is the activation energy ($\mathrm{kJ\,mol^{-1}}$), and k is the reaction rate (h^{-1}).

The staining procedure and the label's activation energy, as determined by the Arrhenius model (23.22–25.77 kcal/mol) provided by Kreyenschmidt et al. [66] and Raab [67], are described in detail. The model and instrument were developed using the data from this study.

11.4.2 Commercialization of TTI

3M Co., CCL Industries Inc., Temptime Corporation, Thin Film Electronics Inc., Bizerba SE & Co. KG, DeltaTrak Inc., Biosynergy, Inc., Freshpoint Quality Assurance Ltd., Insignia Technologies Ltd., and La-Co Industries Inc. are some of the major companies involved in the global temperature indicator label market (Table 11.1).

The color change of an oxidizing chemical system is detected by 3M MonitorMark® (3M Co., St. Paul, Minnesota), a diffusion-based indicator whose migration through the film is dependent on temperature (Figure 11.4a, b). The wick is used to distribute the activation using a fatty acid ester dye in blue color. Light diffuses into the porous matrix at a

Table 11.1 Commercially available TTIs.

TTI	Company
MonitorMark™	3M™
Timestrip®	Timestrip PLC
Fresh-Check®	LifeLines
Checkpoint®	Vitsab

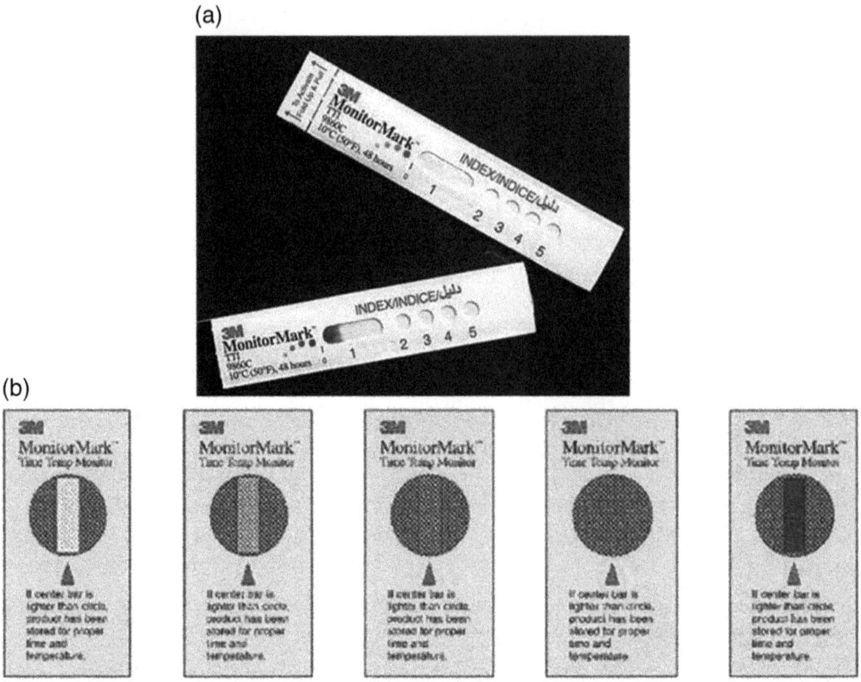

Figure 11.4 (a) Monitor Mark indicator. *Source:* Taoukis [63]/with permission of John Wiley & Sons. (b) Freshness check TTI by 3M. *Source:* Taoukis [63]/with permission of John Wiley & Sons.

rate that depends on temperature, thanks to the viscoelastic material. The label configuration, variation and polymer concentration, glass transition temperature, response rate, and temperature dependence can all be adjusted within a specified range [63, 68]. According to Kuswandi et al. [64], there are two versions of MonitorMark™: a smart label for consumer data and a threshold indicator for industry use in distribution monitoring. When all five windows were blue, the display response was considered complete. It was measured when the blue color flowed along the line. For instance, with a response time of two days, the indicator tag 51 will display a temperature of 5 °C (melting temperature). The time it took for the label to fill in blue over the course of five windows at a constant 2 °C above the response temperature is represented by this response. In a similar vein, response days vary from −17 °C to +48 °C [69, 70].

A smart label called Timestrip® (Timestrip UK Limited, UK) keeps track of how long a product has been used or opened (Figure 11.5a). Maintaining a consistent temperature at home is also crucial for food safety. A smart tag that allows consumers to track the expiration date of perishable goods is called Timestrip® and is intended to enable users to note the amount of time elapsed since tag activation. Packing or labeling perishable goods or goods that need to be replaced or maintained on a regular basis (such as frozen and refrigerated goods) is best suited for this function. Time is automatically tracked for a duration of up to 12 months. When a customer opens the package, it automatically activates. Alternatively, it can be given to them as an external tag that they can manually activate when using the product for the first time [64, 70]. The product has been subjected to excessive heat for

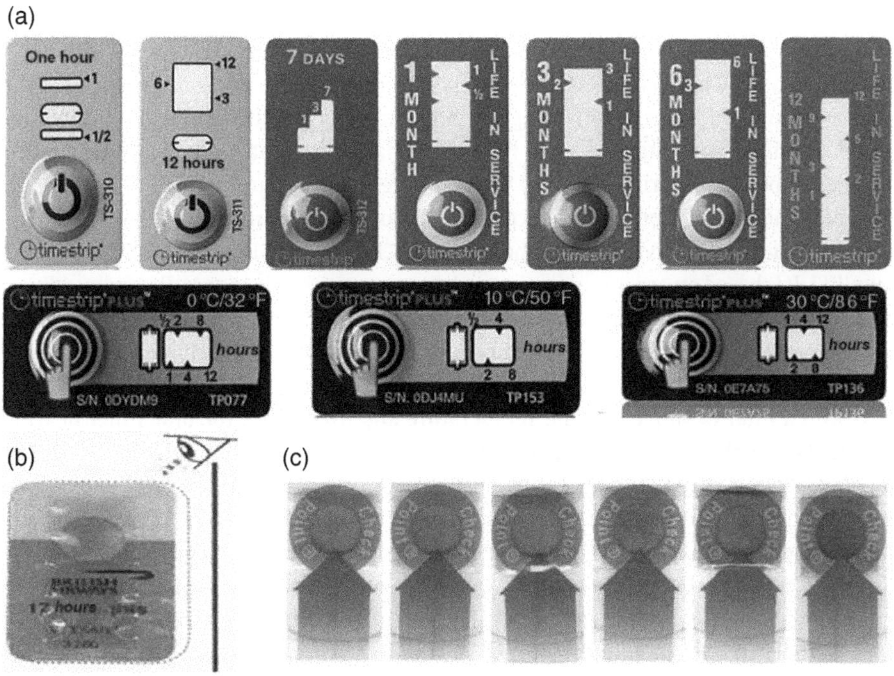

Figure 11.5 (a) Timestrip® time indicator and time–temperature indicators. *Source:* Timestrip UK Ltd. (b) Vistab TTI. *Source:* Taoukis [63]/with permission of John Wiley & Sons. (c) CheckPoint TTI. *Source:* Kuswandi et al. [64]/with permission of Springer Nature.

the entire duration indicated on the Timestrip label. To manually start the timer indicator, just press the button located above it. The product is above the required or desired temperature when it is activated, as indicated by a color change on the printed time scale. Only during inclement weather or periods of high tourist traffic does the color shift occur; otherwise, it ceases. Highly pigmented polymers are formed through solid polymerization reactions, which are the basis for Fresh-Check®TTI (Temptime Corp., Morris Plains, NJ, USA). A decrease in reflectance indicates the TTI response, which is a color shift [63]. A printed ring encircling a tiny polymer circle serves as the indicator. Measured color intensity is compared to label color scale [69] if the inner polymer ring is exposed to unfavorable weather [71]. The polymer's color changes more quickly as the temperature increases. Regardless of the "use by" date, consumers are advised not to purchase or consume the product [72]. To guarantee that the product is still fresh for the customer at the point of purchase and at home, this indicator can be applied to the packaging of perishable goods. This indicator has been found in salads, milk, frozen meals, and fruit cakes.

In an enzyme system, CheckPoint®TTI (VITSAB A. B., Malmö, Sweden) is a straightforward adhesive label (Figure 11.5b, c). These labels show the freshness status and shelf life because they are influenced by time, temperature, and food reactions. The basis of TTI is a change in color brought on by a drop in pH, which is the outcome of carefully regulated enzymatic hydrolysis of the lipid substrate. Acid is released during substrate hydrolysis, and the pH indicator changes from deep green to yellowish red as a result of the pH decrease. Temperature dependence and varying responses can be obtained by varying the concentration and type of enzyme-substrate combination [63]. The CheckPointR I for a single point and the CheckPointR III for three points are the two basic configurations of the device that are offered. Consumer packaging and carton and product unit temperature control employ single-point markers. The three-point label integrates three response levels into a single label and is primarily used in the wholesale distribution chain. Product self-management labels are appropriate for continuous color development [64].

The recently released OnVu™ TTI (Ciba Specialty Chemicals & Freshpoint, SW) is a solid reaction TTI. It is based on organic pigments and photosensitive substances. Benzylpyridines, whose rate of change over time is dependent on temperature. An internal heart shape, or "apple" movement, can be seen in the TTI recording. The image won't change until the UV light from the LED is activated if the inner core turns a deep blue color. The label is then covered with a filter to stop charging. Temperature and time cause the blue inner heart to turn white. The method can be printed directly on the package or used as a label [12, 63, 73]. With pH indicators, eO® TTI measures the pH change that occurs in response to a change in temperature and time, which is brought about by the carefully cultivated growth of specific lactic acid bacteria strains (Figure 11.6a, b). To stop bacteria from growing in the TTI medium, the eO® TTI (CRYOLOG, Gentilly, France) was kept frozen (−18 °C) prior to use. It must be thawed at room temperature for a few minutes in order to activate due to its thinness. The temperature-dependent growth of TTI microorganisms causes the pH of the label to drop after storage in food, and when the temperature drops or the product reaches its use-by date. This results in an irreversible color change, with the medium chromatic index turning red [46, 63]. Biological TTI eO® was assessed by Ellouze and Augustin [11] as a safety and quality indicator for beef and seasoned chicken slices that were packaged in a modified environment.

Diffusion reaction theory serves as the foundation for TT Sensor™ TTI (Avery Dennison Corp., USA). The fluorescence indicator changes from yellow to light pink due to the concentration change caused by the polar compound diffusing between the two polymer layers [63]. Numerous novel forms of TTI have emerged in recent times. Rani and Abraham [40], for instance, created a novel enzymatic reaction for inexpensive TTI.

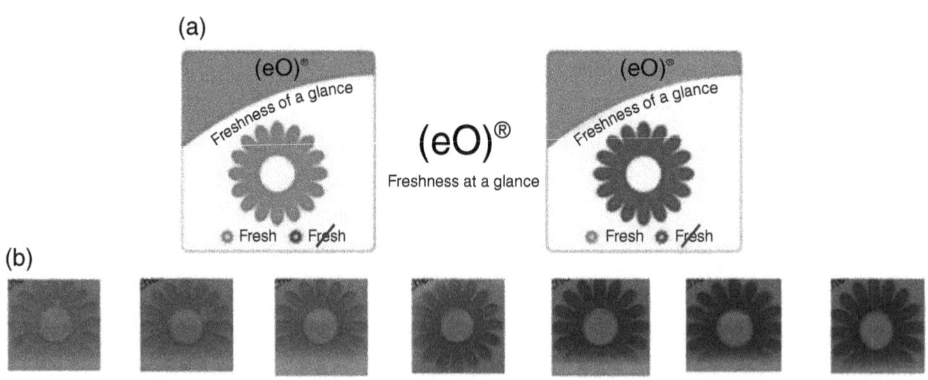

Figure 11.6 (a) eO TTI. *Source:* Ellouze et al. [46]/with permission of Elsevier. (b) Cryolog's eO TTI. *Source:* Ellouze and Augustin [11]/with permission of Elsevier.

Figure 11.7 TTI by Smartdots. *Source:* Insignia.

A prototype lactic acid-based time–temperature indicator for food quality monitoring was created by Wanihsuksombat et al. [74]. Using variable dye, Galagan and Su [75] created a new colorimetric TTI. In order to track food quality in the chilled food chain, Vaikousi et al. [57] created a novel TTI system based on the development and metabolic activity of *Lactobacillus sakei* strains. Based on the interaction between amylase and starch, Yan et al. [25] created a novel amylase-type TTI. Below is a list of additional time–temperature indicators that are available for purchase.

For instance, Insignia's FreshTag is designed for perishables such as meat or fresh produce – items that are particularly susceptible to temperature abuse. The FreshTag is applied to each product and changes colour over the shelf life of the product, with the speed of the colour change accelerating if temperature abuse occurs. This ensures products are used when as fresh as possible and before their shelf life expires (Figure 11.7).

Another is the Freshtag Timer (Figure 11.8a), which is a new type of label for pallets, containers, or individual items that control temperature and time. The label has a color change that happens at a set time. Because the label is more sensitive to temperature and time than the use-by date, it will change color more quickly if the product is stored above the suggested reference temperature. The label also describes the product's freshness more accurately. A helpful stock circulation tool for quick-service restaurants is Freshtag. Check label probes that are frozen and SpotSee are two more TTIs that are sold commercially (Figure 11.8b, c).

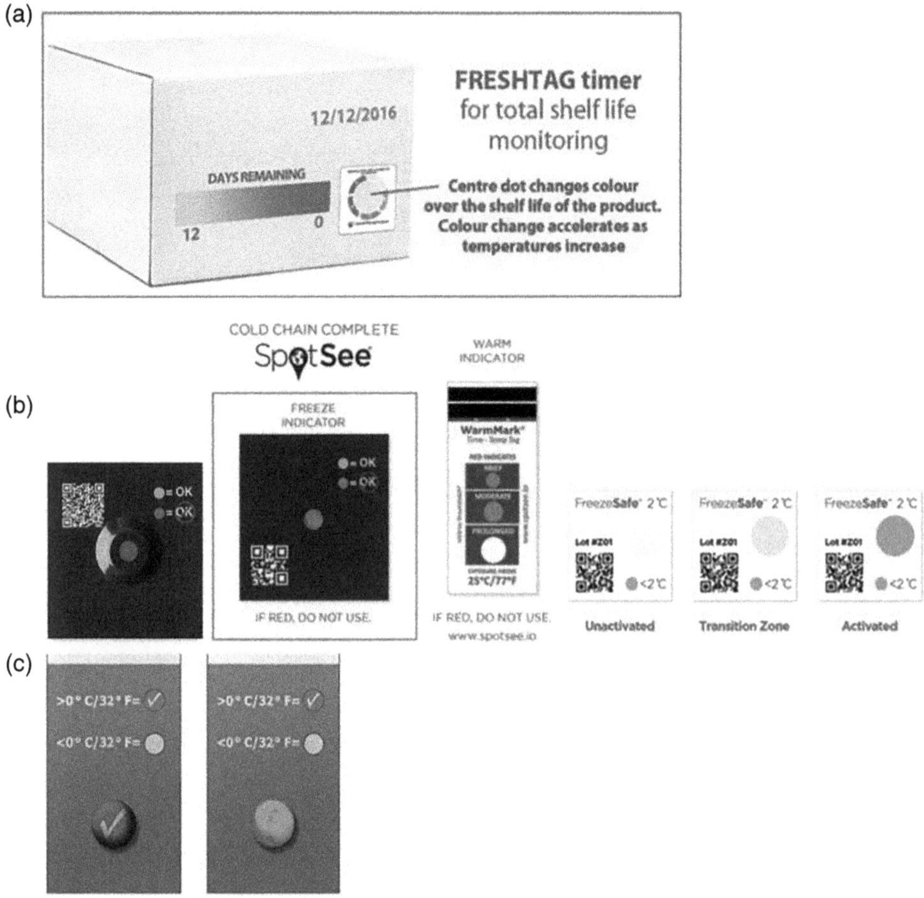

Figure 11.8 (a) Freshtag TTI. *Source:* With permission of Vanprob Solutions [76]. (b) TTI's of SpotSee, USA. *Source:* With permission of SpotSee [77]. (c) Freeze check TTI labels. *Source:* With permission of Vanprob Solutions [76].

11.5 Conclusion and Perspective

Utilizing time–temperature indicators can improve shelf life, optimize product distribution, track and manage the food supply chain, and ultimately cut down on food waste. The requirements for temperature–time indicators to be successful in practice are affordability, dependability, and effective use. According to their specifications, current temperature indicator systems offer dependable and consistent responses. The time–temperature indicator records the effects of temperature and time and shows a visual summary of the product temperature collected over time. Before entering the supply chain, several organizational, functional, and technical requirements must be determined. Numerous studies have demonstrated the value of TTI and the TTI kinetic model as tools for tracking processing temperatures and forecasting shelf life at various supply chain stages. Additionally, a real cold chain with a dynamic shelf life can only be successfully implemented if the TTI dyeing process is executed with high productivity and consistency. TTI's software ought to be expanded to include applications for smartphones. It is advised to measure TTI directly with a smartphone to compute online dynamic latency and enhance performance. It necessitates a thorough adaptation to the unique requirements of clients in accordance with the attributes of the product in the supply chain.

References

1 Taoukis, P.S. and Labuza, T.P. (1989). Reliability of time–temperature indicators as food quality monitors under nonisothermal conditions. *Journal of Food Science* 54 (4): 789–792.

2 Giannakourou, M.C., Koutsoumanis, K., Nychas, G.J.E., and Taoukis, P.S. (2005). Field evaluation of the application of time temperature integrators for monitoring fish quality in the chill chain. *International Journal of Food Microbiology* 102: 323–336.

3 Kerry, J.P., O'Grady, M.N., and Hogan, S.A. (2006). Past, current and potential utilisation of active and intelligent packaging systems for meat and muscle-based products: a review. *Meat Science* 74 (1): 113–130. https://doi.org/10.1016/j.meatsci.2006.04.024.

4 USDA (2023). https://www.fsis.usda.gov/food-safety/safe-food-handling-and-preparation/food-safety-basics/danger-zone-40f-140f

5 USDA (2023). https://www.fsis.usda.gov/food-safety/safe-food-handling-and-preparation/food-safety-basics/how-temperatures-affect-food

6 Karel, M. (1984). Chemical effects in food stored at room temperature. *Journal of Chemical Education* 61 (4): 335–339.

7 Taoukis, P.S. and Labuza, T.P. (2003). 6 – Time-temperature indicators (TTIs). In: *Woodhead Publishing Series in Food Science, Technology and Nutrition, Novel Food Packaging Techniques* (ed. R. Ahvenainen), 103–126. Woodhead Publishing. ISBN: 9781855736757 https://doi.org/10.1533/9781855737020.1.103.

8 Nuin, M., Alfaro, B., Cruz, Z. et al. (2008). Modelling spoilage of fresh turbot and evaluation of a time–temperature integrator (TTI) label under fluctuating temperature. *International Journal of Food Microbiology* 127: 193–199. https://doi.org/10.1016/j.ijfoodmicro.2008.04.010.

9 Shimoni, E., Anderson, E.M., and Labuza, T. (2001). Reliability of time temperature indicators under temperature abuse. *Journal of Food Science* 66: 1337–1340. https://doi.org/10.1111/j.1365-2621.2001.tb15211.x.

10 Coherent Market Insights (2018). Time-temperature-indicator-labels-market. https://www.coherentmarketinsights.com/market-insight/time-temperature-indicator-labels-market-1168.

11 Ellouze, M. and Augustin, J.C. (2010). Applicability of biological time temperature integrators as quality and safety indicators for meat products. *International Journal of Food Microbiology* 138: 119–129.

12 Tsironi, T., Gogou, E., Velliou, E., and Taoukis, P.S. (2008). Application and validation of the TTI based chuill chain management system SMAS (Safety Monitoring and Assurance Sytem) on shelf life optimalization of vacuum packed chilled tuna. *International Journal of Food Microbiology* 128: 108–115.

13 Taoukis, P.S. (2001). Modeling the use of time temperature indicators in distribution and stock rotation. In: *Food Process Modeling* (ed. L.M.M. Tijkskens, M.L.A.T.M. Hertog, and B.M. Nicolaï), 402–432. England: Woodhead Publishing.

14 Giannakourou, M.C. and Taoukis, P.S. (2003). Application of a TTI-based distribution management system for quality optimization of frozen vegetables at the consumer end. *Journal of Food Science* 68 (1): 201–209.

15 Mehauden, K., Cox, P.W., Bakalis, S. et al. (2007). A novel method to evaluate the applicability of time temperature integrators to different temperature profiles. *Innovative Food Science & Emerging Technologies* 8 (4): 507–514. https://doi.org/10.1016/j.ifset.2007.03.001.

16 Bobelyn, E., Hertog, M.L.A.T.M., and Nicolaï, B.M. (2006). Applicability of an enzymatic time temperature integrator as a quality indicator for mushrooms in the distribution chain. *Postharvest Biology and Technology* 42 (1): 104–114. https://doi.org/10.1016/j.postharvbio.2006.05.011.

17 Guiavarc, Y., Loey, A.V., Zuber, F., and Hendrickx, M. (2004). *Bacillus licheniformis* α-amylase immobilized on glass beads and equilibrated at low moisture content: potentials as a time-temperature integrator for sterilization processes. *Innovative Food Science and Emerging Technologies* 5 (3): 317–325.

18 Vistab (2023). https://vitsab.com/?page_id=141 (accessed 29 November 2023).

19 Timestrip (2023). https://timestrip.com/indicator-labels

20 Kim, J.U., Ghafoor, K., Ahn, J. et al. (2016). Kinetic modeling and characterization of a diffusion-based time-temperature indicator (TTI) for monitoring microbial quality of non-pasteurized angelica juice. *LWT- Food Science and Technology* 67: 143–150. https://doi.org/10.1016/j.lwt.2015.11.034.

21 Lim, S., Gunasekaran, S., and Imm, J.Y. (2012). Gelatin-templated gold nanoparticles as novel time-temperature indicator. *Journal of Food Science* 77 (9): N45–N49. https://doi.org/10.1111/j.1750-3841.2012.02872.x.

22 Ramstad, D.E. and Volz, F.E. (1950). Did it thaw? GE will tell. *Food Industries* 1950 22 (84): 182–184.

23 Blixt, K. (1984). The I-Point TTM – a versatile biochemical time-temperature integrator. In: *Thermal Processing and Quality of Foods* (ed. P. Zeuthen, J.C. Cheftel, and C. Eriksson). London: Elsevier Applied Science Publishers.

24 Tucker, S., Brown, H.M., Fryer, P.J. et al. (2007). Sterilisation time–temperature integrator based on amylase from the hyperthermophilic organism Pyrococcus furiosus. *Innovative Food Science & Emerging Technologies* 8: 63–72.

25 Yan, S., Huawei, C., Limin, Z. et al. (2008). Development and characterization of a new amylase type time–temperature indicator. *Food Control* 19 (3): 315–319.

26 Moritz, M., Balasa, A., Jaeger, H. et al. (2012). Investigating the potential of polyphenol oxidase as a temperature–time-indicator for pulsed electric field treatment. *Food Control* 26: 1–5.

27 Jin, Q., Jin, W., Guanglin, Z., and Qin, F. (2012). A time-temperature indicator based on glucoamylase. China Patent 201110394281.X.

28 Kim, K., Kim, E., and Lee, S.J. (2012). New enzymatic time–temperature integrator (TTI) that uses laccase. *Journal of Food Engineering* 113 (1): 118–123.

29 Octavio, L.C.L., Ma, P.P., Ricardo, B.J.R., and Villaseñor, O.F. (2006). Laccase. In: *Enzyme Biotechnology* (ed. R.G. Guevara González and I. Torres-Pacheco), 323–340. Kerala: Advances in Agricultural and Food Biotechnology.

30 Xiao, Y.Z., Chen, Q., Hang, J. et al. (2004). Selective induction, purification and characterization of a laccase isozyme from the basidiomycete Trametes sp. AH28-2. *The Mycological Society of America* 96: 26–35.

31 Xu, F. (1996). Oxidation of phenols, anilines, and benzenethiols by fungal laccases: correlation between activity and redox potentials as well as halide inhibition. *Biochemistry* 35: 7607–7614.

32 Moshtaghioun, S.M., Haghbeen, K., Sahebghadam, A.L. et al. (2011). Direct spectrophotometric assay of laccase using diazo derivatives of guaiacol. *Analytical Chemistry* 83: 4200–4205.

33 Minussi, R.C., Pastore, G.M., and Durán, N. (2002). Potential applications of laccase in the food industry. *Trends in Food Science and Technology* 13: 205–216.

34 Li, A., Zhu, Y., Xu, L. et al. (2008). Comparative study on the determination of assay for laccase of *Trametes* sp. *African Journal of Biochemistry Research* 2: 181–183.

35 Guzman-Maldonadao, H. and Paredes-Lopez, O. (1995). Amylolytic enzymes and products derived from starch: a review. *Critical Reviews in Food Science and Nutrition* 35: 373–403.

36 Pandey, A., Nigam, P., Soccol, C.R. et al. (2000). Advances in microbial amylases. *Applied Biochemistry and Biotechnology* 31: 135–152.

37 Qian J, Zheng GL. and Feng, Q. (2012). Preparation of amylase based time temperature integrator. In Applied Mechanics and Materials. Trans Tech Publications. 2012; 200: 433–436.

38 Wu, D., Wang, Y., Chen, J. et al. (2013). Preliminary study on time–temperature indicator (TTI) system based on urease. *Food Control* 34: 230–234. https://doi.org/10.1016/j.foodcont.2013.04.041.

39 Anbukarasu, P., Sauvageau, D., and Elias, A.L. (2017). Time-temperature indicator based on enzymatic degradation of dye-loaded polyhydroxybutyrate. *Biotechnology Journal* 12 (9): https://doi.org/10.1002/biot.201700050.

40 Rani, D.N. and Abraham, T.E. (2006). Kinetic study of a purified an ionic peroxidase isolated from *Eupatorium odoratum* and its novel application as time temperature indicator for food materials. *Journal of Food Engineering* 77: 594–600.

41 Son, B.H., Lee, Y.J., Jo, H.J., and Hong, K.W. (2014). New enzymatic time-temperature integrator (TTI) using porcine esterase for monitoring food quality. *Journal of Korean Society for Applied Biological Chemistry* 57: 331–334. https://doi.org/10.1007/s13765-013-4309-9.

42 Casanova, C.F., de Souza, M.A., Fischer, B. et al. (2021). Development of enzymatic-colorimetric time-temperature integrator for smart packaging. *Biointerface Research in Applied Chemistry* 11: 9335–9345. https://doi.org/10.33263/BRIAC112.93359345.

43 Park, H.R., Kim, K., and Lee, S.J. (2013). Adjustment of Arrhenius activation energy of laccase-based time–temperature integrator (TTI) using sodium azide. *Food Control* 32 (2): 615–620. https://doi.org/10.1016/j.foodcont.2013.01.046.

44 Labuza, T.P. (1982). Scientific evaluation of shelf-life. In: *Shelf-Life Dating of Foods* (ed. T.P. Labuza), 41–87. Trumbull: Food and Nutrition Press Inc.

45 Poças, M.F.F., Delgado, T.F., and Oliveira, F.A.R. (2008). Smart packaging technologies for fruits and vegetables. In: *Smart Packaging Technologies* (ed. J. Kerr and P. Butler), 151–166. West Sussex: Wiley.

46 Ellouze, M., Pichaud, M., Bonaiti, C. et al. (2008). Modelling pH evolution and lactic acid production in the growth medium of a lactic acid bacterium: application to set a biological TTI. *International Journal of Food Microbiology* 128: 101–107.

47 Cencha, L.G., García, G.F., Budini, N. et al. (2022). Time-temperature indicator based on the variation of the optical response of photonic crystals upon polymer infiltration. *Sensors and Actuators A: Physical* 341: 113571. https://doi.org/10.1016/j.sna.2022.113571.

48 Vaikousi, H., Biliaderis, C.G., and Koutsoumanis, K.P. (2008). Development of a microbial time/temperature indicator prototype for monitoring the microbiological quality of chilled foods. *Applied and Environmental Microbiology* 74 (10): 3242–3250. https://doi.org/10.1128/AEM.02717-07.

49 Grancher, X.V. (2005). Time temperature indicator (TTI) system. EP1725846A1, European Patent Office.

50 Lu, L., Zheng, W., Lv, Z., and Tang, Y. (2013). Development and application of time–temperature indicators used on food during the cold chain logistics. *Packaging Technology and Science* 26: 80–90. https://doi.org/10.1002/pts.2009.

51 Nopwinyuwong, A., Trevanich, S., and Suppakul, P. (2010). Development of a novel colorimetric indicator label for monitoring freshness of intermediate-moisture dessert spoilage. *Talanta* 81 (3): 1126–1132. https://doi.org/10.1016/j.talanta.2010.02.008.

52 Brizio, A.P.D.R. and Prentice, C. (2014). Use of smart photochromic indicator for dynamic monitoring of the shelf life of chilled chicken based products. *Meat Science* 96 (3): 1219–1226. https://doi.org/10.1016/j.meatsci.2013.11.006.

53 Gogou, E., Katapodis, P., Christakopoulos, P., and Taoukis, P.S. (2010). Effect of water activity on the thermal stability of Thermomyces lanuginosus xylanases for process time–temperature integration. *Journal of Food Engineering* 100 (4): 649–655. https://doi.org/10.1016/j.jfoodeng.2010.05.014.

54 Khairunnisa, A., Suyatma, N.E., and Adawiyah, D.R. (2019). Development of smart TTI label based on kinetics diffusion of vegetable oils blends for cold supply chain monitoring. *IOP Conference Series: Earth and Environmental Science* 335: 012015. https://doi.org/10.1088/1755-1315/335/1/012015.

55 Brizio, A.P.B.R., Junior, M.A.G., Fogaca, F.H.D.S., and Prentice, C. (2014). Dynamic monitoring of the shelf life of Cobia (*Rachycentron canadum*): a study on the applicability of a smart photochromic indicator. *International Journal of Food Science and Technology* 1–7. https://doi.org/10.1111/ijfs.12701.

56 Girardeau, A., Biscola, V., Keravec, S., Corrieu, G., Fonseca, F. (2020). Application of lactic acid bacteria in time-temperature integrators, Chapter 14. Marcela Albuquerque Cavalcanti de Albuquerque, Alejandra de Moreno de LeBlanc, Jean Guy LeBlanc, Raquel Bedani. Lactic acid bacteria: a functional approach, CRC Press, pp. 241–260. https://doi.org/10.1201/9780429422591

57 Vaikousi, H., Biliaderis, C.G., and Koutsoumanis, K.P. (2009). Applicability of a microbial Time Temperature Indicator (TTI) for monitoring spoilage of modified atmosphere packed minced meat. *International Journal of Food Microbiology* 133: 272–278.

58 Mai, N.T.T., Gudjónsdóttir, M., Lauzon, H.L. et al. (2011). Continuous quality and shelf life monitoring of retail-packed fresh cod loins in comparison with conventional methods. *Food Control* 22 (6): 1000–1007. https://doi.org/10.1016/j.foodcont.2010.12.010.

59 Sakai, K., Lee, J., Kocharunchitt, C. et al. (2020). Development of a maillard reaction–based time-temperature integrator/indicator (TTI) for visual monitoring of chilled beef during long-term storage and distribution. *Food and Bioprocess Technology* 13: 1–10. https://doi.org/10.1007/s11947-020-02549-z.

60 Jaiswal, R., Mendiratta, S.K., Talukder, S., and Chand, S. (2019). Amylase based enzymatic time temperature indicator (TTI) as thermal abuse marker for frozen chicken meat. *Indian Journal of Poultry Science.* 54 (1): 59. https://doi.org/10.5958/0974-8180.2019.00010.2.

61 Vishnuraj, M.R., Kandeepan, G., Shukla, V. et al. (2017). Development and evaluation of an enzyme based time temperature integrator (TTI) for monitoring meat quality. *Journal of Meat Science* 12 (1): 23–26.

62 Albrecht, A., Ibald, R., Raab, V. et al. (2020). Implementation of time temperature indicators to improve temperature monitoring and support dynamic shelf life in meat supply chains. *Journal of Packaging Technology and Research* 4: 23–32. https://doi.org/10.1007/s41783-019-00080-x.

63 Taoukis, P. (2008). Application of time-temperature integrators for monitoring and management of perishable product quality in the cold chain. In: *Smart Packaging Technologies for Fast Moving Consumer Goods* (ed. J. Kerry and P. Butler), 61–74. London: Wiley. ISBN: 978-0470028025.

64 Kuswandi, B., Wicaksono, Y., Jayus, J. et al. (2011). Smart packaging: sensors for monitoring of food quality and safety. *Sensory and Instrumentation for Food Quality* 5 (3): 137–146.

65 Otles, S. and Yalcin, B. (2008). Intelligent food packaging. *Log Forum* 4 (4): 3. http://www.logforum.net/vol4/issue4/no3.

66 Kreyenschmidt, J., Christiansen, H., Hübner, A. et al. (2010). A novel photochromic time-temperature indicator to support cold chain management. *International Journal of Food Science and Technology* 45: 208–215. https://doi.org/10.1111/j.1365-2621.2009.02123.x.

67 Raab, V. (2011). Assessment of novel temperature monitoring systems for improving cold chain management in meat supply chains. Dissertation, Rheinische Friedrich-Wilhelms-Universität Bonn URN: urn: nbn: de: hbz: 5 N-27540

68 Ahvenainen, R. and Hurme, E. (1997). Active and smart packaging for meeting consumer demands for quality and safety. *Food Additives & Contaminants* 14 (6–7): 753–763. https://doi.org/10.1080/02652039709374586.

69 de Kruijf, N., van Beest, M., Rijk, R. et al. (2002). Active and intelligent packaging: applications and regulatory aspects. *Food Additives & Contaminants* 19: 144–162.

70 Selman, J.D. (1995). Time-temperature indicators. In: *Active Food Packaging* (ed. M.L. Rooney). Boston, MA: Springer https://doi.org/10.1007/978-1-4615-2175-4_10.

71 Summers, L. (1992). Intelligent packaging for quality. *Soft Drinks Management International* 36: 32–33.

72 Han, J.H., Ho, C.H.L., and Rodrigue, E.T. (2005). Intelligent packaging. In: *Innovation in Food Packaging* (ed. J.H. Han), 138–155. London: Elsevier Academic Press.

73 Poców, M. (2001). Innovations in intelligent packaging technologies for perishable foods. In: *Novel processes and Control Technologies in the Food Industry. Series I. Life and Behavioral Sciences*, vol. 228 (ed. F. Bozoglu, B. Deak, and B. Ray). Netherlands: NATO Science Series, IOS Press. ISBN: 1586311856.

74 Wanihsuksombat, C.H., Hongtrakul, V., and Suppakul, P. (2010). Development and characterization of a prototype of a lactic acid– based time–temperature indicators for monitoring food product quality. *Journal of Food Engineering* 100: 427–434.

75 Galagan, Y. and Su, W.F. (2008). Fadable ink for time–temperature control of food freshness: Novel new time –temperature indicator. *Food Research International* 41: 653–657.

76 Vanprob.com (2023). https://www.vanprob.com/time_temperature_monitoring#:~:text=Freeze%20Check%20labels%20provide%20a,C%20%2F%2032%20deg. (accessed 29 November 2023).

77 Spotsee.io (2023). Coldchain complete. https://spotsee.io/ (accessed 29 November 2023).

12

Carbon Dioxide Sensors/Indicators-Based Intelligent Food Packaging

Nitin Gupta[1], *Neelisetty Sesha Sai Baba*[2], *Dimple Singh*[2], *Ayyan P. Borkakoty*[2], *Vinayak Srivastava*[2], *and Vibhore Kumar Rastogi*[2]

[1] *Wood K Plus - Competence Center for Wood Composites and Wood Chemistry, St. Veit an der Glan, Austria*
[2] *Department of Paper Technology, Indian Institute of Technology Roorkee, Roorkee, Uttarakhand, India*

12.1 Introduction

Packaging serves as a means of communication, branding for consumer goods, and protecting the packaged product from the ambient atmosphere and during the supply chain [1]. Packages are needed for different types of products, depending on their intended use, environment, storage, and transportation methods. A perishable product, such as food or beverages, requires packaging that maintains its freshness and quality by protecting it from outside contaminants. Materials such as polyethylene or polypropylene that are airtight and moisture-resistant can be used to achieve this goal [2]. The packaging for products that require physical protection, such as electronics, must provide high levels of shock absorption and cushioning, generally provided by materials such as polyurethane foam or bubble wrap [3]. In this chapter, the main discussion area will be food packaging, whose market needs are growing exponentially. Food packages should protect food from unwanted environmental influences, including light, microbes, moisture, oxygen, mechanical stresses, and dust, which can contribute to the deterioration of food or drink [4]. Traditional packaging solutions are less recommended since consumers and manufacturers are not given any information regarding the quality of food products at any point in the supply chain. A new technology that tracks food degradation from farm to fork is emerging to lessen risks related to foodborne illnesses. Consumers and producers are now more interested than ever in real-time quality monitoring and initiatives to extend the shelf life of food [5]. Increasing food waste, safety, quality insurance, and customer awareness have forced researchers to develop smart packaging solutions [6].

Smart packaging incorporates advanced technology like sensors, microprocessors, microcontrollers, and wireless communication to provide real-time data on the freshness of the product of edible products, which can alert consumers when the product is no longer safe to consume by detecting the presence of contaminants or spoilage microorganisms [7, 8].

Smart Food Packaging Systems: Innovations and Technology Applications, First Edition.
Edited by Avik Mukherjee, Santosh Kumar, Manjusri Misra, and Amar K. Mohanty.
© 2025 John Wiley & Sons Ltd. Published 2025 by John Wiley & Sons Ltd.

Smart packaging can use several indicators to provide real-time information on product freshness and quality [9]. Some of the most common include carbon dioxide (CO_2) indicators, time-temperature indicators (TTIs), oxygen indicators, ethylene indicators, pH indicators, and microbial indicators. CO_2 sensors are typically used in the food and beverage industry to detect and measure carbon dioxide levels within a sealed package [10]. These sensors ensure that products remain fresh and safe for consumption by detecting potential spoilage or contamination. By incorporating these sensors into food packaging, manufacturers can significantly extend the shelf life of the products and reduce food waste. Moreover, the consumer can make more informed purchasing decisions. Overall, smart packaging incorporating CO_2 sensors represents an important step forward in food safety and preservation and has the potential to revolutionize the way food is stored and consumed [11]. Although CO_2 is released after food spoilage, it is commonly used as active gas in food packaging to reduce the microbe metabolism rates; therefore, CO_2 is considered to be a direct indicator of food quality [12–14]. Intelligent food quality control relies heavily on reliable, accurate, quick-responding, and safe CO_2 sensors [15]. Based on the mechanism, these CO_2 sensors can be electrochemical and optical CO_2 sensors. Conventional electrochemical CO_2 sensors work on the principle of the Severinghaus electrode that monitors the change of pH value induced by the dissociation of CO_2 in electrolyte solutions [16]. Using microtechnology, these conventional electrochemical CO_2 sensors were replaced by solid-state electrolyte sensors. However, as these sensors had to be heated to a high temperature of 300–800 °C, they were unstable in the long term [17].

In contrast, optical CO_2 sensors have overcome many drawbacks of electrochemical sensors for severe environments and remote sensing applications [18]. Optical CO_2 sensors are generally classified as infrared detectors, colorimetric-type sensors, and fluorescent-type sensors [19]. Infrared detectors inform about CO_2 levels based on light absorption in the infrared region; colorimetric-type sensors generally use pH-sensitive dyes; and fluorescent-type sensors employ fluorescence change according to the concentration of CO_2 [17]. In this chapter, CO_2 sensors from conventional to state-of-the-art technology are discussed. Furthermore, the application of these sensors in the quality monitoring of fruits, vegetables, dairy products, muscle food, etc. has also been widely discussed. Moreover, to comply with strict regulations, the sensor should also be designed keeping some safety aspects in mind, which are also discussed in this chapter.

12.2 Types of CO_2 Sensors

12.2.1 Conventional CO_2 Sensors

Non-distributive infrared (NDIR) and Severinghaus-type sensors are used conventionally and commercially for CO_2 detection. The NDIR technology, despite its excellent precision, was expensive, bulky, prone to contamination, and interfered with water vapor, giving misleading results [20]. In addition, a sealed package used for gas analysis must be destroyed as part of the procedure, making it a destructive method. Thus, there was a high demand for compact and affordable sensors. Using a gas-permeable membrane for separating the gas sample is very common in conventional CO_2 sensors; one example is Severinghaus electrode CO_2 sensors [21]. Severinghaus electrode consists of a CO_2-sensitive glass electrode surrounded by a bicarbonate film covered by a thin plastic membrane that can permeate CO_2 but not water or electrolytic solvents [16]. It measures CO_2 concentration by monitoring the resultant pH change when ambient CO_2 dissolves into an internal electrolyte to form bicarbonate ions. As an advancement in this kind of sensor, one more study has been done that uses the change in conductivity due to the reaction of carbon dioxide and bicarbonate solution inside a cavity covered with a gas-permeable membrane as a parameter [22]. The response time of this sensor was five seconds, which was an achievement in this area of research. Conventionally, potentiometric CO_2 sensors were used to detect the bacterial growth inside the package, but they involved a destructive analysis [23]. These electrochemical potentiometric CO_2 sensors were compact, highly selective, and low cost. However, because of their higher operating temperature, they are unsuitable for food packaging applications. Environmental factors such as pressure, temperature, and humidity can influence the sensor's performance. Hence, the use of indicative films, sensors, and stickers came into existence to identify the state of food without causing any damage to the package [24].

12.2.2 Innovative Optical CO_2 Sensors

Modified atmosphere packaging (MAP) is a food preservation technique used to extend the shelf life of food [25]. In this technique, a mixture of CO_2 and nitrogen (N_2) is typically added to the packaging to displace the air, reducing oxygen inside the packet to a minimum to limit the growth of aerobic microorganisms. If one can monitor this CO_2 level inside the package, it could be the parameter to determine if the packaged food inside is worth consuming or not. This technique was adapted from the use of controlled atmospheric treatment of archaeological items displayed in museums to control the growth of insect pests, which is referred to as controlled atmosphere treatment (CAT), in which CO_2 gas is used to replace the oxygen within the container for four weeks within which the life cycle of any insect or bacteria lying inside the item ends [26]. Optical CO_2 sensors are used to measure the CO_2 levels inside the food package and confirm whether the food is safe to consume. The mechanism behind the general NDIR CO_2 sensor is light absorbance by CO_2 molecules [27]. It works by shining a light source, such as a laser, through the package. The CO_2 molecules in the mixture absorb some of the light, and the sensor measures the amount of light transmitted through the package. This difference in the amount of light is then used to calculate the concentration of CO_2 in the package. But optical CO_2 sensors in food packaging generally use pH-sensitive dyes and fluorescent dyes, which change their color according to the concentration of CO_2 [20]. It is depicted in Figure 12.1 how CO_2 sensors based on pH-sensitive dyes (Figure 12.1a) and fluorescent dyes (Figure 12.1b) change color when the CO_2 level changes inside the package.

The performance of optical sensors has been improved through several recent innovations. Miniaturization is one of those innovations that completely diversified the application domain of different sensors and allowed them to be integrated into various devices, such as smartphones [28]. Multi-gas sensing, low power consumption, remote sensing, high accuracy, and cost-effectiveness are the other innovations in this era. Numerous research studies have been done in this field, and researchers are continuously developing CO_2

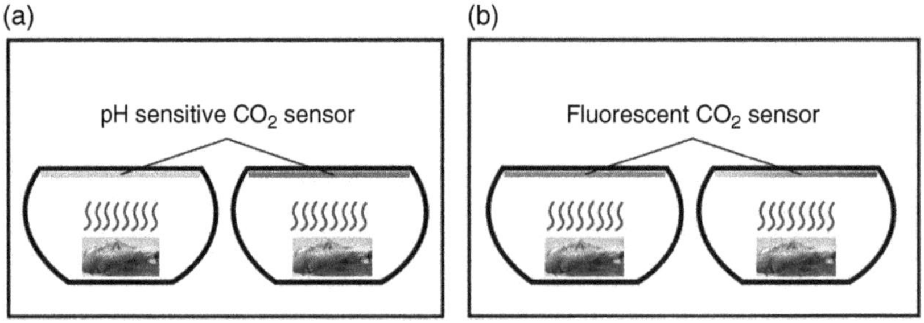

Figure 12.1 Depiction of color change in CO_2 sensors based on (a) pH-sensitive dyes and (b) fluorescent dyes.

sensors using sustainable and safe resources, as shown in Figures 12.2 and 12.3. Agricultural waste material has been used for making the optical CO_2 sensor, and the synthesis process is shown in Figure 12.2a. An indicator hydrogel was developed by Lu et al. from sugarcane bagasse nanocellulose for monitoring the spoilage of chicken breast [29]. This hydrogel was made by employing Zn^{2+} crosslinking to TEMPO-mediated oxidized bagasse fibers. The hydrogel indicated color change based on pH-responsive dyes, which provide this indication by measuring CO_2 levels and microorganism growth in the sample. The color was changed from green to red on the third day, indicating that the chicken was no longer safe for consumption. The application of agricultural waste material into food packaging sensors leads to the circular economy, but at the same time, there are some regulations that these sensors should follow; that is why the final sensor must be of food-grade quality as it may come in contact with the packaged food [32].

Wu et al. have innovated in the miniaturization of CO_2 sensors [17]. They have developed an optical fiber-tip CO_2 sensor that can simultaneously detect CO_2 and temperature. Firstly, a photocrosslinked poly(ionic liquid) (PIL) that is poly(1-allyl-3-vinyl imidazolium bromide) (PAVB) was synthesized in the lab, which was very selective to CO_2 gas because of the large number of imidazolium functional groups on it (Figure 12.2b). Then optical micropatterning technique was incorporated to print Fabry–Pérot interferometers (FPIs) of PAVB on optical fiber. Moreover, commercial SU-8 epoxy resin FPIs were also printed at the end of multicore optical fiber, which worked as a sensor to measure reference temperature. The detection limit of this sensor was nearly 286 ppm. This innovative, tiny sensor was suitable for various applications, including waste gas detection and food quality monitoring. This micropatterning technique can make the sensors at a very high scale in a shorter production time. In another study, an ultrasensitive chemoresistive sensor was made by incorporating functionalized single-walled carbon nanotubes (SWCNTs) and PIL [30]. The detection limit of this sensor was 500 ppt (Figure 12.2c). This sensor was highly selective and resistant to relative humidity interference, making it suitable for real-time monitoring of CO_2 level detection. SWCNTs are known for their eligibility in making gas sensors because they can change their electrical properties drastically due to trace levels of gas analytes because of charge transfer phenomena at CNT/metal electrode contact [33]. But pristine CNTs are insensitive to weak Lewis acids or bases, such as CO_2, CH_4, and H_2, which is why functionalization of CNT is necessary, which facilitates both selectivity and sensitivity of CNT-based gas sensors [34].

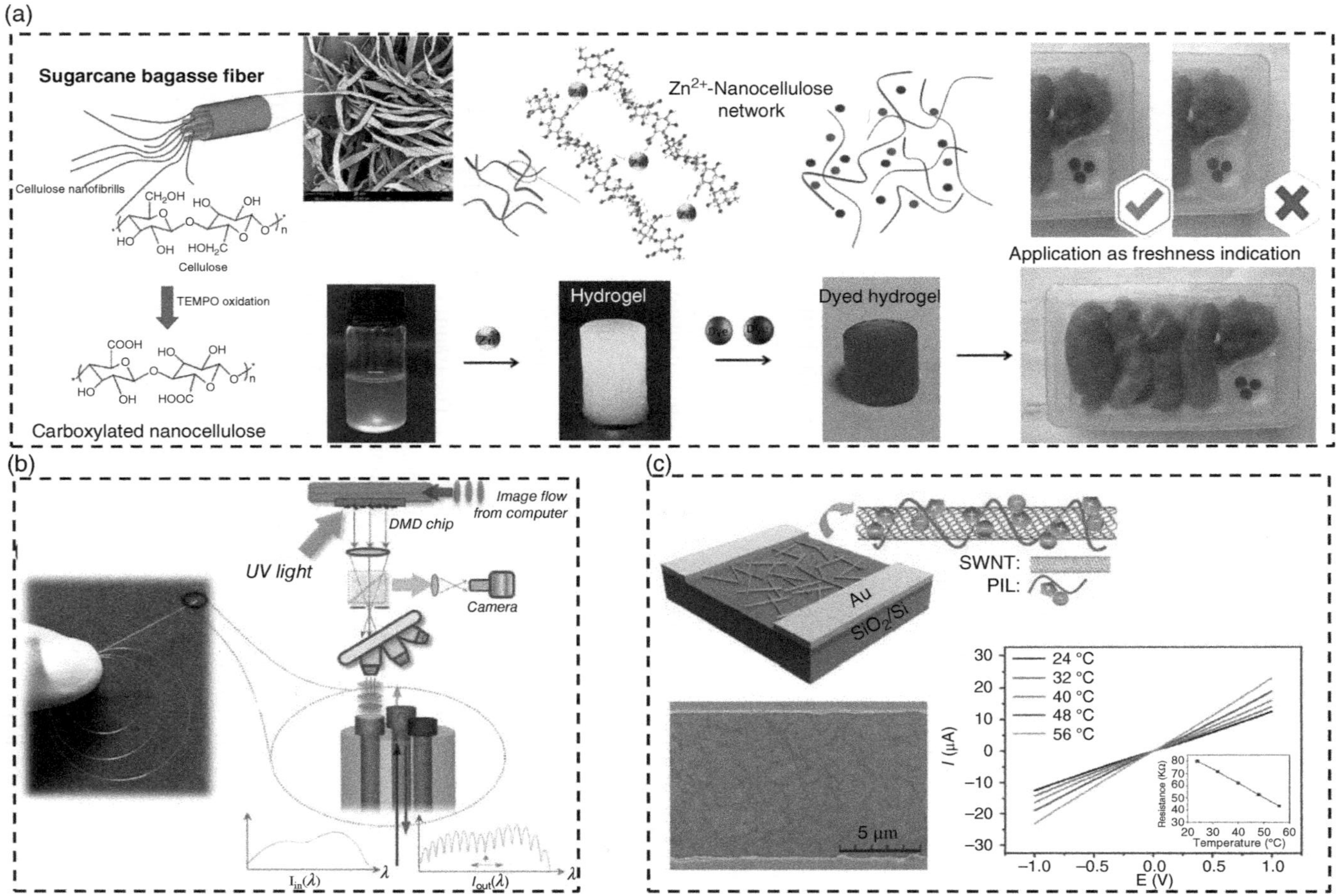

Figure 12.2 Innovations in optical CO_2 sensors: (a) preparation of sugarcane bagasse nanocellulose hydrogel as a colorimetric freshness indicator for intelligent food packaging. *Source:* Reproduced from Lu et al. [29]/with permission of Elsevier. (b) Schematic illustration of the in situ printing of optical fiber-tip CO_2 sensors, photograph of an optical fiber for sensor fabrication, schematic of the optical-printing technology based on a digital-micromirror device, and schematic of the optical fiber-tip CO_2 sensor based on fiber-top Fabry–Pérot interferometers. *Source:* Reproduced from Wu et al. [17]/with permission of Elsevier. (c) A schematic illustration of sensor, SEM image of network across two microfabricated electrodes, and current–voltage characteristic curves of the PIL/SWNTs network-based sensor at different temperatures (24–56 °C). *Source:* Reproduced from Li et al. [30]/with permission of Royal Society of Chemistry.

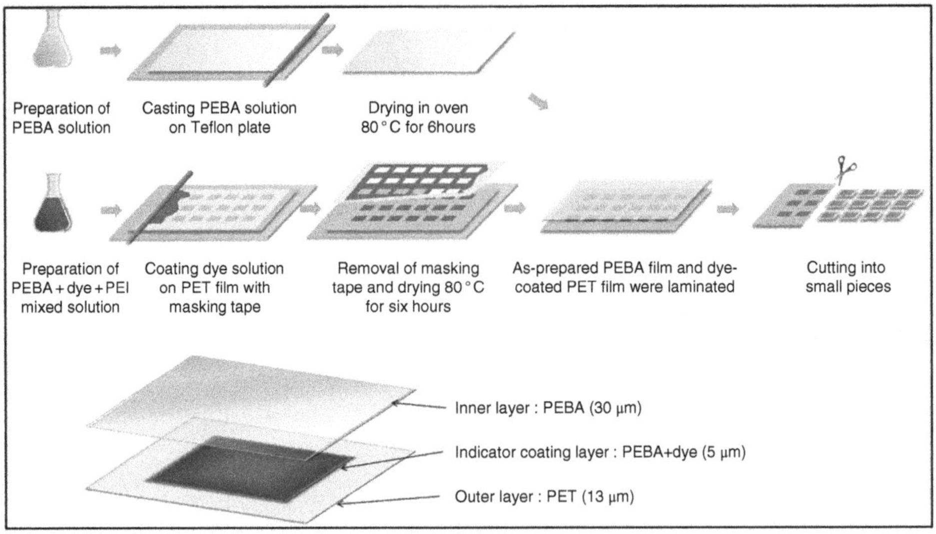

Figure 12.3 Preparation of freshness indicator using PET, PEBA, PEI, and dye for knowing kimchi quality. *Source:* Reproduced from Baek et al. [31]/with permission of Elsevier.

Baek et al. reported the development of a kimchi freshness indicator by using poly(ether-block-amide) (PEBA), polyethylene terephthalate (PET), and pH-sensitive dyes [31], as shown in Figure 12.3. PEBA as the inner layer, PET as the outer layer, and the combination of PEBA, dyes, and polyethyleneimine (PEI) were used as the indicator coating layer for making this sensor. PEI served as a CO_2 molecule adsorbent because of its high basicity and surface area. Different ratios of dyes, namely methyl red (MR) and bromothymol blue (BTB), were used, and the optimum quantity for the best freshness indication was analyzed. CO_2 levels inside the food package can change the pH level, which the pH-sensitive dyes can detect. The most effective film indicator was a mixture of PET/PEBA + dye (MR + BTB) (3 : 7) + 5% PEI/PEBA, considered the best kimchi indicator. Lyu et al. reported an innovative solution that addressed the problems associated with applying pH dyes in food packaging where food has high water content [35]. The traditional pH-sensitive dyes are hydrophilic, which causes them to migrate to food due to the higher water content. To solve this problem, the new BTB−/tetrabutylammonium (TBA+) dye was innovated and used in CO_2 sensing in food. The mechanism behind this dye was employing phase-transfer TBA cation (TBA+) to make the hydrophilic BTB dye hydrophobic. The resultant dye was effectively lipophilic and soluble in polyolefin-based film solutions, which also solved the problem of bad mixing of hydrophilic dyes with hydrophobic packaging materials.

In addition to providing electrical isolation, fluorescent CO_2 sensors reduce noise interference and are very compact, which makes them a better alternative to other types of CO_2 sensors [36]. These sensors frequently employ pyronine (HPTS) because of their unique properties in the visible light region, allowing them to detect CO_2 effectively and resulting in reliable and accurate CO_2 optical sensors [37]. There are many CO_2 sensors reported with the use of HPTS [38], but because of planar sensing films, it was relatively difficult to collect the optical signals [36]. In a recent study, Wang et al. solved this problem by fabricating a CO_2 gas sensor comprising 51 glass capillaries whose channels and the gaps between them have been modified by CO_2-sensitive films [36]. These films were responsible for CO_2 gas sensing and providing optical signals. With this technique, it was possible to have a large sensing area in a small structure.

12.3 Applications of CO_2 Sensors/Indicators in Monitoring Food Quality

CO_2 sensors or indicators can be used in many different areas, including intelligent food packaging. The applications of CO_2 sensors in packaging fruits, vegetables, dairy products, muscle foods, grains, beverages, etc., are explained vastly in this section.

12.3.1 Fruits and Vegetables

Due to technological advancement and changing consumer tastes, fruits and vegetables (F&V) are currently packaged with intelligent packaging instead of conventional gas-regulated packaging. Conventional packaging protects F&V from the outer environment, while intelligent packaging gives information about ripeness, freshness, sensors, labels, etc. It monitors changes including carbon dioxide, humidity, and temperature inside packaged food [39]. This technology reduces a huge amount of fruit and vegetable waste. F&V are extremely vulnerable to physiological degradation, microbiological spoiling, and mechanical harm brought on by endogenous enzymes and exogenous microbes, as well as moisture, oxygen, and mechanical damage during transportation, storage, and marketing. Therefore, fruit and vegetable quality deteriorates, and traits of a lower quality are released [40]. So, the major quality-indicating metabolites, also known as target metabolites, must be identified to create intelligent packaging systems that detect a change in F&V quality. After the harvest of F&V, the food continues to respire, releasing significant amounts of CO_2, the main byproduct of microbial metabolism. The primary byproduct of microorganism metabolism is carbon dioxide. Food degradation and fermentation typically happen due to changes in the external environment. The proliferation of microbes causes carbon dioxide to be produced along with the change in food quality. F&V abruptly increases their respiration rate after being picked, producing significant carbon dioxide, particularly during the transitional phase between growth and age. Consequently, the amount of carbon dioxide in the package correlates with some food freshness [20]. Carbon dioxide is the main component or metabolite that indicates quality in F&V, and some research that detected the CO_2 change using calorimetric dyes is shown in Table 12.1.

In one research, the freshness of freshly cut green bell peppers was monitored using an indicator label that was created using MR and BTB [41]. This work established a correlation between the color variations of the indicator label produced by the 3 : 2 mixture of MR and BTB and the time for which the bell pepper was stored (Figure 12.4a). This correlation helped in the real-time monitoring of bell pepper deterioration. When freshly cut green bell peppers deteriorated, the indication label changed the color to orange from yellow–green (Figure 12.4b). In other investigated work, the MR indicator was used to determine the quality of dragon fruit, which is a non-climatic fruit. The MR

Table 12.1 Monitoring quality changes of fruits and vegetables in intelligent packaging systems using physiological metabolites.

Material	Intelligent packaging strategy	Visible change	Reference
Freshly cut green pepper	Chemical dye-based colorimetric labels	Yellow to green (0–6 days), orange (7–9 days)	[41]
Dragon fruit	Chemical dye-based colorimetric labels	Yellow to orange (0–4 days), red (5–7 days)	[42]
Mushroom	Natural dye-based colorimetric labels	Purple (0–1 day), reddish brown (2–3 days), pink (>4 days)	[43]
Fresh-cut Cantaloupe	CNF hydrogel-based indicator	From dark green to orange on second day	[44]
Kimchi	Switching CO_2 indicator consisting of sodium caseinate solution and pectin	From transparent to opaque after 2 days	[45]

indicator was placed in the dragon fruit package to monitor its freshness. The dragon fruit's freshness was detected by the change in color, which showed a deterioration in fruit quality, as shown in Figure 12.4c. The reason for the color change was an acidic accumulation of CO_2 and volatile acids. The color change occurs from yellow to orange from 0 to 4 days and red from 5 to 7 days [42].

Many visual indicators have been created to convey product quality information by being included in intelligent food packaging systems. For example, indicators help track the quality of agricultural fruit products' ripeness, particularly those with organic dyes mixed into biopolymers. It has been reported that pH dye-based indicators that change color in response to CO_2 levels were embedded in methylcellulose films. Also, the amount of CO_2 inside the container used for packaging assessed the mango fruit's freshness. Moreover, this approach was more feasible as the naked eye can monitor the ripeness stage in mango fruits [46]. In another reported work by Xiang Peng Meng et al., pH dyes with a PEBA film – a polymer that changes color in response to CO_2 levels – were synthesized and characterized. To demonstrate food freshness, color change-based CO_2 indicators could be placed inside the headspace of a kimchi container [31]. To gauge the quality of the kimchi by the partial pressure of CO_2 gas in the headspace of the packing, an irreversible chitosan-based indicator was created [47]. The freshness of mushrooms was also detected in one study ([48]). The study explains the development of an intelligent colorimetric film for determining the freshness of mushrooms. In this study, the major focus was on the usage of nanocomplex and colorimetric films for monitoring the freshness of mushrooms. This colorimetric film was constructed by ovalbumin-carboxymethyl cellulose nanocomplexes loaded with anthocyanins. These were integrated into the starch/polyvinyl alcohol matrix to create the film (OVA-CMC-ACNs). The film was designed to react in such a manner that it detects variations in CO_2 gas and pH buffer, which is an indicator of mushroom spoilage. The color change occurs from purple (0–1 day when the mushroom is fresh) to red–brown

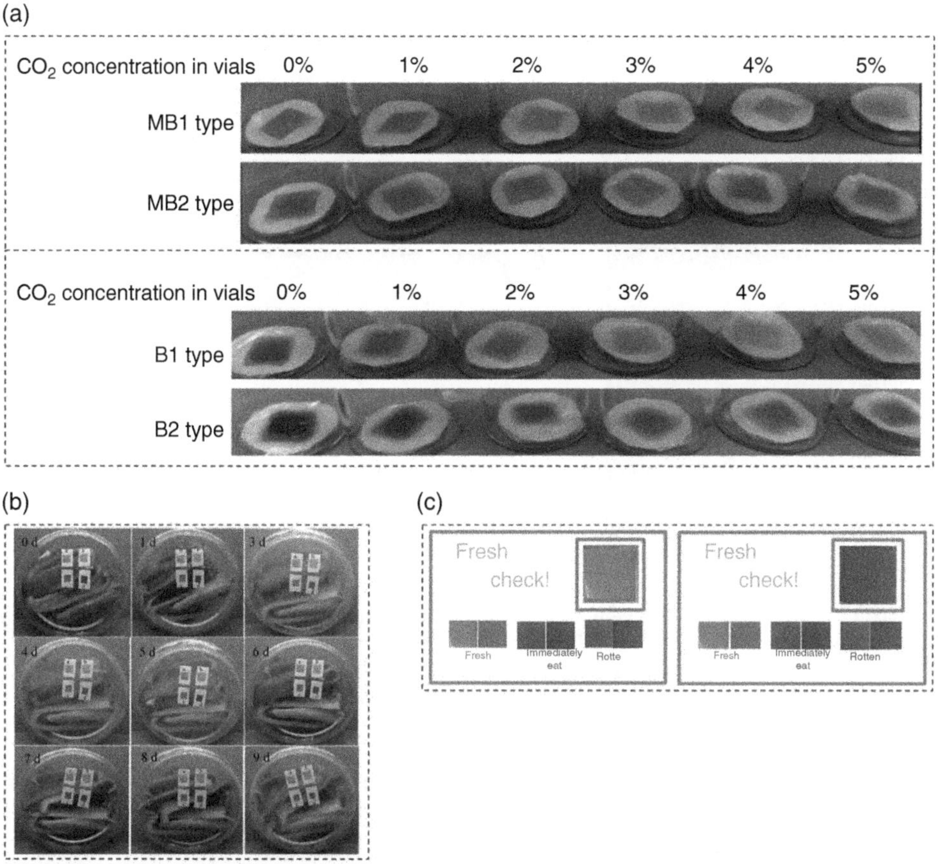

Figure 12.4 (a) Indicator labels' color changes by using different formulas. (b) Changes in label color for fresh-cut pepper deterioration testing at 7 ± 1 °C. *Source:* Reproduced with permission from Chen et al. [41]/Elsevier. (c) Dragon fruit freshness detector with a smart label that is non-climacteric days. *Source:* Reproduced from Warsiki and Rofifah [42]/with permission of IOP Publishing.

(2–3 days when mushroom is subfresh) to pink (when more than 4 days), then mushrooms get spoiled (Figure 12.5a). However, gas indicators are among the intelligent food packaging types that detect changes in gas components in food package headspaces (Figure 12.5b). The goal of the research was to develop a carbon dioxide (CO_2) indicator using sodium caseinate (NaCas) and pectin reactions, which might be used as an intelligent packaging tool to monitor food quality changes [45].

One major factor affecting the consumption and storage of freshly cut products is their freshness. An investigation shows pH-sensitive indicators to build a freshness indicator label for packed fresh-cut green bell peppers [41]. A colorimetric sensor was prepared by immobilizing natural colors taken from red cabbage and rose flowers on filter paper as a carrier. This indicator sensor was built on natural flavonoids (anthocyanins) derived from red cabbage and rose, which changed color from red to green as pH increased [49]. Moreover, the work for detecting the freshness of guava using the bromophenol blue-based color indicator was done in a study. Guavas ripen and produce acetic acid and other volatile

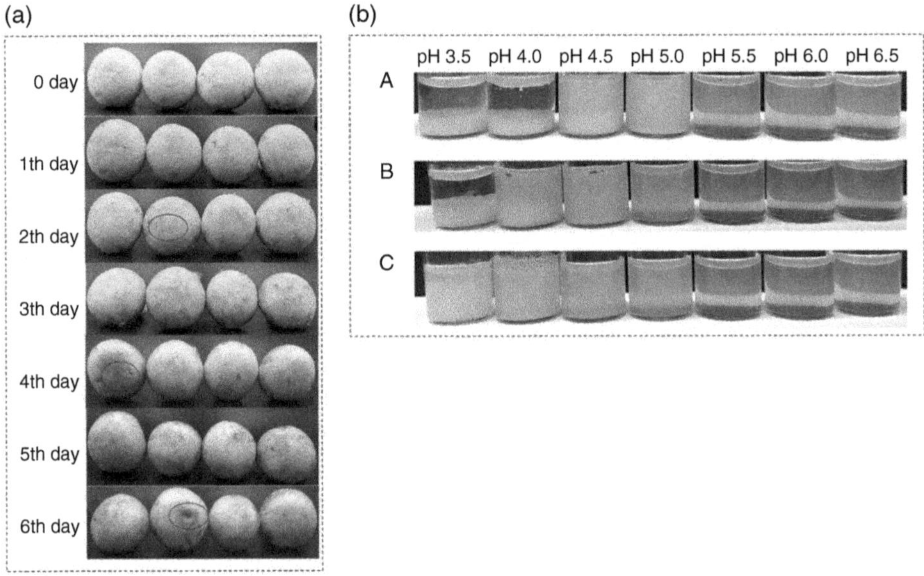

Figure 12.5 (a) Change in appearance and the related color parameter of mushrooms. *Source:* Reproduced with permission from Liu et al. [48]/Elsevier. (b) Appearance changes and agglomeration of NaCas/pectin suspension at various mixing ratios of *A* 9 : 1, *B* 8 : 2, and *C* 7 : 3. *Source:* Reproduced with permission from Choi and Han [45]/Elsevier.

organic chemicals that, over time, build up. The steady accumulation of acetic acid causes the headspace pH to drop gradually. Therefore, in this type of indicator, the color changed from blue to green, which was visible to the naked eye; further, this is useful in determining guava's freshness [50]. It is interesting to notice that grapes undergo metabolite changes during ripening. During the ripening process, the ratio of maleic and tartaric acids changes, leading to a change in the volatile organic acid content. These changes in volatile organic acids were detected by indicator labels based on chlorophenol red, a pH-sensitive dye [51]. However, intelligent packaging helped track food products' freshness during storage by monitoring CO_2 concentration. The possibility of incorrectly interpreting data for packed products using a modified atmosphere is a major issue.

12.3.2 Dairy Products

Milk is the food produced by the mammary glands of mammals, including humans, to feed infants being the primary source of nutrition before they can digest solid food [52]. Humans depend on milk as their first food source; being nutritionally complete, milk provides proteins, fat, carbohydrates, minerals, and vitamins that support growth [53]. Milk is the most consumed food [54] as direct consumption, and various products are also made from milk, which feed humans [55]. Food products made from milk are called dairy products, including yogurt, cheese, butter, cream, and fermented milk products like soured milk, soured cream, frozen foods (ice cream, gelato, ice milk), and custard [56]. The primary functions like quality, safety, cost, and marketing are significantly related to the packaging of dairy products. Hence, the package must preserve the products from physical and environmental hazards, aroma alterations, and moisture loss [57]. Moreover, dairy products are pasteurized, especially milk, so they are extremely perishable. Hence, the packaging must protect dairy throughout the supply chain till it reaches the customer [58, 59]. Intelligent packaging communicates the dairy product history and consumable status [60]. Users can be provided information such as product approach, expiry date, storage conditions, components, packaging material damage, temperature variations, concentration alteration, and microbial growth using intelligent packaging [61]. Indicators are mostly used in packaging to represent the modifications in the food product by interacting with the compounds in the food [62]. The composition of oxygen and carbon dioxide plays a critical role in determining the integrity of the package. The increase in oxygen inside the dairy product package results in the oxidation of lipids and micronutrients. This results in the rise of aerobic microbes, affecting the product's color, flavor, and odor [60]. Hence, MAP, with reduced oxygen levels in the headspace, is preferred, increasing dairy's shelf life by 3–4 times that of normal packaging [63]. The proportion of several gases in the headspace of packaged cheddar cheese is 70% nitrogen gas (N_2) and 30% carbon dioxide (CO_2) [64], for Fior di latte cheese is 50% N_2 and 50% CO_2 [65], for Graviera Agraphon cheese is 50% N_2 and 50% CO_2 [66], for paneer is 50% N_2 and 50% CO_2 [67], for laal peda is 70% N_2 and 30% CO_2 [68], for dietetic rabri is 100% N_2 [69], and for kalakand is 50% N_2 and 50% CO_2 [70].

MAP fails when there is a loss of integrity in the package, the presence of pinholes, or due to oxygen permeability of packaging material affects the composition of gases in the package [71]. Hence, a CO_2 sensor can be utilized to identify the variation in the gas concentration in such cases (usually a decrease in CO_2 with time). Emission of dissolved carbon dioxide due to improper storage conditions is another reason for the usage of CO_2 sensors in milk packaging. The dissolved CO_2 is released when the temperature rises due to improper storage conditions, raising the amount of CO_2 in the package [72]. Hence, a CO_2 sensor can also identify the temperature variation [73, 74]. In addition, the milk contains a high amount of fat consumed by bacteria and emits CO_2, hence increasing the concentration of CO_2 within the package. The need to sense CO_2 in the packaging of dairy foods is represented using a flowchart, as shown in Figure 12.6.

A whey protein isolate (WPI)-based CO_2 indicator was developed to identify the spoilage in cheese packaging. Due to microbial degradation or gradual spoilage, the CO_2 inside the package increases, and when it encounters the indicator placed on the head space of the package, it is dissolved in the WPI suspension, converting into carbonic acid (H_2CO_3) and reducing the pH of the suspension. This results in the indicator's color changing from transparent (neutral pH) to opaque (acidic pH), as shown in Figure 12.7a. The indicator stays transparent above pH 6 and opaque below this pH level [75]. Another CO_2 sensor based on chitosan was developed to identify the microbial growth affecting the freshness of packaged foods by increasing the CO_2 concentration during their storage and transport. When CO_2 comes in contact with the chitosan molecules suspended in aqueous medium of the indicator, it gets dissolved, forming carbonic acid (H_2CO_3) and reducing the pH of the solution. This acidic medium reacts with the sensor transitioning the film from opaque to transparent, as shown in Figure 12.7b. This detector was useful in detecting the increase in the CO_2 concentration in fermented foods like yogurt, cultured buttermilk, cream, meat, and fish [76].

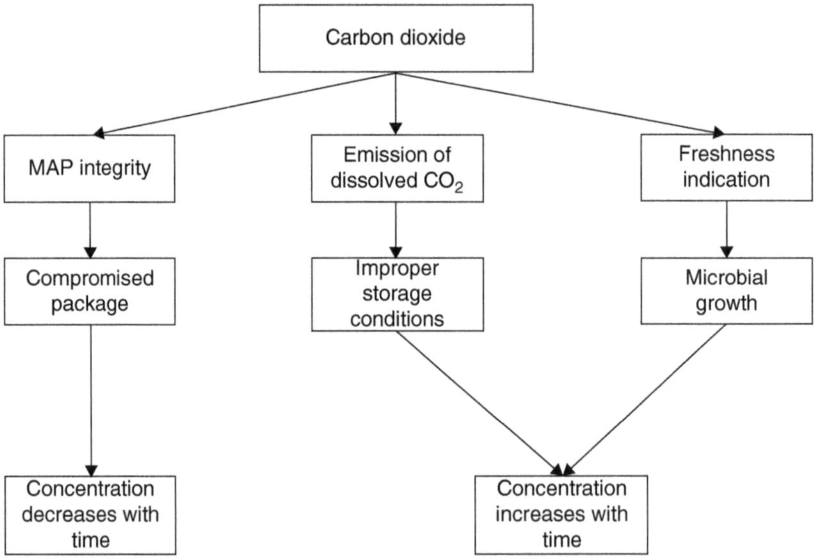

Figure 12.6 Three different routes by which carbon dioxide sensors can sense CO_2 in the dairy package.

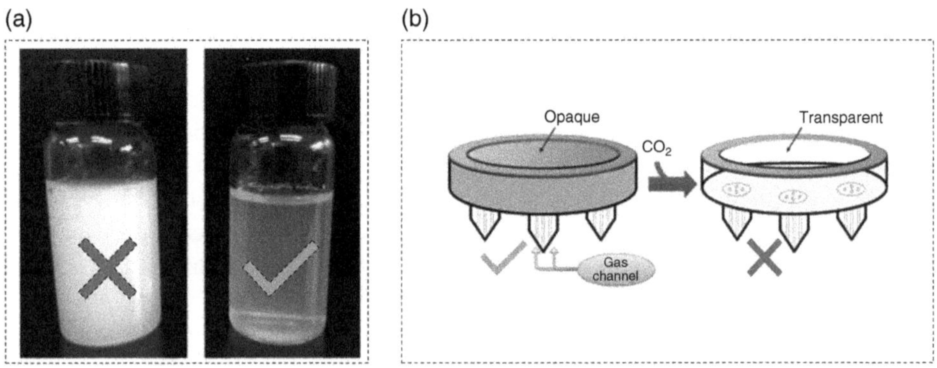

Figure 12.7 Identification of microbial spoilage in dairy products showing (a) whey protein isolate-based dye for CO_2 sensing. *Source:* Reproduced with permission from Lee and Ko [75]/ Springer Nature. (b) Chitosan-based film for CO_2 sensing. *Source:* Reproduced from Jung et al. [76]/ with permission of Elsevier.

The lactococcal bacteria are the reason for the spoilage of milk, converting lactose to glucose, and CO_2 is released during the process. Based on this, a gas sensor utilizing a TGS4161 CO_2 detector with a detection range of 0–4000 ppm was used to measure the concentration in the package containing milk samples. The output voltage was lower for pasteurized milk than fresh milk, indicating that spoilage is slow or less in pasteurized milk. This sensor can be used along with an electronic tongue to assess fresh milk [74]. Cheese is a perishable dairy product that spoils due to the growth of microorganisms. Hence, MAP is mainly preferred in its packaging. Using a CO_2 sensor, the variation in the levels of CO_2

inside the package can be sensed, thereby ensuring that the product is still safe to consume. Whenever the levels of CO_2 inside the package increase, it indicates a growth of microbes inside the cheese, considering it to be spoiled and meant to be discarded [60]. The ripening of any food product leads to the growth of microbes that release gases such as CO_2 or hydrogen, forming small-to-large holes in the cheese mass [77]. The Emmental cheese ripens in two stages: in cold and warm ripening chambers for stabilization and fermentation, respectively; in plastic foil to detect the amount of CO_2 released during maturation. The CO_2 sensor based on nondispersive infrared technology can be used here, whose output changes represent cheese ripening. This analogy was used to develop RFID tags with stretch and pressure sensors to detect cheese ripening [78].

MAP is mostly employed to pack milk products to prevent them from spoilage, and LEAK MASTER is a nondestructive way of identifying leaks in the MAP immediately after packaging. It consists of a test chamber into which the gas will flow continuously, and when there is a leak, the concentration of CO_2 within the chamber rises. By utilizing a highly sensitive infrared sensor, even small leaks in the package can be easily detected [79]. Also, Centec has developed two CO_2 sensors for industrial milk packaging, namely, Carbotec TR–PT and Carbotec NIR, for sensing the amount of dissolved CO_2 in dairy products as well as in carbonated food [80]. Carbotec TR–PT was used for applications with high sensitivity and accuracy, which work on the variation of temperature (TR) and pressure (PT) of the sample [81]. On the other hand, Carbotec NIR works on the principle of attenuated total reflection, where near-infrared (NIR) light is passed through the milk and the wavelength of reflected light is analyzed to identify the change in the percentage of dissolved CO_2 in the product [81] because, with an increase in the CO_2 levels, the infrared absorption also changes [82].

Although CO_2 plays a major role in the packaging and processing of dairy products, the use of CO_2 indicators is not well explored because oxygen, pH sensors, and TTIs are more reliable and convenient. Furthermore, a few studies were conducted based on the combined effect of CO_2 with variation in pH [83], but their use is limited since the pH dyes are highly hydrophilic, resulting in migration into liquid foods and affecting the quality to some extent [84]. Also, these sensors are more respondent to pH rather than CO_2, making the research in developing coupled sensors uncommon [35].

12.3.3 Muscle Foods

Food from animal sources evolved, mainly classified as meat, poultry, and seafood commonly called muscle foods [85]. Muscle foods are desirable, palatable, and nutritious, making them crucial components of a balanced diet [86]. The spoilage of muscle foods is mainly due to microbial contamination, oxidizing lipids, and proteins in meat products, which concerns food safety and quality deterioration [87]. Moreover, the flavor, texture, and color that define the tenderness are meant to be maintained properly, or else it affects the consumer acceptability [88]. So, active packaging systems are used with additives in the package to maintain or extend meat products' quality and shelf life [89]. There are many ways to identify food's chemical and microbial contaminants, but using gas sensors or indicators is still more prominent [10]. Nitrogen inside a MAP develops anaerobic bacteria on a larger scale than in controlled conditions of oxygen and CO_2, so nitrogen alone is less preferred in meat packaging [90]. A few applications include packaging cooked chicken patties with 70% N_2 and 30% CO_2, while the raw and cooked beef is packed at 60% N_2 and 40% CO_2 [91]. As CO_2 is being used in the MAP, the amount of CO_2 present in the package can indicate the quality of the food, which directly symbolizes the freshness of meat [92]. Any variation in the proportion or concentration of CO_2 indicates that the MAP is compromised and there are high chances of bacterial growth [93, 94] over the meat products, which is not tolerable, and the food is considered not safe [26]. Moreover, muscle foods contain various microorganisms that consume oxygen in the headspace and release CO_2 [95]. The increase in CO_2 represents tampering with meat, indicating microbe growth. Hence, the amount of CO_2 and oxygen can indicate the freshness of muscle food [96]. Therefore, an indicator sensitive to CO_2 can be used to identify the variation in the gas proportions [10]. Technological advancement makes using commercially viable sensors practical in the meat industry [97, 98].

Films were made from m-cresol purple and m-cresol red with variations in ethyl cellulose and poly(vinyl butyral). This combination forms a pH-sensitive dye that can be incorporated into a plastic film for the packaging of meat. This film acted as a CO_2 sensor with good sensory properties, like a direct sensor. The films were not water soluble and exhibited long shelf life, but prolonged use for 20–100 hours led to the decomposition of phase-transfer agents, depending on the water content present in the film [99]. Films were also fabricated from silanes and terephthalates to monitor the CO_2 content in the packaged muscle food. It was observed that the sensor responds rapidly to the change in the concentration of CO_2 within the package. The performance characteristics of the dual sensor were analyzed with mutton, chicken, beef, pork, and fish, both fresh and refrigerated, for eight hours with regular monitoring at one-hour intervals. It was found that the CO_2 generation is less in the case of refrigerated food than in fresh food. This implied that the bacterial count in refrigerated meat was less than in fresh meat [100] because refrigeration normally inhibits microbial growth as some bacteria cannot survive at lower temperatures and eventually die [101]. The monitoring of poultry farms using smart sensors was also adapted at several places to increase output. It mainly includes using temperature, humidity, air velocity, ventilation rate, and gas concentrations such as CO_2 and ammonia. The change in the gaseous atmosphere directly affects the humidity, which predominantly impacts the bird's health. Also, increasing the concentration of these gases reduces the farm's production, affecting the livestock's health [102]. Hence, any change in the farm's gas concentrations was monitored so adjustments could be made to obtain optimal conditions [103].

A color sensor was fabricated using a mixture of lysine and anthocyanins with Ɛ-polylysine as an antimicrobial and color stability agent for application in poultry. This sensor can detect even small variations in the gaseous CO_2 and can last several weeks under refrigerated conditions. The evaluation of color was carried out by the naked eye, where the transition is from azure to intense purple with an increase in concentration (Figure 12.8a). It was tested with poultry samples, where a good correlation with the bacterial count was observed [104]. Another research presents a water-based carbon dioxide sensor with ionic liquid that works on the acidity of CO_2 molecules. The ionic liquid enhanced the solubility of CO_2 in water, reducing the response and recovery times and making it more sensitive at lower concentrations. The sensor was stable even at high humidity and refrigerated conditions, hence preferred for packing meat [107]. Another indicator was studied using a spectrophotometer, where the gas concentration was correlated with the hue component.

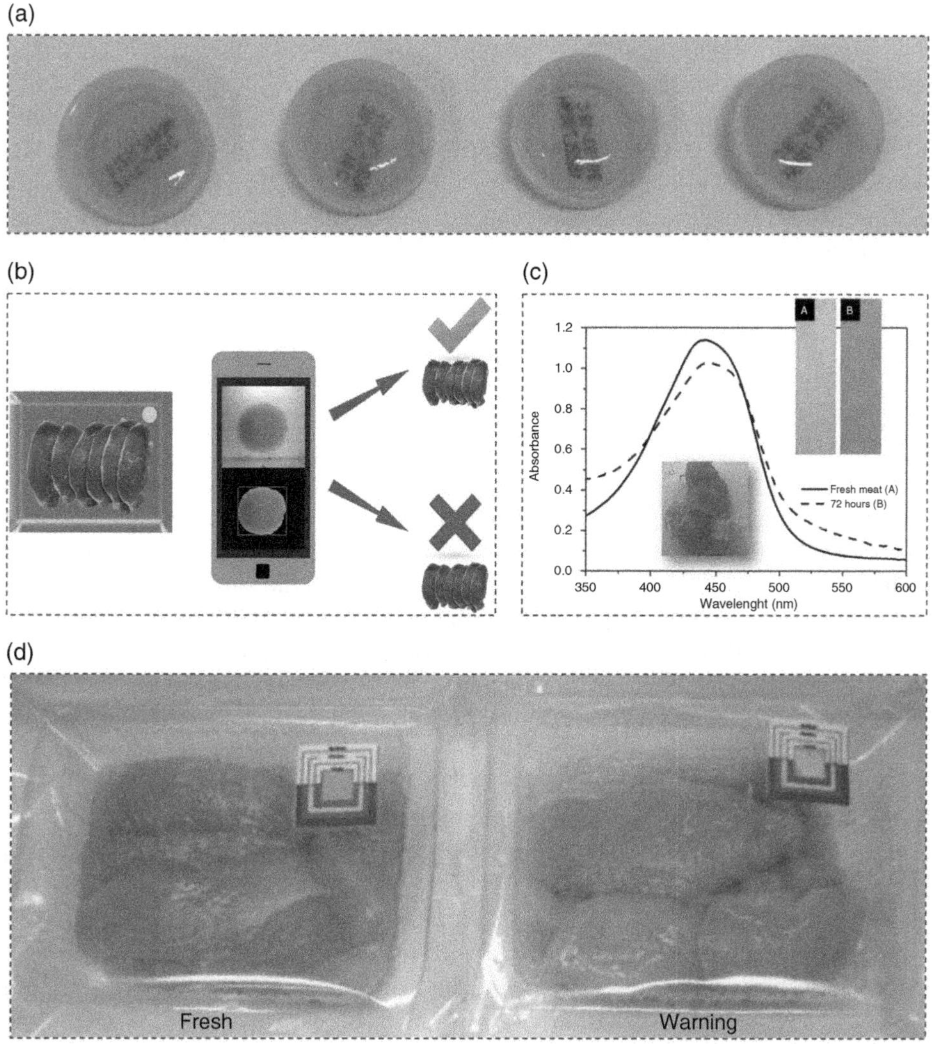

Figure 12.8 (a) Carbon dioxide-sensitive dyes for films. *Source:* Reproduced with permission from Saliu and Della Pergola [104]/Elsevier. (b) Freshness indicator with support of the mobile application. *Source:* Reproduced with permission from Perez de Vargas-Sansalvador et al. [14]/ Elsevier. (c) Absorption spectra of indicator in fresh and rotten meat. *Source:* Reproduced from Pávai et al. [105]/with permission of Springer Nature. (d) Indicator to identify tenderness of chicken. *Source:* Reproduced with permission from Rukchon et al. [106]/Elsevier.

The sensor was utilized in the packaging of pork meat, and it was observed that meat spoilage occurs when the amount of CO_2 within the package exceeds 20% [26]. In addition, a calorimetric freshness sensor was also integrated with pork meat packages. The sensor identifies the increase in CO_2 level in the package and changes the label's color, indicating the rise in the bacterial population within the package. A smartphone application was developed to identify the sensor's color variation (Figure 12.8b), thereby identifying whether the meat was consumable [14].

A curcumin-colored test strip was made from an aqueous solution of Curcuma with the ability to sense changes in pH and CO_2 percentages. This strip was used to sense the rotting process in the fresh chicken breast and meat, where an increase in the percentage of CO_2 inside the package led to the change in the color of the strip from red to yellow, as shown in Figure 12.8c [105]. To monitor the tenderness of the chicken, a chemical barcode was created with BTB and MR based on the acidic properties of CO_2. The indicator was stuck to the inner wall of the package, and the change in color of the indicator represents the variation in the quantity of CO_2, which directly quantifies the number of bacteria present in the food over time (Figure 12.8d). This sensor was found to be stable under all conditions, but a color chart was to be standardized to make commercial use of it [106].

Seafood is considered to be highly perishable, as it deteriorates easily due to the presence of high water content, protein and nonprotein nitrogenous compounds, and polyunsaturated fatty acids. So, the need to preserve for long-term usage is very high [108]. Coating the seer fish meat with sodium acetate and applying MAP alone increases the shelf life from 8 to 28 days. When sensed for gas composition, the sodium acetate-treated meat shows that the CO_2 is maintained at a constant level most of the time, reducing the bacterial count [109]. In sardines (*Sardina pilchardus*), using MAP, vacuum packaging, and air packs provides a shelf life of 12, 9, and 3 days, respectively. The tampering of the package and variation in concentration of CO_2 when sensed were found to be least in MAP and maximum in air packaging. It directly indicates the increases in bacterial count in the package, which also increases the histamine levels [110]. Another study on sardines revealed that the higher CO_2 concentration in the package effectively increases the shelf life by inhibiting bacterial growth [111]. Whereas, in eels (*Anguilla anguilla*), the MAP and vacuum packs provide lower shelf life than that of air packaging, and it was observed that with an increase in the percentage of CO_2 inside the package, the autolysis dominates the microbial growth [112]. Common Carp (*Cyprinus carpio*) was packed under two different MAP conditions, MAP1 (70% N_2: 30% CO_2) and MAP2 (80% O_2: 20% CO_2), analyzing the total viable count and sensory values. The reduction in CO_2 concentration in the package directly affected the bacteriostatic effect. Also, the increase in oxygen content brought an equilibrium of gases in package resulting in inhibiting the total viable count and psychrotrophic count resulting in eight days of shelf life in MAP2, while the same was six days in MAP1 [113].

The paper-based electrical gas sensors (PEGSs) were used for sensing water-soluble gases such as CO_2, ammonia, and trimethylamine at room temperature. These sensors were employed to monitor the quality and freshness of fish and poultry. The results were compared with conventional microbiological testing and were found to be reliable, making their use in integration with near field communication (NFC) tags to indicate the variation in the gas concentration [114]. The latest technology uses machine vision-based colorimetric sensors (MVBCSs) to ensure food safety and quality [115]. Applications include developing a calorimetric sensor nose system from bromocresol purple, bromophenol blue, bromomethyl blue, and thymol blue to detect spoilage in squid, where it detects microbial and physiochemical variations. The sensor could effectively detect the variation of CO_2 and O_2 at the package's headspace along with the microbial growth [116].

An electronic nose can also identify the spoilage rate of vacuum-packaged beef. The beef was stored for eight weeks at 48 °C under vacuum packaging conditions, and an electronic nose equipped with 10 metal oxides, 4 Taguchi, and 1 CO_2 sensor was used to identify the bacterial growth and volatile compounds during the storage period. The mathematical regression analysis was used to determine the relation between spoilage of beef with respect to signal magnitudes by sensors. It was identified that with an increase in the bacterial count, the CO_2 concentration increased [117]. An electronic nose can also be fabricated from metal oxide to identify the volatile compounds behavior in beef. TGS 4160 sensors were used to sense the CO_2 in the package, along with other sensors for methane, hydrogen sulfide, and other gases. The developed nose could distinguish the spoiled and unspoiled beef samples, but further research is necessary to obtain stability and accuracy in the results [118]. These sensors were also applied for quality control of broiler chicken packed under suitable MAP conditions and analyzed at different temperatures for the variation in the proportion of gases inside the package. The results were compared with those obtained by microbiological count values, and it was found that they could also identify the spoilage of broiler chicken [119]. Another electronic nose was developed using discrete wavelet transform and short- and long-term memory to assess beef quality, eliminating the signal noise problem. It can classify beef quality with an accuracy of 94.83%, and the average F-measure was 85.05% [120].

Insignia Technologies has developed a leak detection label for MAP foods, where it can detect leaks even due to pinholes and confirm the integrity of the package. The company states that the label is suitable only with barrier lidding films, and testing must be done before commercial use to confirm the suitability with the product atmosphere [121]. Some of the other commercially available CO_2 sensors are mentioned in Table 12.2, which are extensively used in packaging muscle foods.

Table 12.2 Commercially available CO_2 indicators and mechanisms for packaging of muscle foods.

Manufacturer/patent holder	Compounds detected	Principle of indication	References
Vaisala – CARBOCAP	CO_2	NDIR sensor	[122]
Insignia – after opening freshness label	CO_2	Color change	[121]
Temptime corporation	CO_2	Color change	[123]
PreSens	CO_2	Color change	[124]
Andrew Mills, Grahams Skinner	CO_2	Color change	[125]
James A Kane	CO_2	Color change	[126]
AVL Medical Instruments	CO_2, NH_4, H_2S, amines	Formation of the color of heavy metal sulfides	[127]
Visual Spoilage Indicator Company	CO_2	Color change	[128]
Sealed Air Limited (Tuffex GS)	CO_2	Color change	[129]
Bo Holte	CO_2	Color change	[130]
Mattia et al.	CO_2	Color change of bromothymol blue	[131]
Thomas J. Hooran	CO_2, SO_2, NH_4	Color change of xylenol blue, bromocresol purple, bromocresol green, cresol red, bromothymol blue, neutral red	[132, 133]
Michael P. Neary	CO_2, H_2, NH_4	Color change of liquid crystal indicator	[134]

CO_2 sensors can be utilized to identify the spoilage of all kinds of muscle foods. But there are some challenges in commercializing the sensors. Most of them are destructive technique-based, which is not acceptable in the case of food packaging [23]. Whenever MAP gets tampered with, the CO_2 level decreases. But it also promotes microbial growth, which increases the CO_2 level. So, the transition in most color-based indicators remains unaffected as they are reversible, hence the indication is incorrect. Due to microbial contamination, the food packed in such conditions is no longer safe to consume [135]. Also, there is a need to classify the colors for the color-based indicators. So, integrating color-based sensors with smart packaging identifiers such as RFIDs is essential to track the package's life cycle [136]. Electronic noses can overcome the issue with color-based sensors, but their integration into day-to-day packaging is still lacking [119]. Also, it is to be noted that the variation of CO_2 in a package depends on storage time, microbial respiration, type of food, packaging material, and head space [73].

12.3.4 Other Foods

12.3.4.1 Grains

In response to the growing global market for quality grains, farmers and the grain management industry were urged to invest and work on efficient and reliable grain quality management techniques due to their growing concerns about grain quality [137]. Traditional methods of estimating grain quality, including random grain quality checking and measuring grain temperature and moisture contents, are inadequate to ensure market competitiveness for food-grade grains. The focus is now on the approach of grain quality management using CO_2 measurement. Several grains have been researched, including pinto beans [138] and soybeans [139]. Figure 12.9a shows a grain quality monitoring system in vertical silos for real-time monitoring of quality of stored grains [140]. In a study, CO_2 concentrations in the storage silo were measured and correlated to the presence of insects and mycotoxin levels in the stored grain [143]. The results revealed that at CO_2 concentrations of 400–500 ppm, grains were observed to be safe, but at higher concentrations of CO_2, insect activities inside the grain silo can be seen. The outcomes of this research revealed the potential of CO_2 detection for safe storage limits and grain quality monitoring.

A CO_2 sensor was developed using polyaniline boronic acid (PABA) for detecting spoilage in stored grain [141]. PABA was used as interdigitated conductive electrode region in the sensor (Figure 12.9b). The sensing mechanism was based on the interconversion between conducting emeraldine salt form and the polymer through protonation and deprotonation. When the gaseous CO_2 permeates through the gas-permeable membrane, CO_2 reacts with water from the Nafion electrolyte to create a bicarbonate ion (HCO_3^-), and a proton protonates the polyaniline and converts it to conducting emeraldine salt form. As the CO_2 partial pressure increases, the conductivity increases due to an increase in protonation, which changes the pH of the internal electrolyte. By measuring these changes, the concentration of the CO_2 can be measured. But on the other hand, in the absence of an analyte (gaseous CO_2), the polymers get deprotonated, which converts them to an insulating emeraldine base form [144]. This sensor was tested by measuring CO_2 levels in the range of 380–2400 ppm, temperatures from 10 to 55 °C, and relative humidity of 20–70%. Both quantitative and qualitative loss of grain has occurred because of inadequate management and monitoring of the

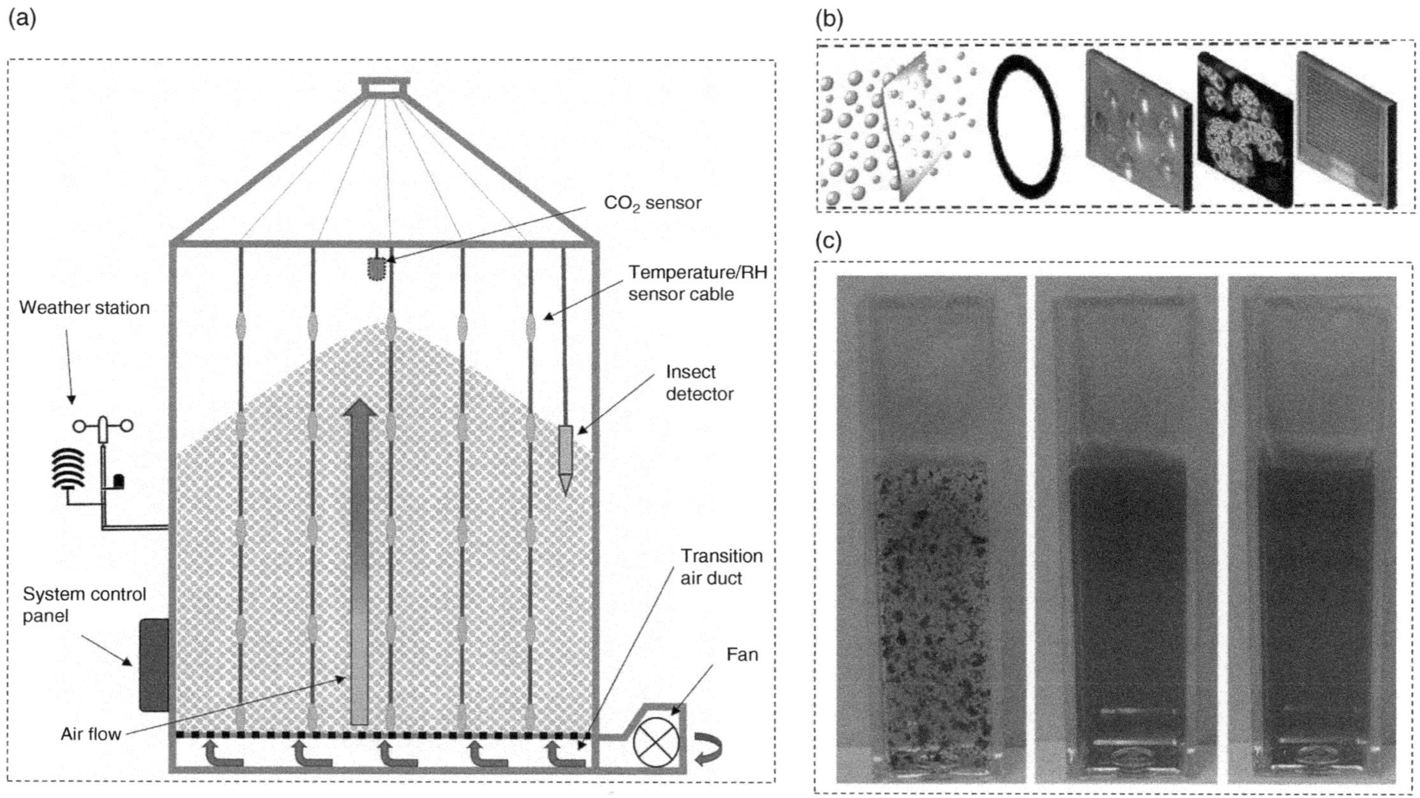

Figure 12.9 (a) Vertical silo used in the storage and monitoring of grain quality. *Source:* Reproduced from Lutz and Coradi [140]/with permission of Elsevier. (b) Pictorial cross-sectional view of the CO_2 sensor from right to left, gold interdigitated electrode PCB array, PABA polymer, Nafion, O-ring, and gas-permeable membrane. *Source:* Reproduced with permission from Neethirajan et al. [141]/Elsevier. (c) Irreversibility of a dye-incorporated chitosan-based CO_2 indicator with changes in pH values of 7, 5, and 7 from left to right. *Source:* Reproduced with permission from Lee et al. [142]/Springer Nature.

grain mass that has been stored [145]. Monitoring the grain's water content, temperature, and relative humidity in the storage environment is essential for maintaining the grain's quality. Thus, researchers are creating smart, intelligent CO_2 detection systems as CO_2 released is a crucial sign of grain quality. Several academic research have demonstrated the effectiveness of preventative monitoring methods in lowering postharvest losses.

12.3.4.2 Beverages

Intelligent packaging extends the shelf life of food products and offers useful information about their quality and safety, helping to manage the cold chain more effectively, reduce food waste, and boost consumer protection. A chitosan-based irreversible CO_2 indicator was developed for Makgeolli (a Korean alcoholic beverage) quality during storage [142]. The dyes were incorporated with chitosan to get a visual change due to the increased CO_2 levels, and to synthesize the CO_2 indicator, 0.3% brilliant blue (BB) was added to chitosan dissolved in 0.1 M HCl solution. During microbial fermentation of food, CO_2 accumulates in the headspace of food packaging, leading to pH change. This phenomenon was used in this study, where chitosan solution changed its visual appearance because of pH changes in the headspace of the packaging. At a low CO_2 concentration, chitosan film in the headspace was opaque (Figure 12.9c), but became transparent at high CO_2 concentrations due to pH-dependent semicrystalline behavior of chitosan molecules in aqueous media [146]. However, the main drawback of this kind of chitosan-based CO_2 indicator is its reversibility. They may again become opaque due to an unintentional decrease in the CO_2 concentration and provide the wrong indication of beverage quality. This study solved this issue using the pH-dependent polarity of chitosan molecules in hydrophobic dye and water. Once the chitosan flocculated particulates dissolved in aqueous conditions, the indicator seemed transparent. However, even if the neutral conditions are back, the chitosan molecules cannot re-encapsulate the color dye since the released color dye molecules have a lyophilic affinity to the acidic aqueous conditions.

12.3.4.3 Kimchi

Several studies have shown a substantial association between the kimchi's properties and the indicator and have emphasized the significance of the CO_2 indicator in food packaging. Korean fermented vegetable dish kimchi is appreciated for its health benefits worldwide [147]. Kimchi is regarded as a symbol of Korean identity and pride because it has been a part of Korean food culture for thousands of years. Kimchi is also a food whose spoilage is often caused by CO_2 and ammonia, and researchers worldwide study it. In a study, by utilizing pH-sensitive dyes like methyl orange, MR, thymol blue, chlorophenol red, mimethyl yellow, phenol red, and m-cresol purple encased in ethyl cellulose matrices, a colorimetric array indicator was designed for the detection of ammonia and CO_2 [148]. In the indicator array, CO_2 gets dissolved in water and forms carbonic acid, which dissociates to produce H^+ ions, resulting in protonation of the dye molecules to cause a color change. The color of each indicator element on the array indicators was converted to RGB scale. The ΔRGB array patterns were determined by subtracting the RGB values before the CO_2 exposure from those after the exposure. The root cause of this color change was the protonation and deprotonation of the dyes in the array indicator, making them sensitive to the detection of CO_2. In a recent study using the same principle, calcium hydroxide was used with different dyes for CO_2 gas monitoring [149]. Here, the selectivity of calcium hydroxide was used as the active material, and color changing dyes assisted it in visual identification. The films were made and used on the kimchi packaging as a ripening indicator.

12.4 Safety Aspects of CO_2 Sensors

According to the reports, one-third of food is wasted each year globally, approximately 1.3 billion tons [150]. Considering these numbers, dealing with this problem using cutting-edge technologies like intelligent food packaging is necessary. At the same time, the components used in these technologies should not affect human health or the environment. Food packaging sensors should be safe because they often come in direct contact with the food. If a sensor is not properly sealed, it could allow pathogens or other contaminants to enter the food package and potentially cause foodborne illness [151]. Furthermore, the food safety regulations and standards, such as those set by the Food and Drug Administration (FDA) and the European Union (EU), must be followed when designing and manufacturing the sensors, which state that all the food-contact materials shall not be hazardous to human health and shall not affect the composition or sensory character of the packed foodstuff in an unacceptable way [152]. With this great technology, some possible challenges like sensor waste recycling also come into the picture that must be addressed. Applying biodegradable materials to CO_2 sensors can contribute to sustainability as they break down naturally and reduce the amount of waste going to landfills [153]. It will fulfill the growing consumer demand for intelligent packaging and reduce environmental impacts. The consumer perspective toward intelligent packaging is as important as developing biodegradable materials for incorporation into this technology. In numerous studies, intelligent packaging has shown its potential, which is sufficient for its application, but sometimes this can undesirably affect the properties such as taste, odor, and texture of packaged food [154]. Therefore, it is required to ensure that the sensory properties of packaged food in intelligent food packaging are appropriate for the consumer. According to some reported studies [155], it is identified that consumer acceptance for intelligent packaging is less than that of active packaging as it was not acceptable for them to have some bio-agents in meat packaging, but on the other hand, some studies also have shown opposite results [156, 157]. In conclusion, safety aspects should be highly considered while designing food packaging sensors like CO_2 sensors, as they belong to sensory characteristics of food.

12.5 Conclusion and Perspective

Intelligent packaging is a kind of packaging that incorporates advanced technology like sensors, microprocessors, microcontrollers, and wireless communication to provide real-time data on the freshness of edible products, which can alert consumers when the product is no longer safe to consume by detecting the presence of contaminants or spoilage microorganisms. After reviewing these studies, it can be concluded that developing CO_2 sensors in intelligent food packaging is a very active area of research and innovation. The new innovative CO_2 sensors have overcome shortcomings like the high response time of conventional CO_2 sensors. While some challenges need to be addressed to make a commercially viable solution, these challenges are easy implementation, reliability, low cost, small size, and food-grade quality of the sensor. The consumer perspective toward intelligent packaging is as important as developing biodegradable materials for incorporation into this technology, which makes it necessary for the sensor to be safe.

Acknowledgment

Nitin Gupta would like to thank Wood K Plus, Austria, for their continuous support during writing of this chapter.

References

1 Stoma, M. and Dudziak, A. (2022). Eastern poland consumer awareness of innovative active and intelligent packaging in the food industry: exploratory studies. *Sustainability (Switzerland)* 14 (20): https://doi.org/10.3390/su142013691.

2 Hess, K.M., Heveran, C.M., and Srubar, W.V. (2020). A computational approach to design moisture-resistant wood polymer composites. *Materials Today Communications* 25 (August): 101594. https://doi.org/10.1016/j.mtcomm.2020.101594.

3 Kim, E., Zhang, H., Lee, J.H. et al. (2021). MXene/polyurethane auxetic composite foam for electromagnetic interference shielding and impact attenuation. *Composites Part A: Applied Science and Manufacturing* 147 (April): 106430. https://doi.org/10.1016/j.compositesa.2021.106430.

4 Sarker, A., Mondal, S.C., Ahmmed, R. et al. (2023). Prospects and challenges of polymer nanocomposites for innovative food packaging. *Smart Polymer Nanocomposites* 355–377. https://doi.org/10.1016/B978-0-323-91611-0.00021-9.

5 Mohebi, E. and Marquez, L. (2015). Intelligent packaging in meat industry: an overview of existing solutions. *Journal of Food Science and Technology* 52 (7): 3947–3964. https://doi.org/10.1007/s13197-014-1588-z.

6 NRDC (2019). Increasing public awareness about food waste prevention guide | NRDC. https://www.nrdc.org/resources/increasing-public-awareness-about-food-waste-prevention-guide (accessed 31 January 2023).

7 Ai, B., Zheng, L., Li, W. et al. (2021). Biodegradable cellulose film prepared from banana pseudo-stem using an ionic liquid for mango preservation. *Frontiers in Plant Science* 12 (February): 1–10. https://doi.org/10.3389/fpls.2021.625878.

8 Alam, A.U., Rathi, P., Beshai, H. et al. (2021). Fruit quality monitoring with smart packaging. *Sensors* 21 (4): 1–30. https://doi.org/10.3390/s21041509.

9 Müller, P. and Schmid, M. (2019). Intelligent packaging in the food sector: a brief overview. *Foods* 8 (1): https://doi.org/10.3390/foods8010016.

10 Shaalan, N.M., Ahmed, F., Saber, O., and Kumar, S. (2022). Gases in food production and monitoring: recent advances in target chemiresistive gas sensors. *Chemosensors* 10 (8): 1–20. https://doi.org/10.3390/chemosensors10080338.

11 Barska, A. and Wyrwa, J. (2017). Innovations in the food packaging market – intelligent packaging – a review. *Czech Journal of Food Sciences* 35 (1): 1–6. https://doi.org/10.17221/268/2016-CJFS.

12 Fuertes, G., Soto, I., Carrasco, R. et al. (2016). Intelligent packaging systems: sensors and nanosensors to monitor food quality and safety. *Journal of Sensors* https://doi.org/10.1155/2016/4046061.

13 Lee, D.S. (2016). Carbon dioxide absorbers for food packaging applications. *Trends in Food Science and Technology* 57: 146–155. https://doi.org/10.1016/j.tifs.2016.09.014.

14 Perez de Vargas-Sansalvador, I.M., Erenas, M.M., Martínez-Olmos, A. et al. (2020). Smartphone based meat freshness detection. *Talanta* 216 (March): 120985. https://doi.org/10.1016/j.talanta.2020.120985.

15 Kuswandi, B. and Jumina (2019). *Active and intelligent packaging, safety, and quality controls*. In: *Fresh-Cut Fruits and Vegetables: Technologies and Mechanisms for Safety Control* (ed. M.W. Siddiqui), 243–294. Elsevier Inc. https://doi.org/10.1016/B978-0-12-816184-5.00012-4.

16 Mills, A. (2009). Optical sensors for carbon dioxide and their applications. *Sensors for Environment, Health and Security* 347–370.

17 Wu, J., Yin, M.-J., Seefeldt, K. et al. (2018). In situ M-printed optical fiber-tip CO2 sensor using a photocrosslinkable poly(ionic liquid). *Sensors and Actuators, B: Chemical* 259: 833–839. https://doi.org/10.1016/j.snb.2017.12.125.

18 Sabri, N., Aljunid, S.A., Salim, M.S. et al. (2013). Toward optical sensors: review and applications. *Journal of Physics: Conference Series* 423 (1): https://doi.org/10.1088/1742-6596/423/1/012064.

19 Struzik, M., Garbayo, I., Pfenninger, R., and Rupp, J.L.M. (2018). A simple and fast electrochemical CO_2 sensor based on $Li_7La_3Zr_2O_{12}$ for environmental monitoring. *Advanced Materials* 30 (44): 1–10. https://doi.org/10.1002/adma.201804098.

20 Puligundla, P., Jung, J., and Ko, S. (2012). Carbon dioxide sensors for intelligent food packaging applications. *Food Control* 25 (1): 328–333. https://doi.org/10.1016/j.foodcont.2011.10.043.

21 Big Chemical Encyclopedia (2019) Severinghaus sensor – big chemical encyclopedia. https://chempedia.info/info/severinghaus_sensor/ (accessed 15 March 2023).

22 Varlan, A.R. and Sansen, W. (1997). Micromachined conductometric p(CO2) sensor. *Sensors and Actuators, B: Chemical* 44 (1–3): 309–315. https://doi.org/10.1016/s0925-4005(97)00223-2.

23 Kress-Rogers, E. (2001). *Sensors for Food Flavour and Freshness: Electronic Noses, Tongues and Testers*, 2001 553–662. Boca Raton, FL: CRC Press. *Instrumentation and Sensors for the Food Industry*, pp. 568–574.

24 Sameen, A., Sahar, A., Khan, S. et al. (2022). Shelf life of foods through nanosensors application. In: *Nanotechnology Interventions in Food Packaging and Shelf Life*, 173–202. Boca Raton: CRC Press https://doi.org/10.1201/9781003207641-11.

25 Winotapun, C., Issaraseree, Y., Sirirutbunkajal, P., and Leelaphiwat, P. (2023). CO_2 laser perforated biodegradable films for modified atmosphere packaging of baby corn. *Journal of Food Engineering* 341 (September 2022): 111356. https://doi.org/10.1016/j.jfoodeng.2022.111356.

26 Perez de Vargas-Sansalvador, I.M., Erenas, M.M., Diamond, D. et al. (2017). Water based-ionic liquid carbon dioxide sensor for applications in the food industry. *Sensors and Actuators B: Chemical* 253: 302–309. https://doi.org/10.1016/j.snb.2017.06.047.

27 UCAR (2023) Carbon dioxide absorbs and re-emits infrared radiation | Center for Science Education. https://scied.ucar.edu/learning-zone/how-climate-works/carbon-dioxide-absorbs-and-re-emits-infrared-radiation (accessed 4 March 2023).

28 Tian, X., Li, Y., Zhao, J. et al. (2022). A silver-based miniature gas sensor for hydrogen sulfide detection. *IEEE Sensors Journal* 23: 1–7. https://doi.org/10.1109/JSEN.2022.3233815.

29 Lu, P., Yang, Y., Liu, R. et al. (2020). Preparation of sugarcane bagasse nanocellulose hydrogel as a colourimetric freshness indicator for intelligent food packaging. *Carbohydrate Polymers* 249 (July): 116831. https://doi.org/10.1016/j.carbpol.2020.116831.

30 Li, Y., Li, G., Wang, X. et al. (2012). Poly(ionic liquid)-wrapped single-walled carbon nanotubes for sub-ppb detection of CO_2. *Chemical Communications* 48 (66): 8222–8224. https://doi.org/10.1039/c2cc33365a.

31 Baek, S., Maruthupandy, M., Lee, K. et al. (2018). Preparation and characterization of a poly(ether-block-amide) film–based CO_2 indicator for monitoring kimchi quality. *Reactive and Functional Polymers* 131 (February): 75–83. https://doi.org/10.1016/j.reactfunctpolym.2018.07.007.

32 Gupta, N., Mahur, B.K., Izrayeel, A.M.D. et al. (2022). Biomass conversion of agricultural waste residues for different applications: a comprehensive review. *Environmental Science and Pollution Research* 29 (49): 73622–73647. https://doi.org/10.1007/s11356-022-22802-6.

33 Li, Y., Hodak, M., Lu, W., and Bernholc, J. (2016). Mechanisms of NH3 and NO_2 detection in carbon-nanotube-based sensors: An ab initio investigation. *Carbon* 101 (2): 177–183. https://doi.org/10.1016/j.carbon.2016.01.092.

34 Han, M., Jung, S., Lee, Y. et al. (2021). PEI-functionalized carbon nanotube thin film sensor for CO_2 gas detection at room temperature. *Micromachines* 12 (9): 1–11. https://doi.org/10.3390/mi12091053.

35 Lyu, J.S., Choi, I., Hwang, K.S. et al. (2019). Development of a BTB−/TBA+ ion-paired dye-based CO_2 indicator and its application in a multilayered intelligent packaging system. *Sensors and Actuators, B: Chemical* 282 (July 2018): 359–365. https://doi.org/10.1016/j.snb.2018.11.073.

36 Wang, J., Wen, Z., Yang, B., and Yang, X. (2017). Optical carbon dioxide sensor based on fluorescent capillary array. *Results in Physics* 7: 323–326. https://doi.org/10.1016/j.rinp.2016.12.044.

37 Kumar, R., Yadav, R., Kolhe, M.A. et al. (2018). 8-Hydroxypyrene-1,3,6-trisulfonic acid trisodium salt (HPTS) based high fluorescent, pH stimuli waterborne polyurethane coatings. *Polymer* 136: 157–165. https://doi.org/10.1016/j.polymer.2017.12.056.

38 Borisov, S.M., Waldhier, M.C., Klimant, I., and Wolfbeis, O.S. (2007). Optical carbon dioxide sensors based on silicone-encapsulated room-temperature ionic liquids. *Chemistry of Materials* 19 (25): 6187–6194. https://doi.org/10.1021/cm7019312.

39 de Oliveira Filho, J.G., Bertolo, M.R.V., Rodrigues, M.Á.V. et al. (2021). Curcumin: a multifunctional molecule for the development of smart and active biodegradable polymer-based films. *Trends in Food Science and Technology* 118: 840–849. https://doi.org/10.1016/J.TIFS.2021.11.005.

40 He, X., Pu, Y., Chen, L. et al. (2023). A comprehensive review of intelligent packaging for fruits and vegetables: target responders, classification, applications, and future challenges. *Comprehensive Reviews in Food Science and Food Safety* (November): 1–40. https://doi.org/10.1111/1541-4337.13093.

41 Chen, H. and zhi, Zhang, M., Bhandari, B. and Guo, Z. (2018). Applicability of a colorimetric indicator label for monitoring freshness of fresh-cut green bell pepper. *Postharvest Biology and Technology* 140: 85–92. https://doi.org/10.1016/J.POSTHARVBIO.2018.02.011.

42 Warsiki, E. and Rofifah, N. (2018). Dragon fruit freshness detector based on methyl red colour indicator. *IOP Conference Series: Earth and Environmental Science* 209 (1): https://doi.org/10.1088/1755-1315/209/1/012016.

43 Liu, M., Zhang, J., Liu, S., and Li, B. (2022). A label-free visual aptasensor for zearalenone detection based on target-responsive aptamer-cross-linked hydrogel and color change of gold nanoparticles. *Food Chemistry* 389 (April): 133078. https://doi.org/10.1016/j.foodchem.2022.133078.

44 Lu, P., Liu, R., Liu, X., and Wu, M. (2018). Preparation of self-supporting bagasse cellulose nanofibrils hydrogels induced by zinc ions. *Nanomaterials* 8 (10): 1–14. https://doi.org/10.3390/nano8100800.

45 Choi, I. and Han, J. (2018). Development of a novel on–off type carbon dioxide indicator based on interactions between sodium caseinate and pectin. *Food Hydrocolloids* 80: 15–23. https://doi.org/10.1016/J.FOODHYD.2018.01.028.

46 Noiwan, D., Suppakul, P., and Rachtanapun, P. (2022). Preparation of methylcellulose film-based CO_2 indicator for monitoring the ripeness quality of mango fruit cv. Nam Dok Mai Si Thong. *Polymers* 14 (17): https://doi.org/10.3390/POLYM14173616.

47 Meng, X., Lee, K., Kang, T.-Y., and Ko, S. (2015). An irreversible ripeness indicator to monitor the CO_2 concentration in the headspace of packaged kimchi during storage. *Food Science and Biotechnology* 24 (1): 91–97. https://doi.org/10.1007/s10068-015-0014-2.

48 Liu, L., Wu, W., Zheng, L. et al. (2022). Intelligent packaging films incorporated with anthocyanins-loaded ovalbumin-carboxymethyl cellulose nanocomplexes for food freshness monitoring. *Food Chemistry* 387: https://doi.org/10.1016/J.FOODCHEM.2022.132908.

49 Shukla, V., Kandeepan, G., Vishnuraj, M.R., and Soni, A. (2016). Anthocyanins based indicator sensor for intelligent packaging application. *Agricultural Research* 5 (2): 205–209. https://doi.org/10.1007/s40003-016-0211-0.

50 Kuswandi, B., Maryska, C., Jayus, A., and A. and Heng, L.Y. (2013). Real time on-package freshness indicator for guavas packaging. *Journal of Food Measurement and Characterization* 7 (1): 29–39. https://doi.org/10.1007/S11694-013-9136-5.

51 Kuswandi, B. and Murdyaningsih, E.A. (2017). Simple on package indicator label for monitoring of grape ripening process using colorimetric pH sensor. *Journal of Food Measurement and Characterization* 11 (4): 2180–2194. https://doi.org/10.1007/S11694-017-9603-5.

52 Van Winckel, M., Vande Velde, S., De Bruyne, R., and Van Biervliet, S. (2011). Clinical practice: vegetarian infant and child. *European Journal of Pediatrics* 170 (12): 1489–1494. https://doi.org/10.1007/s00431-011-1547-x.

53 Pereira, P.C. (2014). Milk nutritional composition and its role in human health. *Nutrition* 30 (6): 619–627. https://doi.org/10.1016/j.nut.2013.10.011.

54 Visioli, F. and Strata, A. (2014). Milk, dairy products, and their functional effects in humans: a narrative review of recent evidence. *Advances in Nutrition* 5 (2): 131–143. https://doi.org/10.3945/an.113.005025.

55 Edgar, S. (2017). Milk as a raw material and food. In: *Milk and Dairy Product Technology* (ed. M. Axel), 9–58. Routledge. https://doi.org/10.1201/9780203747162-2.

56 Chandan, R.C. (2011). Dairy ingredients for food processing: an overview. In: *Dairy Ingredients for Food Processing* (ed. R.C. Chandan and A. Kilara), 3–33. Wiley. https://doi.org/10.1002/9780470959169.ch1.

57 Sohail, M., Sun, D.-W., and Zhu, Z. (2018). Recent developments in intelligent packaging for enhancing food quality and safety. *Critical Reviews in Food Science and Nutrition* 58 (15): 2650–2662. https://doi.org/10.1080/10408398.2018.1449731.

58 Galić, K., Božanić, R., Lisak Jakopović, K. et al. (2019). Packaging perspective of milk and dairy products. *Mljekarstvo* 69 (1): 3–20. https://doi.org/10.15567/mljekarstvo.2019.0101.

59 Karaman, A.D., Özer, B., Pascall, M.A., and Alvarez, V. (2015). Recent advances in dairy packaging. *Food Reviews International* 31 (4): 295–318. https://doi.org/10.1080/87559129.2015.1015138.

60 Mirza Alizadeh, A., Masoomian, M., Shakooie, M. et al. (2022). Trends and applications of intelligent packaging in dairy products: a review. *Critical Reviews in Food Science and Nutrition* 62 (2): 383–397. https://doi.org/10.1080/10408398.2020.1817847.

61 Vanderroost, M., Ragaert, P., Devlieghere, F., and De Meulenaer, B. (2014). Intelligent food packaging: the next generation. *Trends in Food Science & Technology* 39 (1): 47–62. https://doi.org/10.1016/j.tifs.2014.06.009.

62 Han, J.H., Ho, C.H.L., and Rodrigues, E.T. (2005). Intelligent packaging. In: *Innovations in Food Packaging*, 138–155. Elsevier. https://doi.org/10.1016/B978-012311632-1/50041-3.

63 Floros, J.D. and Matsos, K.I. (2005). Introduction to modified atmosphere packaging. In: *Innovations in Food Packaging*, 159–172. Elsevier. https://doi.org/10.1016/B978-012311632-1/50042-5.

64 O'Mahony, F.C., O'Riordan, T.C., Papkovskaia, N. et al. (2006). Non-destructive assessment of oxygen levels in industrial modified atmosphere packaged cheddar cheese. *Food Control* 17 (4): 286–292. https://doi.org/10.1016/J.FOODCONT.2004.10.013.

65 Mastromatteo, M., Conte, A., Lucera, A. et al. (2015). Packaging solutions to prolong the shelf life of Fiordilatte cheese: Bio-based nanocomposite coating and modified atmosphere packaging. *LWT – Food Science and Technology* 60 (1): 230–237. https://doi.org/10.1016/j.lwt.2014.08.013.

66 Fletouris, D.J., Govari, M.A., and Botsoglou, E.N. (2015). The influence of retail display storage on the fatty acid composition of modified atmosphere packaged Graviera Agraphon cheese. *International Journal of Dairy Technology* 68 (2): 218–226. https://doi.org/10.1111/1471-0307.12182.

67 Thippeswamy, L., Venkateshaiah, B.V., and Patil, S.B. (2011). Effect of modified atmospheric packaging on the shelf stability of paneer prepared by adopting hurdle technology. *Journal of Food Science and Technology* 48 (2): 230–235. https://doi.org/10.1007/s13197-010-0155-5.

68 Jha, A., Kumar, A., Jain, P. et al. (2015). Effect of modified atmosphere packaging on the shelf life of lal peda. *Journal of Food Science and Technology* 52 (2): 1068–1074. https://doi.org/10.1007/s13197-013-1064-1.

69 Ghayal, G., Jha, A., Kumar, A. et al. (2015). Effect of modified atmospheric packaging on chemical and microbial changes in dietetic rabri during storage. *Journal of Food Science and Technology* 52 (3): 1825–1829. https://doi.org/10.1007/s13197-013-1171-z.

70 Jain, V., Rasane, P., Jha, A. et al. (2015). Effect of modified atmospheric packaging on the shelf life of Kalakand and its influence on microbial, textural, sensory and physico-chemical properties. *Journal of Food Science and Technology* 52 (7): 4090–4101. https://doi.org/10.1007/s13197-014-1501-9.

71 Papkovsky, D.B., O'Mahony, F.C., Kerry, J.P., and Ogurtsov, V.I. (2005). Quality assessment of packaged foods by optical oxygen sensing. In: (ed. Y.-R. Chen, G.E. Meyer, and S.-I. Tu), 599604.1–599604.10. Bellingham WA: SPIE. https://doi.org/10.1117/12.633474.

72 Fogg, P. and Gerrard, W. (1991). Solubility of gases in liquids: a critical evaluation of gas/liquid systems in theory and practice. *Choice Reviews Online* 29 (2): 29-0903. https://doi.org/10.5860/CHOICE.29-0903.

73 Shao, P., Liu, L., Yu, J. et al. (2021). An overview of intelligent freshness indicator packaging for food quality and safety monitoring. *Trends in Food Science & Technology* 118: 285–296. https://doi.org/10.1016/j.tifs.2021.10.012.

74 Tang, T.B. and Zulkafli, M.S. (2013). Electronic tongue for fresh milk assessment A revisit of using pH as indicator. *2013 IEEE International Conference on Circuits and Systems (ICCAS)*, 18–19 September 2013, Kuala Lumpur, Malaysia, IEEE, pp. 167–171. https://doi.org/10.1109/CircuitsAndSystems.2013.6671633.

75 Lee, K. and Ko, S. (2014). Proof-of-concept study of a whey protein isolate based carbon dioxide indicator to measure the shelf-life of packaged foods. *Food Science and Biotechnology* 23 (1): 115–120. https://doi.org/10.1007/s10068-014-0015-6.

76 Jung, J., Puligundla, P., and Ko, S. (2012). Proof-of-concept study of chitosan-based carbon dioxide indicator for food packaging applications. *Food Chemistry* 135 (4): 2170–2174. https://doi.org/10.1016/j.foodchem.2012.07.090.

77 Law, B.A. (2008). Cheese problems solved – Edited by P.L.H. McSweeney. *International Journal of Dairy Technology* 61 (1): 115–116. https://doi.org/10.1111/j.1471-0307.2008.00356.x.

78 Abdelnour, A., Fonseca, N., Rennane, A., Kaddour, D., and Tedjini, S. (2019). Design of RFID sensor tag for cheese quality monitoring. *2019 IEEE MTT-S International Microwave Symposium (IMS)*, 2–7 June 2019, Boston (Massachusetts), IEEE, pp. 290–292. https://doi.org/10.1109/MWSYM.2019.8700769.

79 Zosel, J., Oelßner, W., Decker, M. et al. (2011). The measurement of dissolved and gaseous carbon dioxide concentration. *Measurement Science and Technology* 22 (7): 072001. https://doi.org/10.1088/0957-0233/22/7/072001.

80 CO_2 and O_2 Measurement for the Dairies (2023). CO_2 and O_2 measurement for the dairy industry, CO_2 and O_2 measurement for the dairies. https://www.instek.co.za/co2-and-o2-measurement-dairy-industri (accessed 26 February 2023).

81 Cantec (2023). CARBOTEC NIR, Cantec. https://www.centec.de/sensors/brewery/carbotec-nir/ (accessed 16 February 2023).

82 Ma, Y., Barbano, D.M., Hotchkiss, J.H. et al. (2001). Impact of CO_2 Addition to Milk on Selected Analytical Testing Methods. *Journal of Dairy Science* 84 (9): 1959–1968. https://doi.org/10.3168/jds.S0022-0302(01)74638-3.

83 Kuswandi, B., Moradi, M., and Ezati, P. (2022). Food sensors: off-package and on-package approaches. *Packaging Technology and Science* 35 (12): 847–862. https://doi.org/10.1002/pts.2683.

84 Soltani Firouz, M., Mohi-Alden, K., and Omid, M. (2021). A critical review on intelligent and active packaging in the food industry: Research and development. *Food Research International* 141: 110113. https://doi.org/10.1016/j.foodres.2021.110113.

85 Claus, J.R., Colby, J.-W., and Flick, G.J. (1994). Processed meats/poultry/seafood. In: *Muscle Foods*, 106–162. Boston, MA: Springer US. https://doi.org/10.1007/978-1-4757-5933-4_5.

86 Kinsman, D.M. (1994). Historical perspective and current status. In: *Muscle Foods*, 1–24. Boston, MA: Springer US https://doi.org/10.1007/978-1-4757-5933-4_1.

87 Fang, Z., Zhao, Y., Warner, R.D., and Johnson, S.K. (2017). Active and intelligent packaging in meat industry. *Trends in Food Science & Technology* 61: 60–71. https://doi.org/10.1016/j.tifs.2017.01.002.

88 Miller, R.K. (1994). Sensory methods to evaluate muscle foods. In: *Muscle Foods*, 333–360. Boston, MA: Springer US. https://doi.org/10.1007/978-1-4757-5933-4_12.

89 Kerry, J.P., O'Grady, M.N., and Hogan, S.A. (2006). Past, current and potential utilisation of active and intelligent packaging systems for meat and muscle-based products: a review. *Meat Science* 74 (1): 113–130. https://doi.org/10.1016/j.meatsci.2006.04.024.

90 Kakouri, A. and Nychas, G.J.E. (1994). Storage of poultry meat under modified atmospheres or vacuum packs: possible role of microbial metabolites as indicator of spoilage. *Journal of Applied Bacteriology* 76 (2): 163–172. https://doi.org/10.1111/j.1365-2672.1994.tb01612.x.

91 Smiddy, M., Papkovskaia, N., Papkovsky, D.B., and Kerry, J.P. (2002). Use of oxygen sensors for the non-destructive measurement of the oxygen content in modified atmosphere and vacuum packs of cooked chicken patties; impact of oxygen content on lipid oxidation. *Food Research International* 35 (6): 577–584. https://doi.org/10.1016/S0963-9969(01)00160-0.

92 Hogan, S.A. and Kerry, J.P. (2008). Smart packaging of meat and poultry products. In: *Smart Packaging Technologies for Fast Moving Consumer Goods*, 33–59. Chichester, UK: Wiley https://doi.org/10.1002/9780470753699.ch3.

93 Von Bültzingslöwen, C., McEvoy, A.K., McDonagh, C. et al. (2002). Sol–gel based optical carbon dioxide sensor employing dual luminophore referencing for application in food packaging technology. *The Analyst* 127 (11): 1478–1483. https://doi.org/10.1039/B207438A.

94 O'Grady, M.N. and Kerry, J.P. (2008). Smart packaging technologies and their application in conventional meat packaging systems. In: *Meat Biotechnology*, 425–451. New York, NY: Springer New York https://doi.org/10.1007/978-0-387-79382-5_19.

95 Andersen, H.J. and Skibsted, L.H. (1991). Oxidative stability of frozen pork patties. Effect of light and added salt. *Journal of Food Science* 56 (5): 1182–1184. https://doi.org/10.1111/j.1365-2621.1991.tb04729.x.

96 Zohora, S.E., Khan, A.M., Srivastava, A.K., and Hundewale, N. (2013). Electronic noses application to food analysis using metal oxide sensors: a review. *International Journal of Soft Computing and Engineering (IJSCE)* 3: 2231–2307.

97 Ghaani, M., Cozzolino, C.A., Castelli, G., and Farris, S. (2016). An overview of the intelligent packaging technologies in the food sector. *Trends in Food Science & Technology* 51: 1–11. https://doi.org/10.1016/j.tifs.2016.02.008.

98 Kerry, J.P. and Papkovsky, D.B. (2002). Development and use of non-destructive, continuous assessment, chemical oxygen sensors in packs containing sensitive foodstuffs. *Research Advances in Food Science* 3: 121–140.

99 Mills, A., Chang, Q., and McMurray, N. (1992). Equilibrium studies on colorimetric plastic film sensors for carbon dioxide. *Analytical Chemistry* 64 (13): 1383–1389. https://doi.org/10.1021/ac00037a015.

100 Anusankari, S., Balaji Ganesh, A., Subasri, R., and Deepa, N. (2019). Optical determination of carbon dioxide and oxygen by a fluorescent membrane to evaluate the freshness of meat products. *Instrumentation Science & Technology* 47 (6): 640–665. https://doi.org/10.1080/10739149.2019.1622132.

101 James Stephen, J. and Christian, J. (2002). *Meat Refrigeration*. CRC Press.

102 David, B., Mejdell, C., Michel, V. et al. (2015). Air quality in alternative housing systems may have an impact on laying hen welfare. Part II—ammonia. *Animals* 5 (3): 886–896. https://doi.org/10.3390/ani5030389.

103 Astill, J., Dara, R.A., Fraser, E.D.G. et al. (2020). Smart poultry management: smart sensors, big data, and the internet of things. *Computers and Electronics in Agriculture* 170: 105291. https://doi.org/10.1016/j.compag.2020.105291.

104 Saliu, F. and Della Pergola, R. (2018). Carbon dioxide colorimetric indicators for food packaging application: applicability of anthocyanin and poly-lysine mixtures. *Sensors and Actuators B: Chemical* 258: 1117–1124. https://doi.org/10.1016/j.snb.2017.12.007.

105 Pávai, M., Mihály, J., and Paszternák, A. (2015). pH and CO_2 sensing by curcumin-coloured cellophane test strip. *Food Analytical Methods* 8 (9): 2243–2249. https://doi.org/10.1007/s12161-015-0102-1.

106 Rukchon, C., Nopwinyuwong, A., Trevanich, S. et al. (2014). Development of a food spoilage indicator for monitoring freshness of skinless chicken breast. *Talanta* 130: 547–554. https://doi.org/10.1016/j.talanta.2014.07.048.

107 Mills, A. and Skinner, G.A. (2010). Water-based colourimetric optical indicators for the detection of carbon dioxide. *The Analyst* 135 (8): 1912. https://doi.org/10.1039/c000688b.

108 Gokoglu, N. (2020). Innovations in seafood packaging technologies: a review. *Food Reviews International* 36 (4): 340–366. https://doi.org/10.1080/87559129.2019.1649689.

109 Yesudhason, P., Lalitha, K.V., Gopal, T.K.S., and Ravishankar, C.N. (2014). Retention of shelf life and microbial quality of seer fish stored in modified atmosphere packaging and sodium acetate pretreatment. *Food Packaging and Shelf Life* 1 (2): 123–130. https://doi.org/10.1016/j.fpsl.2014.04.001.

110 Özogul, F., Polat, A., and Özogul, Y. (2004). The effects of modified atmosphere packaging and vacuum packaging on chemical, sensory and microbiological changes of sardines (*Sardina pilchardus*). *Food Chemistry* 85 (1): 49–57. https://doi.org/10.1016/j.foodchem.2003.05.006.

111 Erkan, N., Özden, Ö., Alakavuk, D.Ü. et al. (2006). Spoilage and shelf life of sardines (Sardina pilchardus) packed in modified atmosphere. *European Food Research and Technology* 222 (5–6): 667–673. https://doi.org/10.1007/s00217-005-0194-8.

112 Arkoudelos, J., Stamatis, N., and Samaras, F. (2007). Quality attributes of farmed eel (*Anguilla anguilla*) stored under air, vacuum and modified atmosphere packaging at 0 °C. *Food Microbiology* 24 (7–8): 728–735. https://doi.org/10.1016/j.fm.2007.03.008.

113 Hudecová, K., Buchtová, H., and Steinhauserová, I. (2010). Effects of modified atmosphere packaging on the microbiological properties of fresh common carp (*Cyprinus carpio* L.). *Acta Veterinaria Brno* 79 (9): S93–S100. https://doi.org/10.2754/avb201079S9S093.

114 Barandun, G., Soprani, M., Naficy, S. et al. (2019). Cellulose fibers enable near-zero-cost electrical sensing of water-soluble gases. *ACS Sensors* 4 (6): 1662–1669. https://doi.org/10.1021/acssensors.9b00555.

115 Jia, X., Ma, P., Tarwa, K., and Wang, Q. (2023). Machine vision-based colorimetric sensor systems for food applications. *Journal of Agriculture and Food Research* 11: 100503. https://doi.org/10.1016/j.jafr.2023.100503.

116 Zaragozá, P., Fuentes, A., Ruiz-Rico, M. et al. (2015). Development of a colorimetric sensor array for squid spoilage assessment. *Food Chemistry* 175: 315–321. https://doi.org/10.1016/j.foodchem.2014.11.156.

117 Blixt, Y. and Borch, E. (1999). Using an electronic nose for determining the spoilage of vacuum-packaged beef. *International Journal of Food Microbiology* 46 (2): 123–134. https://doi.org/10.1016/S0168-1605(98)00192-5.

118 Panigrahi, S., Balasubramanian, S., Gu, H. et al. (2006). Design and development of a metal oxide based electronic nose for spoilage classification of beef. *Sensors and Actuators B: Chemical* 119 (1): 2–14. https://doi.org/10.1016/j.snb.2005.03.120.

119 Rajamäki, T., Alakomi, H.-L., Ritvanen, T. et al. (2006). Application of an electronic nose for quality assessment of modified atmosphere packaged poultry meat. *Food Control* 17 (1): 5–13. https://doi.org/10.1016/j.foodcont.2004.08.002.

120 Wijaya, D.R., Sarno, R., and Zulaika, E. (2021). DWTLSTM for electronic nose signal processing in beef quality monitoring. *Sensors and Actuators B: Chemical* 326: 128931. https://doi.org/10.1016/j.snb.2020.128931.

121 Insignia Technologies (2023). Leak detection label, insignia technologies. https://www.insigniatechnologies.com/leak-detection (accessed 5 March 2023).

122 Food, Beverage and Agriculture | Vaisala. https://www.vaisala.com/en/industries-applications/food-beverage-and-agriculture (accessed 23 January 2024).

123 Food Temperature Indicators | Zebra. https://www.zebra.com/us/en/products/environmental-sensors/ready-to-use-indicators/food-indicators.html (accessed 23 January 2024).

124 CO_2 Sensor Foil SF-CD1R | PreSens. https://www.presens.de/products/detail/co2-sensor-foil-sf-cd1r (accessed 23 January 2024).

125 Mills, A. and Skinner, G. (2013). CO_2 indicator. United States. https://patents.google.com/patent/US20130130399A1/en (accessed 23 January 2024).

126 Kane, J.A. (2013). Carbon dioxide (CO_2) sensor, United States. https://patents.google.com/patent/US9316594B2 (accessed 23 January 2024).

127 Wolfbeis O S L.H. (1993) Method for quality control of packaged organic substances and packaging material for use with this method.

128 Eaton (1975). Visual spoilage indicator for food containers.

129 Balderson, S.N. and Whitwood, R.J. (1994). Gas indicator for a package.

130 Holte, B. (1993). An apparatus for indicating the presence of co2 and a method of measuring and indicating bacterial activity within a container or bag.

131 Mattila, T., Tawast, J., and Ahvenainen, R. (1990). New possibilities for quality control of aseptic packages: microbiological spoilage and seal defect detection using head-space indicators. *Lebensmittel-Wissenschaft & Technologie* 23 (3): 246–251.

132 Horan, T.J. (1996). Method for determining bacteria contamination in food package.

133 Horan, T.J. (1999). Method for determining deleterious bacterial growth in packaged food utilizing hydrophilic polymers.

134 Neary, M.P. (1978). Food spoilage indicator.

135 Ahvenainen, R. and Hurme, E. (1997). Active and smart packaging for meeting consumer demands for quality and safety. *Food Additives and Contaminants* 14 (6–7): 753–763. https://doi.org/10.1080/02652039709374586.

136 Kuswandi, B., Wicaksono, Y., Jayus, A., and A., Heng, L.Y. and Ahmad, M. (2011). Smart packaging: sensors for monitoring of food quality and safety. *Sensing and Instrumentation for Food Quality and Safety* 5 (3–4): 137–146. https://doi.org/10.1007/s11694-011-9120-x.

137 Ramachandran, R.P. (2022). Integrated approach on stored grain quality management with CO2 monitoring—a review. *Journal of Stored Products Research* 96 (February): 101950. https://doi.org/10.1016/j.jspr.2022.101950.

138 Chidananda, K.P., Chelladurai, V., Jayas, D.S. et al. (2014). Respiration of pulses stored under different storage conditions. *Journal of Stored Products Research* 59: 42–47. https://doi.org/10.1016/j.jspr.2014.04.006.

139 Abalone, R., Barreto, A.A., and Yommi, A. (2017). Modelling respiration rate of soybean seeds (*Glycine max* (L.)) in hermetic storage. *Journal of Stored Products Research* 74: 36–45. https://doi.org/10.1016/j.jspr.2017.09.001.

140 Lutz, É. and Coradi, P.C. (2022). Applications of new technologies for monitoring and predicting grains quality stored: sensors, internet of things, and artificial intelligence. *Measurement: Journal of the International Measurement Confederation* 188 (October 2021): https://doi.org/10.1016/j.measurement.2021.110609.

141 Neethirajan, S., Freund, M.S., Jayas, D.S. et al. (2010). Development of carbon dioxide (CO_2) sensor for grain quality monitoring. *Biosystems Engineering* 106 (4): 395–404. https://doi.org/10.1016/j.biosystemseng.2010.05.002.

142 Lee, K., Meng, X., Kang, T.Y., and Ko, S. (2015). A dye-incorporated chitosan-based CO_2 indicator for monitoring of food quality focusing on makgeolli quality during storage. *Food Science and Biotechnology* 24 (3): 905–912. https://doi.org/10.1007/s10068-015-0117-9.

143 Maier, D.E. and Channaiah, L.H. (2010) Monitoring carbon dioxide concentration for early detection of spoilage in stored grain Immmunosorbent Assay (ELISA) (AgraQuant Mycotoxin ELISA Test Kits, Romer Labs Inc. *10th International Working Conference on Stored Product Protection Monitoring*, pp. 505–509. https://doi.org/10.5073/jka.2010.425.332.

144 Khalid, M., Maria, A., and Honorato, B. (2018). Polyaniline: synthesis methods, doping and conduction mechanism. *Polyaniline – From Synthesis to Practical Applications* 1–17. https://doi.org/10.5772/intechopen.79089.

145 Taddese, M., Dibaba, K., Bayissa, W. et al. (2020). Assessment of quantitative and qualitative losses of stored grains due to insect infestation in Ethiopia. *Journal of Stored Products Research* 89: 101689.

146 Yan, K., Xu, F., Li, S. et al. (2020). Ice-templating of chitosan/agarose porous composite hydrogel with adjustable water-sensitive shape memory property and multi-staged degradation performance. *Colloids and Surfaces B: Biointerfaces* 190 (January): 110907. https://doi.org/10.1016/j.colsurfb.2020.110907.

147 Surya, R. and Lee, A.G.Y. (2022). Exploring the philosophical values of kimchi and kimjang culture. *Journal of Ethnic Foods* 9 (1): https://doi.org/10.1186/s42779-022-00136-5.

148 Zhang, Y. and Lim, L.T. (2018). Colorimetric array indicator for NH_3 and CO_2 detection. *Sensors and Actuators, B: Chemical* 255: 3216–3226. https://doi.org/10.1016/j.snb.2017.09.148.

149 Lee, H.G., Jeong, S., and Yoo, S.R. (2023). Development of a calcium hydroxide–dye kimchi ripening indicator and its application in kimchi packaging. *Food Chemistry* 400 (August 2022): 134039. https://doi.org/10.1016/j.foodchem.2022.134039.

150 UNEP (2022) Promoting sustainable lifestyles | North America | UNEP – UN Environment Programme. https://www.unep.org/regions/north-america/regional-initiatives/promoting-sustainable-lifestyles (accessed 18 February 2023).

151 Yousefi, H., Su, H.M., Imani, S.M. et al. (2019). Intelligent food packaging: a review of smart sensing technologies for monitoring food quality. *ACS Sensors* 4 (4): 808–821. https://doi.org/10.1021/acssensors.9b00440.

152 Misko, G.G. (2022). The regulation of active and intelligent food packaging in the U.S. and the EU | PackagingLaw.com. https://www.packaginglaw.com/special-focus/regulation-active-and-intelligent-food-packaging-us-and-eu#FN1 (accessed 18 February 2023).

153 La Fuente, C.I.A., Maniglia, B.C., and Tadini, C.C. (2023). Biodegradable polymers: a review about biodegradation and its implications and applications. *Packaging Technology and Science* 36 (2): 81–95. https://doi.org/10.1002/pts.2699.

154 Zainal Arifin, M.A., Mohd Adzahan, N., Zainal Abedin, N.H., and Lasik-Kurdyś, M. (2023). Utilization of food waste and by-products in the fabrication of active and intelligent packaging for seafood and meat products. *Foods* 12 (3): 456. https://doi.org/10.3390/foods12030456.

155 Nosáľová, M., Loučanová, E., and Parobek, J. (2018). Perception of packaging functions and the interest in intelligent and active packaging. *Problems of Agricultural Economics* 357 (4): 141–152. https://doi.org/10.30858/zer/100715.

156 Aday, M.S. and Yener, U. (2015). Assessing consumers' adoption of active and intelligent packaging. *British Food Journal* 117 (1): 157–177. Available at: https://doi.org/10.1108/BFJ-07-2013-0191.

157 Young, E., Mirosa, M., and Bremer, P. (2020). A systematic review of consumer perceptions of smart packaging technologies for food. *Frontiers in Sustainable Food Systems* 4 (May): 1–20. https://doi.org/10.3389/fsufs.2020.00063.

13

Food Quality Indicator-Based Intelligent Food Packaging

K. P. Chaithra, B. Gayathri, P. S. Aiswarya, and T. P. Vinod

Department of Chemistry, Christ University, Bangalore, Karnataka, India

13.1 Introduction

The United Nations environmental program estimates that ~1.43 billion tons of food is wasted or lost yearly worldwide [1, 2]. The wastage at the consumer stage is significant compared to the production and supply stages due to the inability of the consumers to distinguish the expiration dates and the insufficiency of real-time data of the food quality [3]. Consuming contaminated food has led to 600 million cases of foodborne illnesses, as per the 2017 World Health Organization (WHO) report [1]. Therefore, identifying the real-time status of food quality reduces food poisoning and economic losses from food wastage [4, 5]. Food quality decreases due to various chemical, physical, and biological agents during food storage and supply, resulting in food spoilage [6, 7]. Physical changes, such as odor and taste [8]; chemical changes, such as pH and the evolution of total volatile basic nitrogen (TVB-N) [9]; and microbiological changes [10], indicate the stages of food spoilage. Various quantitative and qualitative indicators that monitor humidity [11], oxygen [12], carbon dioxide [13], pH [14], and microbial [3] are reported for food quality monitoring.

Incorporating these indicators into food packaging helps to monitor and alert the consumer in real time. It also preserves the food quality by preventing moisture, contaminants, and oxygen from reaching the food product [3, 15]. This type of food packaging is known as intelligent packaging (IP), which detects physical, chemical, and biological changes and informs users about the condition of food [16] (Figure 13.1). IP can significantly mitigate food wastage and ensure consumer health by performing tasks such as detecting, recording, sensing, and providing food quality information [17, 18]. It also contributes to one of the sustainable development goals (SDGs), good health and well-being [19]. As per the search results from Scopus, food quality indicators and IP research have taken up interdisciplinary dimensions and significantly increased in the last decade (Figure 13.2). This chapter intends to discuss the physical, chemical, and microbiological changes during food spoilage. Recent advances in food quality indicator-based IP, including humidity, gas, pH, and microbial, are also discussed and summarized in the following sections.

Smart Food Packaging Systems: Innovations and Technology Applications, First Edition.
Edited by Avik Mukherjee, Santosh Kumar, Manjusri Misra, and Amar K. Mohanty.
© 2025 John Wiley & Sons Ltd. Published 2025 by John Wiley & Sons Ltd.

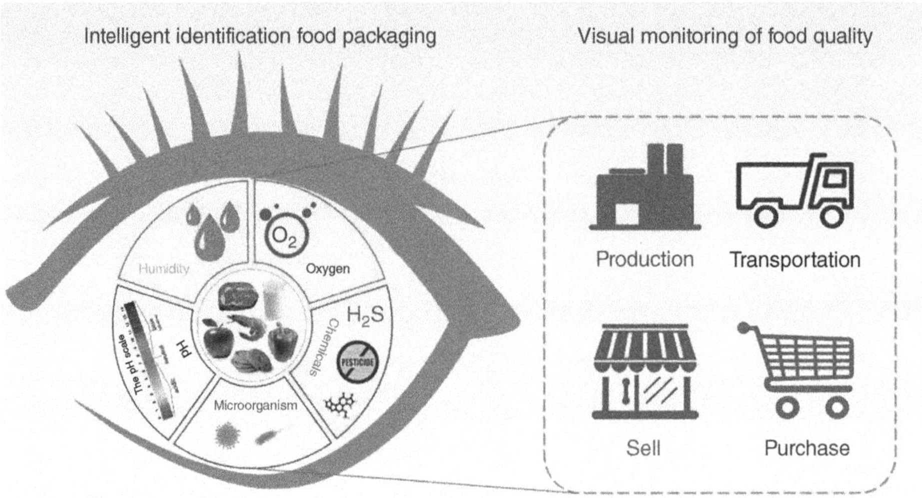

Figure 13.1 Food quality indicator-based IP. *Source:* Reproduced with permission from Cheng et al. [16]/Elsevier.

13.2 Food Quality Indicators

Several indicators that can detect a change in food quality are discussed in this section. These indicators consist of real-time sensing materials showing a qualitative and/or quantitative change in response to physical, chemical, or microbiological changes during food spoilage. These food quality indicators are supported on various substrates, such as polymer, plastic, microfluidic devices, glass, and paper.

13.2.1 Humidity Indicators

The water vapor content present in the atmospheric air is called atmospheric humidity. Analogous to this, a physical measure of moisture in food products is called equilibrium relative humidity (ERH) [20]. ERH is the relative humidity of the air that surrounds the food in equilibrium with the environment. Water activity(a_w) is another quantity related to moisture. It is the ratio between vapor pressure of food in an undisturbed balance with the surrounding media and that of distilled water under similar conditions. At equilibrium, ERH (%) $= a_w \times 100$ [21], and it gives the moisture condition of the food product. Most food items have an ERH value above 95%, providing sufficient moisture to allow the growth of microorganisms like bacteria, yeasts, molds, etc., which can cause spoilage [22, 23]. The excess moisture in food items can promote a suitable environment for microbial growth, whereas lower levels of moisture can cause the edibles to dry [24]. Therefore, maintaining balanced and appropriate humidity values is vital in preserving the condition and texture of food. There exist various ways to alter the humidity within the package. It can be either through the removal of the package seals or by extended exposure to extreme fluctuations in the internal temperature [25].

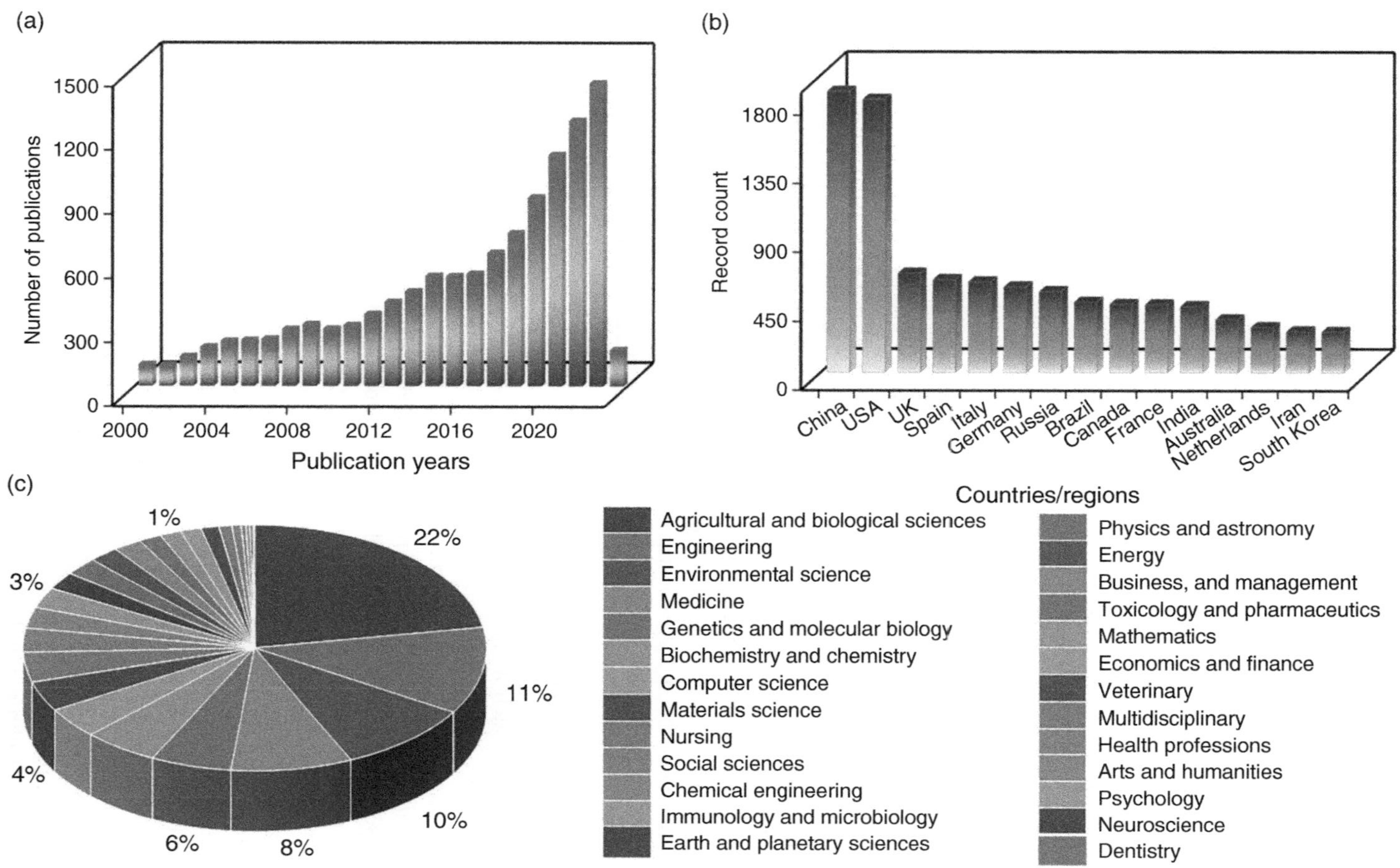

Figure 13.2 Statistics on (a) the number of publications vs. years, (b) the number of publications vs. countries, and (c) research area categories. The literature was focused on search keywords "Food Quality Indicators/Intelligent Packaging/Intelligent Food Packaging" (Data acquired in February 2023 from Scopus).

The humidity level is decisive in products of food industry, including meat products, dairy products, and dried foods, and the most prominent effect is on the lifetime of fruits and vegetables [15]. Hence, these food items are to be maintained at a moderate humidity level. Generally, the use of a refrigerator prevents the humidity from exceeding the limit as the a_w is directly proportional to temperature [26]. The humidity sensor works by sensing the humidity or moisture in the food and sending the data to the microcontroller. Then alerts are generated upon the detection of humidity level [27]. Yi Zhang et al. studied the effect of humidity on pork meat. They showed that pork under high humidity had a fresher color, a lower value for TVB-N, and slower microbial growth than what it had under low and medium environmental humidity [28]. Besides microbial growth, humidity is also the cause of lower shelf-life duration due to spoilage of the food, resulting in softening followed by product dampening. Hence, humidity is a mainstay in the quality assessment of food [3, 15].

13.2.2 Gas Indicators/Detectors

When food in sealed containers spoils under normal atmospheric circumstances, many chemical reactions occur within the packaging atmosphere. Therefore, monitoring the chemical changes in the food can be used to determine its quality. O_2 and CO_2 are the byproducts of bacterial metabolism during food spoilage, and food quality can be monitored using indicators that monitor O_2 and CO_2 levels [29]. IP-based gas indicators/detectors can observe changes in composition and/or concentration of the gases within the food package, which helps in checking the quality of food product. Monitoring spoiled food with a gas indicator is usually done early in the spoiling process to reduce the chance of food wastage [6]. Modified atmosphere packaging (MAP) is food packaging that employs gas barrier materials [30]. It will increase the shelf life and value of food. MAP consists of varying amounts of nitrogen, O_2, and CO_2. MAP is widely used for meat, dairy products, and seafood in the food industry [30–32].

13.2.3 pH Change Indicators/Detectors

Protein-rich foods, such as milk, fish, prawn, meat, and chicken, are easily perishable. Microbial activity during the spoilage of these foods releases chemicals that change the pH. When microbial degradation occurs in milk, the lactic acid produced lowers the pH from ~6.8 (fresh milk) to ~4.0 (spoiled milk) [33]. Chemicals like amines, ammonia, histamine, N-cyclic compounds, etc., are released in various kinds of seafood, increasing the pH [3]. The breakdown of proteins in meat and poultry products releases amino acids. Amino acids result in desulfurization, oxidative deamination, and decarboxylation reactions, releasing TVB-N compounds and CO_2 that change pH [34]. pH changes during food storage are an effective signifier for studying food quality in real time. Indicators that exhibit a color change with pH are called halochromic indicators [14]. Halochromic indicators can have both natural and artificial origins (Figure 13.3a). By incorporating halochromic indicators in/onto the substrates, food quality can be monitored effectively.

Anthocyanins, natural halochromic indicators found in hibiscus, red cabbage, etc., are widely used in food quality monitoring [36]. Chaithra and Vinod fabricated a low-cost, easy-to-prepare paper-based milk quality indicator by rub-coating hibiscus (H-Paper) [5]. The prepared halochromic indicator showed a visible color change from green to pink,

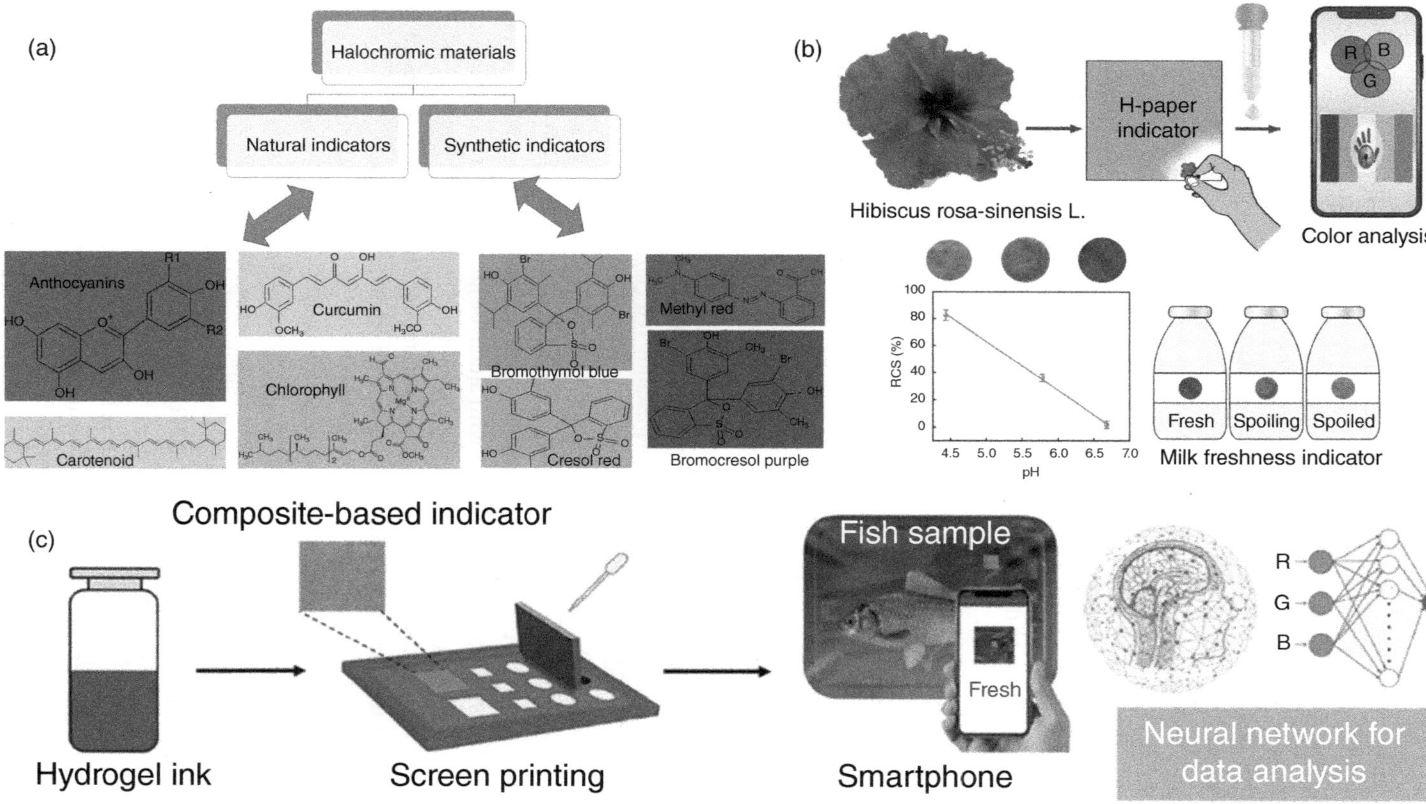

Figure 13.3 (a) Natural and synthetic halochromic indicators. (b) Paper-based indicator for milk freshness monitoring. *Source:* Reproduced from Chaithra and Vinod [5]/with permission of John Wiley & Sons. (c) Anthocyanin-based fish quality monitoring. *Source:* Reproduced from Fang et al. [35]/with permission of Elsevier.

indicating milk freshness (Figure 13.3b). A free Android application was used to quantify the color. Fang et al. fabricated fish quality senor using anthocyanins extracted from red cabbage [35]. The anthocyanins, carboxymethyl chitosan, and oxidized sodium alginate (substrate) were dispersed in silica sol. The sol was used to screen-print colorimetric labels (Figure 13.3c). The label showed a visible color change indicating fish spoilage. Along with natural halochromic dyes, synthetic indicators are also used to monitor food quality. Shukla et al. developed a pH and TVB-N indicator using filter paper and bromophenol blue to monitor the meat quality [37]. When placed inside the buffalo meat package, the indicator changes from yellow to blue, indicating a change in meat quality. Hidayat et al. fabricated indicator paper by soaking filter paper (Whatman No. 1) in pH indicators (Phenol Red and Bromothymol Blue) [38]. The indicator showed a reddish-orange to purple color change, indicating the spoiled state of fresh meat. Liu et al. fabricated a colorimetric indicator by coating filter paper with bromocresol green through a sol–gel matrix that monitors TVB-N levels [39]. The matrix improved moisture resistance and showed a visible and accurate color change in response to different fish freshness states.

13.2.4 Specific Chemical Indicators/Detectors

During food degradation, many chemical changes occur in the food environment, particularly for fish, meat, fruits, and vegetables, with the growth of microorganisms. Several colorimetric assays have been extensively used to check food quality involving chemical indicators. These methods are highly accurate, cost-efficient, simple, and reliable [40]. Any substance that indicates either the presence or absence of a critical concentration of a chemical by changing the observable features, such as color and dimensions, is called a chemical indicator. The chemical indicators are classified as natural, artificial, and olfactory. Examples of natural chemical indicators include red cabbage, turmeric, grape juice, and beets; synthetic indicators include phenolphthalein, methyl orange, and methyl red; and olfactory indicators have unique odors, as in vanilla extract and clove oil [41]. During fish spoilage, several volatile amines and other chemical compounds are evolved. Yujuan Qin et al. fabricated carbon dot-based histamine-responsive indicators, which allowed visual analysis of fish spoilage. The carbon dots with aldehyde groups on their surface provide sites to attach the amino group to the histamine forming a Schiff base, which can enhance the fluorescence. The sensor showed a linear ratiometric fluorescence response with increased histamine concentration and color change from red to blue [42].

Hypoxanthine (Hx) also evolved during fish spoilage. Chen et al. fabricated a sensor based on the etching of gold nanorods (GNRs) that shows the visible color change in response to various levels of Hx [43]. Hx in the presence of xanthine oxidase (XOD) and reaction with dissolved O_2 produces H_2O_2. The GNRs show various colors due to surface plasmon resonance when reacted with H_2O_2 in the presence of Fe^{2+} (Figure 13.4). The freshness of vegetables and fruits must also be detected accurately to utilize their nutrition and benefits and to prevent health hazards. The ethylene (C_2H_4) concentration in the packaging environment measures its freshness since fruits release ethylene during over-ripening or spoilage. Esser et al. reported a chemoresistive sensor with single-walled carbon nanotubes combined with a copper complex placed between gold electrodes, which, when bonded to ethylene, causes a change in resistance (Figure 13.5) [45].

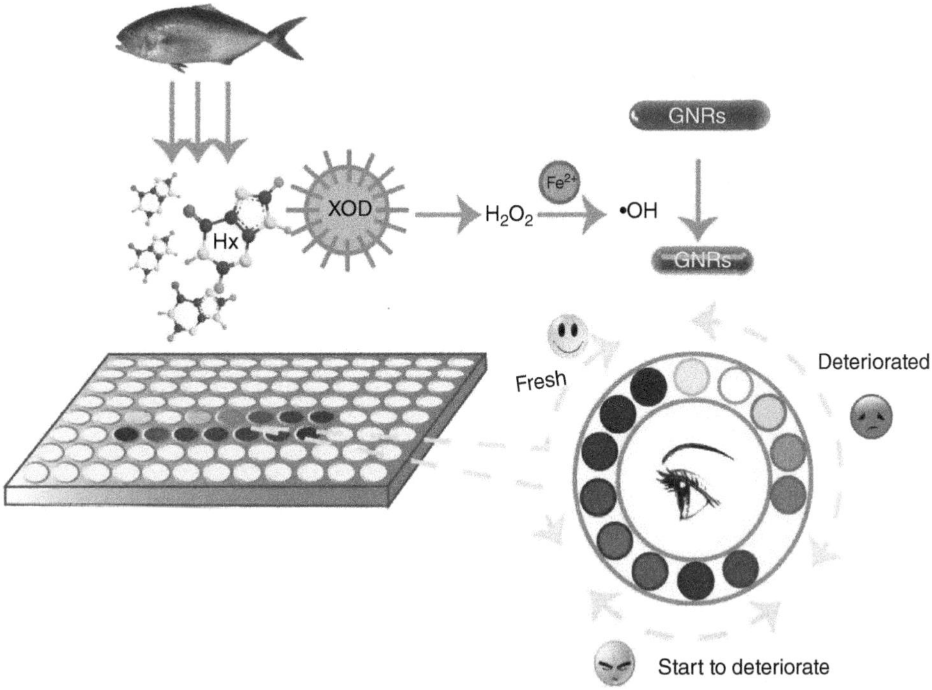

Figure 13.4 Biosensor for detection of fish deterioration. *Source:* Reproduced from Chen et al. [44]/with permission of Elsevier.

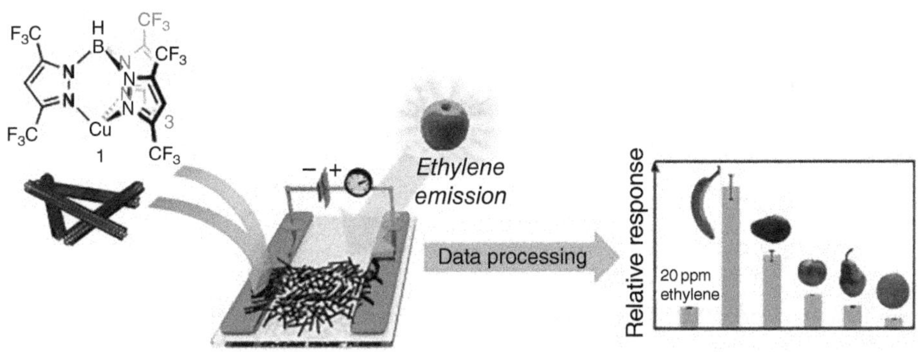

Figure 13.5 Ethylene chemoresistive sensor for detection of fruit deterioration. *Source:* Reproduced from Esser et al. [45]/with permission of John Wiley & Sons.

13.3 Intelligent Packaging (IP) System for Monitoring Food Quality

IP, an advanced form of traditional packaging, senses and records changes in the food environment. IP aims to monitor and indicate packaged food's quality in real time, guaranteeing its safety [6, 16]. IP systems are based on primary technologies that include indicators, sensors, and data carriers (Figure 13.6a) [17]. Indicators provide end users with

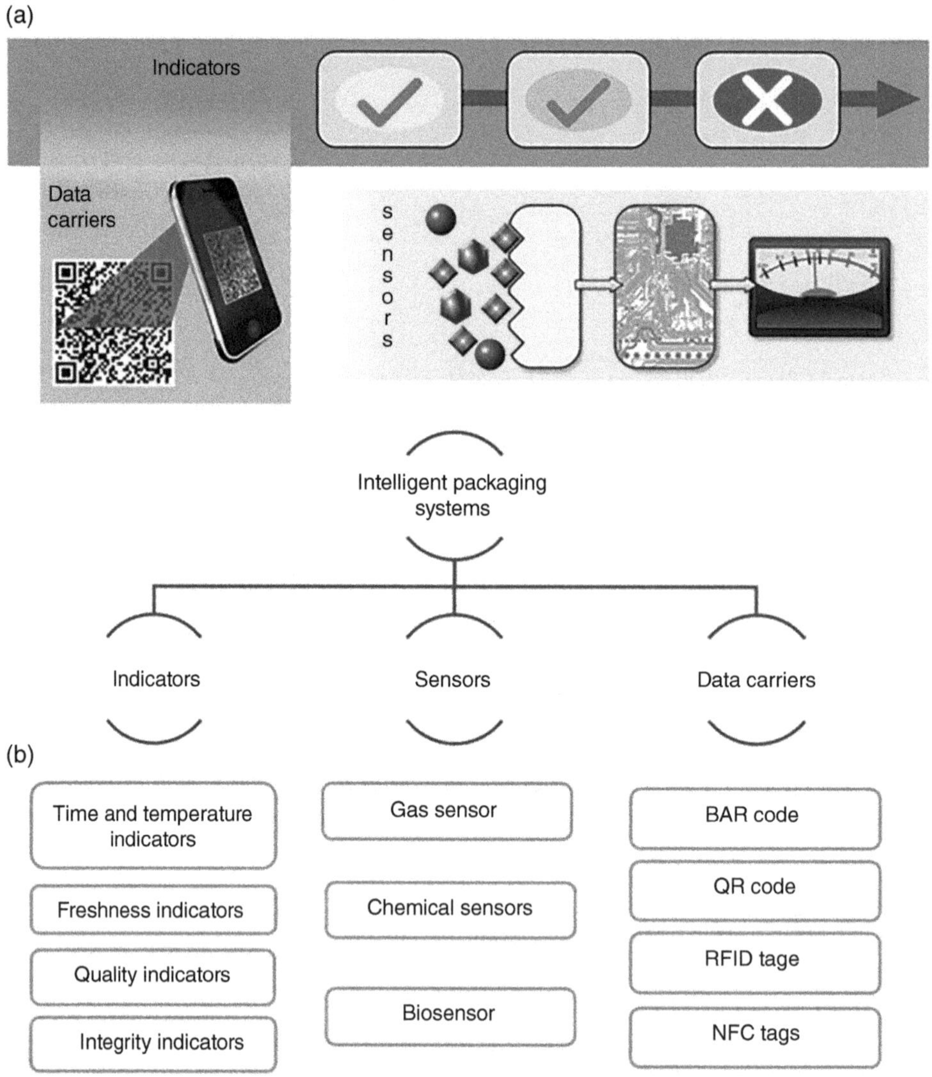

Figure 13.6 (a) Types of intelligent packaging. *Source:* Reproduced from Azeredo et al. [17]/with permission of Elsevier. (b) List of IP systems. *Source:* Adapted from Kumar et al. [46].

real-time data about food safety based on concentration, presence/absence of an analyte, or the extent of a chemical reaction from their interaction with food [47]. Most indicators are colorimetric dyes that indicate a color change in response to food spoilage. Sensors detect analytes that influence changes in food quality by converting information into a quantifiable output signal, which is then processed using software and electronics [17]. The sensor generally consists of the following: (i) a receptor as the sensing part; (ii) a transducer that transforms the information into a helpful signal based on chemical or physical changes; (iii) signal processing electronics; and (iv) a signal display unit. Sensors may be attached inside or outside the package, depending on the conditions monitored [48]. Data carriers such as bar codes and radio frequency identification device (RFID) tags are employed to store and transfer data. It also tracks the movement of food products from manufacturing to consumption [49]. Data carriers display information based on indicators and sensors' responses to the external environment [50]. Various indicators, sensors, and data carriers are listed in Figure 13.6b. The freshness and quality indicators, gas, humidity, and chemical sensors are discussed in the following sessions.

13.3.1 Humidity-Based IP

The construction of humidity-based IP has been carried out through various methods, using measurement of moisture-dependent optical properties of materials like ceramics, hybrid composites, and so on [51]. The latest advancements include biopolymer-based humidity sensors from food-safe transducers and food packaging [11]. Bibi et al. used film from wheat gluten protein as an environmentally friendly packaging material to analyze the relative humidity. The dielectric loss and dielectric permittivity of wheat gluten, which depend on relative humidity, allowed it to be a humidity sensor by pairing it with an RFID tag [52]. Yuan et al. developed a multi-detection system that can analyze NH_3, H_2S, and humidity to monitor food deterioration [53]. The system consists of many silicon-based chemical-sensitive field effect transistors (CSFETs), which quantify the gases and humidity formed due to pork belly and egg spoilage. Tan et al. fabricated a paper-based in situ, wireless, cost-effective sensor with a capacitor and planar inductor attached to the wall of food packaging. As the humidity increases, the substrate absorbs $H_2O(g)$ and changes the sensor's resonant frequency and capacitance of the capacitor [54].

13.3.2 O_2-Based IP

Food quality monitoring can be carried out with the help of O_2 detectors [55, 56]. According to the type of transducer, O_2 sensors are classified as thermomagnetic (based on physical properties of O_2), optical (based on the reduction in fluorescence intensity of an excited fluorescent indicator), and electrochemical (potential electrochemical reaction) [12, 57]. A solid-state material that operates on luminescence quenching by direct interaction with the analyte was reported to constitute optical oxygen sensing devices for IP applications [58, 59]. Since optical sensors are reversible, susceptible, relatively inexpensive, and easy to read and store, several colorimetric oxygen indicators have been developed for IP [58, 60]. Mitsubishi Gas Chemical Corporation manufactures oxygen-scavenging sachets (Ageless^{TM}) that act as optical oxygen indicators. Ageless Eye^{TM} sachets include an oxygen indicator tablet to ensure that Ageless^{TM} absorbers are functioning properly. The indicator becomes pink when no oxygen is in the headspace (0.1%) and turns blue (0.5%) when oxygen is present (Figure 13.7). As a result, without opening the package, any spoilage of food can be assessed by a visual inspection [61, 62].

Wen et al. created a visible colorimetric oxygen indicator comprising an Ag-loaded titanium dioxide nanotube (Ag-TNT), hydroxyethyl cellulose, glycerol, and methylene blue that can effectively develop an entire packaging system to guarantee food quality. Their findings significantly increased the effectiveness of Ag-TNT as an O_2 indicator, and it may now be employed in the packaging of many goods to ensure food safety [63]. A typical electrochemical O_2

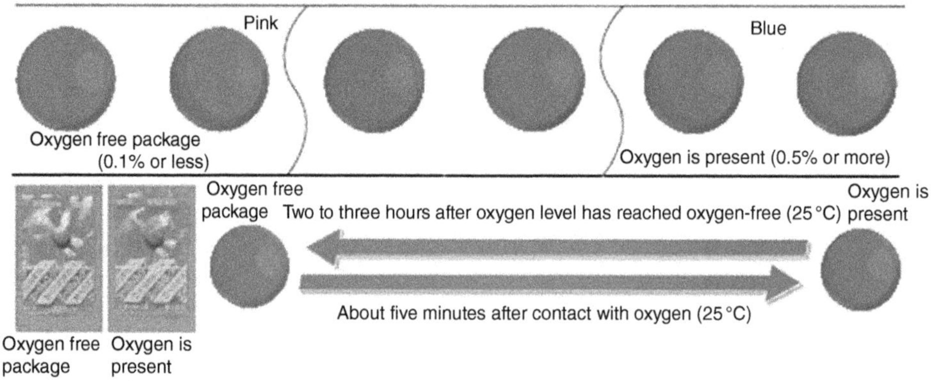

Figure 13.7 The color of the Ageless Eye oxygen indicator changes from pink to blue in the presence of oxygen. *Source:* Wang et al. [61]/Royal Society of Chemistry/CC-BY-4.0.

indicator comprises an alkaline compound (such as sodium hydroxide), a redox dye (such as methylene blue), and a reducing compound (such as reducing sugars). Lee et al. formulated a unique intelligent ink for oxygen indication. It consisted of UV-only absorbing semiconductor photosensitizer (typically titania), a sacrificial electron donor (glycerol), a suitable redox indicator (methylene blue), and a polymer for encapsulating these substances (hydroxyethyl cellulose). The ink can create a variety of colorimetric, UV-active, and nonreversible colors, and they are durable. This indicator can be embedded in paper, glass, plastic, etc. [64].

13.3.3 CO_2-Based IP

Changes in the external environment cause food degradation and fermentation. The principal byproduct of microbial metabolism during food storage is CO_2, and its production accompanies the change in food quality [65, 66]. Optical CO_2 indicators show a visible color change as the percentage of CO_2 gas within the package's headspace rises, providing an observable index of the packed food's shelf life. These indicators can precisely monitor food quality changes visible to the customer, resulting in noticeable alterations [67]. Optical CO_2 sensors are classified as (i) those which use the property of color change of a pH indicator dye (cresol red, phenol red, and thymol blue) [68–70] since CO_2 creates an anaerobic habitat, impedes the microorganism growth, and decreases the pH [71]; and (ii) those which function depending on the CO_2 that causes a change in the fluorescence of a luminescent dye such as ruthenium (II) complexes [72, 73].

Saliu et al. created a CO_2 sensor using polypeptides (poly-L-lysine, PLL), l-lysine, and anthocyanins. When the sensor was exposed to 2.5% CO_2, the sensor's color changed from azure to purple, indicating food spoilage [67]. Nopwinyuwong et al. fabricated IP for significant CO_2 detection using pH dye [74]. This on-package indicator consists of a blend of pH-sensitive pigments, methyl red and bromothymol blue, which change color when exposed to CO_2. Upon examination of the label and indicator solution's features, it indicates a real-time deterioration of food at constant or changing temperatures. Luminescent dyes have remarkable sensing capabilities, but their excitation and sensing require complicated instruments, rendering them unsuitable for broad application. One of this type's devices was entrenching 8-hydroxypyrene-1,3,5-trisulfonic acid within a polymeric film, resulting in a reaction time of fewer than ten seconds [73]. This technology was improved by immobilizing the fluorophore molecule on a substrate such as silica matrix. Regardless, since the chemicals used to make sensors are not food-grade, they must not come into contact with the products [67, 75]. Respiration rates of fruits and vegetables dramatically accelerate after plucking, creating significant CO_2, especially during growth and aging. As a result, the freshness of certain fruits and vegetables is linked to the amount of CO_2 in the container [13]. Chen et al. prepared an indicator label for newly cut green bell peppers with methyl red and bromocresol blue (3 : 2) indicating their quality [76]. In this investigation, the pH reduced from 7.6 to 4.4 as CO_2 increased, causing the label to change from yellow–green (day 0–6) to orange (day 7), as shown in Figure 13.8a.

Lyu et al. created CO_2-based IP by incorporating BTB^-/TBA^+ [bromothymol blue and tetrabutylammonium] ion-paired dye in chemical adsorbent polyethyleneimine. They used it to detect the CO_2 that evolved in kimchi during fermentation [77]. They found a link between the different stages of packaged kimchi fermentation and the sensor's observable color changes, as shown in Figure 13.8b. Furthermore, because the sensor is used in the packaging material, the detection creates continual changes that are easily determined. Hong and Park developed an indicator for detecting CO_2 gas without causing damage to the food packing material [78]. They used methyl red or bromocresol purple mixed into polymeric sheets. Determining carbon dioxide gas concentrations within the package also detects food decomposition during transportation and storage. In modified-environment food packaging, CO_2 is also used as a preservative at high doses. CO_2 inhibits the growth of bacteria and fungi by creating an anaerobic habitat and decreasing the pH of the food environment [71]. Jung et al. found that when an opaque chitosan solution was exposed to CO_2, it became transparent at pH 6. The indicator 2-amino-2-methyl-1-propanol in the chitosan solution can be placed into food packaging in sachets that are permeable to gaseous CO_2. As a result, they are highly likely to be used as sensors for detecting food spoilage and fermentation, as shown in Figure 13.9 [79]. Although significant progress has been made in the development of CO_2 sensors, they are unsuitable for food packaging applications because of drawbacks such as bulkiness, energy input requirements, high equipment costs, and safety concerns [13].

13.3.4 Microbial-Based IP

The leading causes of most foodborne health issues are microorganisms, including pathogenic bacteria like *Escherichia coli (E. coli)* and *Salmonella*. External factors that support these microorganisms' growth are temperature, time, humidity, and anaerobic conditions. Hence, detecting microorganisms is essential in food analysis [80]. The food quality can be indicated by detecting *E. coli* and *Salmonella*. Yousefi et al. fabricated a transparent package by embedding a DNA enzyme (DNAzyme) probe on a cyclo-olefin polymer with fluorescence property that explicitly binds with *E. coli* (Figure 13.10). The package detected *E. coli* in apple juice samples and meat products and assessed their quality [81]. Another detection method for *Salmonella* and *E. coli* is a prototype nanobiosensor and nanoporous silicon-based biosensor. Multiple pathogen detection is more accurate with this than with currently existing devices [82].

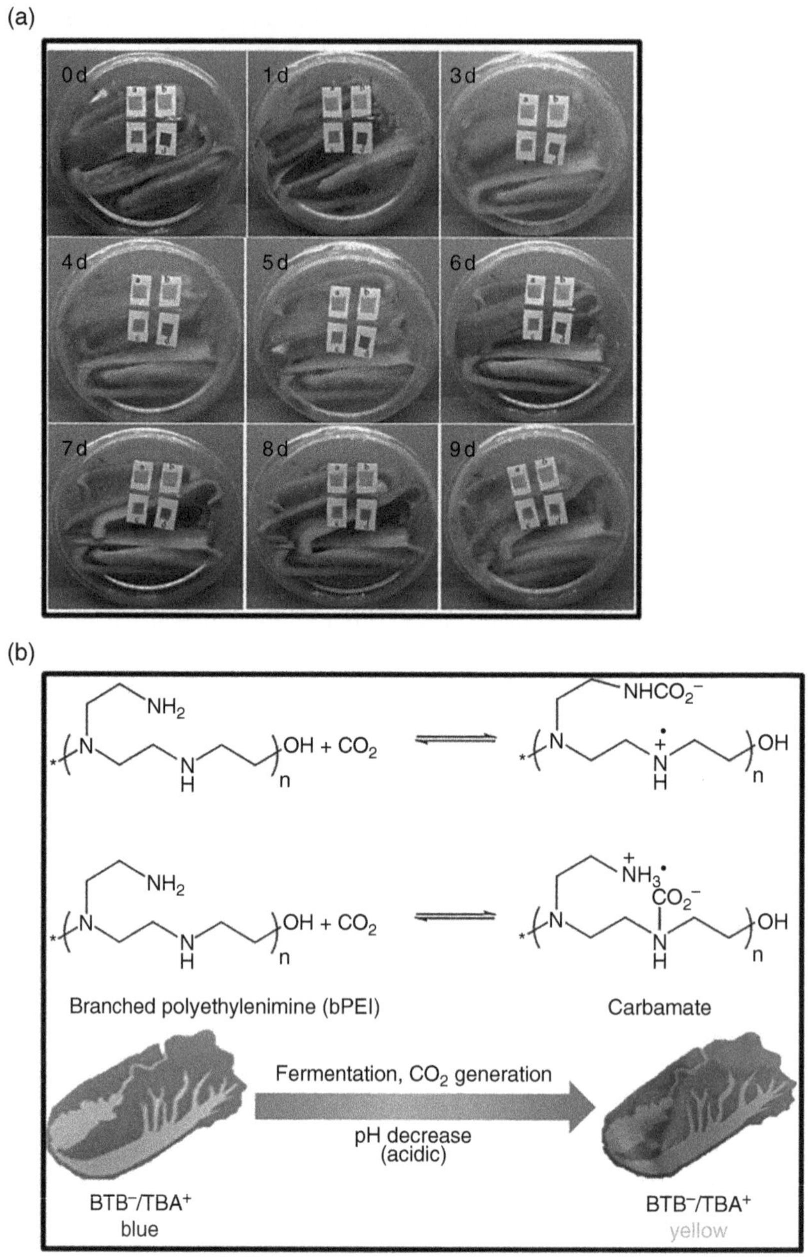

Figure 13.8 (a) Color changes of indicator labels for fresh-cut peppers spoilage trial at 7 ± 1 °C. *Source:* Reproduced with permission from Chen et al. [76]/Elsevier. (b) BTB^-/TBA^+ ion-paired dye-based CO_2 sensor. *Source:* Reproduced from Yousefi et al. [3]/with permission of American Chemical Society.

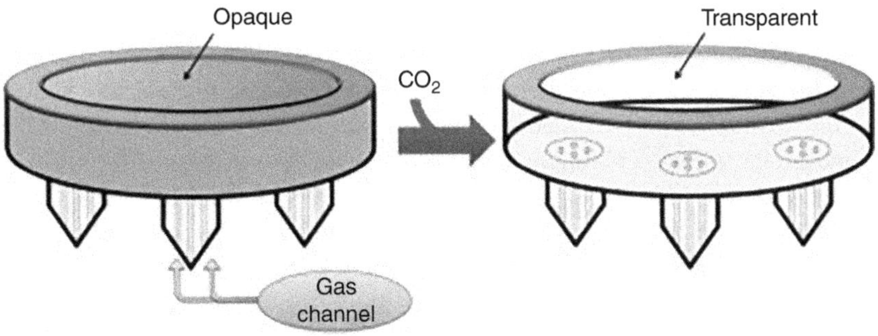

Figure 13.9 Chitosan-based carbon dioxide indicator. *Source:* Reproduced from Jung et al. [79]/ with permission of Elsevier.

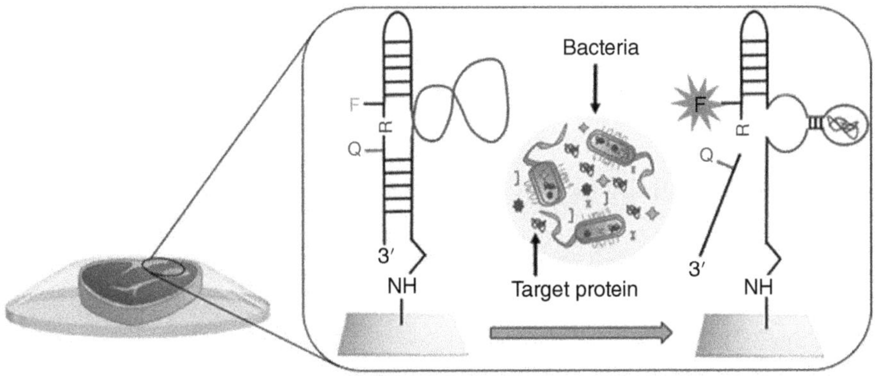

Figure 13.10 Food packaging containing DNAzyme Probes for monitoring food contamination by bacteria. *Source:* Reproduced from Yousefi et al. [81]/with permission of American Chemical Society.

Compared to immunological receptors and DNA-based receptors, synthetic antimicrobial peptides were proposed to inactivate pathogens due to their advantages, including large-scale production and high stability [83]. Sahar Alhogail et al. developed colorimetric sensor strips by immobilizing peptides on a gold chip. This sensor detected *Listeria monocytogenes* responsible for food spoilage, especially meat and milk [84]. Iyer et al. developed ultrathin polymer-based biosensors with RFID chips. The integrated nanocomponents are used to detect foodborne pathogens [85]. Zhiqing Yang et al. developed an electrochemical sensor based on a 3D-ZnO nanorod for detecting and destroying bacteria. This sensor was capable of 50% of bacterial inactivation [86]. The electronic tongue nanosensors and sensing method, which stimulates changes in color in food packages and provides indications of food spoilage, is a recent advancement in microbial detection [87].

13.3.5 pH-Based IP

As discussed in Section 13.2.3, food quality can be indicated by monitoring the pH. pH-responsive IP can be fabricated by immobilizing halochromic dyes into supports such as polymer/biopolymers (Figure 13.11a). Halochromic packaging shows color changes visible to the naked eye, indicating a fresh, spoiling, and spoiled state of food (Figure 13.11b) [88, 89]. Various methods, such as 3D printing, solution casting, sol–gel process, and electrospinning [90–94], are used to prepare halochromic IP systems [18].

Anthocyanins are the widely reported halochromic dyes for food quality monitoring. He et al. used electrospinning to incorporate blueberry anthocyanins and caffeic acid into a polyacrylonitrile polymer [93]. The film showed a colorimetric response to fish spoilage in real time. Anthocyanins undergo chemical transformations in response to TVB-N and pH changes during fish spoilage. These changes showed the color change in the packaging film from pink to light purple and then to dark purple. Caffeic acid enhanced the TVB-N response and the stability of the film through hydrogen bonding. Li et al. used 3D printing to form a sandwich-type composite film of chitosan/mulberry anthocyanin/lemongrass essential oil, and cassava starch [100]. The film in response to pH changes during the spoilage of chilled pork showed a change from red to gray and then blue. Devarajan et al.

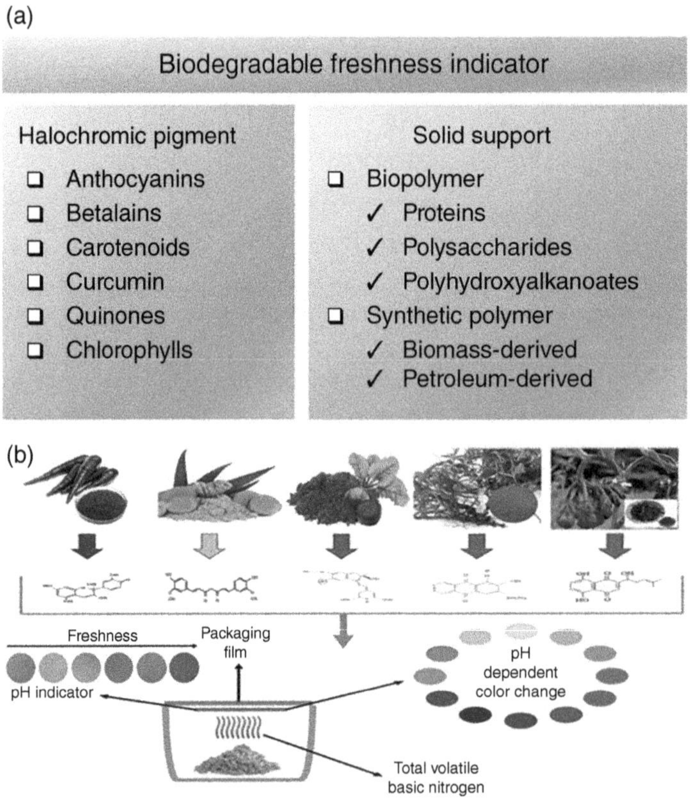

Figure 13.11 (a) Natural halochromic indicators and polymers for intelligent packaging. *Source:* Reproduced from Yu et al. [88]/with permission of American Chemical Society. (b) Schematic illustration of pH-based intelligent packaging. *Source:* Reproduced with permission from Priyadarshi et al. [89]; Yu et al. [88]/American Chemical Society.

reported an IP system for monitoring fish spoilage by casting a thin film of anthocyanins from red cabbage into potato dextrose agar and starch solution. The film showed a color change from brownish to yellowing green, indicating a fresh and spoiled state of the fish [101]. Ghorbani et al. reported an indicator for chicken fillets, fish roe, beef, and shrimp by immobilizing anthocyanins from *Malva sylvestris* into polylactic acid [102]. Along with anthocyanins, betalains and alizarins are also used as halochromic dyes. Apriliyanti et al. incorporated betalains from the dragon fruit peels into glucomannan/ polyvinyl alcohol (PVA) to monitor the fish's freshness [103]. In another study by Sonar et al., betalains extracted from beetroot with polyethylene terephthalate (PET)/Nylon66 as packaging film were used to monitor the freshness of pasteurized vegetable purees [104]. Jaramillo et al. noticed that carotenoids and chlorophyll in green tea and basil extracts change color when exposed to various pH levels, indicating that the materials can be used as food quality indicators [105]. Ezati et al. developed a novel pH-sensitive IP by combining alizarin with starch–cellulose paper to monitor the freshness of rainbow trout fillets [106]. The change of color was noticed from orange to reddish-brown, corresponding to the changes in TVB-N. Due to the narrow color changes, synthetic dyes are usually used to combine different dyes to enable pH indication. Magnaghi et al. reported an ethylene-vinyl alcohol-based pH-sensitive optical array to monitor milk freshness [107]. This work combined various synthetic dyes such as thymol blue, m-Cresol purple, bromothymol blue, bromophenol blue, o-Cresol red, and chlorophenol red with ethylene-vinyl alcohol to obtain a sensor.

13.4 Conclusion and Perspective

The lack of real-time information about food quality puts consumers at risk of food poisoning, a severe problem affecting a large population. Solving these problems requires the proper containment of food, protection from external factors, detection of spoilage, and alerting the consumer about food quality in real time. IP are indicators/sensors that can display food quality in real time based on chemical, physical, and microbiological changes during food spoilage. However, IP is expensive compared to conventional packaging materials, and with their limited ability to be integrated into current packages, such systems are still not widely used. In conclusion, this chapter discussed real-time food quality monitoring based on various food quality indicators, such as humidity, microbial, O_2, CO_2, and pH, which can inform consumers about the freshness of food. Recent advances in these indicators for IP systems have also been summarized.

References

1 Food and Agriculture Organization of the United Nations (2011). *Global Food Losses and food Waste – Extent, Causes and Prevention Food Safety*. World Health Organization.

2 Ma, Y.W., Xia, Y., Xue, W. et al. (2022). Properties and applications of intelligent packaging indicators for food spoilage. *Membranes* 12 (5): 477. https://doi.org/10.3390/membranes12050477.

3 Yousefi, H., Su, H.-M.M., Imani, S.M. et al. (2019). Intelligent food packaging: a review of smart sensing technologies for monitoring food quality. *ACS Sensors* 4 (4): 808–821. https://doi.org/10.1021/acssensors.9b00440.

4 Hernández-Cortez, C., Palma-Martínez, I., Gonzalez-Avila, L.U. et al. (2017). *Food Poisoning Caused by Bacteria (Food Toxins), Poisoning – From Specific Toxic Agents to Novel Rapid and Simplified Techniques for Analysis*. InTech https://doi.org/10.5772/intechopen.69953.

5 Chaithra, K.P. and Vinod, T.P. (2022). A smartphone coupled freshness indicator prepared by rub-coating of hibiscus flowers on paper substrates for visual monitoring of the spoilage of milk. *ChemistrySelect* 7 (41): https://doi.org/10.1002/slct.202201839.

6 Yam, K.L., Takhistov, P.T., and Miltz, J. (2005). Intelligent packaging: concepts and applications. *Journal of Food Science* 70 (1): https://doi.org/10.1111/j.1365-2621.2005.tb09052.x.

7 Weston, M., Geng, S., and Chandrawati, R. (2021). Food sensors: challenges and opportunities. *Advanced Materials Technologies* 6 (5): 1–16. https://doi.org/10.1002/admt.202001242.

8 Kang, W., Lin, H., Jiang, H.,et al. (2021) 'Advanced applications of chemo-responsive dyes based odor imaging technology for fast sensing food quality and safety: a review', *Comprehensive Reviews in Food Science and Food Safety*, 20(5), pp. 5145–5172. https://doi.org/10.1111/1541-4337.12823.

9 Zhang, Z., Sun, Y., Sang, S., Jia, L. and Ou, C. (2022) 'Emerging approach for fish freshness evaluation: principle, application and challenges', *Foods*, 11(13), p. 1897. https://doi.org/10.3390/foods11131897.

10 Lawal, A.T. and Adeloju, S.B. (2012). Progress and recent advances in fabrication and utilization of hypoxanthine biosensors for meat and fish quality assessment: a review. *Talanta* 100: 217–228. https://doi.org/10.1016/j.talanta.2012.07.085.

11 Farahani, H., Wagiran, R., and Hamidon, M. (2014). Humidity sensors principle, mechanism, and fabrication technologies: a comprehensive review. *Sensors* 14 (5): 7881–7939. https://doi.org/10.3390/s140507881.

12 Smiddy, M., Papkovskaia, N., Papkovsky, D. and Kerry, J. (2002) 'Use of oxygen sensors for the non-destructive measurement of the oxygen content in modified atmosphere and vacuum packs of cooked chicken patties; impact of oxygen content on lipid oxidation', *Food Research International*, 35(6), pp. 577–584. https://doi.org/10.1016/S0963-9969(01)00160-0.

13 Puligundla, P., Jung, J., and Ko, S. (2012). Carbon dioxide sensors for intelligent food packaging applications. *Food Control* 25 (1): 328–333. https://doi.org/10.1016/j.foodcont.2011.10.043.

14 Devarayan, K., Pandiyan, P., Nagaraju, K.S. and Anjappan, H. (2020) 'Halochromic sensors for real-time monitoring of spoilage of packed seer fish', *Materials Today: Proceedings*, 33(7), pp. 3961–3966. https://doi.org/10.1016/j.matpr.2020.06.332.

15 Dodero, A., Escher, A., Bertucci, S., Castellano, M. and Lova, P. et al. (2021) 'Intelligent packaging for real-time monitoring of food-quality: current and future developments', *Applied Sciences*, 11(8), p. 3532. https://doi.org/10.3390/app11083532.

16 Cheng, H.et al. (2022) 'Recent advances in intelligent food packaging materials: principles, preparation and applications', *Food Chemistry*, 375, p. 131738. https://doi.org/10.1016/j.foodchem.2021.131738.

17 Azeredo, H.M.C.C. and Correa, D.S. (2021). Smart choices: mechanisms of intelligent food packaging. *Current Research in Food Science* 4: 932–936. https://doi.org/10.1016/j.crfs.2021.11.016.

18 Forghani, S., Almasi, H., and Moradi, M. (2021). Electrospun nanofibers as food freshness and time-temperature indicators: a new approach in food intelligent packaging. *Innovative Food Science & Emerging Technologies* 73: 102804. https://doi.org/10.1016/j.ifset.2021.102804.

19 United, Nations. (2015). THE 17 GOALS| *The UN Sustainable Development Goals*. United Nations, New York Available at: http://www.un.org/sustainabledevelopment/summit/.

20 Taoukis, P.S. and Giannakourou, M.C. (2004). Temperature and food stability: analysis and control. In: *Understanding and Measuring the Shelf-Life of Food*, 42–68. Elsevier. https://doi.org/10.1533/9781855739024.1.42.

21 Ergun, R., Lietha, R., and Hartel, R.W. (2010). Moisture and shelf life in sugar confections. *Critical Reviews in Food Science and Nutrition* 50 (2): 162–192. https://doi.org/10.1080/10408390802248833.

22 Troller, J.A. and Christian, J.H.B. (1978). Control of aw and moisture. In: *Water Activity and Food*, 174–191. Elsevier. https://doi.org/10.1016/B978-0-12-700650-5.50015-9.

23 Kong, F. and Singh, R.P. (2011). *Chemical Deterioration and Physical Instability of Foods and Beverages, Food and Beverage Stability and Shelf Life*. Woodhead Publishing Limited. https://doi.org/10.1533/9780857092540.1.29.

24 McDaniel, C. and Baker, R.C. (1977). Convenience food packaging and the perception of product quality. *Journal of Marketing* 41 (4): 57. https://doi.org/10.2307/1250234.

25 Maddanimath, T., Mulla, I.S., Sainkar, S.R., Vijayamohanan, K. et al. (2002) 'Humidity sensing properties of surface functionalised polyethylene and polypropylene films', *Sensors and Actuators B: Chemical*, 81(2–3), pp. 141–151. https://doi.org/10.1016/S0925-4005(01)00944-3.

26 Jay, J.M. (1992). Spoilage of fruits and vegetables. *Modern Food Microbiology* 187–198. https://doi.org/10.1007/978-94-011-6480-1_8.

27 Sonwani, E., Bansal, U., Alroobaea, R., Baqasah, A.M. and Hedabou, M. et al. (2022) 'An artificial intelligence approach toward food spoilage detection and analysis', *Frontiers in Public Health*, 9(January), pp. 1–13. https://doi.org/10.3389/fpubh.2021.816226.

28 Zhang, Y., Hou, W., Yue, Q., Zhang, Y.,et al. (2021) 'Effects of environmental humidity on microbial community and quality of chilled fresh pork', *SSRN Electronic Journal* https://doi.org/10.2139/ssrn.3964470.

29 Pereira de Abreu, D.A., Cruz, J.M., and Paseiro Losada, P. (2012). Active and intelligent packaging for the food industry. *Food Reviews International* 28 (2): 146–187. https://doi.org/10.1080/87559129.2011.595022.

30 Ali, S., Sattar Khan, A., Ullah Malik, A., Anjum. M.A. et al. (2019) 'Modified atmosphere packaging delays enzymatic browning and maintains quality of harvested litchi fruit during low temperature storage', *Scientia Horticulturae*, 254, pp. 14–20. https://doi.org/10.1016/j.scienta.2019.04.065.

31 Tabassum, N. and Khan, M.A. (2020). Modified atmosphere packaging of fresh-cut papaya using alginate based edible coating: Quality evaluation and shelf life study. *Scientia Horticulturae* 259: 108853. https://doi.org/10.1016/j.scienta.2019.108853.

32 Xiong, Y., Li, S., Warner, R.D. and Fang, Z. (2020) 'Effect of oregano essential oil and resveratrol nanoemulsion loaded pectin edible coating on the preservation of pork loin in modified atmosphere packaging', *Food Control*, 114, p. 107226. https://doi.org/10.1016/j.foodcont.2020.107226.

33 Weston, M., Phan, M.A.T., Arcot, J. and Chandrawati, R. (2020) 'Anthocyanin-based sensors derived from food waste as an active use-by date indicator for milk', *Food Chemistry*, 326, p. 127017. https://doi.org/10.1016/j.foodchem.2020.127017.

34 Yildiz, E., Sumnu, G., and Kahyaoglu, L.N. (2021). Monitoring freshness of chicken breast by using natural halochromic curcumin loaded chitosan/PEO nanofibers as an intelligent package. *International Journal of Biological Macromolecules* 170: 437–446. https://doi.org/10.1016/j.ijbiomac.2020.12.160.

35 Fang, S., Guan, Z., Su, C., Zhang, W. et al.(2022) Accurate fish-freshness prediction label based on red cabbage anthocyanins', *Food Control*, 138(2021), p. 109018. https://doi.org/10.1016/j.foodcont.2022.109018.

36 Chaithra, K.P., Vinod, T.P., and Nagarajan, P. (2023). Smartphone application-based colorimetric fish freshness monitoring using an indicator prepared by rub-coating of red cabbage on paper substrates. *Colloids and Surfaces A: Physicochemical and Engineering Aspects* 132553. https://doi.org/10.1016/j.colsurfa.2023.132553.

37 Shukla, V., Kandeepan, G., and Vishnuraj, M.R. (2015). Development of on-package indicator sensor for real-time monitoring of buffalo meat quality during refrigeration storage. *Food Analytical Methods* 8 (6): 1591–1597. https://doi.org/10.1007/s12161-014-0066-6.

38 Hidayat, S.H., Dirpan, A., Adiansyah, Djalal, M. et al. (2019) 'Sensitivity determination of indicator paper as smart packaging elements in monitoring meat freshness in cold temperature', *IOP Conference Series: Earth and Environmental Science*, 343(1), p. 012076. https://doi.org/10.1088/1755-1315/343/1/012076.

39 Liu, X., Chen, K., Wang, J., Wang, Y. et al. (2020) An on-package colorimetric sensing label based on a sol-gel matrix for fish freshness monitoring', *Food Chemistry*. Edited by Z. Yao, 307(1), p. 125580. https://doi.org/10.1016/j.foodchem.2019.125580.

40 Shao, P., Liu, L., Yu, J., Lin, Y. et al. (2021) 'An overview of intelligent freshness indicator packaging for food quality and safety monitoring', *Trends in Food Science and Technology*, 118(PA), pp. 285–296. https://doi.org/10.1016/j.tifs.2021.10.012.

41 Kulkarni, A. (2022). Preparation and testing of food freshness indicators: an application-oriented learning module integrating basic concepts of microbiology and chemistry laboratory. *Journal of Microbiology & Biology Education*. Edited by D.J. Westenberg 23 (2): 3. https://doi.org/10.1128/jmbe.00063-22.

42 Qin, Y., Huang, P., and Wu, F.-Y. (2022). Histamine-responsive dye-incorporated carbon dots for visual monitoring of food spoilage. *Sensors and Actuators B: Chemical* 365: 131911. https://doi.org/10.1016/j.snb.2022.131911.

43 Karube, I., Matsuoka, H., Suzuki, S., Watanabe, E. and Toyama, K. (1984) 'Determination of fish freshness with an enzyme sensor system', *Journal of Agricultural and Food Chemistry*, 32(2), pp. 314–319. https://doi.org/10.1021/jf00122a034.

44 Chen, Z., Lin, Y., Ma, X., Guo, L. et al. (2017) 'Multicolor biosensor for fish freshness assessment with the naked eye', *Sensors and Actuators B: Chemical*, 252, pp. 201–208. https://doi.org/10.1016/j.snb.2017.06.007.

45 Esser, B., Schnorr, J.M., and Swager, T.M. (2012). Selective detection of ethylene gas using carbon nanotube-based devices: utility in determination of fruit ripeness. *Angewandte Chemie International Edition* 51 (23): 5752–5756. https://doi.org/10.1002/anie.201201042.

46 Kumar, S.B., Jeevamalar, J., Ramu, P. et al. (2021). Evaluation in 4D printing – a review. In: *Materials Today: Proceedings*, vol. 45 (ed. G. Kumaresan), 1433–1437. Elsevier. https://doi.org/10.1016/j.matpr.2020.07.335.

47 Müller, P. and Schmid, M. (2019). Intelligent packaging in the food sector: a brief overview. *Foods* 8 (1): 16. https://doi.org/10.3390/foods8010016.

48 Han, J.H. (2014). A review of food packaging technologies and innovations. In: *Innovations in Food Packaging* (ed. J.H. Han), 3–12. Elsevier https://doi.org/10.1016/B978-0-12-394601-0.00001-1.

49 Ghaani, M., Cozzolino, C.A., Castelli, G., and Farris, S. (2016). An overview of the intelligent packaging technologies in the food sector. *Trends in Food Science & Technology* 51: 1–11. https://doi.org/10.1016/j.tifs.2016.02.008.

50 Heising, J.K., Dekker, M., Bartels, P.V. et al. (2014). Monitoring the quality of perishable foods: opportunities for intelligent packaging. *Critical Reviews in Food Science and Nutrition* 54 (5): 645–654. https://doi.org/10.1080/10408398.2011.600477.

51 Paull, R.E. (1999). Effect of temperature and relative humidity on fresh commodity quality. *Postharvest Biology and Technology* 15 (3): 263–277. https://doi.org/10.1016/S0925-5214(98)00090-8.

52 Bibi, F., Guillaume, C., Vena, A. et al. (2016). Wheat gluten, a bio-polymer layer to monitor relative humidity in food packaging: Electric and dielectric characterization. *Sensors and Actuators, A: Physical* 247: 355–367. https://doi.org/10.1016/j.sna.2016.06.017.

53 Yuan, Z., Bariya, M., Fahad, H.M. et al. (2020). Trace-level, multi-gas detection for food quality assessment based on decorated silicon transistor arrays. *Advanced Materials* 32 (21): 1–8. https://doi.org/10.1002/adma.201908385.

54 Tan, E., Ng, W., Shao, R. et al. (2007). A wireless, passive sensor for quantifying packaged food quality. *Sensors* 7 (9): 1747–1756. https://doi.org/10.3390/s7091747.

55 Hellwig, M. (2020). Analysis of protein oxidation in food and feed products. *Journal of Agricultural and Food Chemistry* 68 (46): 12870–12885. https://doi.org/10.1021/acs.jafc.0c00711.

56 Hu, K., Huyan, Z., Ding, S. et al. (2020). Investigation on food packaging polymers: effects on vegetable oil oxidation. *Food Chemistry* 315: 126299. https://doi.org/10.1016/j.foodchem.2020.126299.

57 Uetake, H., Hirota, N., Nakagawa, J. et al. (2000). Thermal convection control by gradient magnetic field. *Journal of Applied Physics* 87 (9): 6310–6312. https://doi.org/10.1063/1.372689.

58 McNamara, K., Li, X., Stull, A.D. et al. (1998). Fiber-optic oxygen sensor based on the fluorescence quenching of tris (5-acrylamido, 1,10 phenanthroline) ruthenium chloride. *Analytica Chimica Acta* 361 (1–2): 73–83. https://doi.org/10.1016/S0003-2670(97)00703-4.

59 Papkovsky, D.B., Ponomarev, G.V., Trettnak, W., and O'Leary, P. (1995). Phosphorescent complexes of porphyrin ketones: optical properties and application to oxygen sensing. *Analytical Chemistry* 67 (22): 4112–4117. https://doi.org/10.1021/ac00118a013.

60 Chu, C.-S. and Lo, Y.-L. (2011). Highly sensitive and linear calibration optical fiber oxygen sensor based on Pt(II) complex embedded in sol–gel matrix. *Sensors and Actuators B: Chemical* 155 (1): 53–57. https://doi.org/10.1016/j.snb.2010.11.023.

61 Wang, X. and Wolfbeis, O.S. (2014). Optical methods for sensing and imaging oxygen: materials, spectroscopies and applications. *Chem. Soc. Rev.* 43 (10): 3666–3761. https://doi.org/10.1039/C4CS00039K.

62 Ahvenainen, R. and Hurme, E. (1997). Active and smart packaging for meeting consumer demands for quality and safety. *Food Additives and Contaminants* 14 (6–7): 753–763. https://doi.org/10.1080/02652039709374586.

63 Wen, J., Huang, S., Jia, L. et al. (2019). Visible colorimetric oxygen indicator based on Ag-loaded TiO_2 nanotubes for quick response and real-time monitoring of the integrity of modified atmosphere packaging. *Advanced Materials Technologies* 4 (9): 1900121. https://doi.org/10.1002/admt.201900121.

64 Lee, S.-K., Mills, A., and Lepre, A. (2004). An intelligence ink for oxygen. *Chemical Communications* 17: 1912. https://doi.org/10.1039/b406685e.

65 Ha, J.-U., Kim, Y.-M., and Lee, D.-S. (2001). Multilayered antimicrobial polyethylene films applied to the packaging of ground beef. *Packaging Technology and Science* 14 (2): 55–62. https://doi.org/10.1002/pts.537.

66 Kuswandi, B. and Jumina (2020). Active and intelligent packaging, safety, and quality controls. In: *Fresh-Cut Fruits and Vegetables* (ed. M.W. Siddiqui), 243–294. Elsevier. https:doi.org/10.1016/B978-0-12-816184-5.00012-4.

67 Saliu, F. and Della Pergola, R. (2018). Carbon dioxide colorimetric indicators for food packaging application: Applicability of anthocyanin and poly-lysine mixtures. *Sensors and Actuators B: Chemical* 258: 1117–1124. https://doi.org/10.1016/j.snb.2017.12.007.

68 Kawabata, Y., Kamichika, T., Imasaka, T., and Ishibashi, N. (1989). Fiber-optic sensor for carbon dioxide with a ph indicator dispersed in a poly(ethylene glycol) membrane. *Analytica Chimica Acta* 219: 223–229. https://doi.org/10.1016/S0003-2670(00)80353-0.

69 Mills, A., Chang, Q., and McMurray, N. (1992). Equilibrium studies on colorimetric plastic film sensors for carbon dioxide. *Analytical Chemistry* 64 (13): 1383–1389. https://doi.org/10.1021/ac00037a015.

70 Mills, A., Lepre, A., and Wild, L. (1997). Breath-by-breath measurement of carbon dioxide using a plastic film optical sensor. *Sensors and Actuators B: Chemical* 39 (1–3): 419–425. https://doi.org/10.1016/S0925-4005(96)02116-8.

71 Wang, W., Rao, L., Wu, X. et al. (2021). Supercritical carbon dioxide applications in food processing. *Food Engineering Reviews* 13 (3): 570–591. https://doi.org/10.1007/s12393-020-09270-9.

72 Marazuela, M.D., Moreno Bondi, M.C., and Orellana, G. (1995). Enhanced performance of a fibre-optic luminescence CO_2 sensor using carbonic anhydrase. *Sensors and Actuators B: Chemical* 29 (1–3): 126–131. https://doi.org/10.1016/0925-4005(95)01673-2.

73 Mills, A. and Chang, Q. (1993). Fluorescence plastic thin-film sensor for carbon dioxide. *The Analyst* 118 (7): 839. https://doi.org/10.1039/an9931800839.

74 Nopwinyuwong, A., Trevanich, S., and Suppakul, P. (2010). Development of a novel colorimetric indicator label for monitoring freshness of intermediate-moisture dessert spoilage. *Talanta* 81 (3): 1126–1132. https://doi.org/10.1016/j.talanta.2010.02.008.

75 von Bültzingslöwen, C., McEvoy, A.K., McDonagh, C. et al. (2002). Sol–gel based optical carbon dioxide sensor employing dual luminophore referencing for application in food packaging technology. *The Analyst* 127 (11): 1478–1483. https://doi.org/10.1039/B207438A.

76 Chen, H., Zhang, M., Bhandari, B. et al. (2018). Postharvest biology and technology applicability of a colorimetric indicator label for monitoring freshness of fresh-cut green bell pepper. *Postharvest Biology and Technology* 140 (February): 85–92. https://doi.org/10.1016/j.postharvbio.2018.02.011.

77 Lyu, J.S., Choi, I., Hwang, K.-S. et al. (2019). Development of a BTB−/TBA+ ion-paired dye-based CO_2 indicator and its application in a multilayered intelligent packaging system. *Sensors and Actuators B: Chemical* 282: 359–365. https://doi.org/10.1016/j.snb.2018.11.073.

78 Hong, S.-I. and Park, W.-S. (2000). Use of color indicators as an active packaging system for evaluating kimchi fermentation. *Journal of Food Engineering* 46 (1): 67–72. https://doi.org/10.1016/S0308-8146(00)00141-2.

79 Jung, J., Puligundla, P., and Ko, S. (2012). Proof-of-concept study of chitosan-based carbon dioxide indicator for food packaging applications. *Food Chemistry* 135 (4): 2170–2174. https://doi.org/10.1016/j.foodchem.2012.07.090.

80 de Boer, E. (1999). Methodology for detection and typing of foodborne microorganisms. *International Journal of Food Microbiology* 50 (1–2): 119–130. https://doi.org/10.1016/S0168-1605(99)00081-1.

81 Yousefi, H., Ali, M.M., Su, H.-M. et al. (2018). Sentinel wraps: real-time monitoring of food contamination by printing DNAzyme probes on food packaging. *ACS Nano* 12 (4): 3287–3294. https://doi.org/10.1021/acsnano.7b08010.

82 Liu, Y., Chakrabartty, S., and Alocilja, E.C. (2007). Fundamental building blocks for molecular biowire based forward error-correcting biosensors. *Nanotechnology* 18 (42): 424017. https://doi.org/10.1088/0957-4484/18/42/424017.

83 Rydlo, T., Rotem, S., and Mor, A. (2006). Antibacterial properties of dermaseptin S4 derivatives under extreme incubation conditions. *Antimicrobial Agents and Chemotherapy* 50 (2): 490–497. https://doi.org/10.1128/AAC.50.2.490-497.2006.

84 Alhogail, S., Suaifan, G.A.R.Y., and Zourob, M. (2016). Rapid colorimetric sensing platform for the detection of Listeria monocytogenes foodborne pathogen. *Biosensors and Bioelectronics* 86: 1061–1066. https://doi.org/10.1016/j.bios.2016.07.043.

85 Yu, H., Park, J.-Y., Kwon, C.W. et al. (2018). An overview of nanotechnology in food science: preparative methods, Practical Applications, and Safety. *Journal of Chemistry* 2018: 1–10. https://doi.org/10.1155/2018/5427978.

86 Yang, Z., Wang, Y., and Zhang, D. (2017). A novel multifunctional electrochemical platform for simultaneous detection, elimination, and inactivation of pathogenic bacteria based on the Vancomycin-functionalised AgNPs/3D-ZnO nanorod arrays. *Biosensors and Bioelectronics* 98: 248–253. https://doi.org/10.1016/j.bios.2017.06.058.

87 Majid, I., Ahmad Nayik, G., Mohammad Dar, S. et al. (2018). Novel food packaging technologies: innovations and future prospective. *Journal of the Saudi Society of Agricultural Sciences* 17 (4): 454–462. https://doi.org/10.1016/j.jssas.2016.11.003.

88 Yu, Z., Boyarkina, V., Liao, Z. et al. (2023). Boosting food system sustainability through intelligent packaging: application of biodegradable freshness indicators. *ACS Food Science & Technology* 3 (1): 199–212. https://doi.org/10.1021/acsfoodscitech.2c00372.

89 Priyadarshi, R., Ezati, P., and Rhim, J.-W. (2021). Recent advances in intelligent food packaging applications using natural food colorants. *ACS Food Science & Technology* 1 (2): 124–138. https://doi.org/10.1021/acsfoodscitech.0c00039.

90 Duan, M., Sun, J., Huang, Y. et al. (2023). Electrospun gelatin/chitosan nanofibers containing curcumin for multifunctional food packaging. *Food Science and Human Wellness* 12 (2): 614–621. https://doi.org/10.1016/j.fshw.2022.07.064.

91 Goudarzi, J., Moshtaghi, H., and Shahbazi, Y. (2023). Kappa-carrageenan-poly(vinyl alcohol) electrospun fiber mats encapsulated with Prunus domestica anthocyanins and epigallocatechin gallate to monitor the freshness and enhance the shelf-life quality of minced beef meat. *Food Packaging and Shelf Life* 35: 101017. https://doi.org/10.1016/j.fpsl.2022.101017.

92 Han, T. et al. (2023). An electrospun sensing label based on poly (vinyl alcohol)-Ag-grape seed anthocyanidin nanofibers for smart, real-time meat freshness monitoring. *Sensors and Actuators B: Chemical* 376: 132975. https://doi.org/10.1016/j.snb.2022.132975.

93 He, H., Xia, C., Huang, Y. et al. (2022). A novel anthocyanin electrospun film by caffeic acid co-pigmentation for real-time fish freshness monitoring. *Analytical Methods* 15 (2): 228–239. https://doi.org/10.1039/d2ay01434c.

94 Zou, Y., Sun, Y., Shi, W. et al. (2023). Dual-functional shikonin-loaded quaternized chitosan/polycaprolactone nanofibrous film with pH-sensing for active and intelligent food packaging'. *Food Chemistry* 399 (2022): 133962. https://doi.org/10.1016/j.foodchem.2022.133962.

95 Maftoonazad, N. and Ramaswamy, H. (2019). Design and testing of an electrospun nanofiber mat as a pH biosensor and monitor the pH associated quality in fresh date fruit (Rutab). *Polymer Testing* 75: 76–84. https://doi.org/10.1016/j.polymertesting.2019.01.011.

96 Tripathy, S., Reddy, M.S., Vanjari, S.R.K. et al. (2019). A step towards miniaturized milk adulteration detection system: smartphone-based accurate pH sensing using electrospun halochromic nanofibers. *Food Analytical Methods* 12 (2): 612–624. https://doi.org/10.1007/s12161-018-1391-y.

97 Guo, M., Wang, H., Wang, Q. et al. (2020). Intelligent double-layer fiber mats with high colorimetric response sensitivity for food freshness monitoring and preservation'. *Food Hydrocolloids* 101 (2019): 105468. https://doi.org/10.1016/j.foodhyd.2019.105468.

98 Sun, W., Li, W., Jin, Z. et al. (2021). Non-destructive detection of mutton freshness using anthocyanin nanofiber smart label. *Nongye Gongcheng Xuebao/Transactions of the Chinese Society of Agricultural Engineering* 37 (4): 24–30. https://doi.org/10.11975/j.issn.1002-6819.2021.04.004.

99 Aghaei, Z., Emadzadeh, B., Ghorani, B., and Kadkhodaee, R. (2018). Cellulose acetate nanofibres containing alizarin as a halochromic sensor for the qualitative assessment of rainbow trout fish spoilage. *Food and Bioprocess Technology* 11 (5): 1087–1095. https://doi.org/10.1007/s11947-017-2046-5.

100 Li, S., Jiang, Y.Y., Zhou, Y. et al. (2022). Facile fabrication of sandwich-like anthocyanin/chitosan/lemongrass essential oil films via 3D printing for intelligent evaluation of pork freshness. *Food Chemistry* 370: 131082. https://doi.org/10.1016/j.foodchem.2021.131082.

101 Devarayan, K., Motcham, V.V., Kathavarayan, M., and Anjappan, H. (2021). Real-time detection of packaged seer fish spoilage using halochromic optical nose. *Journal of Aquatic Food Product Technology* 30 (4): 484–495. https://doi.org/10.1080/10498850.2021.1897049.

102 Ghorbani, M., Divsalar, E., Molaei, R. et al. (2021). A halochromic indicator based on polylactic acid and anthocyanins for visual freshness monitoring of minced meat, chicken fillet, shrimp, and fish roe. *Innovative Food Science and Emerging Technologies* 74: 102864. https://doi.org/10.1016/j.ifset.2021.102864.

103 Apriliyanti, M.W., Wahyono, A., Fatoni, M. et al. (2018). The Potency of betacyanins extract from a peel of dragon fruits as a source of colourimetric indicator to develop intelligent packaging for fish freshness monitoring. *IOP Conference Series: Earth and Environmental Science* 207 (1): 012038. https://doi.org/10.1088/1755-1315/207/1/012038.

104 Sonar, C.R., Rasco, B., Tang, J., and Sablani, S.S. (2019). Natural color pigments: oxidative stability and degradation kinetics during storage in thermally pasteurized vegetable purees. *Journal of the Science of Food and Agriculture* 99 (13): 5934–5945. https://doi.org/10.1002/jsfa.9868.

105 Medina-Jaramillo, C., Ochoa-Yepes, O., Bernal, C., and Famá, L. (2017). Active and smart biodegradable packaging based on starch and natural extracts. *Carbohydrate Polymers* 176: 187–194. https://doi.org/10.1016/j.carbpol.2017.08.079.

106 Ezati, P., Tajik, H., Moradi, M., and Molaei, R. (2019). Intelligent pH-sensitive indicator based on starch-cellulose and alizarin dye to track freshness of rainbow trout fillet. *International Journal of Biological Macromolecules* 132: 157–165. https://doi.org/10.1016/j.ijbiomac.2019.03.173.

107 Magnaghi, L.R., Zanoni, C., Alberti, G. et al. (2022). Freshness traffic light for fish products: dual-optode label to monitor fish spoilage in sales packages. *ACS Food Science and Technology* 2 (6): 1030–1038. https://doi.org/10.1021/acsfoodscitech.2c00097.

14

Freshness Indicator in Intelligent Food Packaging System

Satti Venu Gopala Kumari[1], *Kannan Pakshirajan*[2], *and G. Pugazhenthi*[1,3]

[1] *Department of Chemical Engineering, Indian Institute of Technology Guwahati, Guwahati, Assam, India*
[2] *Department of Biosciences and Bioengineering, Indian Institute of Technology Guwahati, Guwahati, Assam, India*
[3] *Centre for Sustainable Polymers, Indian Institute of Technology Guwahati, Guwahati, Assam, India*

14.1 Introduction

Public health and social well-being are greatly influenced by the quality and safety of food. As per World Health Organization (WHO) statistics, consumption of contaminated food is known to cause 600 million cases of illness and 420 000 deaths worldwide every year [1]. Traditional methods for the determination of food freshness include microbiological measurements, chemical measurements, and sensory evaluations. However, the aforementioned methods have one or more drawbacks, such as high cost, cumbersome, long detection period, and destruction of food samples [2]. Currently, expiration dates are used to indicate food conditions and decide on food recall times for consumers. However, due to lack of information about real-time condition of food in conventional packaging, even when customers abide by best-before/use-by dates, they are still at risk for foodborne illness. On the other hand, early food recalls due to predetermined/inaccurate expiration dates result in huge wastage of food worldwide [3]. Therefore, developing a facile nondestructive technology for monitoring food freshness throughout the supply chain is at the forefront of contemporary research in the food packaging field. The primary function of food packaging is to contain the product in a suitable form and secure it from external contamination and mechanical damage. In addition, it is used as a marketing tool to communicate brand identity while providing information about food contained in it [4, 5]. For instance, the graphics and text on conventional food packaging provide information about the nature, composition, nutritional quality, and storage conditions required of the food. Whereas, intelligent packaging senses, detects, and records external or internal changes in products, to further provide food quality and safety information, thereby extending the information function of the packaging system [2, 6]. FIs are an indispensable component of intelligent food packaging, primarily fabricated in two forms: indicator film and indicator card [2, 7]. FIs detect quantitative or qualitative changes in the concentration of one or

Smart Food Packaging Systems: Innovations and Technology Applications, First Edition.
Edited by Avik Mukherjee, Santosh Kumar, Manjusri Misra, and Amar K. Mohanty.
© 2025 John Wiley & Sons Ltd. Published 2025 by John Wiley & Sons Ltd.

more substances released inside packaging during food spoilage and transduce those changes into signals that consumers can easily perceive [8]. They are great options for non-destructive testing of freshness of perishable foods, viz. fruits, vegetables, muscle food, and dairy products [9, 10].

In general, FI consists of a solid support typically made of synthetic/bio-based polymer and a pH-sensitive dye that reacts to chemicals released from spoiled food. The subsequent color changes in the indicator caused by characteristic volatile released from the food provide important real-time information on the microbial and chemical quality of the food [11]. The response speed and sensitivity of FIs depend on various factors, including compatibility between support and dye, the amount of colorant loaded, and the fabrication method adapted [11, 12]. Thus, the present chapter begins with enlisting substance releases indicating food freshness, followed by an overview of different types of FIs and their fabrication methods. Thereafter, recent research advancements in the fabrication of FIs for various foods such as fruits and vegetables, dairy products, and muscle food are summarized. Finally, the current status and future prospects of FIs in food packaging field are presented.

14.2 Substance Releases Indicating Food Freshness

During the process of spoilage of various food products, some characteristic substances related to their deterioration are released. As shown in Figure 14.1, an FI indicates the freshness/deterioration status of packaged food through color change by sensing substances released during food spoilage. Characteristic substance releases associated with food deterioration include volatile base nitrogen, carbon dioxide (CO_2), hydrogen sulfide (H_2S), adenosine triphosphate (ATP), lactic acid, and ethylene, just to name a few [8].

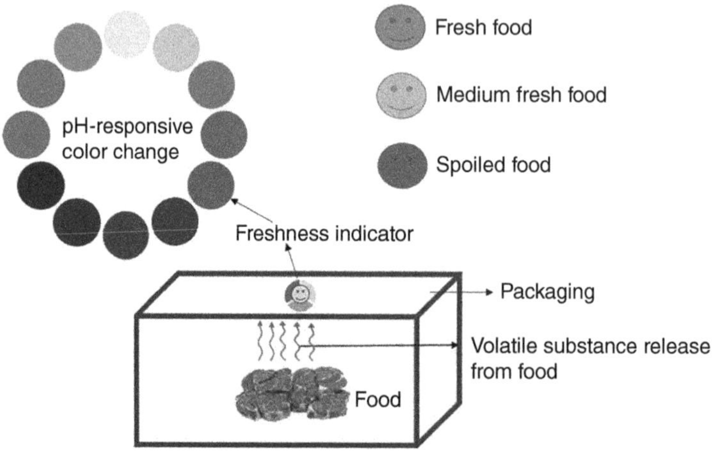

Figure 14.1 Schematic showing working of FI in intelligent food packaging. *Source:* Edward Westmacott/Adobe Systems Incorporated.

14.2.1 Volatile Base Nitrogen

As meat and aquatic food are rich sources of nutrients, spoilage of these food products occurs rapidly due to microbial growth. Under the action of microorganisms, proteins in muscle food decompose and produce biogenic amines such as trimethylamine, dimethylamine, and ammonia, also known as total volatile basic nitrogen (TVBN) [13]. Consequently, TVBN is one of the potential indicators for monitoring the freshness of muscle food. Traditional methods of determination of TVBN are cumbersome and require destruction of food samples. For instance, the TVBN determination method adopted by the European Union involves laborious extraction and steam distillation processes followed by titration of amines with hydrochloric acid [14]. On the other hand, increased TVBN level during spoilage of food increases the pH value of closed packaging environment, which can be detected by pH-responsive colorimetric sensing films [15, 16]. Ghorbani et al. fabricated a TVBN-responsive FI by loading polylactic acid (PLA)/polyethylene glycol/calcium bentonite blend with *Malva sylvestris* anthocyanin for monitoring the real-time status of various meat products such as fish roe, shrimp, chicken fillet, and minced meat [17]. The developed FI successfully discriminated fresh, medium spoiled, and spoiled states of shrimp as well as fish roe, but distinguished only fresh and spoiled states of chicken fillets and beef. This is because, in aquatic food products, less acid production and microbial growth are more favorable for enzyme autolysis. As a result, they are more likely to produce TVBN during the spoilage phase [2]. The FI fabricated using low-density polyethylene and bromophenol blue dye showed the color change from yellow to blue in response to TVBN released during fish spoilage [18]. Moreover, the change in color of the indicator correlated well with the measured bacterial forming units on fish. Several other studies have also reported on the development of TVBN-responsive FIs for muscle food [15, 19, 20].

14.2.2 Carbon Dioxide

Microbial spoilage of food is accompanied by a pronounced release of CO_2. The respiration rate of fruits and vegetables rapidly rises after their harvest and, as a result, a large amount of CO_2 is released between their growth and ripening phases [21, 22]. As CO_2 is an acidic gas, it results in pH changes in the packaging environment. Therefore, it is regarded as a potential indicator for monitoring the freshness of fruits, vegetables, and muscle food. Obaidi et al. fabricated a three-layered CO_2-responsive FI for monitoring the quality of poultry meat and the label changed its color from blue to green upon food spoilage [9]. Wan et al. fabricated a CO_2-responsive indicator label using chitosan derivatives and mixed dye comprising bromocresol blue and methyl red to monitor the freshness of fruits [21]. The developed FI showed excellent color stability in different conditions of humidity, temperature, and CO_2 concentration and was successfully applied to monitor the freshness of mangoes. Several other studies reported on the fabrication of CO_2-responsive FIs for tracking the real-time quality of fruits and vegetables [23–25].

14.2.3 Hydrogen Sulfide

H_2S is generally produced during spoilage of organosulfur-rich foods, viz. fruits, vegetables, meat, and eggs [26, 27]. Hence, H_2S is regarded as a characteristic compound to monitor the spoilage of the aforementioned foods. Zhai et al. developed H_2S-responsive hydrogels using gellan gum and Ag nanoparticles for real-time monitoring of silver carp and chicken breast spoilage [27]. The ultra-strong binding property of Ag with H_2S to form Ag_2S is the working principle of the aforementioned hydrogels. Likewise, a low-cost H_2S-responsive FI for monitoring the spoilage of fish was fabricated by absorbing copper nanoparticles onto Whatman filter paper [28]. The developed FI showed high sensitivity and selectivity to sulfur compounds, and distinct color changes were visible to the naked eye during fish spoilage. H_2S-responsive FIs were also developed for foods such as durian [29] and pork [30, 31].

14.2.4 Other Characteristic Substance Releases

In addition to the aforementioned chemical substances, changes in concentration of organic acids, viz. acetic acid, L-lactic acid, D-lactate, and *n*-butyrate; biogenic amines viz. tyramine, putrescine, cadaverine, and histamine are regarded as potential indicators for monitoring freshness of muscle food [32]. The release of substances such as volatile aldehydes, ethylene, ATP, adenosine diphosphate, and adenosine monophosphate is considered in monitoring the freshness of fruits and vegetables [2]. In particular, ethylene release is linked to various physicochemical changes in fruits including reduction in firmness and rise in the concentration of soluble solids [33]. FIs using ethylene release as a marker for the ripeness of fruits such as bananas and apples have been developed by researchers [33, 34]. Likewise, studies on the fabrication of volatile aldehyde-responsive FIs for fruits such as kiwi and apples were reported in the literature [35, 36]. As it is difficult to detect ATP-related substances, limited studies are available on the development of ATP-responsive FIs for monitoring food quality [37, 38].

14.3 Types of Freshness Indicators

A pH-sensitive FI primarily consists of a chromogenic agent and solid support. Based on the nature of support, FIs can be classified into synthetic polymer-based indicators and natural polymer-based indicators. Based on the type of chromogenic agents, FIs can be categorized into chemical chromogenic formulation indicators and natural chromogenic formulation indicators [2].

14.3.1 Nature of Support

For application in FI, the support material/polymer matrix should meet the subsequent criteria: good film-forming ability; low moisture sensitivity; good barrier and mechanical properties; and low/no toxicity to humans [39].

14.3.1.1 Synthetic Polymer-Based Indicators

Synthetic polymers such as polyethylene terephthalate, low-density polyethylene, and polyaniline were extensively studied as carriers in FIs owing to their advantages such as excellent physicochemical properties and high resistance to physical/chemical/microbial damage [18, 40]. In particular, polyaniline has gained more attention as it changes its color and conductivity with pH due to changes in the degree of protonation in its backbone structure [41, 42]. Consequently, polyaniline can act as both support material and colorant, making it useful as leakage free indicator. Fabrication of polyaniline-based FIs for real-time monitoring of fish spoilage was reported in the literature [40, 42]. Though synthetic polymers are convenient to use, there are numerous environmental and health issues associated with their use. As a result, exploring sustainable substitutes for conventional polymers in the packaging industry is much needed [4].

14.3.1.2 Natural Polymer-Based Indicators

Natural polymers viz. chitosan, agar, starch, cellulose, corn zein, and gelatin were investigated as support materials for the fabrication of FIs owing to their renewable nature, biodegradability, and biocompatibility [43]. In particular, polysaccharide polymers can network with pH-responsive chromogenic agents through ion interactions or hydrogen bonding and augment their color-changing effect [2]. Lu et al. fabricated a FI using sugarcane bagasse nanocellulose hydrogel and pH-responsive dyes (methyl red and bromothymol blue) that displayed a quick color change from green to red in response to spoilage of chicken [44]. Starch films loaded with butterfly pea flower anthocyanin and TiO_2 nanoparticles [45], arrowroot starch/iota-carrageenan films incorporated with kyoho skin extract [46], and starch/polyvinyl alcohol films loaded with red cabbage extract [47] were also studied for monitoring freshness of muscle food. Some plant and animal-based proteins gained attention for application in FIs owing to their nontoxicity and good film-forming ability [48, 49]. Zeng et al. fabricated a wide pH (2–11) sensitive indicator film by incorporating mulberry extract in gelatin/polyvinyl alcohol blend for monitoring fish spoilage [50]. Similarly, FIs were prepared by incorporating alizarin in zein [48], curcumin and cellulose nanocrystals in soy protein [49].

14.3.2 Nature of Chromogenic Agents

In general, colorimetric FI contains one or more dyes that meet the food safety requirements and change color with the change in pH. Besides, the sensitivity of colorimetric FI largely depends on the nature of chromogenic agent used [51].

14.3.2.1 Chemical Chromogen-Based Indicators

Synthetic chromogenic agents viz. methyl red, methyl orange, bromocresol green, bromocresol blue, bromophenol blue, and bromothymol blue were widely applied in the fabrication of FIs owing to their advantages such as low-cost, high color stability, intensity, and versatility [43]. Fabrication of FI using cassia gum and bromothymol blue anchored cellulose fibers for monitoring the freshness of meat products has been reported in the literature [52]. The developed FI showed sensitive in situ response to trimethylamine and a visible color change from pale yellow to blue-green as the meat transitioned from fresh state to medium fresh state. Likewise, polystyrene nanofibrous mats loaded with bromothymol blue [53] and ethylene-vinyl alcohol copolymer composites loaded with bromocresol purple [54] showed their potential in monitoring the condition of wine and milk samples, respectively. Whereas the films fabricated by loading methyl red dye in agar [25], bromophenol blue in bacterial cellulose [55], and methyl red in ethyl cellulose/polyvinyl alcohol blend [35] showed their potential in monitoring the freshness of fruits. However, raising concerns about the carcinogenic or teratogenic impacts of synthetic dyes on humans and other organisms have shifted research as well as consumer interest toward the use of natural colorants in the food industry [43].

14.3.2.2 Natural Chromogen-Based Indicators

Natural pigments are mostly polyphenolic compounds derived from various parts of plants such as fruits, leaves, stems, and roots. Curcumin, alizarin, chlorophyll, carotenoid, betalain, and anthocyanin are examples of some natural pigments widely studied for application in intelligent packaging for monitoring food freshness [56–58]. These natural colorants offer numerous advantages such as safety, easy availability, wide pH indicating range, and good antibacterial and antioxidant properties. The ability to protonate/deprotonate in the acidic/alkaline medium is the basis on which natural colorants work in FIs [59]. For instance, the color variation and chemical structures of curcumin and alizarin at acidic and basic pH conditions are shown in Figure 14.2 [60, 61]. Structurally, curcumin comprises a seven-carbon chain having α, β-unsaturated β-diketone moiety attached to two aromatic rings having ortho-methoxy phenolic groups. In alkaline pH, the α, β-unsaturated β-diketone moiety in curcumin functions as a breakdown point and hydrogen donor leading to hydrolysis and degradation. Consequently, in alkaline pH, curcumin exists in enolic form due to loss of protons from the phenolic group and displays red color. On the other hand, in the pH range of 1–7, it predominately exists in bis-keto form and displays yellow color [43]. Likewise, the color change in alizarin can be attributed to its structural transformations to neutral, monoanionic, and dianionic forms under acidic, neutral, and alkaline

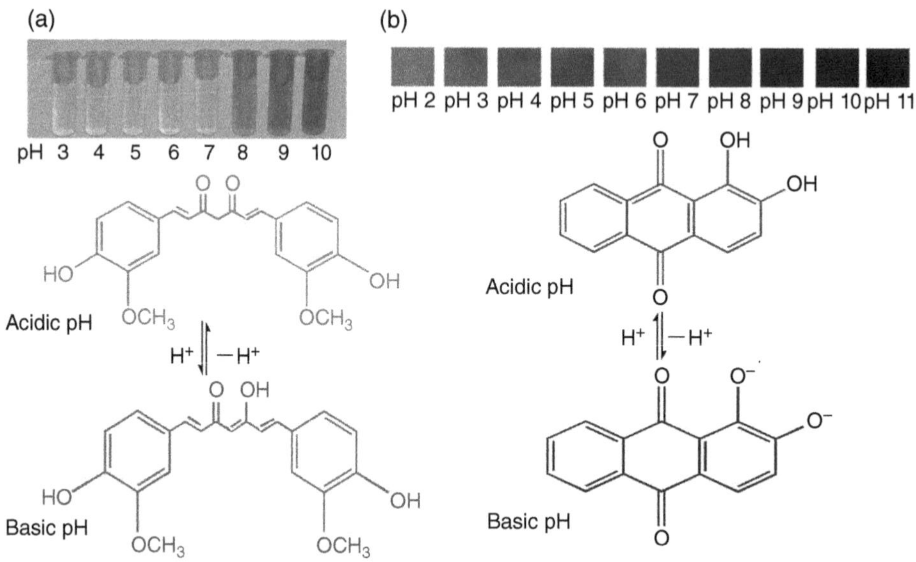

Figure 14.2 Color variation and chemical structures of (a) curcumin and (b) alizarin at different pH conditions. *Source:* (a) Reproduced from Liu et al. [60]/with permission of Elsevier. (b) Reproduced from Ezati et al. [61]/with permission of Elsevier.

pH conditions, respectively [43]. In the early stages of research in this area, paper/synthetic plastics were applied as support materials in FI for loading natural colorants. However, due to concerns about the existence of harmful substances in the carrier, many of those indicators are still not approved for direct contact with food [2]. Consequently, the idea of fabricating FIs with edible matrix and natural colorants is gaining enormous popularity nowadays. Some recent findings on the fabrication and application of edible FIs are summarized in Table 14.1.

Addition of natural pigments such as anthocyanin [56–58, 62], curcumin [59, 63], alizarin [61, 64], and betalain [65] into edible polymers/polymer blends induced not only pH-responsive color-changing property but also improved mechanical, UV-barrier, antimicrobial and antioxidant properties of the composite films. Moreover, the application trails of edible materials reported in Table 14.1 on various food products revealed their potential in intelligent and active food packaging. Nevertheless, low stability of color under

Table 14.1 Summary of recent literature on the fabrication of edible FIs.

Colorant	Source of colorant	Edible matrix	pH indicating range	Foods tested	Color change	Reference
Anthocyanin	Hibiscus flower	Chitosan, gelatin, and starch	2–13	NA	NA	[58]
Anthocyanin	Red cabbage	Bacterial cellulose	1–14	Milk	Blue-gray to pinkish-gray	[57]
Anthocyanin	Rose flower	Polyvinyl alcohol and okra mucilage polysaccharide	2–12	Shrimp	Purple to dark green	[56]
Anthocyanin	Purple sweet potato	Polyvinyl alcohol and agarose	3–10	Pork	Light brown to dark green	[62]
Curcumin	Turmeric	κ-carrageenan and gelatin	3–11	Grass carp fillets	Yellow to dark red	[63]
Curcumin	Turmeric	Chitosan and gelatin	3–11	Pork	Brown to saddle brown	[59]
Alizarin	Madder plant root	Cellulose and chitosan	2–11	Beef	Yellow to purple	[64]
Alizarin	Madder plant root	Cellulose and starch	2–11	Rainbow trout fillet	Orange to reddish brown	[61]
Betalain	Cactus pear	Quaternary ammonium chitosan and polyvinyl alcohol	3–12	Shrimp	Purple to yellow	[65]

NA, not applicable.

harsh light and temperature conditions limits the commercial application of natural chromogenic agents in the food packaging industry [43]. Moreover, hydrophilic natural pigments leach easily from films due to high humidity in the packaging environment. Nano/microencapsulation could be one possible solution to overcome the leaching problems of natural pigments. The findings in the study reported by Liu et al. revealed that compared to starch/polyvinyl alcohol films incorporated only with anthocyanin, the films incorporated with anthocyanin-loaded nanocomplexes showed better color stability in packaging conditions [24]. Starch/butterfly pea anthocyanin/TiO_2 nanocomposite films showed better thermal stability and water vapor barrier properties than starch/butterfly pea anthocyanin composites [45]. Similarly, κ-carrageenan/red grape skin anthocyanin/Ag nanocomposite films demonstrated excellent color stability under UV exposure [15].

14.4 Fabrication of FIs

Fabrication methods of FIs include, but are not limited to, solution casting, extrusion, sol–gel method, absorption, and electrochemical writing [11]. The procedure for fabricating FIs by the aforementioned methods is summarized in Table 14.2 and schematically presented in Figure 14.3. Absorption and solution casting are the most common methods adapted for the fabrication of FIs on a laboratory scale [55, 58, 69, 70]. In particular, cellulose-based FIs are fabricated by absorption method taking advantage of their high water holding capacity [28, 70]. Nonetheless, co-solubility of base material and colorant in the same solvent is a major challenge for adapting solution casting method. For instance, polymers such as PLA and poly (3-hydroxybutyrate) (PHB) are soluble in chloroform [71–73]. On the contrary, natural pigments such as anthocyanins are water-soluble. As a result, it is difficult to achieve a homogeneous solution of PLA or PHB with anthocyanins to adapt the solution casting method for fabricating these composite films [11]. The extrusion method can be used for the fabrication of PLA or PHB-based FIs with natural colorants, but the colorant must be in powder form and also heat resistant. Latos-Brozio et al. successfully fabricated FIs based on two hydrophobic polymers (PLA and PHB) and four natural colorants (lutein, chlorophyll, β-carotene, and curcumin) by extrusion technique [66]. Considering sol–gel method, high porosity of sol–gel matrix facilitates interaction between colorant and food spoilage-related metabolites and strengthens the application of FIs in monitoring food quality [11]. The FIs fabricated using the sol–gel technique [68, 74] demonstrated very good color stability and pH-sensitivity. Techniques such as electrochemical writing [67], electrospinning [75], and 4D printing [76] have gained interest in recent years for the fabrication of FIs.

14.5 Application of FI in Monitoring Food Quality

14.5.1 Fruits and Vegetables

Fruits and vegetables are metabolically active and their biological activities, viz. respiration and ethylene production, persist even after their harvest [77]. They are rich in fiber, antioxidants, vitamins, and minerals and are highly perishable. Deterioration of fruits and vegetables occurs due to environmental, pathological, physiological, and hygienic factors, thus

Table 14.2 Fabrication methods of FIs.

Method	Procedure and application	References
Solution casting	(i) pH-responsive substance (colorant) is dissolved in polymer solution, and then films are obtained by evaporating the solvent. (ii) Simple method and widely used on laboratory scale. (iii) Co-solubility of dye and polymer in the same solvent is a major challenge.	[11, 58]
Extrusion	(i) The polymer and colorant are first loaded into a hopper and then forced into the barrel, where they are heated to the melting temperature of the polymer. Thereafter, the melted mixture enters the die, and the final product is obtained. (ii) This method is easy to scale up; however, it can only be applied to thermoplastic polymers and heat resistant colorants.	[11, 66]
Absorption	(i) This method involves immersing support material in the colorant solution. (ii) It can be used for porous support materials with high absorption capacity. (iii) Dye leaching is a major challenge in this method.	[11, 28]
Electrochemical writing	(i) In this method, the dye-containing film will be placed on a platinum conductive plate which is connected to the cathode of an electrochemical workstation. Subsequently, a platinum pen connected to anode will be controllably moved on the surface of the film to engrave desired pattern. (ii) This method is regarded as green printing technology.	[11, 67]
Sol–gel method	(i) This method involves multiple steps, viz. hydrolysis, polycondensation, gelation, aging, and drying, in which colorant is added into a matrix containing precursor of the inorganic phase. The sol–gel solution containing colorant is then deposited on a different sheet, mostly by dip-coating. (ii) High porosity/low interconnected network reduces mechanical strength and thermal stability of FIs fabricated using this method.	[11, 68]

producing CO_2, aldehydes, ethanol, volatile organic compounds, and other metabolites [78]. Visual criteria are typically used to determine the quality of fruits and vegetables. However, these visual standings may be misleading and difficult to interpret [79]. Techniques such as hyperspectral imaging [80], dielectric sensing [81], fluorescence spectroscopy [82], and near-infrared spectroscopy [83] were investigated for evaluating the freshness of fruits such as kiwi, apple, and apricot. However, the aforementioned techniques cannot provide the real-time condition of fruits throughout the supply chain. Therefore, to overcome the aforementioned limitations, research is directed toward the development of easy-to-use colorimetric FIs for fruits and vegetables to minimize their postharvest loss (Table 14.3). Kim et al. developed an aldehyde-responsive FI label by absorption of methyl red dye on filter paper [36]. The developed labels were observed to undergo periodic color transitions when exposed to ripening apples: yellow to orange

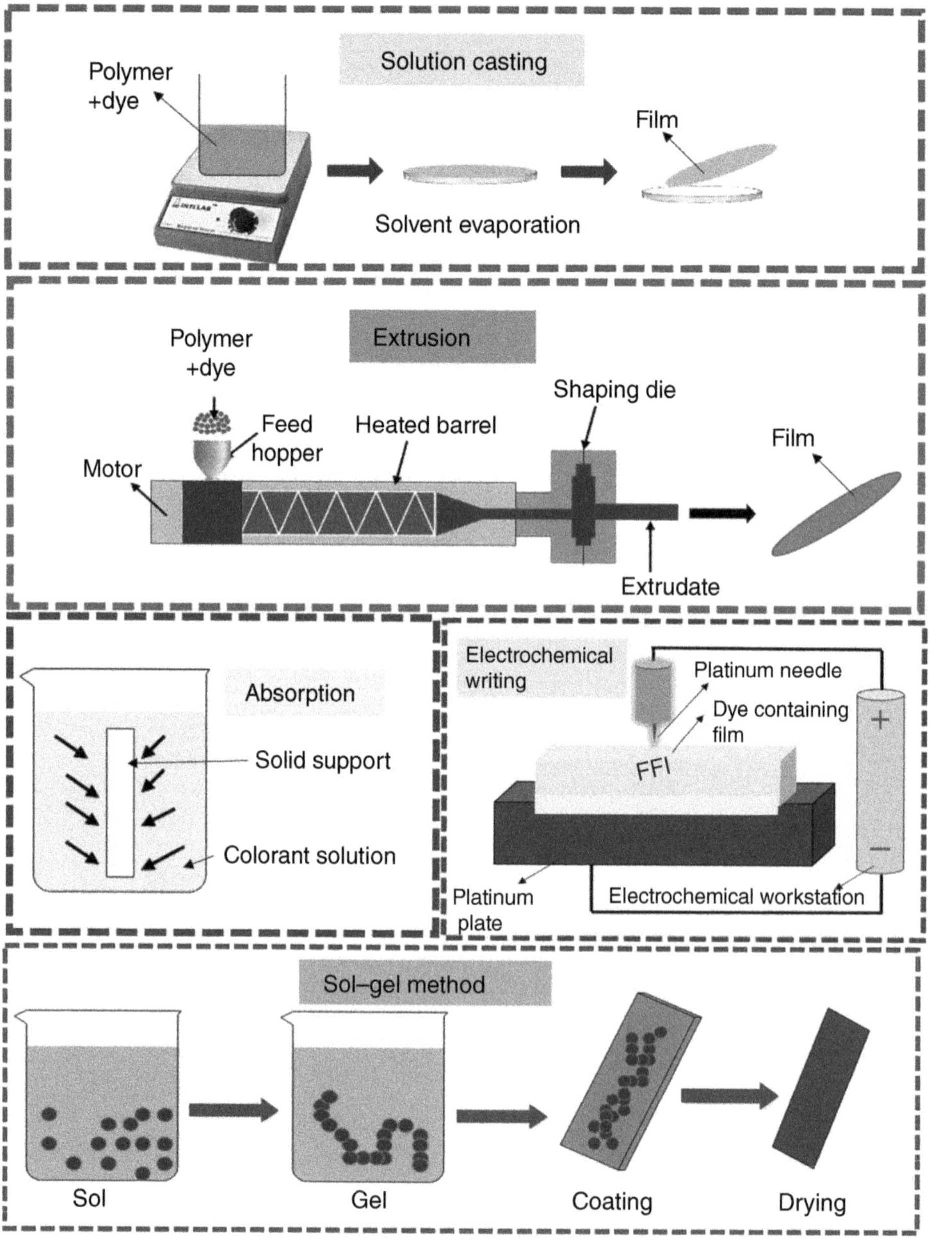

Figure 14.3 Schematic showing different methods to fabricate FIs.

(0–3 days); remained in orange (3–7 days); orange to red (>9 days) (Figure 14.4a). Interestingly, the color transition trend in the labels is in good agreement with the changes in apple freshness parameters such as sweetness, hardness, and soluble solid content. In another study, aldehyde-responsive FI using ethyl cellulose/polyvinyl alcohol blend incorporated with poly (ethylene glycol) bis (3-aminopropyl) and methyl red was developed for

Table 14.3 Application of FIs in monitoring the quality of fruits and vegetables.

Food tested	Support material	Colorant	Fabrication technique	Working pH	Characteristic substance release/responder	Color change	Reference
Apple	Filter paper	Methyl red	Absorption	6.4–10.1	Aldehyde	Yellow to red	[36]
Kiwi	Ethyl cellulose and polyvinyl alcohol	Methyl red	Solution casting	6.6–8.9	Aldehyde	Yellow to red	[35]
Dragon fruit	Agar	Methyl red	Solution casting	4.2–6.4	CO_2	Yellow to red	[25]
Mango	Bacterial cellulose	Bromophenol blue	Absorption	NA	Acetic acid	Blue to green	[55]
Guava	Bacterial cellulose	Bromophenol blue	absorption	NA	Acetic acid	Blue to green	[84]
Avocado	Filter paper	Red cabbage anthocyanin	Absorption	NA	Acetic acid	Light purple to red	[85]
Banana	Polyvinyl alcohol	Sappan wood extract	Solution casting	3–8	CO_2	Yellow to dark red	[69]
Tomato	Carboxymethyl cellulose	Blackberry anthocyanin	Solution casting	2–13	NA	Pink to yellowish green	[86]
Mushroom	Starch and gelatin	Sweet potato anthocyanin	Solution casting	3–10	CO_2	Green to yellow	[23]
Mushroom	Corn starch and polyvinyl alcohol	Blueberry anthocyanin	Solution casting	2–12	CO_2	Purple to pink	[24]
Freshly cut bell pepper	Methylcellulose and polyethylene glycol	Methyl red and bromophenol blue	Solution casting	NA	CO_2	Yellow-green to orange	[87]

NA, not available.

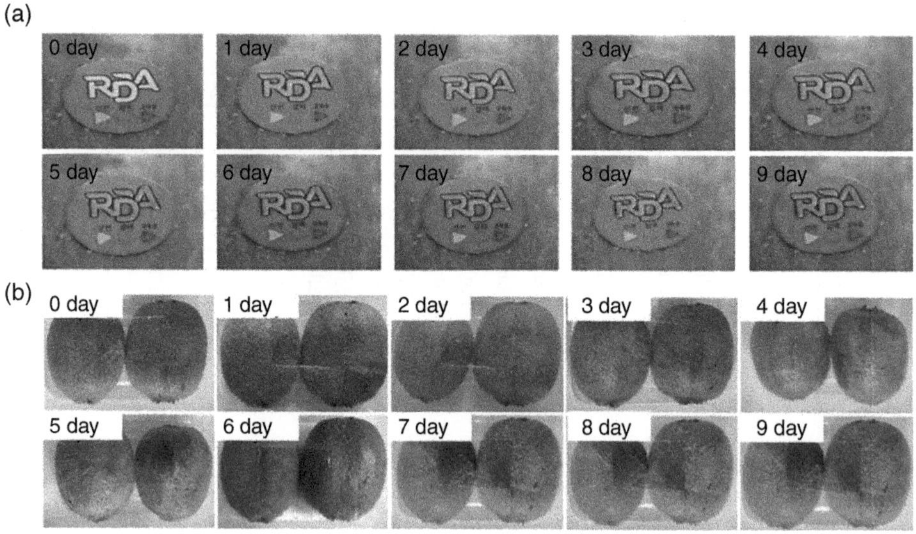

Figure 14.4 Changes in the color of FIs after exposure to (a) apple and (b) kiwi fruits during their storage. *Source:* (a) Kim et al. [36]/Elsevier/CC BY 4.0. (b) Shao et al. [35]/Elsevier/CC BY NC-ND 4.0.

monitoring the condition of kiwi fruits [35]. As the fruits ripened, the film exposed to them transitioned its color from yellow to orange, and then to red (Figure 14.4b). The color change in the aforementioned FI is due to the nucleophilic addition reaction between aldehydes and amine-terminated polymers. Likewise, fabrication of FIs for monitoring the real-time condition of fruits such as dragon fruit [25], mango [55], guava [84], banana [69], and avocado [85] were also reported in the literature. The materials and fabrication methods used, characteristic substance release considered, and color changes observed in FIs in response to various fruit spoilage are summarized in Table 14.3.

Limited studies are available on the fabrication of FIs for vegetables (Table 14.3). In a study reported by Sganzerla et al., a novel bioactive and pH-responsive packaging material was developed for tomatoes by incorporating blackberry extract in carboxymethyl cellulose [86]. Wherein, the authors observed that the fabricated films showed pH-responsive color change which helped in food quality monitoring and indicated controlled release of active components that aid in extending the shelf life of food. Chen et al. observed that compared to methylcellulose/polyethylene glycol films loaded only with bromophenol blue, the films prepared with loading of both methyl red and bromophenol dyes clearly indicated spoilage of bell pepper [87]. Several other studies reported similar observations concerning the use of mixed dyes in the fabrication of FIs [2, 20].

14.5.2 Dairy Products

Milk and its derivatives are rich in nutrients that favor the growth and proliferation of spoilage-causing microorganisms and pathogens. Thus, chemical and microbial tests are regularly performed for dairy products at company level during production and before delivery to ensure their quality and safety [10, 88]. However, this quality testing is typically absent after the products get delivered at distributor/consumer level. Therefore, dairy industry needs rapid, cost-effective, and sensitive technologies for monitoring the quality of products throughout the supply chain. Considering the aforementioned requirements, several studies investigated the fabrication of easy-to-use FIs for milk (Table 14.4). On the other hand, very few studies were available on the fabrication of FIs for milk derivatives such as butter [95] and cheese [94]. pH is considered one of the crucial factors in monitoring the quality of milk in response to microbial growth. pH of Fresh milk is usually in the range of 6.8–6.0. However, during spoilage phase, lactic acid bacteria convert lactose in milk into lactic acid, causing the pH to drop from 6.8 to 4.0 [70, 89]. Fabrication of

Table 14.4 Application of FIs in monitoring the quality of dairy products.

Food tested	Support material	Colorant	Fabrication technique	Working pH	Characteristic substance release/ responder	Color change	Reference
Milk	Agarose	Red cabbage anthocyanin	Solution casting	4–6.8	Lactic acid	Blue to pink	[89]
Milk	Starch and filter paper	Black carrot anthocyanin	Absorption	2–11	Lactic acid	Deep blue to purple	[70]
Milk	Chitosan	Purple tomato anthocyanin	Solution casting	3–11	Lactic acid	Green to pink	[90]
Milk	Starch and polyvinyl alcohol	Purple sweet potato anthocyanin	Solution casting	1–14	Lactic acid	Purple to pink	[91]
Milk	Paper	Hibiscus flowers	Rub coating	NA	NA	Green to pink	[92]
Milk	Chitosan	Black eggplant anthocyanin	Solution casting	2–13	NA	Green to blue	[93]
Cheese	Polyvinyl pyrrolidone, carboxy methyl cellulose, bacterial cellulose, and guar gum	Red cabbage anthocyanin	Solution casting	NA	Lactic acid	Pink to reddish yellow	[94]
Milk	Paper	NA	Printing	NA	Metabolic end products in the glycolytic pathway of *E. coli*	Blue to green	[95]

NA, not available.

agarose-based films loaded with red cabbage anthocyanin, which can discriminate between fresh (pH 6.8–6.0) and spoiled milk (pH 4.5–4.0) by sensing lactic acid concentration, was reported in the literature [89]. In a similar study, FI for milk was developed by immobilizing black carrot anthocyanin in the starch matrix [70]. The fabricated label showed visible color changes as a function of pH and outstanding color stability even after a month of its storage under various conditions. Likewise, fabrication of FIs for milk was also reported using purple tomato anthocyanin loaded in chitosan [90], purple sweet potato anthocyanin loaded in starch/polyvinyl alcohol blend [91], and black eggplant anthocyanin loaded in chitosan [93]. Chaithra et al. developed an FI for milk by rub coating hibiscus flowers onto paper. Interestingly, FI fabrication in the aforementioned study neither entails the extraction of anthocyanin nor requires any trained analyst or laboratory accessories [92]. Thus, it is consumer-friendly and can be prepared easily by anyone who has access to hibiscus flowers and paper.

Fabrication of red cabbage anthocyanin-loaded polyvinyl pyrrolidone/carboxy methyl cellulose/bacterial cellulose/guar gum composite films for monitoring cheese quality is reported in the literature [94]. The developed films showed sensitive in situ response to lactic acid present in cheese and changed its color from pink to reddish yellow as the cheese transitioned from fresh state to spoiled state. *Escherichia coli* is frequently found as contaminating microorganism in dairy products; therefore, its presence is considered as an indicator of deteriorated microbiological quality of milk and its derivatives [96]. Anjali et al. developed an easy-to-use FI for the detection of *E. coli* in milk and milk products by immobilizing polyaniline–pectin nanoparticles on filter paper [95]. The findings in this study revealed that nanoparticles exhibited strong binding with filter paper and no leaching was detected when the FI was immersed in buffer solution of pH 7. Thus, it holds great potential for application in food industry for the detection of *E. coli* in dairy products.

14.5.3 Muscle Foods

Muscle foods, primarily meat and seafood are rich in proteins as well as nutrients, thus, they are highly susceptible to spoilage by both microbial contamination and lipid oxidation. Microbial growth and oxidation of lipids cause discoloration, textural changes, and development of off-flavors, which subsequently decrease shelf life of food and increase risk of foodborne illness [97]. Conventionally, quality of muscle food is checked using sensory tests and experiments such as microbial growth evaluation. Nevertheless, sensory approach involves the destruction of food samples and suffers from low precision due to human errors. Microbiological tests are very time-consuming and are far from real-time monitoring [98]. Therefore, the development of easy-to-use FIs for monitoring muscle food quality is highly desired. As discussed earlier in Section 14.2.1, TVBN content is one of the crucial parameters for evaluating the freshness of meat. As per the National Standard of People's Republic of China, TVBN of fresh meat should be less than or equal to 15 mg per 100 mg of meat [20]. Numerous studies reported on fabrication of easy-to-use TVBN-responsive FIs for muscle foods such as pork, chicken, fish, and shrimp (Table 14.5). For instance, methylcellulose/polyethylene glycol films loaded with mixture of bromothymol blue and methyl red dyes [20] and polyvinylidene fluoride films loaded with roselle anthocyanin and cinnamon oil [99] showed in situ response to TVBN and visible color change as pork

Table 14.5 Application of FIs in monitoring the quality of muscle food.

Food tested	Support material	Colorant	Fabrication technique	Working pH	Characteristic substance release/responder	Storage condition of food	Color change	Reference
Pork	Methylcellulose and polyethylene glycol	Mixture of bromothymol blue and methyl red	Solution casting	NA	TVBN	5 °C	Red to green	[20]
Pork	Polyvinylidene fluoride	Roselle anthocyanin	Electrospinning	2–12	TVBN	4 °C and 50 % RH	Pink to blue	[99]
Chicken fillets	Bovine gelatin and glycerol	Purple basil leaves anthocyanin	Solution casting	4–9	TVBN	25 °C and 75% RH	Purple to green	[100]
Chicken breast	Sodium Alginate	Purple sweet potato anthocyanin	Solution casting	2–13	TVBN	4 and 25 °C	Pink to blue	[101]
Shrimp	Bacterial cellulose	*Echium amoenum* flowers anthocyanin	Absorption	2–12	TVBN	4 °C	Violet to yellow	[102]
Shrimp	Chitosan and polyvinyl alcohol	Jambolan fruit anthocyanin	Solution casting	1–13	TVBN	Several temperatures between −20 °C and 20 °C	Red to blue	[19]
Fish meat	κ-carrageenan	Grape skin anthocyanin	Solution casting	3–10	TVBN	4 °C	Purple to dark green	[15]
Fish meat	Agar	*Arnebia euchroma* root anthocyanin	Solution casting	NA	TVBN	25 °C	Pink to bluish violet	[16]

NA, not applicable; RH, relative humidity; TVBN, total volatile base nitrogen.

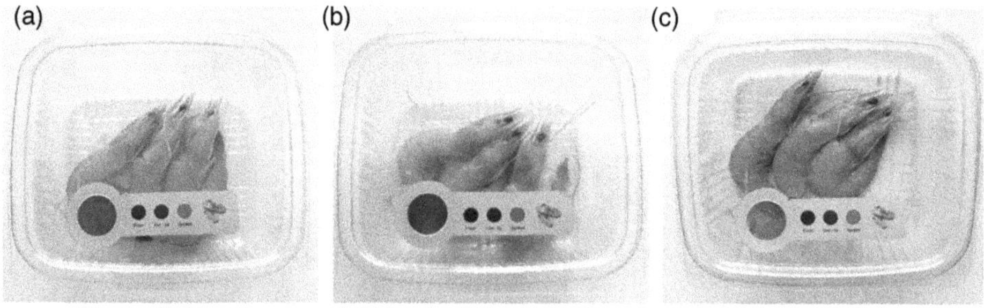

Figure 14.5 Changes in the color of FI attached to package of shrimps: (a) immediately after packaging, (b) after two days of storage at 4 °C, and (c) after four days of storage at 4 °C. *Source:* Mohammadalinejhad et al. [102]/with permission of Elsevier.

transitioned from fresh to spoiled state. Besides, polyvinylidene fluoride/roselle anthocyanin/cinnamon oil composites exhibited antibacterial activity against *E. coli* and *S. aureus* and extended shelf life of pork by two days [99]. The films fabricated by loading purple basil leaf extract in bovine gelatin/glycerol blend [100] and purple sweet potato anthocyanin in sodium alginate [101] showed their potential in monitoring freshness of chicken fillets and chicken breast, respectively.

The FI fabricated by loading *Echium amoenum* flower anthocyanin in bacterial cellulose transitioned its color from violet to gray and then yellow as shrimps stored at 4 °C changed from fresh to medium spoiled and spoiled states (Figure 14.5) [102]. Moreover, the color transitions in the developed FI were in accordance with total viable count and TVBN values of the shrimps. In a similar study, FI was fabricated using chitosan, polyvinyl alcohol, and jambolan fruit extract and tested for monitoring the quality of shrimps stored at different temperatures between −20 and 20 °C for four days [19]. The developed films showed visible color changes for the spoilage of shrimps stored at 4 and 20 °C. On the contrary, the films did not show any color change when the shrimps were stored at −20 °C. This is due to the slower degradation of shrimps stored at −20 °C by microorganisms. Likewise, κ-carrageenan films loaded with grape skin anthocyanin [15] and agar-based composites loaded with *Arnebia euchroma* root anthocyanin [16] showed their potential in monitoring the spoilage of fish meat.

14.5.4 Other Foods

Edible oils are composed of a variety of nutrient elements such as fats, fat-soluble vitamins, and polyunsaturated fatty acids, and are important part of human diet. However, oxidative degradation of vegetable oils produces toxic by-products that not only contribute to off-flavors and bad taste but also pose a serious threat to human health [103]. In this regard, several researchers are working toward development of easy-to-use indicators for visual detection of edible oil oxidation [103, 104]. For instance, the FI developed by incorporating hydroxylamine sulfate and congo red in chitosan films successfully identified the oxidation status of seven different sunflower oil samples [104]. Fabrication of FI label for real-time monitoring of intermediate-moisture dessert (golden drop) spoilage was reported in the literature [105]. The developed labels responded to CO_2 released during dessert spoilage and showed visible color change. Gaikwad et al. fabricated an FI that displayed visible color changes in response to CO_2 released during spoilage of ready-to-cook idli batter, which is widely used in countries such as India and Sri Lanka [106]. Likewise, indicator labels were developed for monitoring the fermentation state of kimchi, a popular side dish in Korean cuisine [107, 108]. In another study, chitosan/polyvinyl alcohol composite films loaded with *Carissa carandas* and *Clitoria ternatea* extracts were prepared using a simple solution casting method for monitoring the quality of orange fruit juice [109]. The developed films exhibited excellent antioxidant activity, biocompatibility, and visible color changes in response to juice spoilage.

14.5.5 Commercial Availability of FIs

Though research on the development of FIs started earlier in 1950s, only few products were commercially available in the market [43]. In 1999, COX Technologies launched Fresh Tag®, a colorimetric FI for monitoring the quality of packaged chicken and seafood. This FI senses volatile amines released in the headspace of packages during food spoilage and changes its color in response to it [110]. But, Fresh Tag® was discontinued later in the year 2004 [43]. SensorQ™, a biogenic amine-responsive FI was developed and launched by Food Quality Sensor International. This indicator works by reacting with biogenic amines released during food spoilage and can be applied to monitor beef and poultry quality [111]. To-Genkyo, a Japanese company, launched an hourglass-shaped label for monitoring the spoilage of meat products. The aforementioned label contains a special ink that changes color upon interacting with ammonia released during food spoilage [111, 112]. Plant & Food Research, Auckland, developed and launched world's first intelligent label "RipeSense™" that can indicate fruit ripeness. This label responds to aromatic compounds released during ripening of fruits and changes its color. Currently, this label is being used only for pears, however, studies are underway for its use for other fruits such as avocado, melon, mango, and kiwi [113].

14.6 Conclusion and Future Perspective

FIs reform traditional "best-serve-period" or "use-by-date" system of indicating food freshness by providing real-time condition of food across the supply chain. Several researchers developed easy-to-use FIs for muscle food, fruits, and milk which work by reacting with substances released during spoilage of food. On the other hand, limited studies are available in literature on the development of FIs for vegetables and milk derivatives. Despite enormous efforts of researchers in developing FIs over many years, very few products are commercially available due to cost issues, safety concerns, poor consumer acceptance due to lack of awareness, and difficulties with their industrial-scale production and incorporation into packaging systems. In recent years, interest in researching edible polymers and natural colorants for the fabrication of FIs has significantly increased due to growing concerns about food intoxication, health, and environmental issues. Use of natural colorants such as curcumin, alizarin, anthocyanins, and betalains in the fabrication of FIs improves/induces their antioxidant and antimicrobial properties and helps in extending the shelf life of food. Nonetheless, edible FIs are still far from commercial-scale application due to the following reasons: poor tolerance of edible polymers to prolonged processing operations, high water solubility, poor thermal resistance, poor color stability, and pH-sensitivity of natural colorants. Thus, subsequent recommendations can be considered to overcome the aforementioned hurdles: modifying carrier material to avoid leaching of natural colorants; improving thermal stability of natural colorants by micro/nanoencapsulation or exploring natural pigments with high heat resistance; improving physicochemical properties and pH-sensitivity of indicator by incorporating new additives or nanomaterials. It is known that most food products get spoiled due to microbial contamination. As the microbial action begins in the food, pH drops to a slightly acidic level. Therefore, researchers should consider developing indicators with distinctive color-changing functions in the pH range of 5–7 for early-stage detection of food spoilage. Solution casting and absorption methods are commonly adapted for the fabrication of FIs on laboratory scale; however, these techniques are not suitable for commercial-scale production and application. Nonetheless, with the advent of advanced fabrication techniques such as 4D printing, electrospinning, and electrochemical writing, a substantial increase in mass production and application of FIs is anticipated in the near future.

References

1 World Health Organization (2022). Estimating the burden of foodborne diseases. https://www.who.int/activities/estimating-the-burden-of-foodborne-diseases (accessed 26 December 2022).

2 Shao, P., Liu, L., Yu, J. et al. (2021). An overview of intelligent freshness indicator packaging for food quality and safety monitoring. *Trends in Food Science & Technology* 118: 285–296. https://doi.org/10.1016/j.tifs.2021.10.012.

3 Yousefi, H., Su, H.M., Imani, S.M. et al. (2019). Intelligent food packaging: a review of smart sensing technologies for monitoring food quality. *ACS Sensors* 4 (4): 808–821. https://doi.org/10.1021/acssensors.9b00440.

4 Kumari, S.V.G., Pakshirajan, K., and Pugazhenthi, G. (2022). Recent advances and future prospects of cellulose, starch, chitosan, polylactic acid and polyhydroxyalkanoates for sustainable food packaging applications. *International Journal of Biological Macromolecules* 221: 163–182. https://doi.org/10.1016/j.ijbiomac.2022.08.203.

5 Vanderroost, M., Ragaert, P., Devlieghere, F., and De Meulenaer, B. (2014). Intelligent food packaging: the next generation. *Trends in Food Science & Technology* 39 (1): 47–62. https://doi.org/10.1016/j.tifs.2014.06.009.

6 Cheng, H., Xu, H., Mcclements, D.J. et al. (2022). Recent advances in intelligent food packaging materials: principles, preparation and applications. *Food Chemistry* 375: 131738. https://doi.org/10.1016/j.foodchem.2021.131738.

7 Ghaani, M., Cozzolino, C.A., Castelli, G., and Farris, S. (2016). An overview of the intelligent packaging technologies in the food sector. *Trends in Food Science & Technology* 51: 1–11. https://doi.org/10.1016/j.tifs.2016.02.008.

8 Becerril, R., Nerín, C., and Silva, F. (2021). Bring some colour to your package: freshness indicators based on anthocyanin extracts. *Trends in Food Science & Technology* 111: 495–505. https://doi.org/10.1016/j.tifs.2021.02.042.

9 Obaidi, A., Karaca, I.M., Ayhan, Z. et al. (2022). Fabrication and validation of CO_2-sensitive indicator to monitor the freshness of poultry meat. *Food Packaging and Shelf Life* 34: 100930. https://doi.org/10.1016/j.fpsl.2022.100930.

10 Mirza Alizadeh, A., Masoomian, M., Shakooie, M. et al. (2021). Trends and applications of intelligent packaging in dairy products: a review. *Critical Reviews in Food Science and Nutrition* 62 (2): 383–397. https://doi.org/10.1080/10408398.2020.1817847.

11 Almasi, H., Forghani, S., and Moradi, M. (2022). Recent advances on intelligent food freshness indicators; an update on natural colorants and methods of preparation. *Food Packaging and Shelf Life* 32: 100839. https://doi.org/10.1016/j.fpsl.2022.100839.

12 Roy, S. and Rhim, J.W. (2020). Anthocyanin food colorant and its application in pH-responsive color change indicator films. *Critical Reviews in Food Science and Nutrition* 61 (14): 2297–2325. https://doi.org/10.1080/10408398.2020.1776211.

13 Tahir, H.E., Hashim, S.B., Mahunu, G.K. et al. (2022). Smart films fabricated from natural pigments for measurement of total volatile basic nitrogen (TVB-N) content of meat for freshness evaluation: a systematic review. *Food Chemistry* 396: 133674. https://doi.org/10.1016/j.foodchem.2022.133674.

14 Dudnyk, I., Janecek, E.R., Vaucher-Joset, J., and Stellacci, F. (2018). Edible sensors for meat and seafood freshness. *Sensors and Actuators B: Chemical* 259: 1108–1112. https://doi.org/10.1016/j.snb.2017.12.057.

15 You, S., Zhang, X., Wang, Y. et al. (2022). Development of highly stable color indicator films based on κ-carrageenan, silver nanoparticle and red grape skin anthocyanin for marine fish freshness assessment. *International Journal of Biological Macromolecules* 216: 655–669. https://doi.org/10.1016/j.ijbiomac.2022.06.206.

16 Huang, S., Xiong, Y., Zou, Y. et al. (2019). A novel colorimetric indicator based on agar incorporated with *Arnebia euchroma* root extracts for monitoring fish freshness. *Food Hydrocolloids* 90: 198–205. https://doi.org/10.1016/j.foodhyd.2018.12.009.

17 Ghorbani, M., Divsalar, E., Molaei, R. et al. (2021). A halochromic indicator based on polylactic acid and anthocyanins for visual freshness monitoring of minced meat, chicken fillet, shrimp, and fish roe. *Innovative Food Science & Emerging Technologies* 74: 102864. https://doi.org/10.1016/j.ifset.2021.102864.

18 Wells, N., Yusufu, D., and Mills, A. (2019). Colourimetric plastic film indicator for the detection of the volatile basic nitrogen compounds associated with fish spoilage. *Talanta* 194: 830–836. https://doi.org/10.1016/j.talanta.2018.11.020.

19 Merz, B., Capello, C., Leandro, G.C. et al. (2020). A novel colorimetric indicator film based on chitosan, polyvinyl alcohol and anthocyanins from jambolan *(Syzygium cumini)* fruit for monitoring shrimp freshness. *International Journal of Biological Macromolecules* 153: 625–632. https://doi.org/10.1016/j.ijbiomac.2020.03.048.

20 Chen, H.Z., Zhang, M., Bhandari, B., and Yang, C.H. (2019). Development of a novel colorimetric food package label for monitoring lean pork freshness. *LWT* 99: 43–49. https://doi.org/10.1016/j.lwt.2018.09.048.

21 Wan, X., He, Q., Wang, X. et al. (2021). Water-soluble chitosan-based indicator label membrane and its response behavior to carbon dioxide. *Food Control* 130: 108355. https://doi.org/10.1016/j.foodcont.2021.108355.

22 Saliu, F. and Della Pergola, R. (2018). Carbon dioxide colorimetric indicators for food packaging application: applicability of anthocyanin and poly-lysine mixtures. *Sensors and Actuators B: Chemical* 258: 1117–1124. https://doi.org/10.1016/j.snb.2017.12.007.

23 Zong, Z., Liu, M., Chen, H. et al. (2023). Preparation and characterization of a novel intelligent starch/gelatin binary film containing purple sweet potato anthocyanins for *Flammulina velutipes* mushroom freshness monitoring. *Food Chemistry* 405: 134839. https://doi.org/10.1016/j.foodchem.2022.134839.

24 Liu, L., Wu, W., Zheng, L. et al. (2022). Intelligent packaging films incorporated with anthocyanins-loaded ovalbumin-carboxymethyl cellulose nanocomplexes for food freshness monitoring. *Food Chemistry* 387: 132908. https://doi.org/10.1016/j.foodchem.2022.132908.

25 Warsiki, E. and Rofifah, N. (2018). Dragon fruit freshness detector based on methyl red colour indicator. *IOP Conference Series: Earth and Environmental Science* 209 (1): 012016. https://doi.org/10.1088/1755-1315/209/1/012016.

26 Wang, B., Leng, J., Wang, X., and Zhao, W. (2022). Reversible AIE-active fluorescent probe with a large emission peak shift for ratiometric detection of food freshness indicator H_2S. *Food Chemistry* 386: 132768. https://doi.org/10.1016/j.foodchem.2022.132768.

27 Zhai, X., Li, Z., Shi, J. et al. (2019). A colorimetric hydrogen sulfide sensor based on gellan gum-silver nanoparticles bionanocomposite for monitoring of meat spoilage in intelligent packaging. *Food Chemistry* 290: 135–143. https://doi.org/10.1016/j.foodchem.2019.03.138.

28 Teymouri, Z. and Shekarchizadeh, H. (2022). A colorimetric indicator based on copper nanoparticles for volatile sulfur compounds to monitor fish spoilage in intelligent packaging. *Food Packaging and Shelf Life* 33: 100884. https://doi.org/10.1016/j.fpsl.2022.100884.

29 Niponsak, A., Laohakunjit, N., Kerdchoechuen, O., and Wongsawadee, P. (2016). Development of smart colourimetric starch-based indicator for liberated volatiles during durian ripeness. *Food Research International* 89: 365–372. https://doi.org/10.1016/j.foodres.2016.08.038.

30 Shi, S., Xu, X., Feng, J. et al. (2023). Preparation of NH_3-and H_2S-sensitive intelligent pH indicator film from sodium alginate/black soybean seed coat anthocyanins and its use in monitoring meat freshness. *Food Packaging and Shelf Life* 35: 100994. https://doi.org/10.1016/j.fpsl.2022.100994.

31 Li, H., Geng, W., Sun, X. et al. (2021). Fabricating a nano-bionic sensor for rapid detection of H_2S during pork spoilage using Ru NPs modulated catalytic hydrogenation conversion. *Meat Science* 177: 108507. https://doi.org/10.1016/j.meatsci.2021.108507.

32 Kerry, J.P., O'grady, M.N., and Hogan, S.A. (2006). Past, current and potential utilisation of active and intelligent packaging systems for meat and muscle-based products: a review. *Meat Science* 74 (1): 113–130. https://doi.org/10.1016/j.meatsci.2006.04.024.

33 Qi, W., Wang, H., Zhou, Z. et al. (2020). Ethylene emission as a potential indicator of fuji apple flavor quality evaluation under low temperature. *Horticultural Plant Journal* 6 (4): 231–239. https://doi.org/10.1016/j.hpj.2020.03.007.

34 Cabanillas-Galan, P., Farmer, L., Hagan, T. et al. (2008). A new approach for the detection of ethylene using silica-supported palladium complexes. *Inorganic Chemistry* 47 (19): 9035–9041. https://doi.org/10.1021/ic800986t.

35 Shao, P., Liu, L., Yu, J. et al. (2022). Novel aldehyde sensitive bio-based colorimetric film for kiwi fruit freshness monitoring. *LWT* 159: 113177. https://doi.org/10.1016/j.lwt.2022.113177.

36 Kim, Y.H., Yang, Y.J., Kim, J.S. et al. (2018). Non-destructive monitoring of apple ripeness using an aldehyde sensitive colorimetric sensor. *Food Chemistry* 267: 149–156. https://doi.org/10.1016/j.foodchem.2018.02.110.

37 Hong, H., Regenstein, J.M., and Luo, Y. (2017). The importance of ATP-related compounds for the freshness and flavor of post-mortem fish and shellfish muscle: a review. *Critical Reviews in Food Science and Nutrition* 57 (9): 1787–1798. https://doi.org/10.1080/10408398.2014.1001489.

38 Kuda, T., Fujita, M., Goto, H., and Yano, T. (2008). Effects of retort conditions on ATP-related compounds in pouched fish muscle. *LWT- Food Science and Technology* 41 (3): 469–473. https://doi.org/10.1016/j.lwt.2007.02.018.

39 Zheng, L., Liu, L., Yu, J., and Shao, P. (2022). Novel trends and applications of natural pH-responsive indicator film in food packaging for improved quality monitoring. *Food Control* 134: 108769. https://doi.org/10.1016/j.foodcont.2021.108769.

40 Liu, X., Wang, Y., Zhu, L. et al. (2020). Dual-mode smart packaging based on tetraphenylethylene-functionalized polyaniline sensing label for monitoring the freshness of fish. *Sensors and Actuators B: Chemical* 323: 128694. https://doi.org/10.1016/j.snb.2020.128694.

41 Ko, Y., Jeong, H.Y., Kwon, G. et al. (2020). pH-responsive polyaniline/polyethylene glycol composite arrays for colorimetric sensor application. *Sensors and Actuators B: Chemical* 305: 127447. https://doi.org/10.1016/j.snb.2019.127447.

42 Kuswandi, B., Restyana, A., Abdullah, A. et al. (2012). A novel colorimetric food package label for fish spoilage based on polyaniline film. *Food Control* 25 (1): 184–189. https://doi.org/10.1016/j.foodcont.2011.10.008.

43 Priyadarshi, R., Ezati, P., and Rhim, J.W. (2021). Recent advances in intelligent food packaging applications using natural food colorants. *ACS Food Science & Technology* 1 (2): 124–138. https://doi.org/10.1021/acsfoodscitech.0c00039.

44 Lu, P., Yang, Y., Liu, R. et al. (2020). Preparation of sugarcane bagasse nanocellulose hydrogel as a colourimetric freshness indicator for intelligent food packaging. *Carbohydrate Polymers* 249: 116831. https://doi.org/10.1016/j.carbpol.2020.116831.

45 Mary, S.K., Koshy, R.R., Daniel, J. et al. (2020). Development of starch based intelligent films by incorporating anthocyanins of butterfly pea flower and TiO_2 and their applicability as freshness sensors for prawns during storage. *RSC Advances* 10 (65): 39822–39830. https://doi.org/10.1039/d0ra05986b.

46 Abdillah, A.A., Lin, H.H., and Charles, A.L. (2022). Development of halochromic indicator film based on arrowroot starch/iota-carrageenan using kyoho skin extract to monitor shrimp freshness. *International Journal of Biological Macromolecules* 211: 316–327. https://doi.org/10.1016/j.ijbiomac.2022.05.076.

47 Zhang, K., Huang, T.S., Yan, H. et al. (2020). Novel pH-sensitive films based on starch/polyvinyl alcohol and food anthocyanins as a visual indicator of shrimp deterioration. *International Journal of Biological Macromolecules* 145: 768–776. https://doi.org/10.1016/j.ijbiomac.2019.12.159.

48 Aghaei, Z., Ghorani, B., Emadzadeh, B. et al. (2020). Protein-based halochromic electrospun nanosensor for monitoring trout fish freshness. *Food Control* 111: 107065. https://doi.org/10.1016/j.foodcont.2019.107065.

49 Xiao, Y., Liu, Y., Kang, S. et al. (2021). Development of pH-responsive antioxidant soy protein isolate films incorporated with cellulose nanocrystals and curcumin nanocapsules to monitor shrimp freshness. *Food Hydrocolloids* 120: 106893. https://doi.org/10.1016/j.foodhyd.2021.106893.

50 Zeng, P., Chen, X., Qin, Y.R. et al. (2019). Preparation and characterization of a novel colorimetric indicator film based on gelatin/polyvinyl alcohol incorporating mulberry anthocyanin extracts for monitoring fish freshness. *Food Research International* 126: 108604. https://doi.org/10.1016/j.foodres.2019.108604.

51 Balbinot-Alfaro, E., Craveiro, D.V., Lima, K.O. et al. (2019). Intelligent packaging with pH indicator potential. *Food Engineering Reviews* 11 (4): 235–244. https://doi.org/10.1007/s12393-019-09198-9.

52 Cao, L., Sun, G., Zhang, C. et al. (2019). An intelligent film based on cassia gum containing bromothymol blue-anchored cellulose fibers for real-time detection of meat freshness. *Journal of Agricultural and Food Chemistry* 67 (7): 2066–2074. https://doi.org/10.1021/acs.jafc.8b06493.

53 Miranda, K.W., Natarelli, C.V., Thomazi, A.C. et al. (2020). Halochromic polystyrene nanofibers obtained by solution blow spinning for wine pH sensing. *Sensors* 20 (2): 417. https://doi.org/10.3390/s20020417.

54 Magnaghi, L.R., Zanoni, C., Alberti, G. et al. (2022). Towards intelligent packaging: BCP-EVOH @ optode for milk freshness measurement. *Talanta* 241: 123230. https://doi.org/10.1016/j.talanta.2022.123230.

55 Dirpan, A., Latief, R., Syarifuddin, A. et al. (2018). The use of colour indicator as a smart packaging system for evaluating mangoes Arummanis *(Mangifera indica L. var. Arummanisa)* freshness. *IOP Conference Series: Earth and Environmental Science* 157 (1): 012031. https://doi.org/10.1088/1755-1315/157/1/012031.

56 Kang, S., Wang, H., Xia, L. et al. (2020). Colorimetric film based on polyvinyl alcohol/okra mucilage polysaccharide incorporated with rose anthocyanins for shrimp freshness monitoring. *Carbohydrate Polymers* 229: 115402. https://doi.org/10.1016/j.carbpol.2019.115402.

57 Kuswandi, B., Asih, N.P., Pratoko, D.K. et al. (2020). Edible pH sensor based on immobilized red cabbage anthocyanins into bacterial cellulose membrane for intelligent food packaging. *Packaging Technology and Science* 33 (8): 321–332. https://doi.org/10.1002/pts.2507.

58 Peralta, J., Bitencourt-Cervi, C.M., Maciel, V.B. et al. (2019). Aqueous hibiscus extract as a potential natural pH indicator incorporated in natural polymeric films. *Food Packaging and Shelf Life* 19: 47–55. https://doi.org/10.1016/j.fpsl.2018.11.017.

59 Liu, J., Li, K., Chen, Y. et al. (2022). Active and smart biomass film containing cinnamon oil and curcumin for meat preservation and freshness indicator. *Food Hydrocolloids* 133: 107979. https://doi.org/10.1016/j.foodhyd.2022.107979.

60 Liu, J., Wang, H., Wang, P. et al. (2018). Films based on κ-carrageenan incorporated with curcumin for freshness monitoring. *Food Hydrocolloids* 83: 134–142. https://doi.org/10.1016/j.foodhyd.2018.05.012.

61 Ezati, P., Tajik, H., Moradi, M., and Molaei, R. (2019). Intelligent pH-sensitive indicator based on starch-cellulose and alizarin dye to track freshness of rainbow trout fillet.

International Journal of Biological Macromolecules 132: 157–165. https://doi.org/10.1016/j.ijbiomac.2019.03.173.

62 He, Y., Li, B., Du, J. et al. (2022). Development of pH-responsive absorbent pad based on polyvinyl alcohol/agarose/anthocyanins for meat packaging and freshness indication. *International Journal of Biological Macromolecules* 201: 203–215. https://doi.org/10.1016/j.ijbiomac.2021.12.171.

63 He, F., Kong, Q., Jin, Z., and Mou, H. (2020). Developing a unidirectionally permeable edible film based on κ-carrageenan and gelatin for visually detecting the freshness of grass carp fillets. *Carbohydrate Polymers* 241: 116336. https://doi.org/10.1016/j.carbpol.2020.116336.

64 Ezati, P., Tajik, H., and Moradi, M. (2019). Fabrication and characterization of alizarin colorimetric indicator based on cellulose-chitosan to monitor the freshness of minced beef. *Sensors and Actuators B: Chemical* 285: 519–528. https://doi.org/10.1016/j.snb.2019.01.089.

65 Yao, X., Hu, H., Qin, Y., and Liu, J. (2020). Development of antioxidant, antimicrobial and ammonia-sensitive films based on quaternary ammonium chitosan, polyvinyl alcohol and betalains-rich cactus pears *(Opuntia ficus-indica)* extract. *Food Hydrocolloids* 106: 105896. https://doi.org/10.1016/j.foodhyd.2020.105896.

66 Latos-Brozio, M. and Masek, A. (2020). The application of natural food colorants as indicator substances in intelligent biodegradable packaging materials. *Food and Chemical Toxicology* 135: 110975. https://doi.org/10.1016/j.fct.2019.110975.

67 Wu, S., Wang, W., Yan, K. et al. (2018). Electrochemical writing on edible polysaccharide films for intelligent food packaging. *Carbohydrate Polymers* 186: 236–242. https://doi.org/10.1016/j.carbpol.2018.01.058.

68 Tirtashi, F.E., Moradi, M., Tajik, H. et al. (2019). Cellulose/chitosan pH-responsive indicator incorporated with carrot anthocyanins for intelligent food packaging. *International Journal of Biological Macromolecules* 136: 920–926. https://doi.org/10.1016/j.ijbiomac.2019.06.148.

69 Kurnianto, M.F., Poerwanto, B., Wahyono, A. et al. (2020). Monitoring of banana deteriorations using intelligent-packaging containing brazilien extract (*Caesalpina sappan L.*). *IOP Conference Series: Earth and Environmental Science* 411 (1): 012043. https://doi.org/10.1088/1755-1315/411/1/012043.

70 Goodarzi, M.M., Moradi, M., Tajik, H. et al. (2020). Development of an easy-to-use colorimetric pH label with starch and carrot anthocyanins for milk shelf life assessment. *International Journal of Biological Macromolecules* 153: 240–247. https://doi.org/10.1016/j.ijbiomac.2020.03.014.

71 Chakraborty, G., Katiyar, V., and Pugazhenthi, G. (2021). Improvisation of polylactic acid (PLA)/exfoliated graphene (GR) nanocomposite for detection of metal ions (Cu^{2+}). *Composites Science and Technology* 213: 108877. https://doi.org/10.1016/j.compscitech.2021.108877.

72 Kumari, S.V.G., Manikandan, N.A., Pakshirajan, K., and Pugazhenthi, G. (2020). Sustained drug release and bactericidal activity of a novel, highly biocompatible and biodegradable polymer nanocomposite loaded with norfloxacin for potential use in antibacterial therapy. *Journal of Drug Delivery Science and Technology* 59: 101900. https://doi.org/10.1016/j.jddst.2020.101900.

73 Valapa, R., Pugazhenthi, G., and Katiyar, V. (2014). Thermal degradation kinetics of sucrose palmitate reinforced poly (lactic acid) biocomposites. *International Journal of Biological Macromolecules* 65: 275–283. https://doi.org/10.1016/j.ijbiomac.2014.01.053.

74 Yari, A. and Dinarvand, M. (2011). Sol-gel film doped with bromopyrogallol red as a highly sensitive sensing element for a new pH optical sensor. *Journal of the Iranian Chemical Society* 8 (4): 1091–1097. https://doi.org/10.1007/bf03246567.

75 Aman Mohammadi, M., Hosseini, S.M., and Yousefi, M. (2020). Application of electrospinning technique in development of intelligent food packaging: a short review of recent trends. *Food Science & Nutrition* 8 (9): 4656–4665. https://doi.org/10.1002/fsn3.1781.

76 He, C., Zhang, M., and Guo, C. (2020). 4D printing of mashed potato/purple sweet potato puree with spontaneous color change. *Innovative Food Science & Emerging Technologies* 59: 102250. https://doi.org/10.1016/j.ifset.2019.102250.

77 Alam, A.U., Rathi, P., Beshai, H. et al. (2021). Fruit quality monitoring with smart packaging. *Sensors* 21 (4): 1509. https://doi.org/10.3390/s21041509.

78 Li, N., Zhou, S., Yang, X., and Lin, D. (2022). Applications of natural polysaccharide-based pH-sensitive films in food packaging: current research and future trends. *Innovative Food Science & Emerging Technologies* 82: 103200. https://doi.org/10.1016/j.ifset.2022.103200.

79 Mustafa, F. and Andreescu, S. (2018). Chemical and biological sensors for food-quality monitoring and smart packaging. *Foods* 7 (10): 168. https://doi.org/10.3390/foods7100168.

80 Pu, Y.Y., Feng, Y.Z., and Sun, D.W. (2015). Recent progress of hyperspectral imaging on quality and safety inspection of fruits and vegetables: a review. *Comprehensive Reviews in Food Science and Food Safety* 14 (2): 176–188. https://doi.org/10.1111/1541-4337.12123.

81 El Khaled, D., Novas, N., Gazquez, J.A. et al. (2015). Fruit and vegetable quality assessment via dielectric sensing. *Sensors* 15 (7): 15363–15397. https://doi.org/10.3390/s150715363.

82 Gao, F., Dong, Y., Xiao, W. et al. (2016). LED-induced fluorescence spectroscopy technique for apple freshness and quality detection. *Postharvest Biology and Technology* 119: 27–32. https://doi.org/10.1016/j.postharvbio.2016.04.020.

83 Camps, C. and Christen, D. (2009). Non-destructive assessment of apricot fruit quality by portable visible-near infrared spectroscopy. *LWT- Food Science and Technology* 42 (6): 1125–1131. https://doi.org/10.1016/j.lwt.2009.01.015.

84 Kuswandi, B., Maryska, C., Abdullah, A., and Heng, L.Y. (2013). Real time on-package freshness indicator for guavas packaging. *Journal of Food Measurement and Characterization* 7 (1): 29–39. https://doi.org/10.1007/s11694-013-9136-5.

85 Dirpan, A., Djalal, M., Rahman, R., and Genisa, J. (2021). The utilization of red cabbage extract (*Brassica oleracea*) in the production of avocado (*Persea americana Mill*) freshness indicator as smart packaging element. *Online Journal of Biological Sciences* 21: 261–270. https://doi.org/10.3844/ojbsci.2021.261.270.

86 Sganzerla, W.G., Ribeiro, C.P.P., Uliana, N.R. et al. (2021). Bioactive and pH-sensitive films based on carboxymethyl cellulose and blackberry (*Morus nigra L.*) anthocyanin-rich extract: a perspective coating material to improve the shelf life of cherry tomato (*Solanum lycopersicum L. var. cerasiforme*). *Biocatalysis and Agricultural Biotechnology* 33: 101989. https://doi.org/10.1016/j.bcab.2021.101989.

87 Chen, H.Z., Zhang, M., Bhandari, B., and Guo, Z. (2018). Applicability of a colorimetric indicator label for monitoring freshness of fresh-cut green bell pepper. *Postharvest Biology and Technology* 140: 85–92. https://doi.org/10.1016/j.postharvbio.2018.02.011.

88 Ziyaina, M., Rasco, B., and Sablani, S.S. (2020). Rapid methods of microbial detection in dairy products. *Food Control* 110: 107008. https://doi.org/10.1016/j.foodcont.2019.107008.

89 Weston, M., Phan, M.A.T., Arcot, J., and Chandrawati, R. (2020). Anthocyanin-based sensors derived from food waste as an active use-by date indicator for milk. *Food Chemistry* 326: 127017. https://doi.org/10.1016/j.foodchem.2020.127017.

90 Li, Y., Wu, K., Wang, B., and Li, X. (2021). Colorimetric indicator based on purple tomato anthocyanins and chitosan for application in intelligent packaging. *International Journal of Biological Macromolecules* 174: 370–376. https://doi.org/10.1016/j.ijbiomac.2021.01.182.

91 Liu, B., Xu, H., Zhao, H. et al. (2017). Preparation and characterization of intelligent starch/PVA films for simultaneous colorimetric indication and antimicrobial activity for food packaging applications. *Carbohydrate Polymers* 157: 842–849. https://doi.org/10.1016/j.carbpol.2016.10.067.

92 Chaithra, K.P. and Vinod, T.P. (2022). A smartphone coupled freshness indicator prepared by rub-coating of hibiscus flowers on paper substrates for visual monitoring of the spoilage of milk. *ChemistrySelect* 7 (41): e202201839. https://doi.org/10.1002/slct.202201839.

93 Yong, H., Wang, X., Zhang, X. et al. (2019). Effects of anthocyanin-rich purple and black eggplant extracts on the physical, antioxidant and pH-sensitive properties of chitosan film. *Food Hydrocolloids* 94: 93–104. https://doi.org/10.1016/j.foodhyd.2019.03.012.

94 Bandyopadhyay, S., Saha, N., Zandraa, O. et al. (2020). Essential oil based PVP-CMC-BC-GG functional hydrogel sachet for 'cheese': its shelf life confirmed with anthocyanin (isolated from red cabbage) bio stickers. *Foods* 9 (3): 307. https://doi.org/10.3390/foods9030307.

95 Anjali, M.K., Bharath, G., Rashmi, H.M. et al. (2022). Polyaniline-Pectin nanoparticles immobilized paper based colorimetric sensor for detection of *Escherichia coli* in milk and milk products. *Current Research in Food Science* 5: 823–834. https://doi.org/10.1016/j.crfs.2022.04.006.

96 Oliver, S.P., Jayarao, B.M., and Almeida, R.A. (2005). Foodborne pathogens in milk and the dairy farm environment: food safety and public health implications. *Foodbourne Pathogens & Disease* 2 (2): 115–129. https://doi.org/10.1089/fpd.2005.2.115.

97 Ahmed, I., Lin, H., Zou, L. et al. (2017). A comprehensive review on the application of active packaging technologies to muscle foods. *Food Control* 82: 163–178. https://doi.org/10.1016/j.foodcont.2017.06.009.

98 Mohebi, E. and Marquez, L. (2015). Intelligent packaging in meat industry: an overview of existing solutions. *Journal of Food Science and Technology* 52 (7): 3947–3964. https://doi.org/10.1007/s13197-014-1588-z.

99 Zhang, J., Huang, X., Zhang, J. et al. (2022). Development of nanofiber indicator with high sensitivity for pork preservation and freshness monitoring. *Food Chemistry* 381: 132224. https://doi.org/10.1016/j.foodchem.2022.132224.

100 Ebrahimi, V., Nafchi, A.M., Bolandi, M., and Baghaei, H. (2022). Fabrication and characterization of a pH-sensitive indicator film by purple basil leaves extract to monitor the freshness of chicken fillets. *Food Packaging and Shelf Life* 34: 100946. https://doi.org/10.1016/j.fpsl.2022.100946.

101 Zhao, M., Nuerjiang, M., Bai, X. et al. (2022). Monitoring dynamic changes in chicken freshness at 4 °C and 25 °C using pH-sensitive intelligent films based on sodium alginate and purple sweet potato peel extracts. *International Journal of Biological Macromolecules* 216: 361–373. https://doi.org/10.1016/j.ijbiomac.2022.06.198.

102 Mohammadalinejhad, S., Almasi, H., and Moradi, M. (2020). Immobilization of *Echium amoenum* anthocyanins into bacterial cellulose film: a novel colorimetric pH indicator for freshness/spoilage monitoring of shrimp. *Food Control* 113: 107169. https://doi.org/10.1016/j.foodcont.2020.107169.

103 Ghohestani, E., Tashkhourian, J., and Hemmateenejad, B. (2023). Colorimetric determination of peroxide value in vegetable oils using a paper based analytical device. *Food Chemistry* 403: 134345. https://doi.org/10.1016/j.foodchem.2022.134345.

104 Liu, X., Li, T., Jiang, J. et al. (2021). Visual detection of edible oil oxidation by using chitin-based colorimetric sensor for aldehydes. *Colloids and Surfaces A: Physicochemical and Engineering Aspects* 628: 127303. https://doi.org/10.1016/j.colsurfa.2021.127303.

105 Nopwinyuwong, A., Trevanich, S., and Suppakul, P. (2010). Development of a novel colorimetric indicator label for monitoring freshness of intermediate-moisture dessert spoilage. *Talanta* 81 (3): 1126–1132. https://doi.org/10.1016/j.talanta.2010.02.008.

106 Gaikwad, P.S., Sarma, C., Rajan, A. et al. (2022). Development of bacterial cellulose film coated with mixed colorimetric indicator for tracking the freshness/spoilage of ready-to-cook idli batter. *Coloration Technology* 138 (1): 28–37. https://doi.org/10.1111/cote.12564.

107 Moon, E.W., Yang, J.S., Yoon, S.R., and Ha, J.H. (2020). Application of colorimetric indicators to predict the fermentation stage of kimchi. *Journal of Food Science* 85 (12): 4170–4179. https://doi.org/10.1111/1750-3841.15532.

108 Kim, S.J., Lee, J.Y., Yoon, S.R. et al. (2019). Regression analysis for predicting the fermentation state of packaged kimchi using a colorimetric indicator. *Journal of Food Engineering* 240: 65–72. https://doi.org/10.1016/j.jfoodeng.2018.06.031.

109 Singh, S., Nwabor, O.F., Syukri, D.M., and Voravuthikunchai, S.P. (2021). Chitosan-poly (vinyl alcohol) intelligent films fortified with anthocyanins isolated from *clitoria ternatea* and *carissa carandas* for monitoring beverage freshness. *International Journal of Biological Macromolecules* 182: 1015–1025. https://doi.org/10.1016/j.ijbiomac.2021.04.027.

110 Pereira de Abreu, D.A., Cruz, J.M., and Paseiro Losada, P. (2012). Active and intelligent packaging for the food industry. *Food Reviews International* 28 (2): 146–187. https://doi.org/10.1080/87559129.2011.595022.

111 Realini, C.E. and Marcos, B. (2014). Active and intelligent packaging systems for a modern society. *Meat Science* 98 (3): 404–419. https://doi.org/10.1016/j.meatsci.2014.06.031.

112 Next Nature (2009). Technology is our next nature. https://www.nextnature.net/story/2009/freshness-label (accessed 27 December 2022).

113 Kalpana, S., Priyadarshini, S.R., Leena, M.M. et al. (2019). Intelligent packaging: trends and applications in food systems. *Trends in Food Science & Technology* 93: 145–157. https://doi.org/10.1016/j.tifs.2019.09.008.

15

Biopolymer-Based Nanocomposites for Intelligent Food Packaging

Himakshi Baishya, Sweety Kalita, Avik Mukherjee, and Santosh Kumar

Department of Food Engineering and Technology, Central Institute of Technology Kokrajhar, Kokrajhar, Assam, India

15.1 Introduction

Food packaging stands as an integral and multifaceted component of food industry, fulfilling critical roles that extend beyond mere containment. Its primary function is to protect and preserve food, shielding it from contamination, physical damage, and environmental factors that could compromise the food quality and safety [1]. Simultaneously, packaging serves as a vital tool for information dissemination, providing consumers with crucial details about nutritional content, ingredients, and various instructions [2]. Moreover, it plays a pivotal role in marketing, as packaging design and branding influence consumer perceptions and purchasing decisions. Compliance with stringent regulatory standards ensures the legality and safety of packaged food products [3]. However, in response to global concerns about environmental sustainability, the food industry is increasingly adopting eco-friendly packaging materials in order to reduce the ecological footprint and food waste fostering a balance between consumer convenience and responsible practices throughout the food supply chain [4–6]. Food packaging also acts as a protective barrier and interface for communication between the product and consumers. In recent times, growing concerns about the ecological footprint have shifted the preference from conventional plastic packaging to biopolymer-based sustainable food packaging [7, 8]. As alternatives to petroleum-based conventional plastic packaging, biopolymer-based packaging led toward sustainable solutions through utilizing materials derived from renewable sources that minimize greenhouse gas emissions, and mitigate synthetic plastic pollution [9]. The different sources for generation of biopolymers include both plants (plant starches and agro-wastes) and animals (animal protein) with the inclusion of microbial sources (pullulan and bacterial cellulose) and chemically synthesized biopolymers derived from natural sources (polylactic acid and polyvinyl alcohol) [10]. The different types of natural biopolymers are illustrated in Figure 15.1a. The most commonly available natural biopolymers include cellulose, chitosan, starch and their derivatives, pectin, zein,

Smart Food Packaging Systems: Innovations and Technology Applications, First Edition.
Edited by Avik Mukherjee, Santosh Kumar, Manjusri Misra, and Amar K. Mohanty.
© 2025 John Wiley & Sons Ltd. Published 2025 by John Wiley & Sons Ltd.

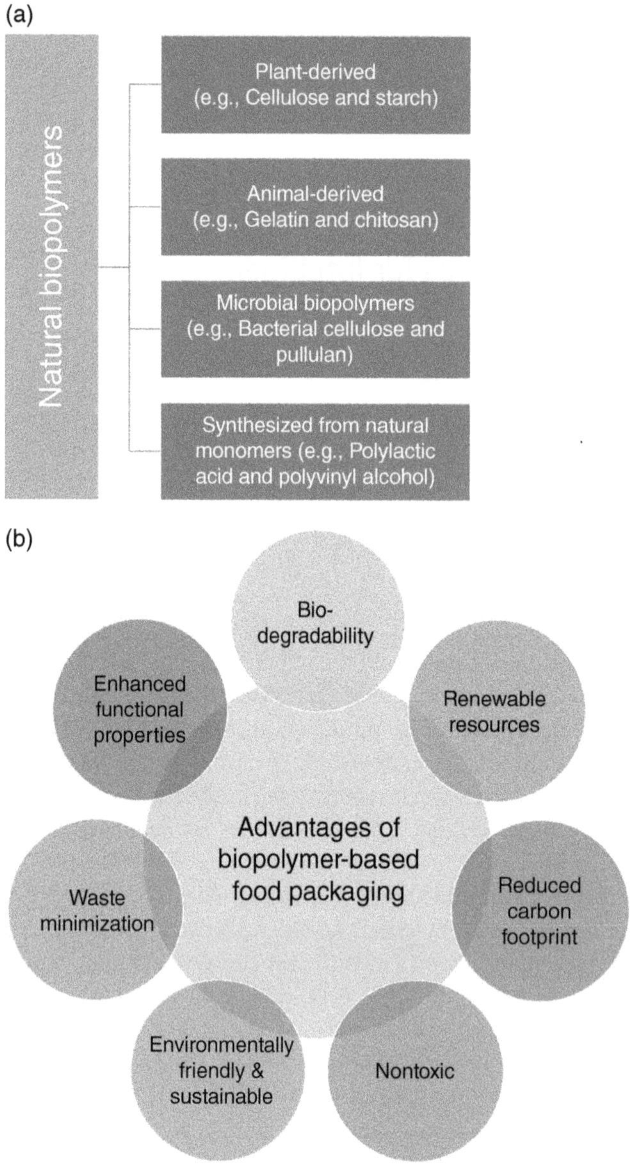

Figure 15.1 (a) Types of natural biopolymers and (b) advantages of biopolymer-based food packaging.

soy protein, gelatin, and collagen [11, 12]. Biopolymers are naturally derived polymers that have several advantages over conventional plastic packaging including being nontoxic in nature, environmentally friendly, biodegradable, biocompatible, and derived from readily available renewable resources, as shown in Figure 15.1b [13]. However, there are some disadvantages of biopolymers, including inferior mechanical characteristics, low heat stability, low vapor and gas permeability, and brittleness. As a result, it is frequently altered or blended with other biopolymers, functional additives, and nanomaterials to mitigate these disadvantages [14]. The application of nanotechnology is revolutionizing the food packaging sector by introducing novel nanomaterials, processes, and features that enhance functionalities of biopolymer-based packaging materials [15]. Nanofillers such as starch nanocrystals, cellulose nanofibers, chitin and chitosan nanoparticles, and mineral nanoparticles, have been used to improve mechanical, barrier, and thermal properties of these composites [16–18]. Similar to synthetic polymer composites, bio-nanocomposites serve a purpose in crafting bio-based sensors, indicators, and data carriers to oversee the quality and safety of food packages, with a focus on sensors utilizing biological and chemical indicators for monitoring food freshness [19, 20].

Intelligent packaging can provide remarkable capabilities by incorporating pH indicators, freshness indicators, humidity sensors, time–temperature sensors, oxygen sensors, and radio frequency identification (RFID) systems sensors [21]. The primary technologies used by intelligent packaging systems are sensors, indicators, and data carriers, as shown in Figure 15.2. The intelligent packaging monitors and interacts with the packaging environment through internal or exterior indicators, and informs the consumer about the change in quality of the packaged food throughout the food supply chain [22]. The integration of sensors, indicators, and data carriers in intelligent packaging systems represents a technological leap toward a more informed, real-time, and efficient packaging ecosystem [23]. Sensors are devices that identify and quantify environmental factors inside the package or surrounding the food that transform electrical signals, which are deciphered and utilized to monitor the state of the packaged food. The major types of sensors used in

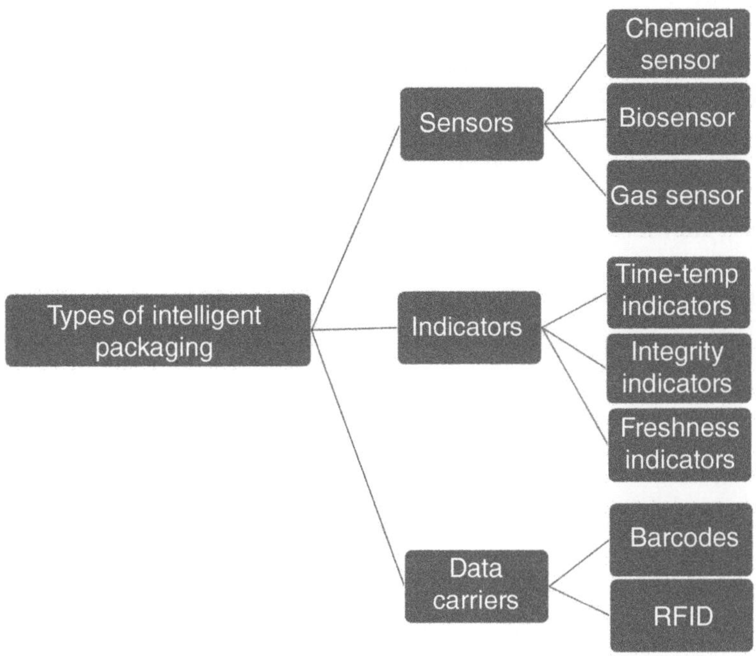

Figure 15.2 Types of intelligent packaging.

intelligent food packaging include biosensors, chemical, and gas sensors [24]. An indicator refers to visual or electronic elements integrated into the packaging system that provides information about the condition or quality of the packaged food product by communicating specific data to consumers, retailers, or other stakeholders in the supply chain, offering insights into the product's freshness, safety, or other environmental conditions. The different types of indicators include time–temperature, integrity, and freshness indicators [25]. Data carriers are technologies that store and transmit information related to the product that enable traceability and access to critical details about the product in the supply chain from production to consumption. The different types of data carriers include barcodes, QR codes, and RFID tags [26]. Nonetheless, it is crucial to summarize the latest research findings to stay informed about recent advancements in bio-nanocomposites for intelligent packaging. This chapter offers a condensed overview of recent innovations, particularly focusing on bio-nanocomposites in intelligent packaging over the past few years.

15.2 Bio-Nanocomposite-Based Intelligent Packaging Films/Labels

As per the EU's legal definition (EC, 2009), intelligent packaging is the novel packaging comprising substances that facilitates the monitoring of the internal environment of the packaged food or the surrounding environment while it is being transported and stored for future use [27]. It provides real-time information about the state of the food, the surroundings, and/or the integrity of the package. Intelligent packaging can feel, detect, or record changes in the product or its surrounding condition, and use that information to communicate with the consumer rather than relying on erroneous expiration dates [28]. This helps to avoid foodborne illnesses brought on by consuming contaminated food as well as the massive food waste that results from over-purchasing, inefficient supply chains, consumer habits, and food product recall systems [29]. Bio-nanocomposite is composed of hybrid material developed by combination of two or more components from renewable sources, which includes a continuous phase called the matrix, and a discontinuous phase at the nanoscale level also known as nanostructure [30]. Inorganic nanomaterials such as Ag, TiO_2, and ZnO are employed as nanofillers due to their advantageous mechanical, barrier, and antibacterial properties [31, 32]. Figure 15.3 shows the improvement in various

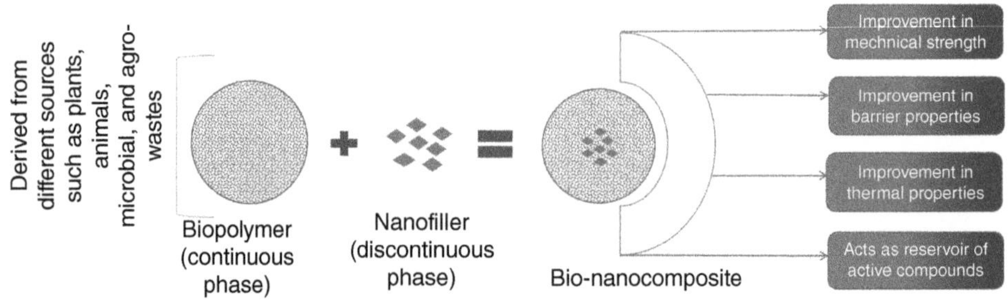

Figure 15.3 Improvement in functional properties of bio-nanocomposites.

functionalities of biopolymers i.e. thermal properties, mechanical strength, and barrier properties by adding nanofillers. The bio-nanocomposites also act as reservoirs for bioactive compounds, which offer antimicrobial, antioxidant, and antibiotic properties to the packaging materials [33]. Consequently, it is imperative to substitute these inorganic nanofillers with natural, nontoxic nanomaterials to uphold the safety standards of bio-nanocomposite packaging materials/films. Chitosan nanofibers, being a polymeric biona-nomaterial, have attracted much attention as nanofillers due to their excellent antibacterial properties, biocompatibility, and biodegradability [34].

Bio-nanocomposite intelligent packaging films are developed by incorporating nanomaterials into biodegradable biopolymer matrix with smart functionalities such as providing real-time information to the consumers [35]. In particular, a pH-sensitive chitosan-based bio-nanocomposite film was created with purple potato anthocyanin extract and surface-deacetylated chitin nanofibers as nanofillers, and the developed films exhibited pH-sensing ability, displaying distinct colors corresponding to different pH ranges (1-pink; 14-light yellow), indicating its potential for intelligent packaging applications [36]. Another chitosan-based bio-nanocomposite film was created with curcumin-loaded mesoporous silica nanoparticles as nanofillers to enhance various functionalities that exhibited good biocompatibility, improved mechanical strength, antibacterial features, as well as pH responsiveness [37]. A chitosan-based film with nano-TiO_2 and red apple pomace extract (APE) exhibited enhanced antimicrobial activity in the chitosan matrix, and consequently induced color changes, serving as an effective freshness indicator for salmon fillets [38]. In a recent study, Shavisi and Shahbazi introduced *Rosa damascena* anthocyanins extract into chitosan/gum Arabic nanofibers, leading to the development of a smart bio-nanocomposite film, which demonstrated noteworthy efficacy as a tool for monitoring the freshness of raw chicken fillet [39]. Several other polysaccharides have been employed to develop intelligent packaging films/labels, for example, corn starch with natural and modified nanoclays, incorporating blueberry extract (anthocyanin) to produce a pH-sensitive bio-nanocomposite film [40], κ-carrageenan/gelatin-based biodegradable nanocomposite film incorporated TiO_2 nanoparticles, anthocyanin from saffron petals, and barberry fruit [41]. In another study, carbon dots (CDs) were integrated into the gelatin/chitosan resulting in a bio-nanocomposite film with outstanding mechanical, barrier, antibacterial, antioxidant, anti-counterfeiting, and pH-responsive properties, attributed to the strong fluorescent emissions of CDs [42]. Gelatin-based intelligent packaging film was further explored using oxidized-chitin nanocrystals and black rice bran anthocyanins (BACNs), displaying notable color changes in various buffer solutions for monitoring shrimp freshness [43]. Gelatin was further effectively employed to create an edible nanocomposite using anthocyanin (*Syzygium cumini*), eggshell membrane, gelatin, and hydroxyapatite that serve as a pH indicator [44]. Zein, a crucial protein, has also been examined in the context of bio-nanocomposite films, incorporating tea tree essential oil-loaded mesoporous nanoparticles as nanofillers and blueberry extract anthocyanin as a pH indicator. The film demonstrated notable color response, transitioning from colorless to brown during packaging and storage with pork samples for nine days [45]. Nanoliposomes, created from phospholipids, represent less-explored biopolymers in the realm of bio-nanocomposite-based intelligent packaging. In recent research, a bio-nanocomposite film was developed using liposomes, gelatin, and chitosan nanofibers loaded with betanin that exhibited enhanced mechanical, antibacterial, antioxidant, as well as high pH-responsive properties, and demonstrated controlled lipid oxidation, improved shelf life, and enhanced color quality of the beef samples [46]. Bio-nanocomposites in packaging primarily focus on active packaging, having limited exploration in intelligent packaging; however, it is noteworthy that the use of bio-nanocomposites for intelligent packaging is a relatively new and emerging research field.

15.3 Natural Active Compounds-Based Intelligent Packaging Systems

The mechanism behind the different types of intelligent packaging is the incorporation of dyes either from natural or synthetic origin that change color with changes in packaging environment such as temperature, pH, humidity, and nitrogen-related compounds. In recent years, with the increase in consumer awareness regarding impact of synthetic dyes on environment and human health, the use of natural dyes or pigments in intelligent packaging is increasing because of their compatibility with food, nontoxicity, and anticarcinogenic properties [47]. Natural active compounds refer to materials obtained from natural sources with active functional properties that play important roles in enhancing shelf life, preventing spoilage, and providing real-time information about the condition of the packaged food [48, 49]. The natural active components are chosen due to their specific properties such as antimicrobial activity, antioxidant activity, and pH sensitivity [50]. Natural pigments such as anthocyanin, betalain, curcumin, and carotene are employed as intelligent agents in food packaging, as their color changes with change in pH (Figure 15.4) [51]. These pigments are derived from different natural sources, such as plants and their various parts such as fruits, vegetables, leaves, seeds, and flowers [52]. For instance, anthocyanins obtained from berries and red cabbage can act as pH indicators, which change color in response to pH changes in food [53]. Chlorophyll, found in green leafy vegetables, can be utilized to monitor food freshness and effects of light on food [54], whereas, curcumin obtained from turmeric, is employed as an indicator for detecting temperature changes and microbial spoilage [55]. These natural pigments not only provide a visual cue to consumers but also align with the growing demand for eco-friendly and clean-label products, contributing to a more sustainable approach in the field of intelligent food packaging [56].

Intelligent packaging, exemplified by food packages featuring time–temperature indicators (TTIs), offers cost-effective labels/tags to record and monitor the change in temperature of a packaged product as it passes through the food chain. TTIs can function as labels/tags on package or be integrated into the packaging film [57]. Anthocyanins demonstrate high sensitivity to temperature changes, and their integration into a biopolymer matrix facilitates the tracking of temperature variations based on food product pH during storage at different temperatures over time [58]. Notably, Chayavanich et al. showed that the starch/gelatin film with red radish anthocyanin exhibited higher color stability at refrigeration temperatures than at room temperature, consistent with the generally known greater stability of anthocyanins at lower temperatures [59]. Furthermore, incorporating humidity sensors on/in packaging film allows for assessing the quality and safety conditions of packaged food [60]. For instance, nanocomposites of poly (vinyl alcohol) (PVA) were crafted by incorporating tannic acid (rGO-TA) coated reduced graphene oxide as a filler, and developed nanocomposites demonstrated good humidity sensing characteristics across an

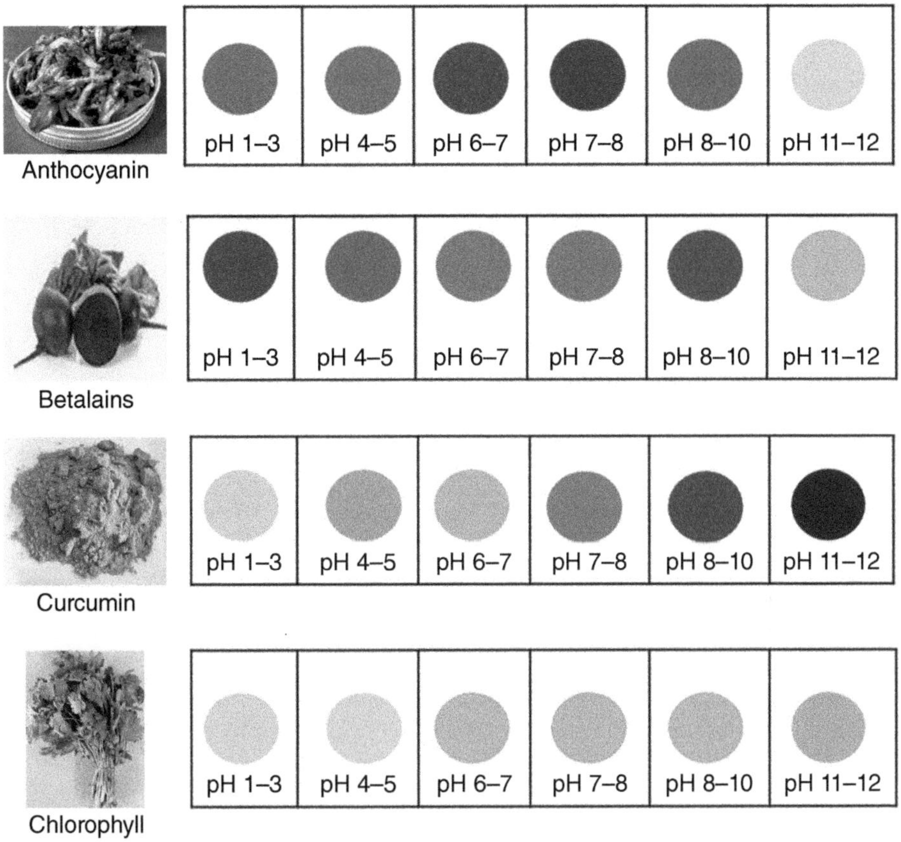

Figure 15.4 Changes in color of various natural pigments with change in pH.

extensive relative humidity range [61]. RFID tags, a barcode type data carrier used in food packaging industry [62, 63]. Intelligent packaging with biosensors tends to monitor microbial growth and other potential contaminants. The electrochemical sensing properties of biosensors in food packaging involve metal oxide ingredients along with receptors such as enzymes, antigens, hormones, and nucleic acids, either individually or in combination [64]. The edible chitosan–agarose hydrogel film, developed by Wu et al. involves anthocyanin, which showed pH sensitivity, and during carp spoilage detection, the hydrogel film exhibited noticeable color change from red to purple, providing a convenient visual indicator of food quality for consumers based on the color shift in the packaging material [65]. Some more recent studies on application of different natural pigments in developing intelligent food packaging films or labels are listed in Table 15.1.

15.4 Bio-Nanocomposite-Based Color Sensor

Color, in the context of food, refers to the visual perception of the various wavelengths of light reflected or emitted by the different substances present in the food. It is a crucial sensory attribute that influences consumers' perception in identifying food quality such as

Table 15.1 Natural pigments-based intelligent food packaging films/labels.

Natural pigment	Source of pigment	Biopolymer/ biocomposite	Stimulating factor	Food packaging application	Reference
Anthocyanin	Butterfly pea flower	Gelatin and chitosan	pH and ammonia	Beef meat	[66]
	Roselle fruit	Starch	pH	Pork and shrimp	[67]
	Red cabbage	Pectin, sodium alginate, and cellulose nanocrystals	pH and ammonia	Shrimp	[68]
	Rose flower	Corn starch, chitosan rose, and potato amylopectin	pH	Shrimp	[69]
	Blueberry fruit	PVA, sodium alginate, and chitosan	pH	Milk	[70]
Betalain	Red prickly pear	Cellulose and alginate	pH	Salmon	[71]
	Amaranthus leaf	PVA and gelatin	pH, microbial counts, and TVB-N	Fish and chicken	[72]
	Beetroot	Starch from oats and cellulose	pH	Chicken	[73]
	Amaranth ammonium	Chitosan and gelatin (from fish)	pH	Shrimp	[74]
	Red pitaya peel	PVA and starch	TVB-N	Shrimp	[75]
Curcumin	Turmeric	Konjac glucomannan and bacterial cellulose nanofibers	pH	Beef	[76]
	Turmeric	Chitosan and gelatin	Ammonia	Animal foods such as meat and seafood	[77]
	Turmeric	Chitosan and polyethylene oxide	pH and TVB-N	Chicken	[78]
	Turmeric	Tara gum and PVA	Ammonia	Shrimp	[79]
	Turmeric	Sodium cellulose sulfate and chitosan	pH	Pork	[80]

freshness and ripening, and it serves as an indicator in the process of decision-making by the consumers [81, 82]. For instance, bright and vibrant colors in fruits and vegetables, are often associated with freshness and high nutritional value. Similarly, the color of meat, fish, and other food products provides cues about their quality and potential spoilage. Overall, color acts as a visual indicator, aiding consumers' choices about the freshness, ripeness, and overall quality of the food [83]. Use of natural pigments as color indicators in intelligent food packaging have increased in recent years [84]. Colorimetric sensors are among the many intelligent packaging technologies that are simple to use, which allow consumers to visually inspect food products without requiring complicated analytical processes [85]. Color sensors made of bio-nanocomposites are designed to monitor quality changes in packaged food products capable of detecting alterations in color [86]. The mechanism involves incorporating responsive materials, such as nanoparticles, pH-sensitive dyes, or colorimetric sensors, into the biopolymeric matrix, when exposed to targeted stimuli such as gases total volatile basic nitrogen (TVB-N) produced during food spoilage or changes in temperature and humidity, these materials undergo color transformations [87]. For instance, if there is a rise in pH due to microbial activity, the color indicator may shift from its original color to signal potential spoilage, while the intensity of the color change can also correlate with the degree of freshness or decay. The incorporation of different pigments such as anthocyanin, alizarin, betalain, and curcumin, derived from various natural sources, into biopolymeric films or labels for intelligent food packaging applications is illustrated in Table 15.2. Thus, intelligent packaging approach has the potential to reduce food waste by allowing consumers and distributors to make informed decisions based on real-time visual cues, contributing to enhanced food safety and quality assurance [105].

Numerous research studies have been explored in this regard including the development of films using starch and nano-clay particles, incorporating methyl orange and bromocresol green indicators for monitoring spoilage of milk. The pH of milk undergoes changes during storage (from 6.7 to 4) and the devised film intelligently demonstrates milk spoilage through color changes in response to the shifting pH levels [106]. Another visually responsive intelligent film, composed of poly (vinyl alcohol), chitosan nanoparticles, and 20% mulberry extracts (MBE) displayed visible color responses across pH 1–13, and film changed color from red to green as the fish deteriorated [107]. Anthocyanin-rich purple eggplant extract and black eggplant extract were combined with chitosan to make pH-sensitive films, which showed notable color changes in a variety of buffer solutions [108]. Anthocyanins particularly are very sensitive to pH changes and show different chemical structures and colors with different polymeric film. Basically, the color of the solution of anthocyanin is different from the color of anthocyanin embedded in the film. In fact, electrostatic interaction between polymer and anthocyanin functional groups can change anthocyanin structure [40, 109]. As an example, anthocyanins from black carrots were used as a chemoresponsive dye to create a colorimetric pH sensor film in a cellulose–chitosan matrix, which allowed for the monitoring of deterioration in pasteurized milk. Across a range of pH values (pH 2–11), the film showed noticeable color shifts from pink to khaki, and even after a month of storage at 20 °C, stability tests showed that the indicator still maintained a respectable level of color stability [110]. Furthermore, fish spoilage was monitored with bilayer colorimetric films made of an outer layer of gelatin combined with ZnO nanoparticles, and an inner layer of gellan gum combined with mulberry anthocyanins. The bilayer films showed a detection limit of 0.01 mM that was attributed to showing discernible color variations in response to crucian spoilage [111]. In addition, anthocyanin-rich films mixed with other natural pigment such as shikonin have shown improved ability to monitor food freshness, which displayed distinctive pH-responsive color change within the pH range of 2–12, showcasing excellent sensitivity to both acidic and basic gas sensing environments [112]. To manufacture pH-sensitive smart reactive packaging films, Zhao et al. used

Table 15.2 Biopolymer and natural pigments-based intelligent packaging films or labels for food packaging applications.

Film formulation	Natural colorant (type and source)	Color change in response to the food		Food application	Reference
		Fresh	Spoiled		
Corn starch and gelatin	Anthocyanin -red radish	Orange	Purple	Meat	[59]
CMC	Anthocyanin – date pits	Yellow	Brown	Beef meat	[88]
Furcellaran	Anthocyanin – beetroot, elderberry, blueberry, green tea yerba mate	Red	Green	Atlantic mackerel	[89]
Pectin, chitosan	Anthocyanin – black rice	Red	Blue	Meat	[90]
Cassava starch	Anthocyanin – Chinese bayberry	Purple	Light brown	Pork	[91]
Chitosan	Anthocyanin – roselle	Light brown	Gray	Fish	[92]
Cassava starch	Anthocyanin – red cabbage, betalain – beetroot	Red	Light brown	Argentine hake fish	[93]
PVA, starch	Anthocyanin – *Lycium ruthenicum*	Violet	Light blue	Largemouth bass fillets	[94]
Chitosan	Alizarin – maddar plant roots	Khaki	Light brown	Fish	[95]
Potato starch	Betacyanin – paper flower	Pink	Yellow	Fish	[96]
Chitosan, PVA	Betacyanin – red beetroot	Purple-red	Pink	Shrimp	[97]
Native oats starch	Betalain – beetroot	Red	Light brown	Chicken	[73]
Locust bean gum, PVA	Betacyanin – cockscomb flower	Reddish purple	Yellow	Shrimp	[98]
Locust bean gum, PVA	Betacyanin – red pitaya fruit	Light pink	Light brown	Shrimp	[99]
Pectin from white-fleshed pitaya peel, montmorillonite	Betacyanin – white-fleshed pitaya peel	Red	Brown	Shrimp	[100]
Lallemantia iberica seed gum	Curcumin – turmeric	Yellow	Red	Shrimp	[101]
κ-carrageenan, gelatin	Curcumin – turmeric	Yellow	Red	Shrimp	[102]
Chitosan, gelatin, starch	Curcumin – turmeric; anthocyanin – red raspberry	Red	Dark gray	Chicken	[103]
κ-carrageenan	Curcumin – turmeric	Bright yellow	Reddish brown	Shrimp, pork	[104]

carboxymethyl cellulose as a substrate, anthocyanin as an indicator, and graphene oxide as an enhancer to monitor pork's freshness. The color of the container changed significantly depending on the pH level, alerting customers in real-time when the food deteriorated [113]. Several other studies have been conducted on anthocyanin as a natural colorant to serve as a sensor within various biopolymer films [108, 114–118].

Betacyanin, extracted from *Bougainvillea glabra* flowers, has been used as a color sensor in potato starch film for real-time fish quality indication, and the developed film exhibited a visual color shift in the packaging label from pink to yellow, aligning with increased TVB-N and microbial count in the fish samples [96]. In a different investigation, lycopene, a natural colorant, was impregnated into a polylactic acid–titanium dioxide nanocomposite film for margarine packaging. The film's color index 'a' decreased (red color diminishes), while 'b' increased (yellow color intensifies), which can be attributed to oxidative deterioration of margarine [119]. Jaramillo et al. found that films based on cassava starch containing green tea and basil extracts (chlorophyll) exhibited good antioxidant properties, and can be used as a food quality indicator when exposed to pH changes [120]. In the commercial realm, freshness indicators, operating as pH indicators or ethylene indicators, have now also been incorporated for detection of volatile compounds produced during food spoilage, such as carbon dioxide, ammonia, and hydrogen sulfide (H_2S) [121]. The development of a real-time monitoring system for meat spoilage using a colorimetric H_2S sensor incorporated into gellan gum-capped silver nanoparticles enables detection of H_2S with 0.81 μM detection limit. The sensor exhibited visible color changes from yellow to colorless, providing effective detection of in situ H_2S generated from chicken breast and silver carp within a packaging system [122].

15.5 Food Packaging Applications of Bio-Nanocomposite Intelligent Films/Labels

15.5.1 Fruits and Vegetables

Fresh fruits and vegetables continue to respire even after harvesting making them inherently perishable. The spoilage of fruits and vegetables is influenced by intrinsic factors such as type or species of crops, maturity, handling, and environmental conditions, resulting in significant losses due to improper storage. The temperature of fruits and vegetables has always been regarded as the primary factor that influences spoilage, respiration rate, production of ethylene, and transpiration. Increase in storage temperature generally increases respiration rates in fresh produce, whereas certain crops, such as bananas and lemons, are also susceptible to physiological damage at low temperatures. This necessitates precise temperature control and gas-permeable monitoring during distribution and storage, which has prompted the exploration of intelligent packaging systems [20, 123, 124]. In recent times, the development and utilization of intelligent packaging incorporated with nanomaterials such as nanocomposites for packaging films, represent a promising approach. These nanocomposites can act as real-time temperature, color, and pH sensors that alert consumers regarding the conditions and safety of the food product [123]. The color-changing indicators, which are responsive to specific gases, moisture, pH alteration,

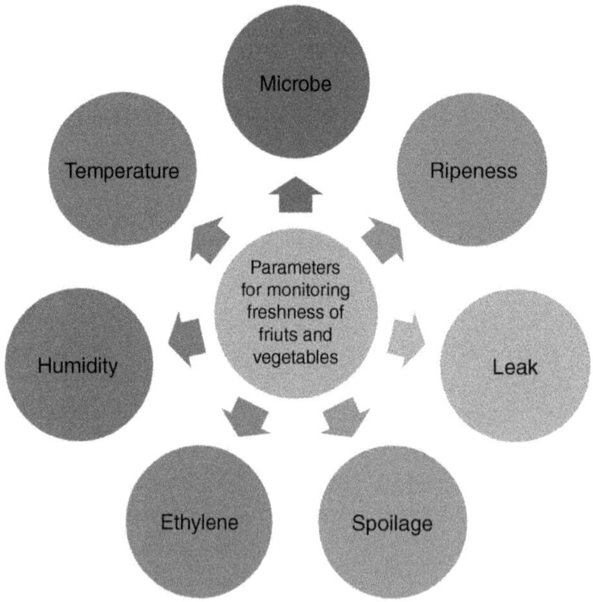

Figure 15.5 Parameters for monitoring freshness of fruits and vegetables.

change in color, temperature change, and microbial contamination have been used to detect and communicate to consumers about the condition of packaged food (Figure 15.5). In a study, a pH sensor-based intelligent packaging system was developed by incorporating polyvinyl alcohol and red cabbage extract (pigment) to serve as a pH biosensor for fresh date fruit, which effectively indicated pH change in the range of 2–12, functioning as a reliable pH biosensor [125]. To evaluate the freshness of bananas, an intelligent film was developed using Brazilian sappan wood extract in combination with polyvinyl alcohol and glucomannan, in which a yellow hue was observed on day 0 (pH 4.5) before it changed to a dark red hue on day 6 (pH 5.1) because of the release of ethylene, water vapor, and carbon dioxide [126]. By incorporation of anthocyanin-loaded ovalbumin–carboxymethyl cellulose nanocomplexes into corn starch/polyvinyl alcohol matrix, an intelligent colorimetric film was prepared to monitor mushroom freshness, and the result showed that the films were sensitive to CO_2 and pH buffer, and change color from purple to pink [127]. Additionally, Chen et al. created a color-changing indicator label to track the freshness of freshly-cut green bell peppers using pH-sensitive indicators [128].

15.5.2 Dairy Products

Dairy products, with their rich nutritional quality, create an environment favorable for the growth and propagation of both spoilage as well as pathogenic microorganisms resulting in heightened risk for quality deterioration and food safety concerns [129]. The limited shelf life and complex handling processes of milk and milk products present challenges associated with microbial and chemical degradation throughout their storage and distribution chains. As a result, it is imperative to establish a real-time, cost-effective monitoring system to ensure that these products meet the highest levels of quality. The pH and acidity of such foods are the most reliable indicators of spoilage, and therefore, attention has shifted to innovative intelligent packaging systems that can provide real-time information about the packaged product [86, 130, 131]. Research in the field of intelligent packaging has focused extensively on pH-based freshness indicators to develop bio-nanocomposites-based intelligent packaging, as a low-cost sustainable solution [118]. A pH-sensitive colorimetric films or labels, were developed using grape skin extracts (GSE), tara gum (TG), and cellulose nanocrystals (CNC), and the developed films showed subtle green hue in alkaline pH compared to red hue in acidic pH. Furthermore, test on milk demonstrated noticeable color changes and suggested that the pH-sensing film can be used to provide information regarding the freshness of packaged milk [132]. Similarly, a freshness indicator was developed by adding anthocyanins from black carrots in starch matrix to detect the spoilage of pasteurized milk, and the developed label displayed a variety of color transformations in pH buffer solutions ranging from 2 to 11, particularly when conditions were mildly acidic, neutral, and mildly basic. During spoilage, changes in the label's color correlated strongly with physical and microbial changes in the milk, indicating that the indicator can differentiate between fresh and moderately fresh milk [118]. The color change of fresh pasteurized milk and spoiled milk was completely distinguished by another indicator fabricated from black carrot anthocyanin in cellulose–chitosan matrix [110].

A pH sensor made from bacterial cellulose and polyvinyl alcohol was developed by adding red cabbage (*Brassica oleracea* [*BO*]) anthocyanins, and the developed film monitored milk freshness by a distinctive color change from blue to pinkish gray after eight hours of storage at ambient temperature [133]. In another study, blueberry anthocyanins, gelatin, and Fe^{2+} were incorporated into a zein matrix using the electrospinning method to produce colorimetric indicator films that indicated freshness of milk by visually discernible color changes from purple-black (fresh milk) to violet-red (spoiled milk), which was strongly correlated with pH/acidity levels of the milk during storage [134]. Zhai et al., crafted composite films incorporating gelatin, gellan gum, and red radish anthocyanins extract that exhibited changes in color as pH changed from 2 to 12, particularly in the context of milk and fish spoilage [135]. The TTI employing PVA/Chitosan polymeric matrix infused with anthocyanins was demonstrated through an activation test conducted on pasteurized milk, which showed observable alterations in the color of the film/label upon changes in the chemical composition of the milk [58]. Some of the other recent developments in bio-nanocomposite-based intelligent packaging systems and their applications for dairy product packaging are summarized in Table 15.3.

15.5.3 Muscle Foods

The spoilage of meat is primarily caused by microbial degradation of the protein due to the high-water activity, and lipid oxidation due to high-fat content, which leads to quality degradation in the meat resulting in consumer rejection and economic losses. The presence of off-odors is one of the most important indicators of meat spoilage, often due to the accumulation of volatile compounds in the headspace of packaging, particularly sulfur-containing compounds. Research efforts have recently concentrated on developing rapid

Table 15.3 Application of biopolymer-based intelligent packaging in dairy products.

Film formulation	Bioactive compound	Dairy product	Indicator type	Reference
PVA and cassava starch	Anthocyanin from chokeberry	Milk	pH indicator	[136]
Chitosan and PVA	Red cabbage anthocyanin	Pasteurized milk	Time–temperature indicator (TTI)	[58]
Chitosan	Anthocyanin from purple eggplant or black eggplant extract	Milk	pH indicator	[108]
Chitosan and cellulose	Anthocyanin obtained from black carrots	Pasteurized milk	pH indicator	[110]
Sodium alginate and konjac glucomannan	*Lycium ruthenicum* anthocyanin and tea polyphenols	Milk	pH indicator	[137]
Sodium alginate and gelatin	Anthocyanin from onion peel	Milk	pH indicator	[22]
Gelatin, sodium alginate, and chitosan	Nonpomace residue of grape juice powder-based anthocyanin	Milk	pH	[138]
Chitosan and konjac glucomannan	Elderberry anthocyanins and oregano essential oil	Cheese	pH	[139]
Potato starch and salicylate-loaded halloysite-layered double hydroxide filler	Anthocyanin from freeze-dried onion extract	Milk	pH	[140]
Starch	Anthocyanin extracted from black carrot	Milk	pH	[118]
PVA and cassava starch	Ethyl lauroyl arginate and mulberry anthocyanin	Milk	pH	[140]

methodologies for monitoring microbial breakdown in muscle foods and assessing their freshness in real-time using bio-nanocomposites in intelligent packaging [85, 141, 142]. For example, Shavisi and Shahbazi integrated anthocyanin extract from *Rosa damascene* into a nanofiber matrix of chitosan and gum Arabic to create an intelligent bio-nanocomposite film/label. This film provided visual hues from pale pink at pH 1, green at pH 9–10, and ultimately brown at pH 11–12, while monitoring the freshness of raw chicken fillet [39]. Blending barberry anthocyanin with methylcellulose/chitosan nanofiber composite film produced another pH-responsive color indicator film that was suitable for identifying changes in food pH by changing color from reddish-pink to yellow [143].

As a breakthrough to advance the industrial application of innovative eco-friendly packaging, another study explored the potential of a portable starch/butterfly pea extract-based intelligent material, which exhibited a noticeable shift in color, transitioning from red to blue-violet and ultimately to green-yellow as the pH increased from 2 to 13. When applied to monitor the freshness of chicken stored at room temperature, the film displayed a conspicuous color transformation, indicating the chicken's deterioration [144]. Another novel smart packaging film, possessing pH sensitivity along with antibacterial and antioxidant properties, was formulated using highly cross-linked zein along with tea tree essential oil-loaded mesoporous silica nanoparticles. Additionally, blueberry anthocyanin was introduced as a colorimetric indicator in the film, enabling monitoring of pork meat product safety via a mobile phone during the nine-day storage, and the color of the composite film transitioned from colorless to brown, indicating pork meat spoilage [45]. Several successful studies have been carried out recently concerning the pH indicator for different meat products [78, 145, 146]. Wagh et al. also explored the efficacy of CD-based intelligent packaging films having real-time sensing properties made from cellulose nanofiber, anthocyanins from BO, and CDs. Results showed real-time monitoring of the freshness of minced pork, fish, and shrimp, as evident through a distinct visual transition from red to colorless/light yellow during storage at 25 °C for 48 hours [147]. The colorimetric H_2S sensor employing gellan gum-capped silver nanoparticles for real-time monitoring of meat spoilage facilitated H_2S analysis with a limit of detection of 0.81 μM and demonstrated exceptional selectivity for H_2S over other volatile components produced during storage. Capitalizing on these attributes, the sensor exhibited visible color alterations from yellow to colorless, effectively detecting H_2S generation from chicken breast and silver carp within a packaging system [122].

Perishable foods, such as seafood, are particularly susceptible to spoilage during storage and transportation that necessitates advanced packaging solutions such as intelligent packaging. In such cases, the integration of nanomaterials into biocomposite has facilitated the creation of diverse sustainable sensing platforms capable of monitoring food quality [148, 149]. A novel intelligent film was developed through integration of 1%, 3%, and 5% BACNs into an oxidized-chitin nanocrystals/chitosan matrix, which showed shifts in color from red to green across a pH range of 2–12. Experimentation revealed that films with 3% BACNs effectively tracked fish and shrimp spoilage through observable changes in color [114]. Another research used tangerine peel and resazurin along with CDs to develop biocompatible food freshness indicators, which displayed pH-responsive transition from yellow to orange with increasing pH. The CDs demonstrated ammonia sensing capability with a detection limit of 0.84 μM, facilitated detection of biogenic amines (BAs) by changing hue from yellow to brown, and demonstrated the ability to monitor shrimp freshness [150]. A pH-sensitive film based on soluble soybean polysaccharide, grape skin anthocyanin extract, and graphene oxide was investigated for intelligent packaging of salmon fish, and result demonstrated distinguishable color variations in response to TVB-N release and pH changes [151]. There has been a substantial amount of research conducted recently on muscle foods to monitor their freshness over the storage period in relation to nanobiocomposites-based intelligent packaging (Table 15.4).

Table 15.4 Nanocomposite-based intelligent packaging for monitoring muscle foods.

Indicator	Biopolymer	Active agents	Food product	Function	Reference
pH-responsive indicator	Pectin	Curcumin and sulfur nanoparticles	Shrimp	Using a composite film for shrimp packaging resulted in a pH-sensitive color shift from yellow to orange, indicating changes in shrimp quality.	[152]
	Bacterial nanocellulose	Black carrot anthocyanins	Rainbow trout and common carp fillet	This pH-sensitive indicator demonstrated clear color changes indicating the fresh stage (carmine color), optimal consumption stage (pink color), and spoiled stage (blue color) of fish fillets.	[153]
	Gelatin and κ-carrageenan	Red barberry anthocyanins and TiO_2 nanoparticles	Fish	Adding TiO_2 nanoparticles and natural anthocyanins greatly enhanced the mechanical strength and moisture resistance of the films. Additionally, the films changed color as the freshness of the fish samples changed, a response correlated with the production of ammonia during fish degradation.	[41]
	Pullulan/chitosan/chitin nanofibers	Red cabbage extracts and nisin	Sea bass (*Lateolabrax japonicas*)	The color of the nanocomposite film shifts from red to blue as the pH increases from 2 to 12, allowing for the monitoring of sea bass freshness during storage.	[154]
	Oxidized-chitin nanocrystals (O-ChNCs)/gelatin	Black rice bran anthocyanins (BACNs)	Shrimp and hairtail	The O-ChNCs/gelatin/BACNs films were sensitive to pH changes, displaying notable color variations in various buffer solutions, which were utilized to monitor the freshness of shrimp and hairtail based on visible color changes.	[43]
	Polyvinyl alcohol, chitosan, xylan, and hydroxyapatite	Curcumin	Indian oil sardine (*Sardinellalongiceps*) fish	UV studies revealed that higher relative humidity favored a more pronounced color response. The colorimetric properties of the composite film exhibited visible and distinct color changes, providing a means to assess the real-time freshness of fish.	[155]

	Chitosan / oxidized-chitin nanocrystals	Red cabbage anthocyanins	Shrimp and hairtail	The intelligent films displayed noticeable color alterations in response to ammonia vapor and acidic/alkaline conditions in short time frames, easily visible to the naked eye, effectively distinguishing between three stages of product freshness (fresh, medium fresh, and spoiled).	[156]
	Carboxymethyl cellulose/gelatin	Ge-montmorillonite (OM) nanofiller, red cabbage anthocyanins (ATH), and *pistacia* leaves extract (PE)	Shrimp	Color-changing pH-sensitive films, enhanced with improved light, showed superior performance with PE/ATH/OM additives. These ternary indicator films demonstrated high physical and functional capabilities, enabling the detection and prevention of shrimp spoilage.	[157]
Time–temperature indicator (TTI)	-	Tyrosinase and tyrosine	Turbot sashimi	The potential of employing this TTI as a quality gauge for turbot sashimi was evaluated, with an activation energy of 0.409 kJ mol−1 and an indication time of 50 hours at 4 °C.	[158]
	-	Lactic acid bacteria	Vacuum-packed grouper fish fillets	In varying temperature conditions, the TTI response can exhibit an irreversible color transformation from red to yellow, effectively indicating three levels of fish freshness (very fresh, fresh, and spoilage).	[159]

15.6 Conclusion and Perspective

The development and utilization of biopolymer-based nanocomposites for intelligent food packaging represent a remarkable stride toward ensuring the safety, quality, and sustainability of the global food supply chain. The integration of biopolymers derived from renewable resources, with nanomaterials has ushered in a new era of intelligent packaging solutions. These nanocomposites offer multifaceted functionalities that contribute to the preservation and monitoring of various food products. Integrating bio-nanocomposites as biosensors in food packaging could revolutionize the industry, emphasizing the need to understand biosensors and bio-nanocomposites for sustainable and affordable intelligent packaging materials. This chapter discusses research findings and articles on intelligent packaging technology and its applications in packaging of perishable food products. Various commercial packaging systems incorporating intelligent concepts, such as freshness, temperature–time, gas, and radio frequency detectors, have recently been introduced to the food market. Researchers can delve deeper into the design of intelligent packaging systems that adapt to specific food types and evolving industry demands. Efforts should be directed toward optimizing the scalability and cost-effectiveness of the production processes for biopolymer-based nanocomposites to facilitate widespread adoption in the food packaging industry. Collaboration between academia, industry, and regulatory bodies is essential to establish standardized testing methods and safety guidelines for these innovative materials, ensuring their seamless integration into mainstream food packaging practices.

References

1 Ares, G., Velázquez, A.L., Vidal, L. et al. (2022). The role of food packaging on children's diet: insights for the design of comprehensive regulations to encourage healthier eating habits in childhood and beyond. *Food Quality and Preference* 95: 104366.

2 Upadhyay, A., Agbesi, P., Arafat, K.M.Y. et al. (2024). Argyropoulos, D Bio-based smart packaging: fundamentals and functions in sustainable food systems. *Trends in Food Science & Technology* 145: 145.

3 Gordon, A. and Williams, R. (2020). The role and importance of packaging and labeling in assuring food safety, quality and regulatory compliance of export products II: packaging & labeling considerations. In: *Food Safety and Quality Systems in Developing Countries* (ed. A. Gordon), 285–341. Elsevier.

4 Etxabide, A., Maté, J.I., and Kilmartin, P.A. (2021). Effect of curcumin, betanin and anthocyanin containing colourants addition on gelatin films properties for intelligent films development. *Food Hydrocolloids* 115: 106593.

5 Yan, M.R., Hsieh, S., and Ricacho, N. (2022). Innovative food packaging, food quality and safety, and consumer perspectives. *Processes* 10 (4): 747.

6 Basumatary, I.B., S. Kumar, A. Mitra et al., Edible film and coating on food products: progress in deposition methods in Comrehensive Materials Processing. S. Hasmi, Editor 2023. p. 244–261, Elsevier.

7 Nemat, B., Razzaghi, M., Bolton, K. et al. (2019). The role of food packaging design in consumer recycling behavior – a literature review. *Sustainability* 11 (16): 4350.

8 Aliabbasi, N., Faraji, B., Emam-Djomeh, Z. et al. (2022). Biopolymer-based active and intelligent packaging for food applications. In: *Biodegradable Polymer-Based Food Packaging* (ed. A.D. Tripathi, K.K. Darani, D.C. Rai, and V. Paul), 245–275. Springer.

9 Chia, M.R., Phang, S.-W., and Ahmad, I. (2023). Influence of polyaniline and cellulose nanocrystals on starch biopolymer film for intelligent food packaging. *Food Bioscience* 56: 103212.

10 Perera, K.Y., Jaiswal, A.K., and Jaiswal, S. (2023). Biopolymer-based sustainable food packaging materials: challenges, solutions, and applications. *Foods* 12 (12): 2422.

11 Singh, R., Gautam, S., Sharma, B. et al. (2021). Biopolymers and their classifications. In: *Biopolymers and their Industrial Applications* (ed. S. Thomas, S. Gopi, and A. Amalraj), 21–44. Elsevier.

12 Purkayastha, M.D. and Kumar, S. (2022). Protein-based films and coatings. In: *Biopolymer-Based Food Packaging* (ed. S. Kumar, A. Mukherjee, and J. Dutta), 178–224. Wiley.

13 Torres-Giner, S., K. J. Figueroa-Lopez, B. Melendez-Rodriguez, et al., Emerging trends in biopolymers for food packagingin *Sustainable Food Packaging Technology*. A. Athanassiou, Editor. 2021. Wiley. p. 101–33.

14 Majumder, R., Das, A., Mukherjee, A. et al. (2022). Biopolymers: The Chemistry of Food and Packaging. In: *Biopolymer-Based Food Packaging* (ed. S. Kumar, A. Mukherjee, and J. Dutta), 29–65. Wiley.

15 Ahmad, K., Li, Y., Tu, C. et al. (2024). Nanotechnology in food packaging with implications for sustainable outlook and safety concerns. *Food Bioscience* 58: 103625.

16 Raghuvanshi, S., Khan, H., Saroha, V. et al. (2023). Recent advances in biomacromolecule-based nanocomposite films for intelligent food packaging – a review. *International Journal of Biological Macromolecules* 253: 127420.

17 Sobhan, A., Muthukumarappan, K., and Wei, L. (2021). Biosensors and biopolymer-based nanocomposites for smart food packaging: challenges and opportunities. *Food Packaging and Shelf Life* 30: 100745.

18 Kumar, S., Mukherjee, A., Kalita, S. et al. (2022). Nanomaterials in food packaging. In: *Biopolymer-Based Food Packaging* (ed. S. Kumar, A. Mukherjee, and J. Dutta), 336–367. Wiley.

19 Ramos, Ó.L., Pereira, R.N., Cerqueira, M.A. et al. (2018). Bio-based nanocomposites for food packaging and their effect in food quality and safety. In: *Food packaging and preservation* (ed. A.M. Grumezescu and A.M. Holban), 271–306. Elsevier.

20 Pirsa, S., Sani, I.K., and Mirtalebi, S.S. (2022). Nano-biocomposite based color sensors: investigation of structure, function, and applications in intelligent food packaging. *Food Packaging and Shelf Life* 31: 100789.

21 Jayakumar, A., Radoor, S., Kim, J.T. et al. (2022). Recent innovations in bionanocomposites-based food packaging films – a comprehensive review. *Food Packaging and Shelf Life* 33: 100877.

22 Devi, D., Kumar, S., and Mukherjee, A. (2024). Onion peel extract as milk freshness indicator in biopolymer-based intelligent packaging films. *Food and Humanity* 2: 100223.

23 Salgado, P.R., Giorgio, L.D., Musso, Y.S. et al. (2021). Recent developments in smart food packaging focused on biobased and biodegradable polymers. *Frontiers in Sustainable Food Systems* 5: 630393.

24 Rodrigues, C., Souza, V.G.L., Coelhoso, I. et al. (2021). Bio-based sensors for smart food packaging – current applications and future trends. *Sensors* 21 (6): 2148.

25 Rahimah, S., Malinda, W., Zaida et al. (2020). Betacyanin as bioindicator using time-temperature integrator for smart packaging of fresh goat milk. *The Scientific World Journal* 2020: 1–9.

26 Azeredo, H.M. and Correa, D.S. (2021). Smart choices: mechanisms of intelligent food packaging. *Current Research in Food Science* 4: 932–936.

27 Vanderroost, M., Ragaert, P., Devlieghere, F. et al. (2014). Intelligent food packaging: the next generation. *Trends in Food Science & Technology* 39 (1): 47–62.

28 Chaudhary, V., Bangar, S.P., Thakur, N. et al. (2022). Recent advancements in smart biogenic packaging: reshaping the future of the food packaging industry. *Polymers* 14 (4): 829.

29 Chen, L., Guttieres, D., Levi, R. et al. (2021). Public health risks arising from food supply chains: challenges and opportunities. *Naval Research Logistics (NRL)* 68 (8): 1098–1112.

30 Basavegowda, N. and Baek, K.-H. (2021). Advances in functional biopolymer-based nanocomposites for active food packaging applications. *Polymers* 13 (23): 4198.

31 Jagadeesh, P., Puttegowda, M., Rangappa, S.M. et al. (2021). Influence of nanofillers on biodegradable composites: a comprehensive review. *Polymer Composites* 42 (11): 5691–5711.

32 Nesic, A.R. and Seslija, S.I. (2017). The influence of nanofillers on physical–chemical properties of polysaccharide-based film intended for food packaging. In: *Food Packaging* (ed. A.M. Grumezescu), 637–697. Elsevier.

33 Abdullah, Cai, J., Hafeez, M.A. et al. (2022). Biopolymer-based functional films for packaging applications: a review. *Frontiers in Nutrition* 9: 1000116.

34 Bakshi, P.S., Selvakumar, D., Kadirvelu, K. et al. (2020). Chitosan as an environment friendly biomaterial–a review on recent modifications and applications. *International Journal of Biological Macromolecules* 150: 1072–1083.

35 Shen, X., Zheng, Q., and Kim, J.-K. (2021). Rational design of two-dimensional nanofillers for polymer nanocomposites toward multifunctional applications. *Progress in Materials Science* 115: 100708.

36 Li, Y., Ying, Y., Zhou, Y. et al. (2019). A pH-indicating intelligent packaging composed of chitosan-purple potato extractions strength by surface-deacetylated chitin nanofibers. *International Journal of Biological Macromolecules* 127: 376–384.

37 Wu, C., Zhu, Y., Wu, T. et al. (2019). Enhanced functional properties of biopolymer film incorporated with curcurmin-loaded mesoporous silica nanoparticles for food packaging. *Food Chemistry* 288: 139–145.

38 Lan, W., Wang, S., Zhang, Z. et al. (2021). Development of red apple pomace extract/chitosan-based films reinforced by TiO_2 nanoparticles as a multifunctional packaging material. *International Journal of Biological Macromolecules* 168: 105–115.

39 Shavisi, N. and Shahbazi, Y. (2022). Chitosan-gum Arabic nanofiber mats encapsulated with pH-sensitive Rosa damascena anthocyanins for freshness monitoring of chicken fillets. *Food Packaging and Shelf Life* 32: 100827.

40 Gutiérrez, T.J. and Alvarez, V.A. (2018). Bionanocomposite films developed from corn starch and natural and modified nano-clays with or without added blueberry extract. *Food Hydrocolloids* 77: 407–420.

41 Sani, M.A., Tavassoli, M., Salim, S.A. et al. (2022). Development of green halochromic smart and active packaging materials: TiO_2 nanoparticle-and anthocyanin-loaded gelatin/κ-carrageenan films. *Food Hydrocolloids* 124: 107324.
42 Fu, B., Liu, Q., Liu, M. et al. (2022). Carbon dots enhanced gelatin/chitosan bio-nanocomposite packaging film for perishable foods. *Chinese Chemical Letters* 33 (10): 4577–4582.
43 Ge, Y., Li, Y., Bai, Y. et al. (2020). Intelligent gelatin/oxidized chitin nanocrystals nanocomposite films containing black rice bran anthocyanins for fish freshness monitorings. *International Journal of Biological Macromolecules* 155: 1296–1306.
44 Priyadharshee, M. and Preetha, R. (2023). Fabrication and characterization of gelatin-based nanocomposite edible film prepared from eggshell with anthocyanin as pH indicator to assure quality of food. *Journal of Food Science and Technology* 60 (4): 1389–1401.
45 Kong, J., Ge, X., Sun, Y. et al. (2023). Multi-functional pH-sensitive active and intelligent packaging based on highly cross-linked zein for the monitoring of pork freshness. *Food Chemistry* 404: 134754.
46 Amjadi, S., Nazari, M., Alizadeh, S.A. et al. (2020). Multifunctional betanin nanoliposomes-incorporated gelatin/chitosan nanofiber/ZnO nanoparticles nanocomposite film for fresh beef preservation. *Meat Science* 167: 108161.
47 Phan, K., Raes, K., Speybroeck, V.V. et al. (2021). Non-food applications of natural dyes extracted from agro-food residues: a critical review. *Journal of Cleaner Production* 301: 126920.
48 Azman, N., Khairul, W.M., and Sarbon, N. (2022). A comprehensive review on biocompatible film sensor containing natural extract: active/intelligent food packaging. *Food Control* 141: 109189.
49 Yu, D., Basumatary, I.B., Kumar, S. et al. (2023). Chitosan modified with bio-extract as an antibacterial coating with UV filtering feature. *International Journal of Biological Macromolecules* 230: 123145.
50 Parlak, M.E., Sahin, O.I., Dundar, A.N. et al. (2023). Natural colorant incorporated biopolymers-based pH-sensing films for indicating the food product quality and safety. *Food Chemistry* 439: 138160.
51 Ndwandwe, B.K., Malinga, S.P., Kayitesi, E. et al. (2024). Recent developments in the application of natural pigments as pH-sensitive food freshness indicators in biopolymer-based smart packaging: challenges and opportunities. *International Journal of Food Science & Technology* https://doi.org/10.1111/ijfs.16990.
52 Ribeiro, J.S. and Veloso, C.M. (2021). Microencapsulation of natural dyes with biopolymers for application in food: a review. *Food Hydrocolloids* 112: 106374.
53 Choi, I., Choi, H., Lee, J.S. et al. (2023). Novel color stability and colorimetry-enhanced intelligent CO_2 indicators by metal complexation of anthocyanins for monitoring chicken freshness. *Food Chemistry* 404: 134534.
54 Manzoor, M.F., Afraz, M.T., Yılmaz, B.B. et al. (2024). Recent progress in natural seaweed pigments: green extraction, health-promoting activities, techno-functional properties and role in intelligent food packaging. *Journal of Agriculture and Food Research* 15: 100991.
55 Duan, M., Yu, S., Sun, J. et al. (2021). Development and characterization of electrospun nanofibers based on pullulan/chitin nanofibers containing curcumin and anthocyanins

for active-intelligent food packaging. *International Journal of Biological Macromolecules* 187: 332–340.

56 Yao, Q.-B., Huang, F., Lu, Y. et al. (2024). Polysaccharide-based food packaging and intelligent packaging applications: a comprehensive review. *Trends in Food Science & Technology* 147: 104390.

57 Tsironi, T., Ronnow, P., Giannoglou, M. et al. (2017). Developing suitable smart TTI labels to match specific monitoring requirements: the case of *Vibrio* spp. growth during transportation of oysters. *Food Control* 73: 51–56.

58 Pereira, V.A. Jr., de Arruda, I.N.Q., and Stefani, R. (2015). Active chitosan/PVA films with anthocyanins from Brassica oleraceae (Red Cabbage) as time–temperature indicators for application in intelligent food packaging. *Food Hydrocolloids* 43: 180–188.

59 Chayavanich, K., Thiraphibundet, P., and Imyim, A. (2020). Biocompatible film sensors containing red radish extract for meat spoilage observation. *Spectrochimica Acta Part A: Molecular and Biomolecular Spectroscopy* 226: 117601.

60 Yousefi, H., Su, H.-M., Imani, S.M. et al. (2019). Intelligent food packaging: a review of smart sensing technologies for monitoring food quality. *ACS Sensors* 4 (4): 808–821.

61 Lim, M.-Y., Shin, H., Shin, D.M. et al. (2016). Poly (vinyl alcohol) nanocomposites containing reduced graphene oxide coated with tannic acid for humidity sensor. *Polymer* 84: 89–98.

62 Fernandez-Salmeron, J., Rivadeneyra, A., Rodríguez, M.A.C. et al. (2015). HF RFID tag as humidity sensor: two different approaches. *IEEE Sensors Journal* 15 (10): 5726–5733.

63 Deng, F., He, Y., Li, B. et al. (2018). Design of a slotted chipless RFID humidity sensor tag. *Sensors and Actuators B: Chemical* 264: 255–262.

64 Yang, J., Shen, M., Luo, Y. et al. (2021). Advanced applications of chitosan-based hydrogels: from biosensors to intelligent food packaging system. *Trends in Food Science & Technology* 110: 822–832.

65 Wu, S., Wang, W., Yan, K. et al. (2018). Electrochemical writing on edible polysaccharide films for intelligent food packaging. *Carbohydrate Polymers* 186: 236–242.

66 Li, R., Wang, S., Feng, H. et al. (2024). An intelligent chitosan/gelatin film via improving the anthocyanin-induced color recognition accuracy for beef sub-freshness differentiation monitoring. *Food Hydrocolloids* 146: 109219.

67 Lv, H., Wang, C., He, D. et al. (2024). Intelligent food tag: a starch-anthocyanin-based pH-sensitive electrospun nanofiber mat for real-time food freshness monitoring. *International Journal of Biological Macromolecules* 256: 128384.

68 Lei, Y., Yao, Q., Jin, Z. et al. (2023). Intelligent films based on pectin, sodium alginate, cellulose nanocrystals, and anthocyanins for monitoring food freshness. *Food Chemistry* 404: 134528.

69 Zheng, L., Liu, L., Yu, J. et al. (2023). Intelligent starch/chitosan-based film incorporated by anthocyanin-encapsulated amylopectin nanoparticles with high stability for food freshness monitoring. *Food Control* 151: 109798.

70 Zhang, Q., Lin, G., Wang, H. et al. (2024). Development of smart packaging film incorporated with sodium alginate-chitosan quaternary ammonium salt nanocomplexes encapsulating anthocyanins for monitoring milk freshness. *International Journal of Biological Macromolecules* 263: 130336.

71 Halloub, A., Raji, M., Essabir, H. et al. (2023). Stable smart packaging betalain-based from red prickly pear covalently linked into cellulose/alginate blend films. *International Journal of Biological Macromolecules* 234: 123764.

72 Kanatt, S.R. (2020). Development of active/intelligent food packaging film containing Amaranthus leaf extract for shelf life extension of chicken/fish during chilled storage. *Food Packaging and Shelf Life* 24: 100506.

73 Surendhiran, D., Roy, V.C., Ho, T.C. et al. (2023). Smart packaging film prepared from subcritical water-modified oat starch and betalain of beetroot extract reinforced with cellulose nanofibrils. *Sustainable Chemistry and Pharmacy* 36: 101349.

74 Hu, H., Yao, X., Qin, Y. et al. (2020). Development of multifunctional food packaging by incorporating betalains from vegetable amaranth (Amaranthus tricolor L.) into quaternary ammonium chitosan/fish gelatin blend films. *International Journal of Biological Macromolecules* 159: 675–684.

75 Qin, Y., Liu, Y., Zhang, X. et al. (2020). Development of active and intelligent packaging by incorporating betalains from red pitaya (Hylocereus polyrhizus) peel into starch/polyvinyl alcohol films. *Food Hydrocolloids* 100: 105410.

76 Li, N., Yang, X., and Lin, D. (2022). Development of bacterial cellulose nanofibers/konjac glucomannan-based intelligent films loaded with curcumin for the fresh-keeping and freshness monitoring of fresh beef. *Food Packaging and Shelf Life* 34: 100989.

77 Duan, M., Sun, J., Huang, Y. et al. (2023). Electrospun gelatin/chitosan nanofibers containing curcumin for multifunctional food packaging. *Food Science and Human Wellness* 12 (2): 614–621.

78 Yildiz, E., Sumnu, G., and Kahyaoglu, L.N. (2021). Monitoring freshness of chicken breast by using natural halochromic curcumin loaded chitosan/PEO nanofibers as an intelligent package. *International Journal of Biological Macromolecules* 170: 437–446.

79 Ma, Q., Du, L., and Wang, L. (2017). Tara gum/polyvinyl alcohol-based colorimetric NH3 indicator films incorporating curcumin for intelligent packaging. *Sensors and Actuators B: Chemical* 244: 759–766.

80 Tang, C., Zhao, Z., Yang, M. et al. (2022). Preparation and characterization of sodium cellulose sulfate/chitosan composite films loaded with curcumin for monitoring pork freshness. *Current Research in Food Science* 5: 1475–1483.

81 Spence, C. (2023). On the manipulation, and meaning (s), of color in food: a historical perspective. *Journal of Food Science* 88 (S1): A5–A20.

82 Kim, J., Spence, M.T., and Marshall, R. (2018). The color of choice: the influence of presenting product information in color on the compromise effect. *Journal of Retailing* 94 (2): 167–185.

83 Mead, J.A. and Richerson, R. (2018). Package color saturation and food healthfulness perceptions. *Journal of Business Research* 82: 10–18.

84 Bhatlawande, A.R., Ghatge, P.U., Shinde, G.U. et al. (2024). Unlocking the future of smart food packaging: biosensors, IoT, and nano materials. *Food Science and Biotechnology* 33 (5): 1075–1091.

85 Luo, X., Zaitoon, A., and Lim, L.T. (2022). A review on colorimetric indicators for monitoring product freshness in intelligent food packaging: indicator dyes, preparation methods, and applications. *Comprehensive Reviews in Food Science and Food Safety* 21 (3): 2489–2519.

86 Alizadeh, S., Pirsa, S., and Amiri, S. (2023). Development of a colorimetric sensor based on nanofiber cellulose film modified with ninhydrin to measure the formalin index of fruit juice. *International Journal of Biological Macromolecules* 253: 127035.

87 Escher, A., Megahd, H., Tavella, C. et al. (2023). Colorimetric polymer sensors for smart packaging. *Macromolecular Chemistry and Physics* 224 (14): 2300022.
88 Elhadef, K. et al. (2024). pH-sensitive films based on carboxymethyl cellulose/date pits anthocyanins: a convenient colorimetric indicator for beef meat freshness tracking. *Food Bioscience* 57: 103508.
89 Jamróz, E., Chaari, M., Akermi, S. et al. (2019). The verification of intelligent properties of furcellaran films with plant extracts on the stored fresh Atlantic mackerel during storage at 2 C. *Food Hydrocolloids* 97: 105211.
90 Zeng, F., Ye, Y., Liu, J. et al. (2023). Intelligent pH indicator composite film based on pectin/chitosan incorporated with black rice anthocyanins for meat freshness monitoring. *Food Chemistry: X* 17: 100531.
91 Yun, D., Cai, H., Liu, Y. et al. (2019). Development of active and intelligent films based on cassava starch and Chinese bayberry (Myrica rubra Sieb. et Zucc.) anthocyanins. *RSC Advances* 9 (53): 30905–30916.
92 Khezerlou, A., Tavassoli, M., Sani, M.A. et al. (2023). Smart packaging for food spoilage assessment based on Hibiscus sabdariffa L. anthocyanin-loaded chitosan films. *Journal Of Composites Science* 7 (10): 404.
93 Otálora González, C.M., Schelegueda, L.I., Ruiz-Henestrosa, V.M.P. et al. (2022). Cassava starch films with anthocyanins and betalains from agroindustrial by-products: their use for intelligent label development. *Foods* 11 (21): 3361.
94 Qin, Y., Yun, D., Xu, F. et al. (2021). Smart packaging films based on starch/polyvinyl alcohol and Lycium ruthenicum anthocyanins-loaded nano-complexes: functionality, stability and application. *Food Hydrocolloids* 119: 106850.
95 Ezati, P. and Rhim, J.-W. (2020). pH-responsive chitosan-based film incorporated with alizarin for intelligent packaging applications. *Food Hydrocolloids* 102: 105629.
96 Naghdi, S., Rezaei, M., and Abdollahi, M. (2021). A starch-based pH-sensing and ammonia detector film containing betacyanin of paperflower for application in intelligent packaging of fish. *International Journal of Biological Macromolecules* 191: 161–170.
97 Yao, X., Qin, Y., Zhang, M. et al. (2021). Development of active and smart packaging films based on starch, polyvinyl alcohol and betacyanins from different plant sources. *International Journal of Biological Macromolecules* 183: 358–368.
98 Wu, Y., Tang, P., Quan, S. et al. (2021). Preparation, characterization and application of smart packaging films based on locust bean gum/polyvinyl alcohol blend and betacyanins from cockscomb (Celosia cristata L.) flower. *International Journal of Biological Macromolecules* 191: 679–688.
99 Yao, X., Liu, J., Hu, H. et al. (2022). Development and comparison of different polysaccharide/PVA-based active/intelligent packaging films containing red pitaya betacyanins. *Food Hydrocolloids* 124: 107305.
100 Jiang, H., Zhang, W., Pu, Y. et al. (2023). Development and characterization of a novel active and intelligent film based on pectin and betacyanins from peel waste of pitaya (Hylocereus undatus). *Food Chemistry* 404: 134444.
101 Taghinia, P., Abdolshahi, A., Sedaghati, S. et al. (2021). Smart edible films based on mucilage of lallemantia iberica seed incorporated with curcumin for freshness monitoring. *Food Science & Nutrition* 9 (2): 1222–1231.

102 Yang, X., Zhang, M., Zhao, B. et al. (2024). Development, characteristics and biological activities of bilayer composite film based on gelatin/κ-carrageenan loading curcumin and ε-polylysine. *Food Hydrocolloids* 150: 109646.

103 Duan, A., Yang, J., Wu, L. et al. (2022). Preparation, physicochemical and application evaluation of raspberry anthocyanin and curcumin based on chitosan/starch/gelatin film. *International Journal of Biological Macromolecules* 220: 147–158.

104 Liu, J., Wang, H., Wang, P. et al. (2018). Films based on κ-carrageenan incorporated with curcumin for freshness monitoring. *Food Hydrocolloids* 83: 134–142.

105 Versino, F., Ortega, F., Monroy, Y. et al. (2023). Sustainable and bio-based food packaging: a review on past and current design innovations. *Foods* 12 (5): 1057.

106 Pirsa, S., Karimi Sani, I., and Khodayvandi, S. (2018). Design and fabrication of starch-nano clay composite films loaded with methyl orange and bromocresol green for determination of spoilage in milk package. *Polymers for Advanced Technologies* 29 (11): 2750–2758.

107 Ma, Q., Liang, T., Cao, L. et al. (2018). Intelligent poly (vinyl alcohol)-chitosan nanoparticles-mulberry extracts films capable of monitoring pH variations. *International Journal of Biological Macromolecules* 108: 576–584.

108 Yong, H., Wang, X., Zhang, X. et al. (2019). Effects of anthocyanin-rich purple and black eggplant extracts on the physical, antioxidant and pH-sensitive properties of chitosan film. *Food Hydrocolloids* 94: 93–104.

109 Koosha, M. and Hamedi, S. (2019). Intelligent Chitosan/PVA nanocomposite films containing black carrot anthocyanin and bentonite nanoclays with improved mechanical, thermal and antibacterial properties. *Progress in Organic Coatings* 127: 338–347.

110 Tirtashi, F.E., Moradi, M., Tajik, H. et al. (2019). Cellulose/chitosan pH-responsive indicator incorporated with carrot anthocyanins for intelligent food packaging. *International Journal of Biological Macromolecules* 136: 920–926.

111 Yang, Z., Zhai, X., Zou, X. et al. (2021). Bilayer pH-sensitive colorimetric films with light-blocking ability and electrochemical writing property: application in monitoring crucian spoilage in smart packaging. *Food Chemistry* 336: 127634.

112 Roy, S., Kim, H.-J., and Rhim, J.-W. (2021). Effect of blended colorants of anthocyanin and shikonin on carboxymethyl cellulose/agar-based smart packaging film. *International Journal of Biological Macromolecules* 183: 305–315.

113 Zhao, Y., Du, J., Zhou, H. et al. (2022). Biodegradable intelligent film for food preservation and real-time visual detection of food freshness. *Food Hydrocolloids* 129: 107665.

114 Wu, C., Sun, J., Zheng, P. et al. (2019). Preparation of an intelligent film based on chitosan/oxidized chitin nanocrystals incorporating black rice bran anthocyanins for seafood spoilage monitoring. *Carbohydrate Polymers* 222: 115006.

115 Wang, X., Yong, H., Gao, L. et al. (2019). Preparation and characterization of antioxidant and pH-sensitive films based on chitosan and black soybean seed coat extract. *Food Hydrocolloids* 89: 56–66.

116 Qin, Y., Liu, Y., Yong, H. et al. (2019). Preparation and characterization of active and intelligent packaging films based on cassava starch and anthocyanins from Lycium ruthenicum Murr. *International Journal of Biological Macromolecules* 134: 80–90.

117 Uranga, J., Etxabide, A., Guerrero, P. et al. (2018). Development of active fish gelatin films with anthocyanins by compression molding. *Food Hydrocolloids* 84: 313–320.

118 Goodarzi, M.M., Moradi, M., Tajik, H. et al. (2020). Development of an easy-to-use colorimetric pH label with starch and carrot anthocyanins for milk shelf life assessment. *International Journal of Biological Macromolecules* 153: 240–247.

119 Pirsa, S. and Asadi, S. (2021). Innovative smart and biodegradable packaging for margarine based on a nano composite polylactic acid/lycopene film. *Food Additives & Contaminants: Part A* 38 (5): 856–869.

120 Medina-Jaramillo, C., Ochoa-Yepes, O., Bernal, C. et al. (2017). Active and smart biodegradable packaging based on starch and natural extracts. *Carbohydrate Polymers* 176: 187–194.

121 Liu, J., Zhuang, Y., Hu, Y. et al. (2020). Improving the color stability and antioxidation activity of blueberry anthocyanins by enzymatic acylation with p-coumaric acid and caffeic acid. *LWT* 130: 109673.

122 Zhai, X., Li, Z., Shi, J. et al. (2019). A colorimetric hydrogen sulfide sensor based on gellan gum-silver nanoparticles bionanocomposite for monitoring of meat spoilage in intelligent packaging. *Food Chemistry* 290: 135–143.

123 Xing, Y., Li, W., Wang, Q. et al. (2019). Antimicrobial nanoparticles incorporated in edible coatings and films for the preservation of fruits and vegetables. *Molecules* 24 (9): 1695.

124 He, X., Pu, Y., Chen, L. et al. (2023). A comprehensive review of intelligent packaging for fruits and vegetables: target responders, classification, applications, and future challenges. *Comprehensive Reviews in Food Science and Food Safety* 22 (2): 842–881.

125 Maftoonazad, N. and Ramaswamy, H. (2019). Design and testing of an electrospun nanofiber mat as a pH biosensor and monitor the pH associated quality in fresh date fruit (Rutab). *Polymer Testing* 75: 76–84.

126 Kurnianto, M., Kurnianto, M.F., Poerwanto, B. et al. (2020). Monitoring of banana deteriorations using intelligent-packaging containing brazilien extract (Caesalpina sappan L.). In: *IOP Conference Series: Earth and Environmental Science*, 012043. IOP Publishing.

127 Liu, L., Wu, W., Zheng, L. et al. (2022). Intelligent packaging films incorporated with anthocyanins-loaded ovalbumin-carboxymethyl cellulose nanocomplexes for food freshness monitoring. *Food Chemistry* 387: 132908.

128 Chen, H.-Z., Zhang, M., Bhandari, B. et al. (2018). Applicability of a colorimetric indicator label for monitoring freshness of fresh-cut green bell pepper. *Postharvest Biology and Technology* 140: 85–92.

129 Mirza Alizadeh, A., Masoomian, M., Shakooie, M. et al. (2021). Trends and applications of intelligent packaging in dairy products: a review. *Critical Reviews in Food Science and Nutrition* 62 (2): 383–397.

130 Sohail, M., Sun, D.-W., and Zhu, Z. (2018). Recent developments in intelligent packaging for enhancing food quality and safety. *Critical Reviews in Food Science and Nutrition* 58 (15): 2650–2662.

131 Biegańska, M. (2022). Packaging of Dairy Products: Emerging Strategies. In: *Food Packaging: The Smarter Way* (ed. A.K. Shukla), 127–164. Springer.

132 Ma, Q. and Wang, L. (2016). Preparation of a visual pH-sensing film based on tara gum incorporating cellulose and extracts from grape skins. *Sensors and Actuators B: Chemical* 235: 401–407.

133 Kuswandi, B., Asih, N.P.N., Pratoko, D.K. et al. (2020). Edible pH sensor based on immobilized red cabbage anthocyanins into bacterial cellulose membrane for intelligent food packaging. *Packaging Technology and Science* 33 (8): 321–332.

134 Gao, R., Hu, H., Shi, T. et al. (2022). Incorporation of gelatin and $Fe2^+$ increases the pH-sensitivity of zein-anthocyanin complex films used for milk spoilage detection. *Current Research in Food Science* 5: 677–686.

135 Zhai, X., Li, Z., Zhang, J. et al. (2018). Natural biomaterial-based edible and pH-sensitive films combined with electrochemical writing for intelligent food packaging. *Journal of Agricultural and Food Chemistry* 66 (48): 12836–12846.

136 Wu, X., Yan, X., Zhang, J. et al. (2024). Preparation and characterization of pH-sensitive intelligent packaging films based on cassava starch/polyvinyl alcohol matrices containing Aronia melanocarpa anthocyanins. *LWT* 194: 115818.

137 Wang, S., Li, R., Han, M. et al. (2023). Intelligent active films of sodium alginate and konjac glucomannan mixed by Lycium ruthenicum anthocyanins and tea polyphenols for milk preservation and freshness monitoring. *International Journal of Biological Macromolecules* 253: 126674.

138 de Azevedo, E.S. and Noreña, C.P.Z. (2024). Anthocyanin-based indicators design by polyelectrolyte complexation: a study on structural and thermodynamic properties, and application for milk freshness assessment. *Food Hydrocolloids* 147: 109389.

139 Zhang, D., Cao, G., Bu, N. et al. (2023). Multi-functional konjac glucomannan/chitosan bilayer films reinforced with oregano essential oil loaded β-cyclodextrin and anthocyanins for cheese preservation. *International Journal of Biological Macromolecules* 244: 125365.

140 Li, N., Zhou, Z., Wu, F. et al. (2022). Development of pH-indicative and antimicrobial films based on polyvinyl alcohol/starch incorporated with ethyl lauroyl arginate and mulberry anthocyanin for active packaging. *Coatings* 12 (10): 1392.

141 Khodaei, S.M., Gholami-Ahangaran, M., Sani, I.K. et al. (2023). Application of intelligent packaging for meat products: a systematic review. *Veterinary Medicine and Science* 9 (1): 481–493.

142 Ahmed, I., Lin, H., Zou, L. et al. (2018). An overview of smart packaging technologies for monitoring safety and quality of meat and meat products. *Packaging Technology and Science* 31 (7): 449–471.

143 Alizadeh-Sani, M., Tavassoli, M., Mohammadian, E. et al. (2021). pH-responsive color indicator films based on methylcellulose/chitosan nanofiber and barberry anthocyanins for real-time monitoring of meat freshness. *International Journal of Biological Macromolecules* 166: 741–750.

144 Wang, L., Yang, C., Deng, X. et al. (2023). A pH-sensitive intelligent packaging film harnessing Dioscorea zingiberensis starch and anthocyanin for meat freshness monitoring. *International Journal of Biological Macromolecules* 245: 125485.

145 Huang, X., Xu, F., Yun, D. et al. (2023). Development and application of intelligent packaging films based on guar gum, polyvinyl alcohol and hyacinth bean (Lablab purpureus (L.) sweet) anthocyanins. *International Journal of Biological Macromolecules* 251: 126369.

146 Fathi, M., Babaei, A., and Rostami, H. (2022). Development and characterization of locust bean gum-Viola anthocyanin-graphene oxide ternary nanocomposite as an efficient pH indicator for food packaging application. *Food Packaging and Shelf Life* 34: 100934.

147 Wagh, R.V., Khan, A., Priyadarshi, R. et al. (2023). Cellulose nanofiber-based multifunctional films integrated with carbon dots and anthocyanins from Brassica oleracea for active and intelligent food packaging applications. *International Journal of Biological Macromolecules* 233: 123567.

148 Mustafa, F. and Andreescu, S. (2020). Nanotechnology-based approaches for food sensing and packaging applications. *RSC Advances* 10 (33): 19309–19336.

149 Idumah, C., Mohamad, Z., Ogbu, J. et al. (2019). A review on innovations in polymeric nanocomposite packaging materials and electrical sensors for food and agriculture. *Composite Interfaces* 27: 1–72.

150 Ezati, P., Khan, A., Rhim, J.-W. et al. (2023). pH-Responsive strips integrated with resazurin and carbon dots for monitoring shrimp freshness. *Colloids and Surfaces B: Biointerfaces* 221: 113013.

151 Kafashan, A., Joze-Majidi, H., Kazemi-Pasarvi, S. et al. (2023). Nanocomposites of soluble soybean polysaccharides with grape skin anthocyanins and graphene oxide as an efficient halochromic smart packaging. *Sustainable Materials and Technologies* 38: e00755.

152 Ezati, P. and Rhim, J.-W. (2020). pH-responsive pectin-based multifunctional films incorporated with curcumin and sulfur nanoparticles. *Carbohydrate Polymers* 230: 115638.

153 Moradi, M., Tajik, H., Almasi, H. et al. (2019). A novel pH-sensing indicator based on bacterial cellulose nanofibers and black carrot anthocyanins for monitoring fish freshness. *Carbohydrate Polymers* 222: 115030.

154 Wu, C., Jiang, H., Zhao, J. et al. (2022). A novel strategy to formulate edible active-intelligent packaging films for achieving dynamic visualization of product freshness. *Food Hydrocolloids* 133: 107998.

155 Vadivel, M., Sankarganesh, M., Raja, J.D. et al. (2019). Bioactive constituents and bio-waste derived chitosan/xylan based biodegradable hybrid nanocomposite for sensitive detection of fish freshness. *Food Packaging and Shelf Life* 22: 100384.

156 Chen, M., Yan, T., Huang, J. et al. (2021). Fabrication of halochromic smart films by immobilizing red cabbage anthocyanins into chitosan/oxidized-chitin nanocrystals composites for real-time hairtail and shrimp freshness monitoring. *International Journal of Biological Macromolecules* 179: 90–100.

157 Sadi, A. and Ferfera-Harrar, H. (2023). Cross-linked CMC/Gelatin bio-nanocomposite films with organoclay, red cabbage anthocyanins and pistacia leaves extract as active intelligent food packaging: colorimetric pH indication, antimicrobial/antioxidant properties, and shrimp spoilage tests. *International Journal of Biological Macromolecules* 242: 124964.

158 Xu, F., Ge, L., Li, Z. et al. (2017). Development and application of a tyrosinase-based time-temperature indicator (TTI) for determining the quality of turbot sashimi. *Journal of Ocean University of China* 16: 847–854.

159 Hsiao, H.I. and Chang, J.N. (2017). Developing a microbial time–temperature indicator to monitor total volatile basic nitrogen change in chilled vacuum-packed grouper fillets. *Journal of Food Processing and Preservation* 41 (5): e13158.

16

Life Cycle Assessment of Smart Food Packaging Systems

Sweety Kalita[1], Debarshi Nath[2,3], Ehsan Pesaranhajiabbas[2,3], Manjusri Misra[2,3], Amar K. Mohanty[2,3], Avik Mukherjee[1], and Santosh Kumar[1]

[1] *Department of Food Engineering and Technology, Central Institute of Technology Kokrajhar, Kokrajhar, Assam, India*
[2] *Department of Plant Agriculture, Bioproducts Discovery and Development Center, Crop Science Building, University of Guelph, Guelph, Ontario, Canada*
[3] *School of Engineering, Thornbrough Building, University of Guelph, Guelph, Ontario, Canada*

16.1 Introduction

A sustainable food chain ensures that minimal food goes to waste, food packaging is the most effective and extensively used technology to mitigate food loss. Synthetic plastic packaging has revolutionized food packaging since the first half of the twentieth century; however, in the context of environmental and consumer health concerns, extensive research and development efforts have been dedicated to developing biodegradable alternatives. It has been estimated that synthetic plastics were responsible for almost 2 gigaton CO_2-eq of greenhouse gas emission in 2015, and it is expected to rise more than threefold to 6.5 gigaton CO_2-eq by 2050 [1]. Life cycle assessment (LCA) of a packaging material is an effective technique to quantify the hazardous substance emission, the amount of energy consumed, and the volume of greenhouse gas emissions throughout the life cycle of plastic packaging [2]. It can also be useful to compare the environmental sustainability of various synthetic plastics as well as biopolymer-based food packaging materials throughout their entire lifespan, i.e. from raw material extraction to end-of-life (EoL). In simplest terms, the LCA of a packaging material is the principal measurement of the "environmental footprint" of the packaging material that includes CO_2 released into the atmosphere, toxic substances leached into aquatic environments, and other such impacts. Globally, these LCA methods are regulated by ISO 14040 and are used to evaluate the impacts of packaging production, the EoL assessment, the impact(s) of packaging material, and their release into the natural environment. However, traditional LCA methods did not allow measurement of the impacts of plastic exposure on living organisms because the effects of synthetic plastic wastes on living organisms in various habitats such as soil, freshwater, seas, and oceans may arise after several years or even decades of exposure. In recent years, the application of mathematical

Smart Food Packaging Systems: Innovations and Technology Applications, First Edition.
Edited by Avik Mukherjee, Santosh Kumar, Manjusri Misra, and Amar K. Mohanty.
© 2025 John Wiley & Sons Ltd. Published 2025 by John Wiley & Sons Ltd.

modeling in LCA analysis has opened new avenues to estimate impact(s) of food packaging wastes on the environment and ecosystem(s), e.g. climate change, ozone depletion, scarcity of drinking water, and so on. In 2019, the Marine Impacts on LCA (MarILCA) project was launched to assess the impacts of microplastic litter on the life cycle impact assessment (LCIA) of marine habitats [3]. After a comprehensive gap analysis, MarILCA has published characterization factors such as size, shape, and polymer type for LCIA calculations. Among the impacts assessed in LCIA, human toxicity and ecotoxicity are traditionally included, whereas physical effects on biota, such as entanglement, smothering, and any other type of external or internal injuries, have been added subsequently in the calculation.

As the world attempts to transition from a linear to a circular economy, research and technology developments are crucial along with the necessary synergy between the stakeholders for achieving the common goal of circularity in food packaging systems (Figure 16.1). To achieve social sustainability in the LCA, the social life cycle assessment (S-LCA) approach has been developed [5]. The United Nations Environment Program (UNEP) has developed guidelines to use S-LCA to improve social conditions and the overall socio-economic performance of a product (including food packaging) throughout its life cycle for all of its stakeholders [6]. Numerous studies have used a LCA of plastic used for food packaging to evaluate direct effects on the environment and circularity [7]. Also, the effects of production and EoL stages of trays made of polystyrene (PS), polypropylene (PP), and polylactic acid (PLA) on the environment and circularity are also studied [8]. Only a few studies have reported S-LCA assessment of food packaging wastes, in developing countries, such analyses are even more challenging due to informal collection, segregation, and/or weak management [6]. This chapter discusses the important contributions of

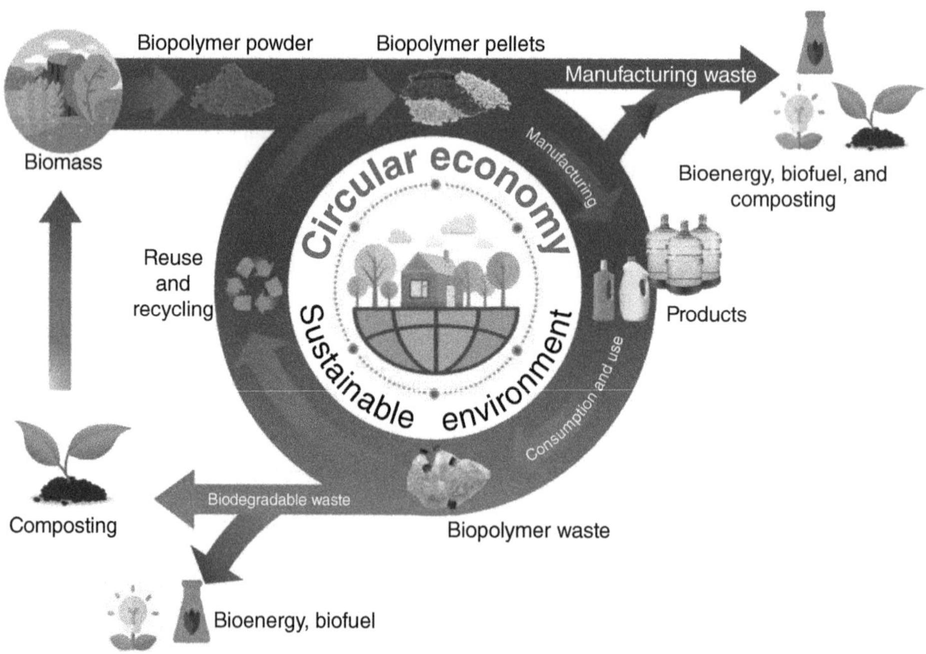

Figure 16.1 Circular economy for bio-polymeric composites manufacturing. *Source:* Reproduced from Shanmugam et al. [4]/with permission of Elsevier.

bio-nanocomposites as sustainable food packaging materials towards achieving a circular economy framework. Additionally, this chapter also details the significance of LCA analyses to quantify the impacts of applications of these biodegradable materials on the environment.

16.2 Biodegradability in the Circular Economy Framework

Several forms of packaging are currently available, ranging from passive packaging to more sophisticated options such as smart, active, interactive, ergonomic, and intelligent packaging [9]. Considering these factors, packaging has become an integral part of corporate business strategies, not only from a marketing perspective but also as an opportunity to rethink an organization's ability to attain successful performance and to transition to a circular economy from a linear economy. Circular economy perspectives tend to place a greater emphasis on the features and functions of packaging; and in the practical realm of business, companies are embracing the circular economy model and leveraging advanced technologies to maximize its advantages [10, 11]. A leading multinational corporation with a diverse portfolio of personal care and food brands, such as Dove and Lipton (Unilever), strives to ensure that all of its plastic packaging is reusable, compostable, or recyclable by 2025 [12], while Nestle targets 100% recyclability or reusability of its packaging by the same year [13]. In recent times, conventional materials used in food packaging, such as standard synthetic plastics and untreated or plastic-coated paper, have been substituted with innovative alternatives like bioplastics, natural fiber-reinforced composites, and barrier papers with natural coatings [14–16]. These replacements have resulted in reduced ecological footprints and helped diminish dependence on fossil fuels, microplastic pollution, and greenhouse gas emissions. In this framework, the primary benefit of utilizing renewable materials for food packaging lies in their potential to enhance the circular economy by cleansing waste streams, redirecting food waste away from landfills, and providing mechanically nonrecyclable packaging options that complement the mechanosphere cycle with the biosphere [14, 17].

Strategies such as reduce, reuse, and recycle, also known as the "3Rs," embody circular approaches by enhancing product design and utilization, prolonging the utility of products and components, and efficient utilization of materials. These approaches constitute vital parts of the resource and disposal hierarchy and are intimately related to the life cycle of a product [18, 19]. The Circular Design Guide, created by the Ellen MacArthur Foundation (EMF), listed six essential circular design strategies, out of which two of these are crucial for food packaging; selecting materials wisely by considering how they can be reused, composted, or recycled at the end of their life; and, implementing closed-loop systems that recover materials from old products for recycling or reuse in new ones [14, 19]. Implementing the smart material selection strategy in the pre-production stage of food packaging is pivotal, which affects the packaging integration into circular material flows by directly influencing EoL choices regarding treatment. Consequently, there is a growing global demand for biopolymer-based packaging or bioplastic solutions due to their biodegradability and environment-friendly nature [14]. Bioplastics can undergo degradation via three primary mechanisms: microbial breakdown, aerobic degradation, and anaerobic decomposition [20]. A biodegradable microbial polymer, known as polyhydroxyalkanoates (PHA), was discovered to be an effective alternative to petroleum-based synthetic polymers responsible for negative effects on the environment. In contrast to other bioplastics, such as poly(ε-caprolactone) (PCL) and PLA, PHAs are fully biodegradable in both natural and marine environments [21]. García-Depraect et al. compared the biodegradability of poly(3-hydroxybutyrate) (PHB) and poly(3-hydroxybutyrate-co-3-hydroxyvalerate) (PHBV), which are common types of PHAs, along with PCL. Under standard conditions, PHB and PHBV biodegraded anaerobically (PHB: 83.9%, PHBV: 81.2%) in 77 days or aerobically (PHB: 83.0%, PHBV: 87.4%) in 117 days. In contrast, PCL only aerobically biodegraded (77.6%) in over 177 days. This study emphasizes the importance of analyzing biodegradation kinetics and carbon fate to improve the development and testing of biodegradable materials and waste treatments in a circular economy context [22].

Various techniques for evaluating the biological breakdown of bioplastics during aerobic digestion and anaerobic metabolizing were covered by Ruggero et al., who pointed out that while international criteria are followed, experimental parameters such as test time, composition of waste, and humidity, frequently differ between investigations [23]. Poly(butylene adipate-co-terephthalate) (PBAT) is also a bioplastic that can be used in flexible packaging, and despite being completely biodegradable and compostable, PBAT is not fully aligned with the principles of a circular economy due to its dependence on fossil fuels [24]. PLA on the other hand, as a biodegradable polymer, consumes two-thirds less energy during manufacturing than conventional polymers, it also produces no net increase in CO_2 during biodegradation, it emits fewer greenhouse gases when degrading in landfills, and it reduces greenhouse gas emissions by 25%; therefore, PLA can be considered as one of the most suitable candidates to replace conventional plastics [25, 26]. According to a study, PLA films (0–3% by weight) incorporated with nanostructure silver, titania, and graphene were tested to determine their preservation activity when wrapped around curd cheese. Consequently, the used composite-PLA sheets were subjected to another round of curd cheese wrapping to determine their reusability after being washed and disinfected. In the case of 0.5% and 3% PLA, the authors concluded that the PLA film could be reused with reproducibility of 100% and 85%, respectively [27]. Furthermore, polyesters break down in the environment either by hydrolyzing ester bonds (hydrolytic degradation) or by an enzymatic process by microbes (biodegradation) [28]. Thus, bioplastics and biopolymer-based products and materials are sustainable alternatives to petroleum-based synthetic plastics. Consequently, policymakers, local government officials, practitioners, and academics have increased their efforts and emphasis on bio-based materials for smart packaging (SP) applications [29].

16.3 Sustainability of Food Packaging: Rationales, Importance, and Approaches

As the global population continues to grow, there is an escalating demand for food necessitating a proportional increase in food preservation and packaging technologies as well as materials. Modern food packaging plays a pivotal role in enhancing human life quality by preserving the integrity of food products. Its primary function is to protect foodstuffs from spoilage induced by environmental factors such as oxygen, humidity, and light, while also serving as a conduit for essential information to consumers [30, 31]. Presently, the predominant materials in food packaging are derived from petroleum-based plastics, notably polyethylene (PE), PP, and PS. These polymers offer a spectrum of advantages for food packaging applications, including thermal stability, mechanical resilience, effective barrier properties against gases, durability, lightweight, and cost-effectiveness. Their versatility empowers manufacturers to design and provide the market with efficient packaging solutions to meet diverse market demands effectively. Despite all the advantages that petroleum-based polymer packaging materials have brought for food packaging, there are some serious concerns about the broad application of these polymers in the packaging industry. These materials are mainly dependent on nonrenewable fossil fuel resources. Another concern regarding the application of these polymers relates to their adverse effect on our planet and its natural resources. The production process of fossil-based polymers is one of the main contributors to greenhouse gases released into the atmosphere resulting in accelerated global warming [32]. On the other hand, such polymers are highly stable chemically and can remain in the environment for hundreds of years after being disposed. Considering the concise lifespan from manufacturing to disposal for polymers utilized in the food packaging sector, this means that a huge amount of polymer wastes is generated and discarded everyday. Currently, from all this waste generated, only a small portion can be recycled and returned to the application. Figure 16.2 indicates the global production of plastics and their fate from 1950 to 2015. As it can be observed, only a small share of these polymers were recycled, while most of the plastic wastes were discarded or incinerated [33]. Such polymer wastes gradually decompose into micro-size pieces, nano-size particles, and diffuse into the human food chain resulting in health problems. The use of sustainable polymers, such as

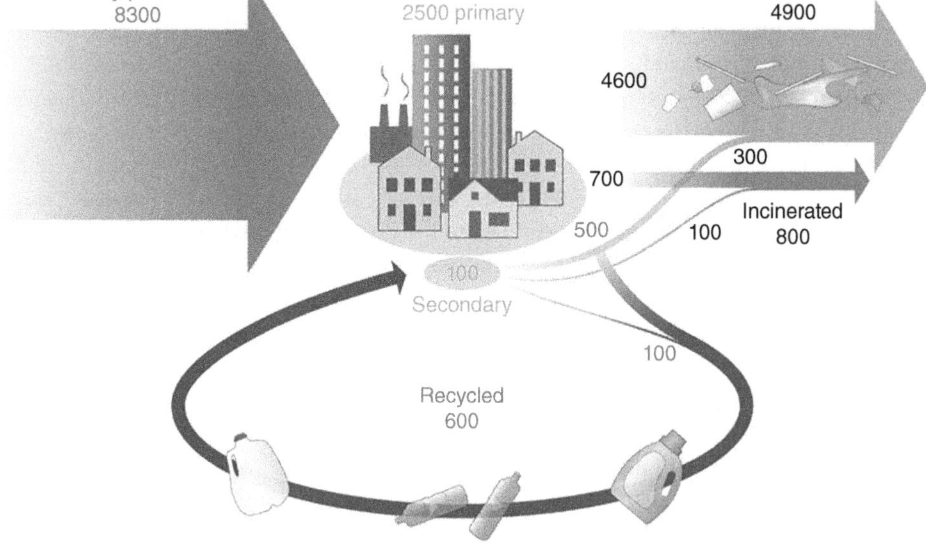

Figure 16.2 The global production of plastics and their fate from 1950 to 2015. *Source:* Reproduced from Geyer et al. [33]/with permission of American Association for the Advancement of Science.

biopolymers, can be a promising solution. Also, several pieces of legislation passed by governments and authorities in support of such sustainable materials around the world can support this cause. The different approaches that can be taken to achieve long-term sustainability in food packaging are discussed in this section.

16.3.1 Approaches to Achieve Sustainability in Packaging

In response to these concerns, there has been a significant movement toward sustainable solutions in food packaging, and both academia and industry are intensifying efforts to develop sustainable and eco-friendly alternatives for food packaging. These practices are aimed at mitigating the issues of landfill waste generation, water usage, CO_2 emissions, greenhouse gas emissions, and depleting fossil resources while protecting human health from harmful additives [34]. Moreover, the reduction in fossil raw materials for packaging and the growing shortage of energy resources over the past century has resulted in the exploration of sustainable materials and production techniques [35].

16.3.2 Use of Biodegradable Materials

Sustainable polymers, derived from bio-based resources or recycled materials, offering promising solutions to petroleum-based synthetic non-degradable plastics [36]. These materials exhibit similar functions to traditional plastics while enabling more environmentally responsible EoL scenarios, such as recycling or biodegradation [37]. Agro-waste-derived bioplastics have gained popularity as a sustainable and nonpolluting substitute for nonbiodegradable plastics. Therefore, "bioplastic" refers to plastics that (i) are biodegradable (ii) are derived from renewable resources, or (iii) possess both characteristics. As such, bioplastics are a very diverse family of materials rather than a single type of polymer [38, 39]. Bioplastics are blended with other biopolymers, natural fillers, and nanoparticles to develop sustainable packaging materials having properties like most commercial products [40]. The bioplastic-based composites and nanocomposites can be used for a variety of food packaging applications ranging from flexible packaging and food wraps to one-time takeaway containers [41, 42]. These products obtained from bioplastic can be termed as sustainable since they are considered to be biodegradable under specific conditions which lead to lesser waste generation. The term "biodegradable" refers to the chemical breakdown of materials into simple compounds such as CO_2, water, and methane with the help of microorganisms. Nowadays, home compostable products are also gaining popularity since they can degrade under ambient conditions and do not require a specific biodegradation facility. The term "compostability" here refers to the aerobic degradation of organic matter into compost (humus-like product), water, CO_2, and heat. Therefore, to achieve sustainability in packaging, bioplastic-based packaging products need to demonstrate biodegradability [38, 43].

Packaging materials can also be developed from biopolymers that are directly derived from plant or animal biomass. These biopolymers (such as carbohydrates, protein, and lipids) can be mixed with bioplastic to get the desired features for packaging materials. Edible films are developed from food-grade biopolymers with film-forming ability. Occasionally, the edible films and coating are used synonymously where the food surface is covered with relatively thin layers [44, 45]. Sometimes, plasticizers and other food-grade additives (such as antimicrobial agents) are mixed with the film-forming solutions to enhance the functionality of the films such as antioxidant and antimicrobial properties. The rationality for using edible films in food packaging is its protective function by preventing moisture loss and maintaining the original organoleptic properties of packaged food. Moreover, edible films with antimicrobial additives can delay microbial spoilage significantly thereby increasing the shelf life of the packaged product [46, 47]. This will help reduce food spoilage, which is a global concern due to higher food insecurity in present times. Traditional plastic food packaging may not be immediately replaced by edible films, but they can lead the way towards the commercialization of sustainable biopolymers in the long run as an alternative to plastic packaging [48].

16.3.2.1 Circular Economy and Recycling

Circular economy is a closed-loop economic system that aims for zero waste and pollution across the entire life cycle of materials and is considered a pathway towards a sustainable future [37]. The circular economy concept emphasizes recycling plastic waste in order to reduce the ecological effects of waste, where packaging waste is the main contributor. In recent times there has been an exponential growth of packaging waste resulting in several calls from the United Nations and European Union (EU) to mitigate this problem [49]. The European Commission mandated a 55% recycling rate for waste plastic packaging by 2030. This approach will trigger increased recycling of plastic packaging also specifically for food, as this sector poses the largest fraction of all plastic packaging in the EU. In the same way, China and the USA encourage greater recycling rates to promote the circular economy [32]. Closed-loop recycling, often referred to as direct or primary recycling, involves collecting postconsumer plastic and physically reprocessing it for reuse. Open-loop recycling, which is also called secondary recycling, involves the recycling of plastic packaging wastes as materials for applications other than food and beverage packaging. Chemical recovery/recycling and biodegradable plastic composting are included in tertiary recycling, whereas energy recovery (incineration) is included in quaternary recycling [50–52]. Chemical recycling is one of the suitable technologies for achieving a circular economy to effectively recycle large quantities of plastic waste. Chemical recycling diverts the requirement for fossil fuels by transforming waste plastic packaging into new chemical products. Thus, chemical recycling will help reduce greenhouse gas emissions and the demand for fossil resources. Moreover, this process helps obtain chemical compounds that are chemically similar to the original product but a slight reduction in the performance might take place, which is also known as downcycling [32]. Similarly, mechanical recycling is a method of transforming waste plastic into new goods without significantly changing the chemical structure of the material, except for adding new additives. It is crucial to the circular economy and may be classified as either primary or secondary recycling [37]. Although recycling contributes largely towards a sustainable system, it also has challenges. The product derived from plastic waste needs to be analyzed on a case-to-case basis because of the specific product requirements. Also, recycled products might contain hazardous components making them unsuitable for packaging applications [49].

16.3.2.2 Reducing and Lightweighting

Minimizing the usage of packaging materials and other resources in food product manufacture is crucial. That does not suggest we should get rid of packing altogether; on the contrary, it is obvious that it plays a crucial role in maintaining the quality of food products and that ready-to-eat food products would be inconceivable without it. Packaging needs to be designed in such a way as to utilize the least number of materials required to provide the appropriate protection. The concept of lightweight packaging, known as using lighter and/or thinner materials for packaging, lowers resource requirements and can be used for sustainable packaging goals [53]. The problem of overpackaging, or the excessive use of packing materials, is severe but frequently ignored. Several studies show that significant thickness reductions are possible without sacrificing package performance, leading to substantial economic and environmental advantages. Using composites and reducing the thickness of the materials can be looked upon as a viable way toward lightweighting which can significantly improve the sustainability of food packaging. Nevertheless, to ensure eco-friendliness and higher recyclability in EoL scenarios, it is suggested wherever possible to use mono-materials over composite multilayer materials [53].

16.3.2.3 Regulatory Initiatives and Role of Customers

Regulations and policies play a crucial part in influencing and directing the adoption of environmentally conscious practices in the food packaging industry to develop a sustainable manufacturing model for packaging systems. This regulation can promote sustainability by promoting innovation in packaging and setting standards for manufacturers. The Sustainable Packaging Coalition (SPC) is a voluntary industry association headquartered in the United States. Its primary objective is "to collectively strengthen and advance the business case for more sustainable packaging" [7]. Walmart published the Sustainable Packaging Playbook (SPP) to advise dealers about ways to boost their Sustainability Index Score by enhancing the packaging of their products. The SPP suggests the use of LCA to evaluate material carbon footprint, greenhouse gas emissions, and water consumption [26]. The Australian Packaging Covenant Organisation requires all enterprises to follow the Sustainable Packaging Guidelines (SPG), which is governed by the National Environment Protection Measure 2011 [7, 54]. As per the SPG, sustainability in packaging is resource efficient, developed from low-carbon footprint material, and can be reused or recycled at the EoL phase. To promote a circular economy, the updated Directive 94/62/EC on Packaging and Packaging Waste [55] prohibits packaging waste generation and promotes reuse and recycling. The recycling rates for various packaging materials are established by the regulation (Article 6) [7]. The EU, in its plastic strategy and Circular Economy Action Plan, addresses the issue of plastic food packaging. It advocates for shifting towards a circular economy [56]. Furthermore, the United Nations General Assembly established the 17 Sustainable Development Goals (SDGs) in 2015 with a view to reduce waste and pollution and promote sustainability by 2030. The Stockholm Convention (2004), Basel Convention (1989), and Rotterdam Convention (2004) involve the management and safe disposal of hazardous materials related to the disposal of plastics [57, 58].

Sustainable packaging can only help the environment if buyers are also committed to buying eco-friendly items, rather than relying just on their biodegradable characteristics to attract consumers. The desires of consumers have a significant impact on promoting biodegradable materials to act as an alternative to prevalent nonbiodegradable materials. Moreover, innovation in packaging is determined by the requirements of the consumers and global trends such as sustainability, convenience, and product safety [59]. Moreover, environmental knowledge and the realization of the magnitude of problems involved with nonbiodegradable packaging must be instilled among the general population, which can motivate them to purchase green products [60, 61]. This suggests that the ecological behavior of consumers may be positively influenced by environmental awareness. Persuading customers to switch from conventionally packaged items to sustainably packaged ones is a major hurdle on the path to substantial ecological advantages, although in recent times consumers are heading in this direction. Additionally, improved communication from businesses that are prepared to assume social responsibility might motivate customers to choose and act in ways that are more environmental friendly and sustainable [61].

16.4 Case Studies on Life Cycle Assessment of Smart Packaging Systems

Conventional food packaging acts as an inert shield intended to prolong the adverse impacts of the external surroundings on the enclosed food item. However, the rising focus on reducing food wastage and the carbon footprint of products, coupled with the escalating demands of consumers for enhanced food quality and safety, indicates that conventional packaging (CP) is inadequate [62, 63]. SP offers a novel packaging solution that enables distributors and sellers to minimize food wastage while also meeting the current preference of consumers for higher-quality foods [64, 65]. This pioneering packaging system integrates the principles of intelligent and active packaging by having the ability to detect alterations in both the product and its surrounding environment and subsequently manage and respond to these alterations [66, 67]. In recent years, significant advancements have occurred in this field, primarily centered on the utilization of biodegradable packaging materials, bio-derived and ingestible sensors and coatings, and natural antimicrobials with enhanced and consistent characteristics (Figure 16.3) [68–71]. Smart food packaging must meet regulatory requirements and consider various sustainability factors, such as waste reduction and recycling, overall environmental benefit, and economic viability [72]. LCA identifies environmental issues and suggests improvements by considering all stages of packaging, from production to disposal for enhanced environmental sustainability [73, 74]. LCA has two main approaches: attribution methods, which detail the environmental attributes of a current life cycle and its subsystems, and consequential methods, which predict the outcomes of changes in established processes [75].

A variety of impact assessment methods are employed worldwide, such as Tool for the Reduction and Assessment of Chemical and Other Environmental Impacts (TRACI) in the United States and Ecoindicator; ReCiPe Endpoint (ReCiPe) in Europe, alongside the widely adopted Center of Environmental Science (CML) method. This method gauges a product's environmental impact across diverse categories such as eutrophication, ionizing radiation, aquatic ecotoxicity, land use, and human toxicity [76]. A study using ISO 14040 series, ReCiPe methodology, and economic evaluation compared the environmental impact of SP to CP across production, use phase, and end-of-life phases (EoLP) for both packaging and

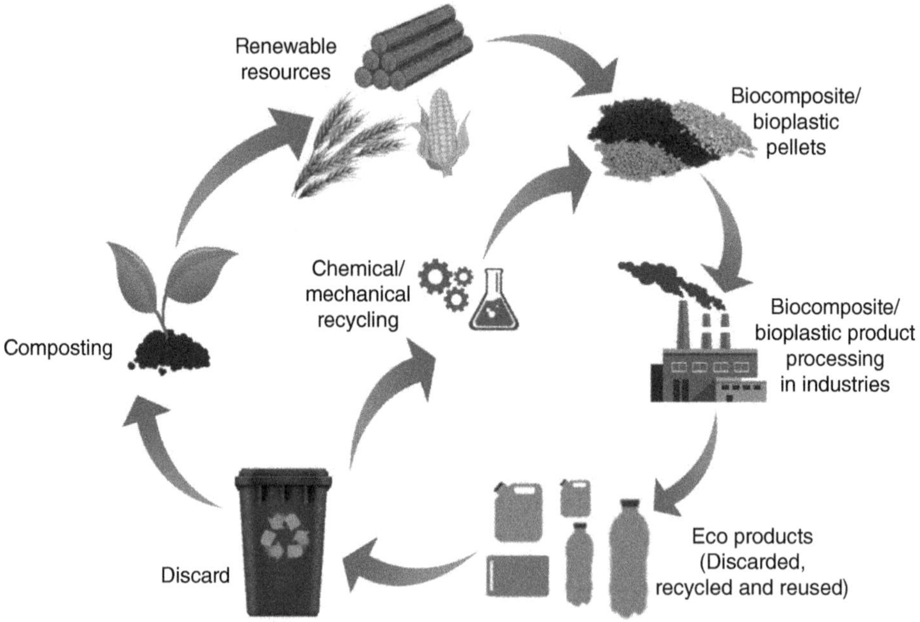

Figure 16.3 Recyclability of bio-nanocomposites. *Source:* Reproduced from Shanmugam et al. [4]/ with permission of Elsevier.

packaged food products. The LCA showed that SP had a 1.6 times higher environmental footprint in the climate change category compared to CP. However, when factoring in the usage and EoLP of both packaging systems and their contained food product, SP performed better environmentally. In the optimistic scenario of SP producing 5% food waste, SP outperformed CP in 14 out of 18 total impact categories. In the realistic scenario (10% food waste) and the conservative scenario (20% food waste), SP surpassed CP in 11 and 7 impact categories, respectively [77]. Using LCA with ReCiPe indicators, Vigil et al. discovered that active packages with active agents surpassed traditional packaging in extending shelf life and reducing waste of fresh-cut produce. They assessed biodegradable PLA and nonbiodegradable polypropylene packages, both coated with zinc oxide nanoparticles (ZnONPs), as new technologies. Two LCA methods were used, including an attributional approach comparing materials in production and EoLP, and a consequential approach considering the entire fresh produce life cycle, focusing on produce loss implications. In both cases, scenarios with PLA and PP active packages with ZnONPs exhibited superior profiles to traditional methods, particularly in the consequential LCA for the supply chain of fresh-cut produce [78].

16.5 Conclusion

SP integrates active and intelligent packaging to prolong the shelf life of food products, reduce food waste, and communicate real-time information on product quality during the supply chain. This chapter presents a preliminary conceptual framework for assessing the environmental impacts associated with food packaging, drawing upon principles from LCA and sustainability theories. The implementation of a circular design strategy based on the selection of smart material during the pre-production phase has a significant impact on the final phase of the packaging life cycle, the disposal phase. As the disposal phase of food packaging approaches, closed-loop design strategies emerge as a potent tool for reducing packaging waste and enhancing resource recovery. Utilizing bio-based plastics in food packaging can be an effective method of closing the loop, whether it is through reusing or recycling. In accordance with the 3R circular strategy and the resource management flowchart, the goal is to conserve energy and resources, while reducing waste production and preventing littering by prioritizing reuse over recycling. In a nutshell, SP offers a sustainable solution that combines the advantages of preserving food quality and extending shelf life, all while being environmentally friendly and economically viable. This realization is poised to impact both industry trends, leading to increased production of SP systems for various applications, and consumer attitudes, as they become more aware of the positive impacts of SP across its life cycle.

References

1 Boone, L., Préat, N., Nhu, T.T. et al. (2023). Environmental performance of plastic food packaging: life cycle assessment extended with costs on marine ecosystem services. *Science of the Total Environment* 894: 164781.
2 Vignali, G. (2016). Life-cycle assessment of food-packaging systems. In: *Environmental Footprints of Packaging. Environmental Footprints and Eco-design of Products and Processes* (ed. S.S. Muthu). Singapore: Springer.
3 Corella-Puertas, E., Hajjar, C., Lavoie, J. et al. (2023). MarILCA characterization factors for microplastic impacts in life cycle assessment: physical effects on biota from emissions to aquatic environments. *Journal of Cleaner Production* 418: 138197.
4 Shanmugam, V., Mensah, R.A., Försth, M. et al. (2021). Circular economy in biocomposite development: state-of-the-art, challenges and emerging trends. *Composites Part C: Open Access* 5: 100138.
5 Reinales, D., Zambrana-Vasquez, D., and Saez-De-Guinoa, A. (2020). Social life cycle assessment of product value chains under a circular economy approach: a case study in the plastic packaging sector. *Sustainability* 12: 6671.
6 Yıldız-Geyhan, E., Altun-Çiftçioğlu, G.A., and Kadırgan, M.A.N. (2017). Social life cycle assessment of different packaging waste collection system. *Resources, Conservation and Recycling* 124: 1–12.
7 Pauer, E., Wohner, B., Heinrich, V. et al. (2019). Assessing the environmental sustainability of food packaging: an extended life cycle assessment including packaging-related food losses and waste and circularity assessment. *Sustainability* 11: 925.
8 Maga, D., Hiebel, M., and Aryan, V. (2019). A comparative life cycle assessment of meat trays made of various packaging materials. *Sustainability* 11: 5324.
9 Rundh, B. (2016). The role of packaging within marketing and value creation. *British Food Journal* 118: 2491–2511.
10 Sumrin, S., Gupta, S., Asaad, Y. et al. (2021). Eco-innovation for environment and waste prevention. *Journal of Business Research* 122: 627–639.

11 Meherishi, L., Narayana, S.A., and Ranjani, K.S. (2019). Sustainable packaging for supply chain management in the circular economy: a review. *Journal of Cleaner Production* 237: 117582.

12 McMullen, C. (2017). Unilever calls for a circular economy in the consumer goods industry. https://www.waste360.com/plastics/unilever-calls-for-a-circular-economy-in-the-consumer-goods-industry (accessed March 2024).

13 Meier, C. (2018). Nestlé aiming at 100% recyclable or reusable packaging by 2025. Nestlé good food, good life. https://www.nestle.com/media/pressreleases/allpressreleases/nestle-recyclable-reusable-packaging-by-2025 (accessed March 2024).

14 Markevičiūtė, Z. and Varžinskas, V. (2022). Smart material choice: the importance of circular design strategy applications for bio-based food packaging preproduction and end-of-life life cycle stages. *Sustainability* 14: 6366.

15 Basumatary, I.B. et al. (2023). Edible film and coating on food products: progress in deposition methods in *Comprehensive Materials Processing*, S. Hashmi, Editor. 2023. p. 244–261, Elsevier

16 Basumatary, I.B. et al. (2022). Edible films and coatings. In: *Biopolymer-Based Food Packaging: Innovations and Technology Applications* (ed. S. Kumar, A. Mukherjee, and J. Dutta), 445–475. Wiley.

17 Spierling, S. et al. (2018). Bio-based plastics – a review of environmental, social and economic impact assessments. *Journal of Cleaner Production* 185: 476–491.

18 Blomsma, F. et al. (2018). Exploring circular strategy combinations – towards understanding the role of PSS. *Procedia CIRP* 69: 752–757.

19 Nunes, B. et al. (2018). University contributions to the circular economy: professing the hidden curriculum. *Sustainability* 10: 2719.

20 Chong, J.W.R., K.S. Khoo, G.Y. Yew, et al., Advances in production of bioplastics by microalgae using food waste hydrolysate and wastewater: a review. *Bioresource Technology*, 2021. 342: p. 125947.

21 Bugnicourt, E. et al. (2014). Polyhydroxyalkanoate (PHA): review of synthesis, characteristics, processing and potential applications in packaging. *eXPRESS Polymer Letters* 8: 791–808.

22 Garcia-Depraect, O. et al. (2022). Biodegradation of bioplastics under aerobic and anaerobic aqueous conditions: kinetics, carbon fate and particle size effect. *Bioresource Technology* 344: 126265.

23 Ruggero, F., Gori, R., and Lubello, C. (2019). Methodologies to assess biodegradation of bioplastics during aerobic composting and anaerobic digestion: a review. *Waste Management & Research* 37: 959–975.

24 Odegard, I. et al. (2017). Biobased plastics in a circular economy policy suggestions for biobased and biobased biodegradable plastics. *Ministry of Infrastructure and Environment* https://cedelft.eu/wp-content/uploads/sites/2/2017/10/CE_Delft_2J66_Bioplastics_in_a_circular_economy_DEF.pdf (accessed March 2024).

25 Elsawy, M.A., Kim, K.H., Park, J.W. et al. (2017). Hydrolytic degradation of polylactic acid (PLA) and its composites. *Renewable and Sustainable Energy Reviews* 79: 1346–1352.

26 Iwata, T. (2015). Biodegradable and bio-based polymers: future prospects of eco-friendly plastics. *Angewandte Chemie International Edition* 54: 3210–3215.

27 Peter, A., Cozmuta, L.M., Nicula, C. et al. (2021). Chemical and organoleptic changes of curd cheese stored in new and reused active packaging systems made of Ag-graphene-TiO(2)-PLA. *Food Chemistry* 363: 130341.

28 Rydz, J., Sikorska, W., Kyulavska, M. et al. (2015). Polyester-based (bio)degradable polymers as environmentally friendly materials for sustainable development. *International Journal of Molecular Sciences* 16: 564–596.

29 Moshood, T., Nawanir, G., Mahmud, F. et al. (2021). Expanding policy for biodegradable plastic products and market dynamics of bio-based plastics: challenges and opportunities. *Sustainability* 13: 6170.

30 Wu, F., Misra, M., and Mohanty, A. (2021). Challenges and new opportunities on barrier performance of biodegradable polymers for sustainable packaging. *Progress in Polymer Science* 117: 101395.

31 Chakraborty, P., Nath, D., Hoque, M. et al. (2022). Biopolymer-based antimicrobial coatings for aquatic food products: a review. *Journal of Food Processing and Preservation* 46: 16465.

32 Meys, R., Frick, F., Westhues, S. et al. (2020). Towards a circular economy for plastic packaging wastes – the environmental potential of chemical recycling. *Resources, Conservation and Recycling* 162: 105010.

33 Geyer, R., Jambeck, J., and Law, K. (2017). Production, use, and fate of all plastics ever made. *Science Advances* 3: e1700782.

34 Russell, D. (2014). Sustainable (food) packaging – an overview. *Food Additives & Contaminants. Part A, Chemistry, Analysis, Control, Exposure & Risk Assessment* 31: 396–401.

35 Oliveira, A. and Melo, N. (2019). Market and sustainability of food packaging – a review. *Boletim do Centro de Pesquisa de Processamento de Alimentos* 36: 1–10.

36 Mohanty, A., Vivekanandhan, S., Pin, J.M. et al. (2018). Composites from renewable and sustainable resources: challenges and innovations. *Science* 362: 536–542.

37 Mohanty, A., F. Wu, R. Mincheva, et al., *Sustainable polymers*. Nature Reviews Methods Primers, 2022. 2: p. 46.

38 Babaremu, K., Oladijo, O.P., and Akinlabi, E. (2023). Biopolymers: a suitable replacement for plastics in product packaging. *Advanced Industrial and Engineering Polymer Research* 6: 333–340.

39 Devi, L., Purkayastha, M.D., Mukherjee, A. et al. (2021). Biopolymer-based films and coatings: emerging technologies to extend shelf-life of fruits and vegetables. *Pray Rasayan* 3: 82–90.

40 Yu, D., Basumatary, I.B., Liu, Y. et al. (2024). Chitosan-photocatalyst nanocomposite on polyethylene films as antimicrobial coating for food packaging. *Progress in Organic Coatings* 186: 108069.

41 Nayak, A., Mukherjee, A., Kumar, S. et al. (2024). Exploring the potential of jujube seed powder in polysaccharide based functional film: characterization, properties and application in fruit preservation. *International Journal of Biological Macromolecules* 260: 129450.

42 Yu, D., Basumatary, I.B., Kumar, S. et al. (2023). Chitosan modified with bio-extract as an antibacterial coating with UV filtering feature. *International Journal of Biological Macromolecules* 230: 123145.

43 Samantaray, P., Little, A., Wemyss, A.M. et al. (2021). Design and control of compostability in synthetic biopolyesters. *ACS Sustainable Chemistry & Engineering* 9: 28.

44 Mellinas, C., Valdés, A., Ramos, M. et al. (2015). Active edible films: current state and future trends. *Journal of Applied Polymer Science* 133: 42631.

45 Devi, L.S., Mukherjee, A., Dutta, D. et al. (2023). Carnauba wax-based sustainable coatings for prolonging postharvest shelf-life of citrus fruits. *Sustainable Food Technology* 1 (3): 415–425.

46 Campos, C.A., Gerschenson, L.N., and Flores, S.K. (2011). Development of edible films and coatings with antimicrobial activity. *Food and Bioprocess Technology* 4: 849–875.

47 Purkayastha, M.D. and Kumar, S. (2022). Protein-based films and coatings. In: *Biopolymer-Based Food Packaging: Innovations and Technology Applications* (ed. S. Kumar, A. Mukherjee, and J. Dutta), 178–224. Wiley.

48 Kumar, L., Ramakanth, D., Akhila, K. et al. (2021). Edible films and coatings for food packaging applications: a review. *Environmental Chemistry Letters* 20: 26.

49 Ajwani-Ramchandani, R., Figueira, S., de Oliveira, R.T. et al. (2021). Towards a circular economy for packaging waste by using new technologies: the case of large multinationals in emerging economies. *Journal of Cleaner Production* 281: 125139.

50 Hopewell, J., Dvorak, R., and Kosior, E. (2009). Plastics recycling: challenges and opportunities. *Philosophical Transactions of the Royal Society B* 364: 2115–2126.

51 Vanderroost, M., Ragaert, P., Verwaeren, J. et al. (2017). The digitization of a food package's life cycle: Existing and emerging computer systems in the logistics and post-logistics phase. *Computers in Industry* 87: 15–30.

52 Sundqvist-Andberg, H. and Åkerman, M. (2021). Sustainability governance and contested plastic food packaging – An integrative review. *Journal of Cleaner Production* 306: 127111.

53 Licciardello, F. and Piergiovanni, L. (2020). Packaging and food sustainability. In: *The Interaction of Food Industry and Environment* (ed. C. Galanakis), 191–222. Elsevier Academic Press.

54 National Environment Protection Measures (NEPMs) (2011). National environment protection (used packaging materials) measure. https://www.nepc.gov.au/nepms/used-packaging (accessed March 2024).

55 Cakal, H.S. (2016). Mapping and analysing the current waste paper recycling system in Europe for a new fiber-based packaging material. In: *Strategies for Sustainable Development* (ed. A. Papageorgiou). Stockholm: KTH Royal Institute of Technology.

56 European Commission (2011). Communication from the Commission to the European Parliament, the Council, the European Economic and Social Committee and the Committee of the Regions Youth Opportunities Initiative. https://eur-lex.europa.eu/legal-content/en/TXT/?uri=CELEX%3A52011DC0933 (accessed March 2024).

57 Reichert, C.L., Bugnicourt, E., Coltelli, M.B. et al. (2020). Bio-based packaging: materials, modifications, industrial applications and sustainability. *Polymers (Basel)* 12: 1558.

58 Perera, K.Y., Jaiswal, A.K., and Jaiswal, S. (2023). Biopolymer-based sustainable food packaging materials: challenges, solutions, and applications. *Foods* 12: 2422.

59 Guine, R.P.F., Florença, S.G., Barroca, M.J. et al. (2020). The link between the consumer and the innovations in food product development. *Foods* 9: 1317.

60 Khan, M.S., Saengon, P., Alganad, A.M.N. et al. (2020). Consumer green behaviour: an approach towards environmental sustainability. *Sustainable Development* 28: 1168–1180.

61 Chirilli, C., Molino, M., and Torri, L. (2022). Consumers' awareness, behavior and expectations for food packaging environmental sustainability: influence of socio-demographic characteristics. *Foods* 11: 2388.

62 Drago, E., Campardelli, R., Pettinato, M. et al. (2020). Innovations in smart packaging concepts for food: an extensive review. *Foods* 9: 1628.

63 Schaefer, D. and Cheung, W.M. (2018). Smart packaging: opportunities and challenges. *Procedia CIRP* 72: 1022–1027.

64 Ganeson, K., Mouriya, G.K., Bhubalan, K. et al. (2023). Smart packaging – a pragmatic solution to approach sustainable food waste management. *Food Packaging and Shelf Life* 36: 101044.

65 Fernandez, C.M., Alves, J., Gaspar, P.D. et al. (2023). Innovative processes in smart packaging. A systematic review. *Journal of the Science of Food and Agriculture* 103: 986–1003.

66 Yu, Z., Boyarkina, V., Liao, Z. et al. (2023). Boosting food system sustainability through intelligent packaging: application of biodegradable freshness indicators. *ACS Food Science & Technology* 3: 199–212.

67 Roopa, H. et al. (2022). Active packaging in food industry. In: *Novel Technologies in Food Science* (ed. N. Chhikara, A. Panghal, and G. Chaudhary), 375–404. Wiley.

68 Sani, M.A., Azizi-Lalabadi, M., Tavassoli, M. et al. (2021). Recent advances in the development of smart and active biodegradable packaging materials. *Nanomaterials (Basel)* 11: 1331.

69 Guo, H., Shao, C., Ma, Y. et al. (2023). Development of active and intelligent pH food packaging composite films incorporated with litchi shell extract as an indicator. *International Journal of Biological Macromolecules* 226: 77–89.

70 Kong, J., Ge, X., Sun, Y. et al. (2023). Multi-functional pH-sensitive active and intelligent packaging based on highly cross-linked zein for the monitoring of pork freshness. *Food Chemistry* 404: 134754.

71 Wang, F., Xie, C., Tang, H. et al. (2023). Development, characterization and application of intelligent/active packaging of chitosan/chitin nanofibers films containing eggplant anthocyanins. *Food Hydrocolloids* 139: 108496.

72 Siracusa, V. and Lotti, N. (2019). Intelligent packaging to improve shelf life. In: *Food Quality and Shelf Life* (ed. C.A. Galanakis), 261–279. Academic Press, Elsevier.

73 Stramarkoua, M., Boukouvalas, C., Eleni, P. et al. (2021). Comparative life cycle assessment of polyethylene terephthalate (PET) and multilayer tetra pak juice packaging systems. *Chemical Engineering Transactions* 87: 103–108.

74 Del Borghi, A., Strazza, C., Magrassi, F. et al. (2018). Life cycle assessment for eco-design of product-package systems in the food industry – the case of legumes. *Sustainable Production and Consumption* 13: 24–36.

75 Schaubroeck, T., Schaubroeck, S., Heijungs, R. et al. (2021). Attributional & consequential life cycle assessment: definitions. Conceptual characteristics and modelling restrictions. *Sustainability* 13: 7386.

76 Dong, Y., Hossain, M.U., Li, H. et al. (2021). Developing conversion factors of LCIA methods for comparison of LCA results in the construction sector. *Sustainability* 13: 9016.

77 Stramarkou, M., Boukouvalas, C., Koskinakis, S.E. et al. (2022). Life cycle assessment and preliminary cost evaluation of a smart packaging system. *Sustainability* 14: 7080.

78 Vigil, M., Pedrosa-Laza, M., Cabal, J.V.A. et al. (2020). Sustainability analysis of active packaging for the fresh cut vegetable industry by means of attributional & consequential life cycle assessment. *Sustainability* 12: 7207.

Index

a

Absorbent pads 11, 12, 20
Absorbers 3, 4, 6–8, 10, 16, 18
Absorption 369, 370, 373, 375
Acetylmuramic acid 83
Acid compounds 201, 202
Activated carbon 173, 176
Active agents 1, 3, 11–13, 19, 20, 22, 23
Active compounds 1, 12, 20
Active materials 4, 22
Active packaging 1–3, 7, 11–13, 16, 18–23
Active packaging films 174
Additives 39–41, 51, 52, 54, 56
Aldehyde dehydrogenase 199, 201
Aldehydes 196, 197, 199, 201, 202, 205, 208, 210, 212
Aluminosilicates 155, 156
2-Amino-2-methyl-1-propanol 347
Anthocyanin 312, 318, 340, 350–352, 366–368, 373, 374, 376, 377, 391, 392, 394–400, 402
Antibacterial 43–46, 48, 49, 51–56
Anticancer 48, 52, 53
Anti-inflammatory 46–48, 51–53
Antimicrobial 224, 228, 231, 233, 235, 238, 240–243
Antimicrobial agents 11, 12, 18–20
Antioxidants 141, 144, 225, 229, 231, 233, 234
Ascorbic acid 5, 11, 15
Autocatalytic 149, 151, 153, 156
Avenanthramides 47, 48

b

Bactericidal 83, 84
Bacteriocins 49, 50, 55
Betalin 391, 392, 394–396
Bioactive substance 52, 61
Biochemical 149, 150, 152, 153
Biocide action 82, 85
Biodegradable 417, 418, 420, 421, 423
Bio-nanocomposites 91, 390–394, 399, 400, 404
Biopolymer 9, 14, 15, 18, 20, 219–221, 223, 225, 226, 231, 232, 289, 387, 388, 391, 395, 397, 404
Biopolymer-based film 8, 20
Biopreservative 83, 84
Biosynthesis 153
Blending 42, 43, 116–119, 122
Bromothymol blue and tetrabutylammonium ion-paired dye 347, 348

c

Calcium hydroxide 133, 135, 142, 145
Capsaicinoids 47, 48
Carbon dioxide 2, 10, 11, 18
Carbon dioxide indicator 306–311, 313, 315, 317, 320, 322, 324–326
Carbotec TR–PT 317
Carboxymethyl cellulose 7, 14, 15, 170, 174

Smart Food Packaging Systems: Innovations and Technology Applications, First Edition.
Edited by Avik Mukherjee, Santosh Kumar, Manjusri Misra, and Amar K. Mohanty.
© 2025 John Wiley & Sons Ltd. Published 2025 by John Wiley & Sons Ltd.

Cardiovascular diseases 46, 47
Carrageenan 12, 19
Cellulose 3, 4, 7, 228, 229, 233, 235, 236
Cell walls 44, 49
Chelating agents 207, 210
Chemical absorbers 135
Chemical indicators 342
Chemical preservatives 73
Chemical-sensitive field effect transistors (CSFETs) 345
Chili peppers 47
Chitin 55
Chitosan 228, 229, 233, 234, 236–238, 241, 242, 395, 396, 399–401
Chlorophyll 152, 161,
Citric acid 201, 202, 206, 209, 219
Circular economy 417, 418, 421, 422
Climacteric 149, 150, 152, 153, 159, 160
Climacteric fruits 4, 10
Coating 111, 116–118, 120, 122
Colorimetric-type sensors 306, 309, 311–313, 319
Compatibility 110, 116, 118
Composite film 1, 4–6, 8, 9, 12, 13, 17, 19, 20
Conjugation 202, 210
Contaminants 172–174
Conventional plastic 13, 20
CO_2-responsive indicator 363, 376, 377
Covalent bonding 84, 86
Critical temperature/time (CTI) 283–285
Curcumin 48, 55, 366, 367, 377, 391, 392, 394–396, 402
Cyclodextrin 85, 90, 199, 201, 210, 211
Cytoplasm 82, 85
Cytoplasmic 50, 59
Cytoplasmic membrane 85

d

Dairy products 120, 121
Decomposition 173, 180, 185
Discoloration 120, 121
Dry food 169, 184

e

Ecotoxicology 57
Ellen MacArthur Foundation (EMF) 417
Encapsulation 55, 56, 61
Environmental impact 10, 22
Enzymatic TTIs 285, 286, 289, 292, 295
Enzymes 195–197, 201, 204, 211
EoL assessment 415, 417, 420, 422–424
Ethylene exposure 150, 153
Ethylene production 150, 152, 153, 156, 159–162
Ethylene receptors 153, 156–158
Ethylene removal 131
Ethylene scavengers 3, 4
Eugenol 52, 53, 57
Exothermic 135, 136

f

Fatty foods 213
Ferrous salt 199, 202, 206, 208–210
Film-forming 42, 61
Flavonoids 45–47
Flavor absorbents 195–199, 201–204, 206, 208, 209, 211, 213, 214
Foodborne diseases 40
Foodborne parasites 74
Food contamination 39
Food matrix 87, 88
Food packaging 1–4, 7–13, 15, 17–22, 257, 260, 263, 265, 271–274, 416–422, 425
Food products 39, 40, 44, 45, 47, 49, 50
Food quality indicator 238, 337, 344–346
Food rotting 73
Food spoilage 40, 50
Fresh-Check® 293, 295, 297, 298
Freshness indicators 262–264, 272, 361–364, 368, 369, 372, 374, 376–378, 389, 390, 404

g

Gallic acid 106, 108, 109, 119
Gas indicators 340–342, 344, 345, 347, 352
Gelatin 230, 233, 237, 242
Generally Recognized as Safe (GRAS) 77
Glucose oxidase 106, 111, 118, 121
Greenhouse gas emissions 417–420

h

Halochromic indicators 340, 342, 349–352
Heme protein 83

Histidine 50
H_2S-responsive indicator 364
H_2S sensor 397, 401
Humidity indicators 338, 340, 344, 345
Humidity sensors 389, 390, 392
Hydrocarbon 48, 51, 196
Hydro-distillation 50, 51
Hydrophilic 83
Hydrophilicity 156, 170
Hydroxyl groups 113
Hygroscopic substance 6, 7

i
Immobilization 115–119
Inhibitory effect 73
Intelligent packaging 257–259, 266, 268, 270, 272, 274, 282, 284, 337, 343–345, 352, 388–395, 397–399, 401, 404
Iron powder 203, 213

k
Kimchi 310, 312, 324, 325

l
Laccases 112
Lactic acid 133
Lactoferrin 49, 83
Lactoperoxidase 83
Leghemoglobin 106, 108, 110
Life cycle assessment 415, 416, 422–425
Lipid oxidation 5, 13, 16, 17, 40, 41, 45, 48, 51
Lipopolysaccharide 82, 83
Lysozyme 83, 84, 86

m
Machine vision-based colorimetric sensors (MVBCSs) 320
Magnesium 133, 136, 145
Matrix 40, 43, 45, 50, 53–55, 61
Mechanical properties 174, 175, 179
Metal oxides 4, 12, 106, 107
Methylcellulose 87, 89, 91
Microbial activity 202, 209
Microbial growth 1, 5, 6, 10–12, 14, 16, 18, 19
Microbial TTI 287, 288, 291
Microorganisms 3, 4, 20, 131, 134, 139–141, 143, 144, 347
Migration 85, 90, 92
Mitochondria 201, 210
Modified atmospheric packaging 307, 316–318, 320–322
Moisture absorber 6, 7, 18, 170, 171, 173, 174, 176–178
Moisture migration 1, 3–5, 15, 16
MonitorMark™ 293, 294
Myoglobin 211

n
Nanocellulose fiber 89
Nanocomposite 155, 159–161, 227–230, 232, 296, 297
Nanocomposite film 4, 9, 15
Nanocrystals 88, 90
Nanogels 56
Nanomaterials 222–224, 226–230, 240, 241
Nano-packaging 86
Nanoparticle-based TTI 285
Nanoparticles 3, 4, 6–8, 10, 16, 18
Nanospheres 54
Nanostructure 107
Nisin 49, 50, 56, 85
Non-climacteric 149, 150, 152, 153
Non-distributive infrared (NDIR) technology 306, 307, 321
Nonmetallic scavenger 5
Nonmigratory 131

o
Off-flavor 195–199, 201–214
Organic substances 108
Oxalate oxidase 106, 111, 118
Oxidative deteriorations 4, 10
Oxygen scavenger 3–6, 11, 103–106, 108, 110–113, 115–123
Oxygen sensors 389, 390, 404

p
Passive packaging 103
Packaging film 119, 120, 139
Paper-based electrical gas sensors (PEGSs) 320

Pasteurization 50
Pediocin 49, 50
Perishable foods 3, 144
Phenolic compounds 171, 186
pH indicators 389–392, 395, 398, 399, 401, 402
Phospholipid 44
Photocatalysis 44
Photochromic TTI 289, 291, 292
pH sensitive dye/ flim 362, 363, 365, 367, 368, 372–374, 378
pH-sensitive indicators 310, 312–314
pH sensors 267, 268, 271, 272
Physical absorbers 136
Physiological disorders 203, 211
Plant hormone 203, 211
Pleurocidin 84, 88
Polyacrylic acid 174–176, 178, 179
Poly(1-allyl-3-vinyl imidazolium bromide) (PAVB) 308
Polyaniline boronic acid 322
Polyester films 6
Poly(etherblock-amide) (PEBA) 310
Polyethylene 5, 7, 10, 14, 15
Polyethylene glycol 6
Poly (3-hydroxybutyrate) (PHB) 368
Polylactic acid (PLA) 228, 232, 233, 235, 237
Polymeric absorbents 174, 175
Polyolefin 196, 199
Polyphenols 47, 48, 61
Polysaccharides 3, 12
Polyunsaturated fatty acids 110, 120
Proline 84
Prolonged shelf life 121
Pseudomonas 77, 83

q

Quercetin 55

r

Radio frequency identification (RFID) 389, 390, 393, 404
Ready-to-eat 132, 144
Recycling 419–423, 425
Redox reactions 206
Reducing 417, 421–423, 425
Resins 198–201, 204, 205, 207, 210, 211, 214
Reusing 417, 422
Reuterin 85
RFID chips 340, 344, 348, 349
RGB scale 324
RipeSense™ 377, 390

s

Sachets 106, 116, 117
Scavengers 1, 3–5, 15, 16, 131–134, 139, 142–145
Scavenging oxygen 111, 112
Scavenging systems 3, 5
Senescence 203
Sensor 259, 262, 263, 271
SensorQ™ 377
Severinghaus-type sensors 306, 307
Shelf-life 150, 159–162
Silica gel 170, 171, 173, 176, 183, 185
Single-walled carbon nanotubes (SWCNTs) 308, 309
Smart packaging 305, 306, 324
Sodium bicarbonate 138
Sodium glycinate 135
Sol-gel 44
Solution casting 42, 43, 369–371, 373, 375
Sorbitol 6, 7, 172–174, 183
Spoilage bacteria 73
Sterilization 50
Sulfhydryl oxidase 201, 213
Superabsorbent 170, 179, 184, 185
Sustainability 419–421, 423, 425
Sustainable development goals (SDGs) 337, 422
Sustainable polymers 419–420
Synthetic preservatives 11, 16
Synthetic rubbers 112, 113

t

Tannins 47
Thermal properties 39
Thymol 53, 54, 56

Timestrip® 293–295
Time temperature indicator 259–262, 266, 281–287, 289–292, 297–299
Time–temperature sensors 389, 390, 392, 399, 403
TiO_2 nanocomposite 361, 368
TiO_2 nanoparticles 232, 235, 237–240
Titanium oxide 154, 157
Tocopherols 199, 202, 213
Tool for the Reduction and Assessment of Chemical and Other Environmental Impacts (TRACI) 423
Total volatile basic nitrogen 363, 374, 375
Toxicity 50, 56–61
Traditional packaging systems 39

u
Unsaturated polymer 106, 113
UV-blocking 226, 229, 230, 233, 236, 238, 240
UV irradiation 107, 108, 113, 119

v
Vacuum-packaged 135

w
Water vapor barrier 170, 181
Water vapor sorption isotherm 6
World Health Organization (WHO) 337

z
Zeolite 133, 136, 137, 145